T0074131

Bijaya Ketan Panigrahi, Yuhui Shi, and Meng-Hiot Lim (Eds.)

Handbook of Swarm Intelligence

Adaptation, Learning, and Optimization, Volume 8

Series Editor-in-Chief

Meng-Hiot Lim
Nanyang Technological University, Singapore
E-mail: emhlim@ntu.edu.sg

Yew-Soon Ong
Nanyang Technological University, Singapore
E-mail: asysong@ntu.edu.sg

Further volumes of this series can be found on our homepage: springer.com

Vol. 1. Jingqiao Zhang and Arthur C. Sanderson
Adaptive Differential Evolution, 2009
ISBN 978-3-642-01526-7

Vol. 2. Yoel Tenne and Chi-Keong Goh (Eds.)
Computational Intelligence in
Expensive Optimization Problems, 2010
ISBN 978-3-642-10700-9

Vol. 3. Ying-ping Chen (Ed.)
Exploitation of Linkage Learning in Evolutionary Algorithms, 2010
ISBN 978-3-642-12833-2

Vol. 4. Anyong Qing and Ching Kwang Lee
Differential Evolution in Electromagnetics, 2010
ISBN 978-3-642-12868-4

Vol. 5. Ruhul A. Sarker and Tapabrata Ray (Eds.)
Agent-Based Evolutionary Search, 2010
ISBN 978-3-642-13424-1

Vol. 6. John Seiffertt and Donald C. Wunsch
Unified Computational Intelligence for Complex Systems, 2010
ISBN 978-3-642-03179-3

Vol. 7. Yoel Tenne and Chi-Keong Goh (Eds.)
Computational Intelligence in Optimization, 2010
ISBN 978-3-642-12774-8

Vol. 8. Bijaya Ketan Panigrahi, Yuhui Shi, and Meng-Hiot Lim (Eds.)
Handbook of Swarm Intelligence, 2011
ISBN 978-3-642-17389-9

Bijaya Ketan Panigrahi, Yuhui Shi,
and Meng-Hiot Lim (Eds.)

Handbook of Swarm Intelligence

Concepts, Principles and Applications

Dr. Bijaya Ketan Panigrahi
Electrical Engineering Department
Indian Institute of Technology, Delhi
New Delhi, India - 110016
E-mail: bijayaketan.panigrahi@gmail.com

Dr. Meng-Hiot Lim
School of Electrical & Electronic Engineering
Nanyang Technological University
Singapore 639798
E-mail: EMHLIM@ntu.edu.sg

Prof. Yuhui Shi
Director of Research and Postgraduate Office
Xi'an Jiaotong-Liverpool University,
Suzhou, China 215123
E-mail: Yuhui.Shi@xjtlu.edu.cn

ISBN 978-3-642-17389-9 e-ISBN 978-3-642-17390-5

DOI 10.1007/978-3-642-17390-5

Adaptation, Learning, and Optimization ISSN 1867-4534

Typeset & Cover Design: Scientific Publishing Services Pvt. Ltd., Chennai, India.

Printed on acid-free paper

9 8 7 6 5 4 3 2 1

springer.com

Preface

Swarm Intelligence is a collection of nature-inspired algorithms under the big umbrella of evolutionary computation. They are population based algorithms. A population of individuals (potential candidate solutions) cooperating among themselves and statistically becoming better and better over generations and eventually finding (a) good enough solution(s).

Researches on swarm intelligence generally fall into three big categories. The first category is on algorithm. Each swarm intelligence algorithm has been studied and modified to improve its performance from different perspectives such as the convergence, solution accuracy, and algorithm efficiency, *etc*. For example, in particle swarm optimization (PSO) algorithm, the impacts of inertia weight, acceleration coefficients, and neighborhood on the performance of the algorithm have been studied extensively. Consequently, all kind of PSO variants have been proposed. The second category is on the types of problems that the algorithms are designed and/or modified to solve. Generally speaking, most of the swarm intelligence algorithms are originally designed to solve unconstraint single-objective optimization problems. The algorithms are then studied and modified to suit for solving other types of problems such as constraint single-objective optimization problems, multi-objective optimization problems, constraint multi-objective optimization problems, and combinatorial optimization problems, etc.. The third category is on algorithms' applications. Swarm intelligence algorithms have been successfully applied to solve all kinds of problems covering a wide range of real-world applications. Due to algorithms' characteristics such as that they usually do not require continuity and differentiability which are critical for traditional optimization algorithms to have, swarm intelligence algorithms have been able to solve a lot of real-world application problems which are very difficult, if not impossible, for traditional algorithms to solve. Therefore, swarm intelligence algorithms have been attracting more and more attentions from engineers in all industrial sectors. It is the successful real-world applications that are the impetus and vitality that drives the research on swarm intelligence forward.

A research trend of swarm intelligence is on hybrid algorithms. A swarm intelligence algorithm is combined with other swarm intelligence algorithms, evolutionary algorithms, or even other traditional algorithms to take advantage of strengths from each algorithm to improve algorithms' performance for more effective and efficient problem-solving. Initially and intuitively, a swarm intelligence algorithm such as particle swarm optimization algorithm is used as a global search method to find good initial starting point(s), and then a local method such as greedy algorithm, hill-climbing algorithms, etc. is employed to search for the better solution from the initial starting point(s). Currently a new concept of memetic computation is becoming popular. Within the memetic computation framework, memetic algorithms is arguably the most popular. In memetic algorithms, two levels of learning processes are involved: one on learning capability, another on knowledge (individual) learning. Over generations, each individual evolves to have better learning capability. Within each generation, each individual learns knowledge in its neighborhood. Currently, the individuals at two different learning levels generally use the same representation. But it is reasonable to expect different representation for individuals at different level of learning.

Another research trend is on solving large-scale problems. Similar to the traditional algorithms for solving large scale problems, swarm intelligence algorithms also suffer the "dimensionality curse". Researchers are actively engaging in applying swarm intelligence algorithms to solve large-scale optimization problems.

In recent years, swarm intelligence algorithms have become a very active and fruitful research area. There are journals dedicated to swarm intelligence. There are also international conferences held annually that are dedicated to swarm intelligence. The number of papers published in both journals and conferences are increasing exponentially. It is therefore the right time to commit a book volume dedicated to swarm intelligence. Due to the limited space, this book does not intend to cover all aspects of research work on swam intelligence. More appropriately, it is a snapshot of current research on swarm intelligence. It covers particle swarm optimization algorithm, ant colony optimization algorithm, bee colony optimization algorithm, bacterial foraging optimization algorithm, cat swarm optimization algorithm, harmony search algorithm, and their applications.

This book volume is primarily intended for researchers, engineers, and graduate students with interests in swarm intelligence algorithms and their applications. The chapters cover different aspects of swarm intelligence, and as a whole should provide broad perceptive insights on what swarm intelligence has to offer. It is very suitable as a graduate level text whereby students may be tasked with the study of the rich variants of swarm intelligence techniques coupled with a project type assignment that involves a hands-on implementation to demonstrate the utility and applicability of swarm intelligence techniques in problem-solving.

The various chapters are organized into 4 parts for easy reference. Part A grouped together 12 chapters, forming the most substantial part of this volume. The focus is primarily on particle swarm optimization. Besides the first two chapters that focus on general issues pertaining to particle swarm optimization, the remaining of the chapters in Part A offer a good mix of work, with some focusing on analytical issues and the remaining on applications. Part B has 3 chapters on honey bee algorithms, one of which describes the application of honey bee algorithm in vehicle routing problem. Part C covers ant colony optimization, a long established swarm technique. The two chapters in this part should serve as a gentle reminder to the fact that established or matured swarm techniques such as ant colony optimization can be combined innovatively with other swarm or conventional techniques to further enhance its problem-solving capability. In Part D, we grouped together the various swarm techniques, including work that combines two swarm techniques, otherwise referred to as hybrid swarm intelligence. This serves as a glimpse of the possibilities of swarm techniques inspired by the metaphors of swarming behavior in nature. Glowworm optimization, bacterial foraging optimization and cat swarm optimization are examples to this effect.

We are grateful to the authors who have put in so much effort to prepare their manuscripts, sharing their research findings as a chapter in this volume. Based on compilation of chapters in this book, it is clear that the authors have done a stupendous job, each chapter directly or indirectly hitting on the spot certain pertinent aspect of swarm intelligence. As editors of this volume, we have been very fortunate to have a team of Editorial Board members who besides offering useful suggestions and comments, helped to review manuscripts submitted for consideration. Out of the chapters contributed and finally selected for the volume, 4 were honored with the Outstanding Chapter Award. The chapters selected for the awards were decided with the help of the Editorial Board members. The original intention was to choose only 3. However, due to a tie in ranking scores submitted by the Editorial Board members, it was deemed as more appropriate to grant 4 awards instead. We must emphasize that this had not been easy as many of the chapters contributed are just as deserving. Finally, we would like to put in no definitive order our acknowledgements of the various parties who played a part in the successful production of this volume:

- Authors who contributed insightful and technically engaging chapters in this volume;
- To the panel of experts in swarm intelligence that form Editorial Board members, a big thank you;
- Our heartiest Congratulations to the authors of the 4 chapters honored with the Outstanding Chapter Award;
- The publisher, in particular Tom Ditzinger and his team have been very helpful and accommodating, making our tasks a little easier.

Editors

Yuhui Shi
Meng-Hiot Lim
Bijaya Ketan Panigrahi

Contents

Part A
Particle Swarm Optimization

From Theory to Practice in Particle Swarm Optimization

Maurice Clerc

Independent Consultant
Maurice.Clerc@WriteMe.com

Summary. The purpose of this chapter is to draw attention to two points that are not always well understood, namely, a) the "balance" between exploitation and exploration may be not what we intuitively think, and b) a mean best result may be meaningless. The second point is obviously quite important when two algorithms are compared. These are discussed in the appendix. We believe that these points would be useful to researchers in the field for analysis and comparison of algorithms in a better and rigorous way, and help them design new powerful tools.

1 Introduction

Like any other optimization algorithm, Particle Swarm Optimization (PSO) sometimes work well on a complex problem and sometimes not so well on another. Therefore, it makes sense to investigate if there are ways to "improve" the algorithm. To do that, we need to first understand why the algorithm works at all. If we know why and how the algorithm works, then it may be possible to modify some of its parameters or introduce some novel ideas so as to improve its performance on a practical problem.

In fact, this approach is already the main topic of several works, and the reader is encouraged to look those up ([5, 7, 8, 10, 11, 15, 16, 18, 21, 23]). However, there are certain points which are not well understood and sometimes even misinterpreted. In this chapter, we focus on a few such points and try to explain those.

This chapter is organized as follows. First, we discuss the concept of "nearer is better" (NisB) that implicitly lies under stochastic algorithms like PSO. We define the class of NisB functions, estimate a NisB value for every function, and describe some of its properties. We also try to give some motivation regarding why this property is interesting, and show that a very large class of problems including many real life ones satisfy this property.

This leads to a NisB PSO variant which can be shown to be theoretically excellent on average on this huge class. However, it should be noted that this class also contains many uninteresting or strange functions, and the variant may not work very well on those.

B.K. Panigrahi, Y. Shi, and M.-H. Lim (Eds.): Handbook of Swarm Intelligence, ALO 8, pp. 3–36.
springerlink.com

On the other hand, if we consider only classical problems, all if which are reasonably structured, NisB PSO's performance may again be quite bad. The reason is that the number of "classical well structured problems" is negligible compared to the total number of possible functions, and so they count for almost nothing on the average. As the performance on classical problems is important for the user, this also means that the "average performance" is almost of no interest for him/her.

It is possible to partly handle this weakness of NisB PSO by modifying the probability distribution of the next possible positions (we call this version the Gamma NisB PSO). However, both these variants need to define a metric on the search space, as for Standard PSO (SPSO) the movement equations on each dimension are independent and hence do not need any concept of distance between two positions.

For this reason, we next try to find reasonable definitions of "exploration" and "exploitation", which considers each dimension separately. According to the definition we propose, SPSO does not really exploit. So, it is natural to try to modify it in order to explicitly manipulate the exploitation rate. That was the first purpose of the Balanced PSO presented here in section 6 . Later, it became a modular tool that can help us to design a specific version of PSO, assuming we know what kind of problems we have to solve. To illustrate this, we build a PSO variant better than SPSO (in the sense 6.1) for the problems that are moderately multimodal and of low dimensionality (typically smaller than 10).

In the appendix, we show with an example why we must be very careful when using the classical "mean best" performance criterion. If not interpreted carefully, it may give completely meaningless results.

Before we can start our discussion, we need to state precisely what we mean by "Standard PSO." Let us begin with that.

2 Standard PSO in Short

There is no "official" definition of Standard PSO, and that is why there are different descriptions. However, it seems reasonable to choose the one that is supported by the main researchers in the field. It has been put online since 2006 and was slightly modified in 2007. In this chapter, we will use the terms "Standard PSO" or simply "SPSO" to indicate this "Standard PSO 2007", as available on the Particle Swarm Central[24][1]. The main features are the following.

The swarm size S is automatically computed according to the formula

$$S = Int\left(10 + 2\sqrt{D}\right) \tag{1}$$

where D is the dimension of the problem, and Int the integer part function. As this formula is partly empirical, this is not a crucial element of standard PSO though.

[1] SPSO 2006 does not specifically describe the pseudo-random number generator to be used. For reproducible results, we will use the generator KISS throughout this chapter[17].

Far more important is how the neighbourhood is defined. It is modified at each time step, if there has been no improvement of the best solution found by the swarm. Each particle chooses K others at random (using a uniform distribution) to inform. It means that each particle is informed by n others, according to the formula

$$proba\left(nbOfNeighbours = n\right) = \binom{S-1}{n-1}\left(\frac{K}{S}\right)^{n-1}\left(1-\frac{K}{S}\right)^{S-n} \tag{2}$$

The suggested value for K is 3. Observe that each particle informs itself. Two distributions are shown on the figure 1. Most of the time a particle has about K informants, but may have far less or far more with a low probability.

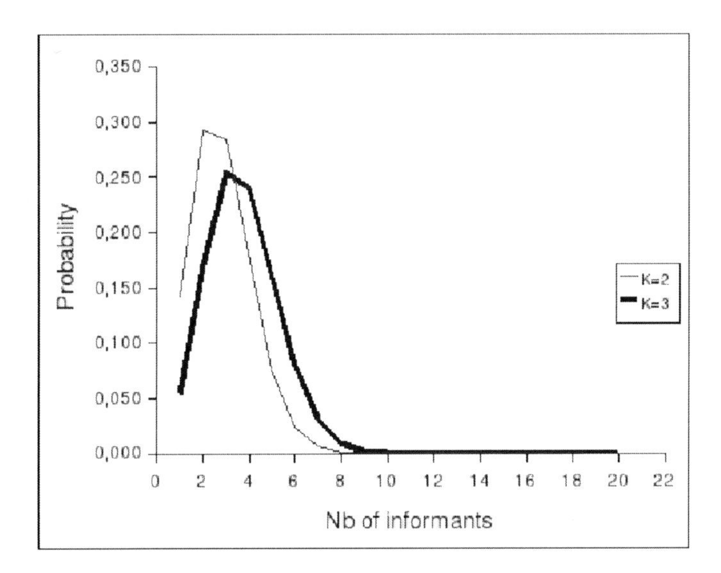

Fig. 1. Distribution of the neighbourhood size (number of informants) of a given particle, for a swarm of 20 particles

For the positions, the initialisation is at random on each dimension d according to an uniform distribution:

$$x\left(0\right)_d = \mathcal{U}\left(x_{min,d}, x_{max,d}\right) \tag{3}$$

For the velocities, the initialisation is the so called "half-diff" method:

$$v\left(0\right)_d = 0.5\left(\mathcal{U}\left(x_{min,d}, x_{max,d}\right) - x\left(0\right)_d\right) \tag{4}$$

Note that the intermediate variable v is wrongly called "velocity" for historical reasons, as it is in fact a displacement during a time step equal to 1.

The general movement equations are the well known ones. For each dimension and each particle we have

$$\begin{cases} v(t+1) = wv(t) + \widetilde{c}(p(t) - x(t)) + \widetilde{c}(g(t) - x(t)) \\ x(t+1) = x(t) + v(t+1) \end{cases} \quad (5)$$

where t is the time step. As usual, $p(t)$ is the best position known by the particle at time t (sometimes called "previous best") and $g(t)$ is the best "previous best" found by its neighbours. In the case the particle is its own best neighbour, the component $\widetilde{c}(g(t) - x(t))$ is removed.

We point out that a very common mistaken notion often occurs here, that g is a "global best". In fact, g is not anymore the so called "global best" or "gbest", as it was in the very first PSO version (1995), but in fact a "local best". As noted in 2006 by one of the inventors of PSO, James Kennedy, on his blog Great Swarmy Speaks:

> *and it might be time to mount a sword-swinging crusade against any use of the gbest topology. How did this happen? It is the worst way to do it, completely unnecessary. I will not tolerate anybody who uses that topology complaining about "premature convergence."*

Note that, though, this is valid only for a mono-objective PSO. For a multiobjective one, the global best method is sometimes still acceptable. This is also true for uni-modal functions, but in that case, there are better specific algorithms, for example the pseudo-gradient descent.

The coefficient w is constant[2], and $\widetilde{c}$ means "a number drawn at random in $[0, c]$, according to a uniform distribution." The values given below are taken from SPSO 2007 and are slightly different from the ones that were used in Standard PSO 2006. They come from a more complete analysis of the behaviour of the particle [5](see also [22]):

$$\begin{cases} w = \frac{1}{2ln(2)} \simeq 0.721 \\ c = 0.5 + ln(2) \simeq 1.193 \end{cases} \quad (6)$$

Last but not least, the following clamping method is used:

$$\begin{cases} x(t+1) < x_{min} \Rightarrow \begin{cases} x(t+1) = x_{min} \\ v(t+1) = 0 \end{cases} \\ x(t+1) > x_{max} \Rightarrow \begin{cases} x(t+1) = x_{max} \\ v(t+1) = 0 \end{cases} \end{cases} \quad (7)$$

Why and how PSO in general and Standard PSO in particular do work has been explained several times, with more and more details emerging year after year. Two not too theoretical recent explanations can be found in [6] and [13]. However, there is an important underlying assumption that is almost never explicit. Let us try to explain what it is.

[2] A classical variant is to use a coefficient decreasing over the time t, assuming you give the maximum number of time steps T. For example $w = w_{max} - t(w_{max} - w_{min})/T$. There is a risk of inconsistency, though. Let us say you have $T_1 < T_2$. Let us suppose that for $T = T_1$ the mean best result is f_1. Now you run the algorithm with $T = T_2$, and consider the mean best result for $t = T_1$. Usually, it is different of f_1. Moreover, the final result may perfectly be worse than f_1.

3 When Nearer Is Better

3.1 Motivation

Most iterative stochastic optimisation algorithms, if not all, at least from time to time look around a good point in order to find an even better one. We can informally call it "nearer is better" (NisB), but this can also be mathematicaly defined, and summarised by a single real value for any function. This NisB truth value can be estimated (or even exactly calculated for not too large problems). The functions for which this value is positive define a huge class, which in particular contains most of classical and less classical tests functions, and apparently also all real problems. Although it is out of the scope of this paper, it is worth noting that the No Free Lunch Theorem (NFLT,[28]) does not hold on this class, and therefore we can explicitly define an algorithm that is better than random search. Therefore it makes sense to look for the best possible algorithm.

When negative, this NisB value indicates that the corresponding function is very probably deceptive for algorithms that implicitly assume the NisB property to be true. PSO is typically such an algorithm, and an example of the design of a function on which it is worse than random search is given in B.

3.2 Notations and Definitions

3.2.1 Search Space, Problems and No Free Lunch Theorem (NFLT)

On a computer all sets of numbers are finite, so without loss of generality, we can consider only discrete problems. As in [28], we consider a finite search space X, of size $|X|$, and a finite set of fitness values Y, of size $|Y|$.

An optimisation problem f is identified with a mapping $f : X \rightarrow Y$ and let $\mathcal{F}$ be the space of all problems. It may be useful to quickly recall under which conditions NFLT holds.

Condition 1: *For any position in X all values of Y are possible*

In such a case the size of $\mathcal{F}$ is obviously $|Y|^{|X|}$. An optimisation algorithm A generates a time ordered sequence of points in the search space, associated with their fitnesses, called a *sample.*

Condition 2: *The algorithm A does not revisit previously visited points*

So an optimisation algorithm A can be seen as a permutation of the elements of X.

The algorithm may be stochastic (random), but under Condition 2.

For this reason, Random Search (let us call it R) in the context of NFLT is not exactly similar to the usual one, in which each draw is independent of the previous draws. Here R is defined not only by "drawing at random according to an uniform distribution" but also by "... amongst points not already drawn". It means that under Condition 2 any algorithm, including R, is in fact an exhaustive search.

3.2.2 Performance and Characteristic Distribution

In the context of NFLT, the performance depends only on the fitness values, and not on the positions in the search space. It means, for example, that NFLT does not hold if the performance measure takes the distance to the solution point into account. Let f be a problem, and A an algorithm that samples x_{α_t} at "time step" t. There is a probability $p(f,t,A)$ that the sampled point is a solution point. We compute the following expectation

$$r(f,A) = \sum_{t=1}^{|X|} p(f,t,A)\,t \tag{8}$$

Roughly speaking, it means that the algorithm finds a solution, "on average" after $r(f,A)$draws. Then we say that algorithm A is better than algorithm B for the problem f if $r(f,A) < r(f,B)$, i.e. if on average A finds a solution quicker than B.

On a set of functions $\mathcal{E}$ we can then define the global performance by the following mean

$$\boldsymbol{r}(\mathcal{E},A) = \frac{1}{|\mathcal{E}|}\sum_{f\in\mathcal{E}} r(f,A) \tag{9}$$

NFLT claims that when averaged over *all* $|Y|^{|X|}$ functions, i.e. over $\mathcal{F}$, "Any algorithm is equivalent to Random Search". It means that $\boldsymbol{r}(\mathcal{F},A)$ is the same for all algorithms. Therefore its value is $|X|/2$.

Another point of view is to consider the characteristic distribution. For each point x_i of X, let $n(x_i)$ be the number of functions for which x_i is the position of at least one global optimum. Then we define

$$s(x_i) = \frac{1}{\sum_{k=1}^{|X|} n(x_k)} n(x_i) \tag{10}$$

which can be seen as the probability that there is at least one function whose global optimum is on x_i. The characteristic distribution is $S = (s(x_1), ..., s(x_{|X|}))$. An example is given in the A. An interpretation of the NFLT is that the characteristic distribution is completely "flat" (all $s(x_i)$ have the same value). However, it is easy to define subsets of functions on which the characteristic distribution is not uniform anymore. We are now going to study such a subset which has interesting practical applications.

3.3 The Nearer Is Better (NisB) Class

We assume here that we are looking for a minimum. Let f be a function; x_b, x_w two positions in its definition domain, so that $f(x_b) \leq f(x_w)$; and δ a distance defined on this domain. Let us define two subdomains

$$\begin{aligned} N_{b,w} &= \{x, 0 < \delta(x_b, x) < \delta(x_b, x_w)\} \\ B_{w,b} &= \{x, x \neq x_b, x \neq x_w, f(x) \leq f(x_w)\} \end{aligned} \tag{11}$$

That is, $N_{b,w}$ is the set of points to which x_b is closer than to x_w, with a nonzero distance, and $B_{w,b}$the set of positions that are better than x_w (or equivalent to), except the two points already drawn. The figure 2 illustrates the NisB principle for a two dimension search space and the Euclidean distance.

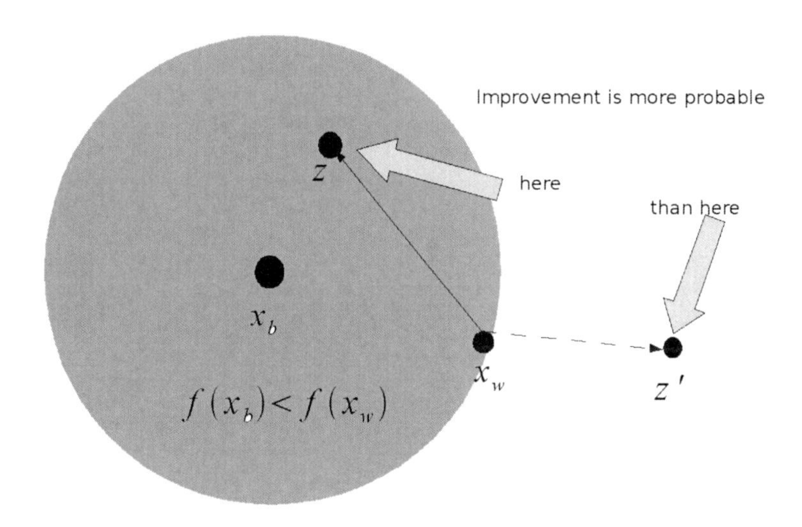

Fig. 2. Illustration of the NisB principle

If we choose x at random in $X - \{x_b, x_w\}$, according to a uniform distribution, the probability of finding a position better than x_w is simply $|B_{w,b}| / (|X| - 2)$. Now, what happens if we choose x in $N_{b,w}$(assuming it is not empty)? The probability of finding a better position than x_w is $|N_{b,w} \bigcap B_w, b| / |N_{b,w}|$. If this probability is greater than the previous one, this strategy may be interesting. We do not require that it happens always though, but just that it happens "on average". Let us call this method "NisB strategy". To formalise this we define the set of acceptable pairs by

$$\Omega = \{(x_b, x_w), x_b \neq x_w, f(x_b) \leq f(x_w), |N_{b,w}| > 0\} \tag{12}$$

and the two mean probabilities

$$\begin{cases} \xi(f) = \dfrac{1}{|\Omega|} \displaystyle\sum_{\Omega} \dfrac{|N_{b,w} \bigcap B_{w,b}|}{|N_{b,w}|} \\ \rho(f) = \dfrac{1}{|\Omega|} \displaystyle\sum_{\Omega} \dfrac{|B_{w,b}|}{|X| - 2} \end{cases} \tag{13}$$

Table 1. "Nearer is Better" truth value $\nu(f)$ for some functions, and how much a pure NisB algorithm is better than Random Search. The numbering of the first three functions is referring to the 3125 possible functions from $X = \{1, 2, 3, 4, 5\}$ into $Y = \{1, 2, 3, 4, 5\}$

Function f	X, granularity	Landscape	$\xi(f)$	$\nu(f)$
F1188	$[1, 5]$,1	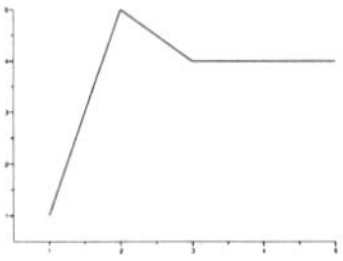	0.67	-0.10
F2585	$[1, 5]$, 1		0.86	0.00
Step, plateau 10%	$[-100, 100]$, 5	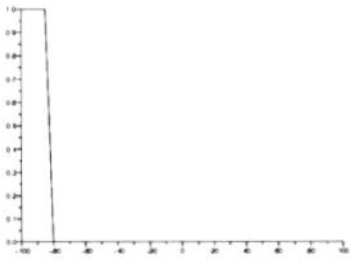	0.995	0.05
Alpine 2 $f(x) = \|x \sin(x) + 0.1x\|$	$[-100, 100]$, 5		0.78	0.09
Rosenbrock $f(x, y) = 100(x^2 - y)^2 + (x - 1)^2$	$[-10, 10]^2$, 2	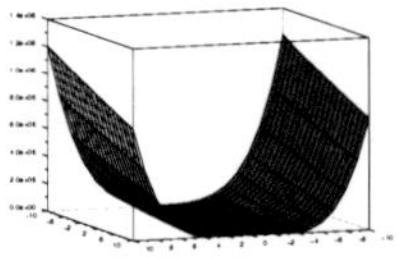	0.80	0,13
Rastrigin $f(x) = 10 + x^2 - 10\cos(2\pi x)$	$[-10, 10]$, 0.5	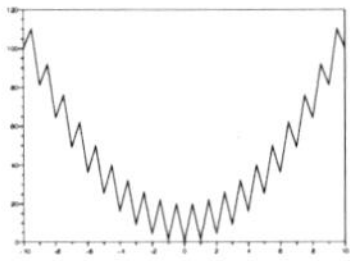	0.83	0.15
Parabola $f(x) = x^2$	$[0, 200]$, 5	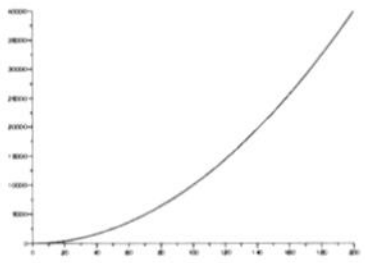	1.00	0.33

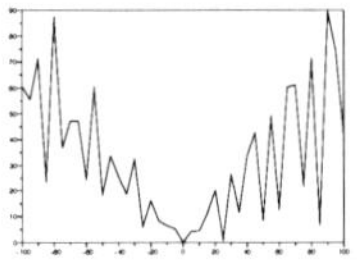

Then we say "nearer is better" is true for the function f iff

$$\nu(f) = \xi(f) - \rho(f) > 0 \tag{14}$$

It can be seen as an estimation of how much the NisB strategy is better than Random Search on f. Or, intuitively, it can be seen as an estimation of how much "nearer is better, not just by chance". Note that $\nu(f)$ may be null even for non-constant functions. For example, in dimension 1, that is the case with $X = (1, 2, 3, 4)$and $f(x) = (4, 2, 3, 4)$. This is because if we consider the "contribution" of a given pair

$$\eta_{b,w}(f) = \frac{|N_{b,w} \bigcap B_{w,b}|}{|N_{b,w}|} - \frac{|B_{w,b}|}{|X| - 2} \tag{15}$$

it is obviously equal to zero when $f(x_w)$ is maximum, and in this example all four possible pairs (i.e. $(x_1, x_4), (x_2 x_4), (x_3, x_1), (x_4, x_1)$) contain a x_w so that $f(x_w) = 4$.

From now on we call $\mathcal{F}^+$ the set of functions of $\mathcal{F}$ for which $\nu(f) > 0$, $\mathcal{F}^-$ the set of functions for which $\nu(f) < 0$, and $\mathcal{F}^=$ the set of functions for which $\nu(f) = 0$. When $|X|$ is not too big this truth value can be computed by exhaustively considering all pairs of points, as in Table 1, but most the time we just estimate it by sampling. For simplicity, the discrete domain X is defined by a continuous real interval and a "granularity", and the discrete domain Y by the $|X|$ fitness values. To build the table, we used the following distance measure between two points $x = (x_1, ..., x_D)$ and $x' = (x'_1, ..., x'_D)$, which is the well known L_∞ norm:

$$\delta(x, x') = max(|x_i - x'_i|) \tag{16}$$

It is clear that on dimension one it is equal to the Euclidean distance. The table shows that $\mathcal{F}^-$ is not empty: we can have a "nearer is worse" effect, i.e. a deceptive function. For such deceptive functions (for example F1188 in the table[3]) Random Search may be better than a more sophisticated algorithm.

4 NisB PSOs

It is tempting to design a PSO variant that explicitly takes advantage of the NisB property. The simplest way is to replace equation 5 by "draw $x(t+1)$ at random "around" $g(t)$." Let us consider two ways to do that.

4.1 Pure NisB PSOs

We can apply exactly the definition of the NisB truth value. It means that $g(t)$ has to be the best known estimation of the solution point, i.e. the global best. The next point $x(t+1)$ is then just drawn at random (uniform distribution) in

[3] The numbering is quite arbitrary. Actually, it has been generated by a small program that finds all the $5^5 = 3125$ permutations.

the hypersphere $H(g,r)$ with centre $g(t)$ and radius $r = distance(x(t), g(t))$. This can be summarised by a formula[4] like

$$x(t+1) = randUnif(H(distance(x(t), g(t)))) \tag{17}$$

As we can see from table 3 in the D, this variant is good not only on classical easy functions (like Sphere), but also on some other functions that are considered difficult, but have in fact also a high NisB truth value (Rosenbrock, Rastrigin). For Rastrigin, it is even significantly better than SPSO. However, it is clearly bad on the four quasi-real-world problems (Pressure vessel, Compression spring, Gear train, and Cellular phone).

So, we can try to "cheat" a bit. To be completely consistent with the NisB approach, $g(t)$ has to be the best estimation of the solution point, i.e. the "global best." However, in what follows, $g(t)$ is not anymore the global best, but the local best as defined in SPSO (here with $K = 3$), i.e. the best position known in the neighbourhood of the particle that is moving. Not surprisingly, the results are a little better, but still can not compete with the ones of SPSO for the quasi-real-world problems (in passing, it suggests that some classical artificial problems may not be very relevant for comparing algorithms). This is why we want to try a variant that is significantly more different from pure NisB strategy.

4.2 Gamma NisB PSO

We can now try a "radius" that may be greater than just $distance(x(t), g(t))$, and moreover with a non uniform distribution. This is typically the idea that lies under the 2003 Bare Bones PSO [12], with its Gaussian distribution. Although here we do not take each dimension independently into account, we can consider the same kind of distribution, with a standard deviation γ. This can be summarised by a formula like

$$x(t+1) = randGauss(g, \gamma \times distance(x(t), g(t))) \tag{18}$$

For example, if we call "exploitation" the fact that $x(t+1)$ is strictly inside the hypersphere H, and if we want the probability of this "exploitation" to be approximately equal to 0.5, then we can set $\gamma = 0.67$. The results (not detailed here) are not significantly better than the previous variant. It suggests that simply manipulating the exploitation/exploration rate is not a so good strategy by itself. What we did here is that we added the NisB variation on exploitation strategy, but the results did not improve much. What if we take SPSO and tune the exploitation rate? We investigate this in the next section.

[4] In order to avoid any bias, and without lost of generality, we assume here that the search space is a hypercube. If it is given as a hyperparallelepid, it is easy to convert it into a hypercube before the run, and to convert back the solution point after the run.

5 The Mythical Balance, or When PSO Does Not Exploit

5.1 A Ritual Claim

In many PSO papers (see for example [1][14][20][25]), we find a claim similar to this: "this algorithm ensures a good balance between exploration and exploitation". Sometimes the authors use the words "diversification" and "intensification", but the idea is the same. However, a precise definition of "exploitation" is rarely given, or, if it is, it may be inconsistent. That is. we can not even be sure that the defined "exploitation area" contains the best position found by now.

Therefore, here we use a rigorous and "reasonable" definition, directly usable by PSO versions that consider each dimension independently. We show that, according to this definition, Standard PSO does not really exploit.

5.2 A Bit of Theory

Roughly speaking, exploitation means "searching around a good position". In PSO, good positions are already defined: each particle memorises the best position it has found. So, it seems natural to define a *local exploitation area (LEA)* around each of these positions. As we want something defined for each dimension, we can not use a hypersphere like we did for NisB PSO.

Let us consider a search space $\otimes_{d=1}^{D}[x_{min,d}, x_{max,d}]$, the Cartesian product of D intervals, and the following points $(p_0 = x_{min}, p_1, \ldots, p_N, p_{N+1} = x_{max})$, where $(p_1, ..., p_N)$ are the local best positions found by the swarm. Each p_i is a vector $p_i = (p_{i,1}, \ldots, p_{i,D})$. On each dimension d, we sort the $p_{i,d}$s in increasing order. We have then $p_{\varsigma(0),d} \leq p_{\varsigma(1),d} \leq \cdots \leq p_{\varsigma(N),d} \leq p_{\varsigma(N+1),d}$, where ς is a permutation on $\{0, 1, \ldots, N+1\}$, with $\varsigma(0) = 0$, and $\varsigma(N+1) = N+1$. Then we define define the intervals

$$\begin{array}{l} e_{i,d} = [p_{\varsigma(i),d} - \alpha(p_{\varsigma(i),d} - p_{\varsigma(i-1),d}), p_{\varsigma(i),d} + \alpha(p_{\varsigma(i+1),d} - p_{\varsigma(i),d})] \\ \qquad i \in \{1, \cdots N\}, d \in \{1, \ldots D\} \end{array} \tag{19}$$

where $\alpha > 0$ is a parameter, called *relative local size*. The local exploitation area around each best position p_i is defined by the Cartesian product

$$E_i = e_{i,1} \otimes \ldots \otimes e_{i,D}, i \in \{1, \cdots N\} \tag{20}$$

Figure 3 shows an example for a two dimensional search space, and two known best positions.

Now, if we consider a particle, what is the probability that its next position is within a given LEA? For the d-th coordinate, let us say the partial probability is $\pi(d)$. As each coordinate is updated independently, the total probability is then

$$\boldsymbol{\pi} = \prod_{d=1}^{D} \pi(d) \tag{21}$$

It is clear that this quantity can very easily be extremely small as the dimension D becomes large. Let us see it on a simple example.

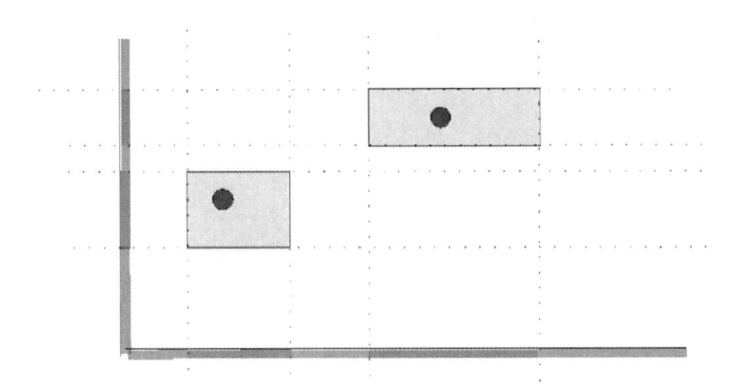

Fig. 3. 2D example of Local exploitation areas, with $\alpha = 1/3$

5.3 Checking the Exploitation Rate

At each iteration t, let $S_E(t)$ be the number of particles that are inside a local exploitation area. We can define the *exploitation rate* by

$$r(t) = \frac{S_E(t)}{S} \tag{22}$$

For an algorithm like SPSO, we can observe the evolution of the exploitation rate on some test functions. A typical example is shown in figure 4: there is no clear cut tendency, except a very slight increase when the swarm has converged. This is not surprising. A more important observation is that the mean value of r quickly decreases when the dimension increases, as expected. This is shown in figure 5. In practise, as soon as the dimension becomes greater than 6, there is no exploitation at all, And this holds no matter what the value of α is (except of course very big ones, but in that case, speaking of "exploitation" is meaningless). That is why, for some PSO variants, the "good balance" often claimed is just a myth.

6 Balanced PSO: A Tool to Design a Specific Algorithm[5]

As SPSO does not really exploit, it seems natural to slightly modify it in order to ensure a better exploitation rate, and to really define a "balanced" PSO, in which it is kept more or less constant, or, at least, significantly greater than zero for any number of dimensions of the search space. The basic idea here is to "force" some particles to move to a given local exploitation area (LEA). It can be seen as a cheap and rudimentary kind of local search. This is discussed in detail in sub-section 6.2.1. We will see that it is not enough, though.

Indeed, this rate is not the only component of the algorithm to consider when designing a PSO variant. There are also the probability distributions, the parameters w and c, the topology of the information links etc. For example, it

[5] A C source code is available on [3].

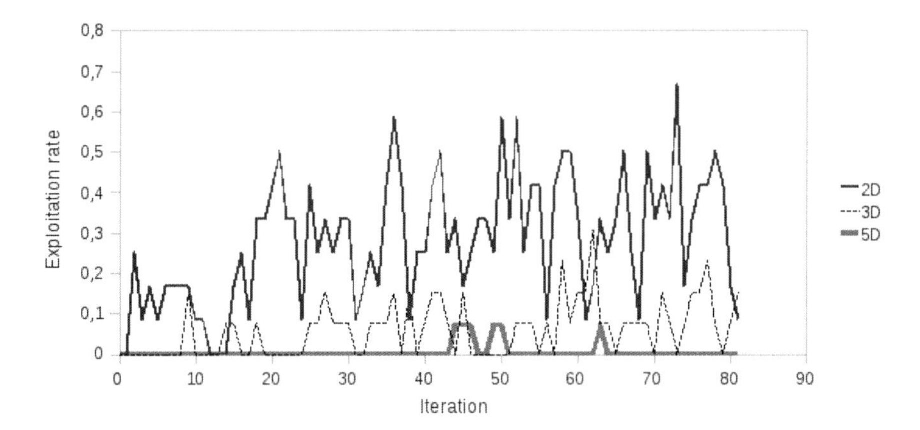

Fig. 4. Rosenbrock 2D. Evolution of the exploitation rate for different dimensions, for $\alpha = 0.5$

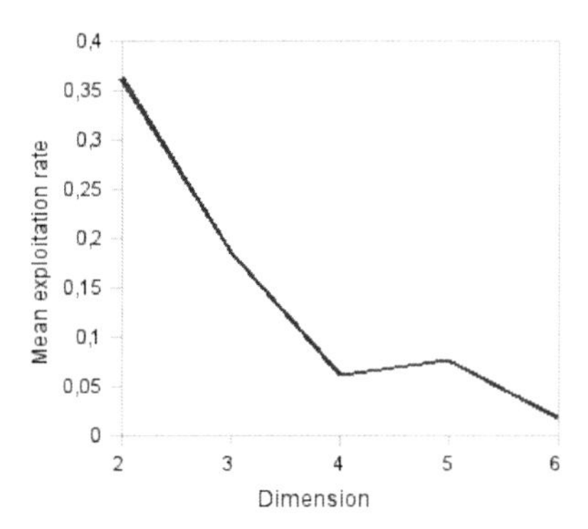

Fig. 5. Rosenbrock 2D. Mean exploitation rate for different dimensions, for $\alpha = 0.5$

has been proved that the classical random initialisations for the positions and the velocities have an important drawback: many particles may leave the search space in the beginning, with a probability that increases with the dimension [11].

So, assuming we know what kind of problems we have to solve, it is possible to design a specific and more accurate version of PSO. To illustrate the method, we consider the class of problems that are moderately multimodal (but not unimodal) and of low dimensionality (typically smaller than 30), and, of course, NisB. A modified velocity update equation is used, adapted to the considered class of problems. Then, in order to improve the robustness of the algorithm, more general modifications are added (in particular, better initialisations, and the use of two kind of particles). On the whole, the resulting algorithm is indeed better than SPSO, though the improvement is not always very impressive. However, first we have to define what "better than" means.

6.1 My Algorithm Is Better than Yours

When comparing two algorithms, we have to agree on a common benchmark, and on a given criterion. For example, for each function, the criterion may be Cr = "success rate over N runs, for a given accuracy ε, after at most F fitness evaluations for each run, or, in case of a tie, mean of the best results". Note that the criterion may be a probabilistic one (for example Cr'= "the probability that A is better than B on this problem according to the criterion Cr, is greater than 0.5"). What is important is to clearly define the meaning of "A is better than B on *one* problem".

Frequently, one uses such an approach, and then perform statistical analyses (null hypothesis, p-test, and so on), in order to decide "in probability" whether an algorithm is better than another one, on the whole benchmark. However the method is not completely satisfying for there is no total order in R^D compatible with its division ring structure on each dimension[6], for all $D > 1$. Why is that important? It may be useful to express it more formally.

Let us say the benchmark contains P problems. We build two comparison vectors. First $C_{A,B} = (c_{A,B,1}, \ldots, c_{A,B,P})$with $c_{A,B,i} = 1$ if A is better than B on problem i (according to the unique criterion defined), $c_{A,B,i} = 0$ otherwise. Second $C_{B,A} = (c_{B,A,1}, \ldots, c_{B,A,P})$ with $c_{B,A,i} = 1$ if B is better than A on the problem i, and $c_{B,A,i} = 0$ otherwise. We have to compare the two numerical vectors $C_{A,B}$ and $C_{B,A}$. Now, precisely because there is no total order on R^D compatible with its division ring structure, we can say that A is better than B if and only if for any i we have $c_{A,B,i} \geq c_{B,A,i}$ for all i, and if there exists j so that $c_{A,B,j} > c_{B,A,j}$.

This is similar to the definition of the classical Pareto dominance. As we have P values of one criterion, the process of comparing A and B can be seen as a multicriterion (or multiobjective) problem. It implies that most of the time no comparison is possible, except by using an aggregation method. For example, here, we can count the number of 1s in each vector, and say that the one with the larger sum "wins". But the point is that any aggregation method is arbitrary, i.e. for each method there is another one that leads to a different conclusion[7].

Let us consider an example:

- the benchmark contains 5 unimodal functions f_1 to f_5, and 5 multimodal ones f_6 to f_{10}

---

[6] A complete discussion of this is outside the scope of this article, but the basic idea is as follows. A total order is a order defined for every pair of elements of a set. Evidently, it is easy to define such an order for R^D, $D > 1$, for example the lexicographic order. However, what we truly need is not any arbitrary order but an order that is compatible with the addition and multiplication defined on the algebraic division ring structure of R^D with D>1, and this does not exist. For example, in R^2, for classical complex number like addition and multiplication, no total order exists.

[7] For example, it is possible to assign a "weight" to each problem (which represents how "important" is this kind of problem for the user) and to linearly combine the $c_{A,B,i}$ and $c_{B,A,i}$. However, if for a set of (non identical) weights, A is better than B, then there always exists another set of weights for which B is better than A.

- the algorithm A is extremely good on unimodal functions (very easy, say a pseudo-gradient method)
- the algorithm B is quite good for multimodal functions, but not for unimodal ones.

Suppose you find that $c_{A,B,i} = 1$ for $i = 1, 2, 3, 4, 5$, and also for 6 (a possible reason is that the attraction basin of the global optimum is very large, compared to the ones of the local optima), and $c_{B,A,i} = 1$ for $i = 7, 8, 9, 10$. You say then "A is better than B". A user trusts you, and chooses A for his/her problems, and as most interesting real world problems are multimodal, s(h)e will be very disappointed.

So, we have to be both more modest and more rigorous. That is why the first step in our method of designing a specific improved PSO is to choose a small benchmark. We will say that A is better than B only if it is true for all the problems of this benchmark. Now, let us see if we can build an algorithm better than SPSO by choosing different parameters associated only with the exploitation strategy. We stress here that this is different from what we did with NisB PSOs. In NisB PSO variants, one single hypersphere was enough, which can not be done for SPSO, the dimensions being independent. Here, we need to take into account the local exploitation area for each dimension, and therefore need to change several parameters. We will try to achieve this by selecting several parameters in a modular way.

6.2 A Modular Tool

Starting from SPSO, and in order to have a very flexible research algorithm., we add a lot of options, which are the following:

- two kind of randomness (KISS [17], and the standard randomness provided in LINUX C compiler). With KISS the results can be more reproducible
- six initialisation methods for the positions (in particular a variant of the Hammersley's one [29])
- six initialisation methods for the velocities (zero, completely random, random "around" a position, etc.)
- two clamping options for the position (actually, just "clamping like in SPSO", or "no clamping and no evaluation")
- possibility to define a search space larger than the feasible space. Of course, if a particle flies outside the feasible space, its fitness is not evaluated.
- six local search options (no local search as in SPSO, uniform in the best local area, etc.). Note that it implicitly assumes a rigorous definition of a "local area"
- two options for the loop over particles (sequential or according to a random permutation)
- six random distributions$R$
- six strategies

The random distributions R and the strategies are related to the velocity update formula that can be written in a more general form

$$v(t+1) = wv(t) + R(c_1)(p(t) - x(t)) + R(c_2)(g(t) - x(t)) \quad (23)$$

A strategy defines the coefficients w, c_1, and c_2, constant or variable. Different particles may have different strategies, and different random distributions. As the initial purpose of Balanced PSO was to explicitly modify the exploitation rate, let us try it this way.

6.2.1 Example: Manipulating the Exploitation Rate

It was the first purpose of Balanced PSO. Let us see what happens when we just control the exploitation rate. To find a good approach experimentally, we consider four parameters:

- *force_LEA*, which is the probability that a particle is forced to a LEA
- *choose_LEA*, which defines what kinds of LEAs are taken into account. This has two possible "values": the LEA of the particle, or the best LEA (i.e. around the global best position)
- *size_LEA*, which is in fact the parameter α that we have seen above
- *draw_LEA*, which defines how the new position is drawn in the LEA. This has three possible "values": the centroid of the LEA, at random (uniform distribution), and at random (truncated Gaussian distribution).

Also, a variable *size_LEA* may be interesting to experiment with. We tested this with four methods:

- adaptive. After each iteration t, the exploitation rate $r(t)$ is calculated, and *size_LEA* is modified according to a formula (if $r(t)$ increases then *size_LEA* decreases, and vice versa). Several formulae were tested
- alternative, random choice between two values $1/S$ and a given λ
- random, uniform distribution in a given interval $[0, \lambda]$
- random, Gaussian distribution, mean $1/S$, standard deviation λ

Of course, more options are obviously possible, but anyway, the experiments confirm what we have seen with our Gamma NisB PSO: the idea that a good balance always implies good performance is simply wrong. Let us consider two very different variants. Balanced H2 D1 is a quite complicated one[8] , with the following options:

- *force_LEA* $= 0.5$, i.e. about half of the particles are forced to a LEA
- *choose_LEA* = the best LEA
- *size_LEA* = alternative α, chosen randomly in $\{1/S, 0.5\}$
- *draw_LEA* = uniform in the LEA

On the contrary, Balanced H2 D0 is very simple: at each iteration, in addition to the normal moves, the middle of the best local area is sampled. Note that there is a theoretical reason for this method, if we assume that the function is NisB inside this LEA (see A).

[8] Codes like H2, D1, P2 etc. are the names of the options in the C source.

From the results given in table 3, we can see that more complicated does not necessary means more efficient: Balanced L4 is significantly worse than Balanced H2 D1 only for three functions (Ackley F8, Rastrigin F9, and Cellular phone), and it is far better for some others (Pressure vessel, Compression spring). More importantly, the two variants do not beat Standard PSO for all problems. Again, although the definition of "exploitation" is different from the one used in NisB PSO, we see that just manipulating the exploitation rate is not enough.

We need something more sophisticated. However, on the other hand, we may also have to be more specific. That is why the approach needs three steps.

6.3 Step 1: A Small "Representative" Benchmark

Let us consider the following class of problems:

- moderately multimodal, but not unimodal (although, of course, we do hope that our method will be not too bad on these kinds of problems)
- low dimensionality D (say no more than 30)

For this class, to which a lot of real problems belong, a good small benchmark may be the following one[9]:

- Rosenbrock F6 ($D = 10$. Note that it is multimodal as soon as the dimension is greater than 3 [27])
- Tripod ($D = 2$)
- Compression spring ($D = 3$)
- Gear train ($D = 4$)

These functions are supposed to be "representative" of our class of problems. Now, our aim is to design a PSO variant that is better than SPSO for these three functions. Our hope is that this PSO variant will indeed also be better than SPSO on more problems of the same class, and if it is true even for some highly multimodal problems, and/or for higher dimensionality, well, we can consider that as a nice bonus!

6.4 Step 2: A Specific Improvement Method

First of all, we examine the results found using SPSO. When we consider the surfaces of the attraction basins, the result for Tripod is not satisfying (the success rate is only 51%, and any rudimentary pseudo-gradient method would give 50%). What options/parameters could we modify in order to improve the algorithm? Let us call the three attraction basins B_1, B_2, and B_3 . The problem is deceptive because two of them, say B_2 and B_3, lead to only local optima. If, for a position x in B_1 (i.e. in the basin of the global optimum) the neighbourhood

[9] Unfortunately, there is no clear guideline, just experiments. Rosenbrock and Tripod are well known to be difficult. Compression spring and Gear train are also largely used, and supposed to be more "similar" to a lot of real problems. But there is no precise definition of what "similar" means.

best g is either in B_2 or in B_3, then, according to equation 23, even if the distance between x and g is high, the position x may get easily modified such that it is not in B_1 any more. This is because in SPSO the term $R(c_2)(g(t) - x(t))$ is simply $U(0, c_2)(g(t) - x(t))$, where U is the uniform distribution.

However, we are interested here on functions with just a few local optima, and therefore we may suppose that the distance between two optima is usually not very small. So, in order to avoid the above behaviour, we use the idea that the further an informer is, the smaller is its influence (this can be seen as a kind of niching). We may then try an $R(c_2)$ that is in fact an $R(c_{2,} |g - x|)$, and decreasing with $|g - x|$. The formula used here is

$$R(c_{2,} |g - x|) = U(0, c_2)\left(1 - \frac{|g - x|}{x_{max} - x_{min}}\right)^{\lambda} \quad (24)$$

Experiments suggest that λ should not be too high, because in that case, although the algorithm becomes almost perfect for Tripod, the result for Sphere becomes quite bad. In practice, $\lambda = 2$ seems to be a good compromise. This PSO is called Balanced R2. The result for Compression spring is just slightly bettert (success rate: 36% vs 33%), but the result for Gear train is significantly improved (24% vs 3%). Of course, the one for Tripod is also largely improved (69% vs 51%). For Rosenbrock, the result becomes just slightly worse (67% vs 70%). So, on the whole keeping this specific option seems to be a good idea. Anyway, we may now try to improve our PSO variant further.

6.5 Step 3: Some General Improvement Options

The above option was specifically chosen in order to improve what seemed to be the worst result, i.e. the one for the Tripod function. Now, we can successively trigger some options that are often beneficial, at least for moderately multimodal problems:

- the modified Hammersley method for the initialisation of the positions x (option P2)
- another velocity initialisation (option V1), defined by $v(0)_d = \mathcal{U}(x_{min,d}, x_{max,d}) - x(0)_d$
- the sampling of the centroid of the best LEA (options H2 D0. Recall that we already have seen Balanced H2 D0 in 6.2.1)

The results (table 3) show that we now have indeed something very interesting, although results for some of the problems wich are not in the specific class defined by us get worse (see Cellular phone). It is difficult to decide between the two variants, though. Also, these results are just for a given number of fitness evaluations, for each function. Is the improvement still valid for less or more fitness evaluations?

6.6 Success Rate vs. "Search Effort"

Here, on our small specific benchmark, we simply consider different maximum numbers of fitness evaluations (FE_{max}), and we evaluate the success rate over

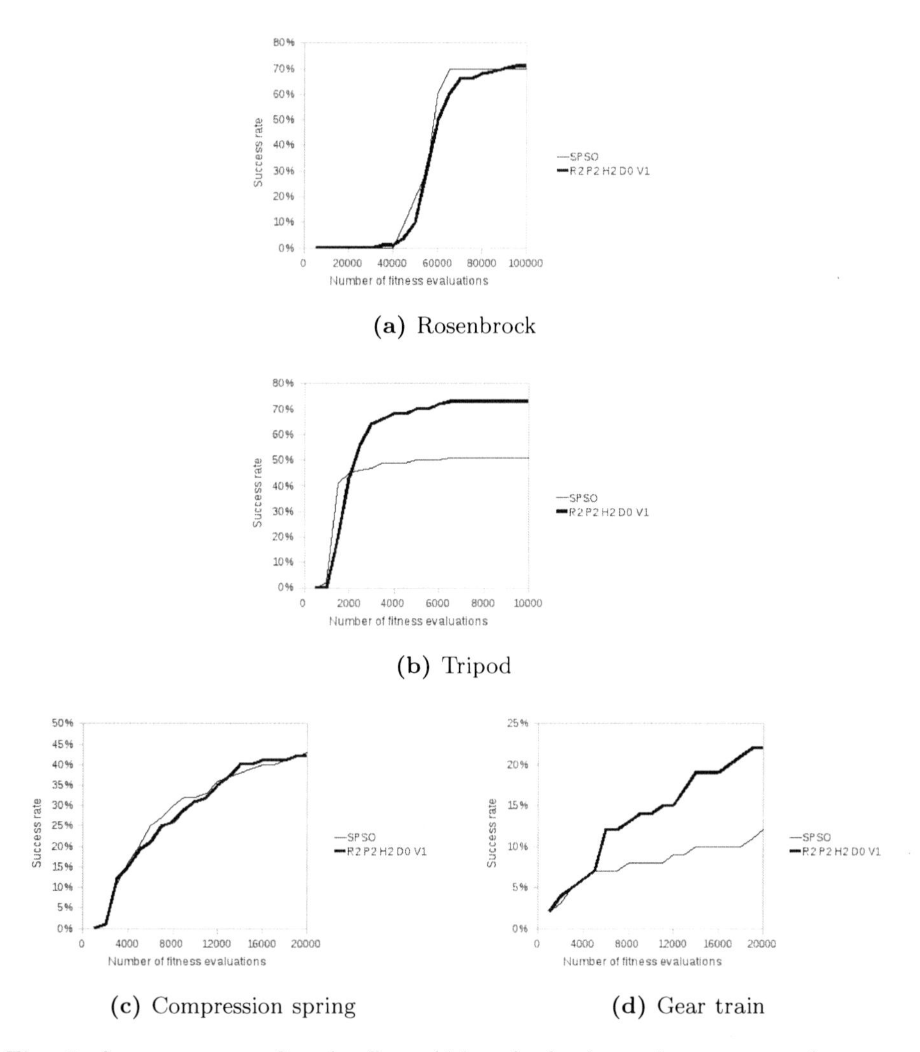

(a) Rosenbrock

(b) Tripod

(c) Compression spring

(d) Gear train

Fig. 6. Success rate vs Search effort. Although clearly an improvement, for some problems the R2 P2 H2 D0 V1 variant may perform better than SPSO or not, depending on the maximum number of fitness evaluations

100 runs. As we can see from figure 6, for some problems, and for any FE_{max}, the success rate of our variant is greater than that of Standard PSO. In such a case, we can safely say that it is really better on such problems. However, it is not always so obvious. Sometimes the variant is better than SPSO for some values but not for some others. When such a situation arises, any comparison that does not specifically mention FE_{max} is not technically sound, as explained in 6.1.

7 Conclusion

Let us try to summarise what we have seen. First, we defined a class of functions (the ones that have the Nearer is Better (NisB) property) on which not all algorithms are equivalent, and therefore a best one exists. We gave a measure of this NisB property, and a candidate algorithm (NisB PSO) is proposed. This is the theoretical part. However, as this class is huge, being the best algorithm for the average performance does not mean being the best one for the small subsets of benchmark problems and real problems. As a future direction of research, it may be an interesting idea to define a NisB(ν) PSO variant that works well on all functions that have a NisB property value greater than ν.

Fortunately, in practice, the user does know some things about the problem to be solved. In particular, the dimension is known, and also, quite often, an estimate of the number of local optima. Such information define a subclass of problems for which it is possible to design a more specific and "better" algorithm. The second main contribution of this chapter is to propose a three steps method to do that. To be useful, such a method needs a modular algorithm, with a lot of options that can be chosen according to the kind of problem under consideration. We propose such a tool, Balanced PSO, based on Standard PSO, and whose code is freely available. Finally, we explain how to use it with one example (the class of low dimensional moderately multimodal functions).

Acknowledgement

I want to thank here Abhi Dattasharma who looked closely at the final version of this chapter for English style and grammar, correcting both and offering suggestions for improvement.

References

1. Ben Ghalia, M.: Particle swarm optimization with an improved exploration-exploitation balance. In: Circuits and Systems, MWSCAS, pp. 759–762. Circuits and Systems, MWSCAS (2008)
2. CEC. Congress on Evolutionary Computation Benchmarks (2005), http://www3.ntu.edu.sg/home/epnsugan/
3. Clerc, M.: Math Stuff about PSO, http://clerc.maurice.free.fr/pso/
4. Clerc, M.: Particle Swarm Optimization. ISTE (International Scientific and Technical Encyclopedia) (2006)
5. Clerc, M.: Stagnation analysis in particle swarm optimization or what happens when nothing happens, Technical report (2006), http://hal.archives-ouvertes.fr/hal-00122031
6. Clerc, M.: Why does it work? International Journal of Computational Intelligence Research 4(2), 79–91 (2008)
7. Clerc, M., Kennedy, J.: The Particle Swarm-Explosion, Stability, and Convergence in a Multidimensional Complex Space. IEEE Transactions on Evolutionary Computation 6(1), 58–73 (2002)

8. Fernandez-Martinez, J.L., Garcia-Gonzalo, E., Fernandez-Alvarez, J.P.: Theoretical analysis of particle swarm trajectories through a mechanical analogy. International Journal of Computational Intelligent Research (this issue, 2007)
9. Gacôgne, L.: Steady state evolutionary algorithm with an operator family. In: EISCI, Kosice, Slovaquie, pp. 373–379 (2002)
10. Helwig, S., Wanka, R.: Particle swarm optimization in high-dimensional bounded search spaces. In: IEEE Swarm Intelligence Symposium (SIS 2007) (2007)
11. Helwig, S., Wanka, R.: Theoretical analysis of initial particle swarm behavior. In: Rudolph, G., Jansen, T., Lucas, S., Poloni, C., Beume, N. (eds.) PPSN 2008. LNCS, vol. 5199, pp. 889–898. Springer, Heidelberg (2008)
12. Kennedy, J.: Bare Bones Particle Swarms. In: IEEE Swarm Intelligence Symposium, pp. 80–87 (2003) (Fully informed PSO)
13. Kennedy, J.: How it works: Collaborative Trial and Error. International Journal of Computational Intelligence Research 4(2), 71–78 (2008)
14. Koduru, P., et al.: A particle swarm optimization-nelder mead hybrid algorithm for balanced exploration and exploitation in multidimensional search space (2006)
15. Langdon, W., Poli, R.: Evolving problems to learn about particle swarm and other optimisers. In: Congress on Evolutionary Computation, pp. 81–88 (2005)
16. Li, N., Sun, D., Zou, T., Qin, Y., Wei, Y.: Analysis for a particle's trajectory of pso based on difference equation. Jisuanji Xuebao/Chinese Journal of Computers 29(11), 2052–2061 (2006)
17. Marsaglia, G., Zaman, A.: The kiss generator. Technical report, Dept. of Statistics, U. of Florida (1993)
18. Mendes, R.: Population Topologies and Their Influence in Particle Swarm Performance. PhD thesis, Universidade do Minho (2004)
19. Onwubolu, G.C., Babu, B.V.: New Optimization Techniques in Engineering. Springer, Berlin (2004)
20. Parsopoulos, K.E., Vrahatis, M.N.: Parameter selection and adaptation in unified particle swarm optimization. Mathematical and Computer Modelling 46, 198–213 (2007)
21. Poli, R.: The Sampling Distribution of Particle Swarm Optimisers and their Stability. Technical report, University of Essex (2007); Poli, R.: On the moments of the sampling distribution of particle swarm optimisers. GECCO (Companion), 2907–2914 (2007)
22. Poli, R.: Dynamics and stability of the sampling distribution of particle swarm optimisers via moment analysis. Journal of Artificial Evolution and Applications (2008)
23. Poli, R., Langdon, W.B., Clerc, M., Stephen, C.R.: Continuous Optimisation Theory Made Easy? Finite-Element Models of Evolutionary Strategies, Genetic Algorithms and Particle Swarm Optimizers. In: Stephens, C.R., et al. (eds.) Foundations of Genetic Algorithms, Mexico, vol. 9, pp. 165–193. Springer, Heidelberg (2007)
24. PSC. Particle Swarm Central, `http://www.particleswarm.info`
25. Richards, M., Ventura, D.: Dynamic Sociometry and Population Size in Particle Swarm Optimization. C'est juste un extrait de la thèse (2003)
26. Sandgren, E.: Non linear integer and discrete programming in mechanical design optimization (1990) ISSN 0305-2154
27. Shang, Y.-W., Qiu, Y.-H.: A note on the extended rosenbrock function. Evolutionary Computation 14(1), 119–126 (2006)
28. Wolpert, D.H., Macready, W.G.: No free lunch theorems for optimization. IEEE Transactions on Evolutionary Computation 1(1), 67–82 (1997)

29. Wong, T.-T., Luk, W.-S., Heng, P.-A.: Sampling with Hammersley and Halton points. Journal of Graphics Tools 2(2), 9–24 (1997)

Appendix

A Examples of Characteristic Distributions

An important conjecture is that the characteristic distribution of the $\mathcal{F}^+$ class is unimodal (possibly with a plateau) with the minimum on the boundary. If this conjecture is true, then it directly implies that there exists a best algorithm on F^+ which can be explicitly described. Note that if the search space is symmetrical, and has a "centre", like in the examples below, then the characteristic distribution also is obviously symmetrical around this centre.

This conjecture is well supported by experiments. Let us give some examples in dimension 1 and 2.

Example 1. One dimension

$X = (1,2,3,4,5)$, $Y = (1,2,3,4,5)$ $|\mathcal{F}| = 3125$, $|\mathcal{F}^+| = 1090$

i	1	2	3	4	5
$n(x_i)$	205	411	478	411	205
S^+	0.19	0.38	0.44	0.38	0.19

Example 2. Two dimensions

$X = ((1,1),(2,1),(3,1),(1,2),(2,2),(3,2))$, $Y = (1,2,3,4,5,6)$ $|\mathcal{F}| = 46656$, $|\mathcal{F}^+| = 18620$

i	1	2	3
	4	5	6
$n(x_i)$	3963	6580	3963
	3963	6580	3963
S^+	0.212	0.352	0.212
	0.212	0.352	0.212

Computing a characteristic distribution is quite time consuming, if the search space is not very small. However, $\mathcal{F}$ can be divided into equivalence classes, on which most algorithms have the same behaviour. In short, f and g are said to be equivalent if for any x_i and any x_j we have $f(x_i) < f(x_j) \Leftrightarrow g(x_i) < g(x_j)$. All the functions in a class have the same NisB truth value, and the same "profile".

For example,in SPSO, we use rules which only use comparisons between f values and therefore SPSO would behave similarly across an equivalence class. Some versions of PSO, however, do use the f values directly; that is why we say "most algorithms" have the same behaviour on such a class and not "all algorithms".

The point is that as soon as we have defined these equivalence classes, we can work on them to compute the characteristic distribution. We indeed have two methods to compute $n(x_i)$.

First method:

- consider all the functions. For each point x_i, count how many have a a global minimum on x_i

Second method:

- define the equivalence classes and choose a representative for each class
- consider all the representatives and their "weight" (size of the class). For each point x_i, count how many have a a global minimum on x_i. Multiply it by the size of the class

Experimentally, the computer time needed by the second method is significantly smaller than the one needed by the first one. That is why even the small tables for Example 1 and Example 2 have been built by using this method.

B Designing a Deceptive Continuous Function

The NisB notion gives us a nice way to design a function that is deceptive for most algorithms, and particularly for SPSO. We can start from any discrete function that belongs to $\mathcal{F}^-$, say $f = (0,3,2,4,2)$ on $X = (1,2,3,4,5)$. Here we have $\nu(f) = -0.17$. Then we can derive a piece-wise function g (more precisely an union of plateaus) on say $[0,5[$ by:

$$x \in [i-1, i[\Rightarrow g(x) = f(i)$$

where i is in X. On this function the probability of success of Random Search R after at most n attempts is given by $p(n) = 1 - (1 - 1/5)^n$. We can compare this to the result obtained by a classical PSO with, say, five particles (swarm size $S = 5$), as shown in Table 2. Of course, when the number of attempts is precisely equal to the number of particles, PSO is equivalent to R because only the random initialisation phase is performed.

Table 2. Comparison of Random Search and PSO on a piece-wise deceptive function. For PSO the success rate is estimated over 5000 runs

Number of attempts	Random search	PSO 5 particles
5	0.67	0.67
20	0.99	0.73

C Problems

The problems with an offset are taken from the CEC 2005 benchmark[2].

C.1 Sphere/Parabola F1

The function to be minimised is

$$f = -450 + \sum_{d=1}^{30} (x_d - o_d)^2 \tag{25}$$

The search space is $[-100, 100]^{30}$. The offset vector $O = (o_1, \cdots, o_{30})$ is defined by its C code below. It is also the solution point, where $f = -450$.

Offset (C code)

static double offset_0[30] =
{ -3.9311900e+001, 5.8899900e+001, -4.6322400e+001, -7.4651500e+001, -1.6799700e+001, -8.0544100e+001, -1.0593500e+001, 2.4969400e+001, 8.9838400e+001, 9.1119000e+000, -1.0744300e+001, -2.7855800e+001, -1.2580600e+001, 7.5930000e+000, 7.4812700e+001, 6.8495900e+001, -5.3429300e+001, 7.8854400e+001, -6.8595700e+001, 6.3743200e+001, 3.1347000e+001, -3.7501600e+001, 3.3892900e+001, -8.8804500e+001, -7.8771900e+001, -6.6494400e+001, 4.4197200e+001, 1.8383600e+001, 2.6521200e+001, 8.4472300e+001 };

C.2 Schwefel F2

$$f = -450 + \sum_{d=1}^{10} \left(\sum_{k=1}^{d} x_k - o_k \right)^2 \tag{26}$$

The search space is $[-100, 100]^{10}$. The solution point is the offset $O = (o_1, \ldots, o_{10})$, where $f = -450$.

Offset (C code)

static double offset_4[30] =
{ 3.5626700e+001, -8.2912300e+001, -1.0642300e+001, -8.3581500e+001, 8.3155200e+001, 4.7048000e+001, -8.9435900e+001, -2.7421900e+001, 7.6144800e+001, -3.9059500e+001};

C.3 Rosenbrock F6

$$f = 390 + \sum_{d=2}^{10} \left(100 \left(z_{d-1}^2 - z_d \right)^2 + \left(z_{d-1} - 1 \right)^2 \right) \tag{27}$$

with $z_d = x_d - o_d + 1$. The search space is $[-100, 100]^{10}$. The offset vector $O = (o_1, \cdots, o_{30})$ is defined by its C code below. It is the solution point where $f = 390$. There is also a local minimum at $(o_1 - 2, \cdots, o_{30})$, where $f = 394$[10].

Offset (C code)

static double offset_2[10] =
{ 8.1023200e+001, -4.8395000e+001, 1.9231600e+001, -2.5231000e+000, 7.0433800e+001, 4.7177400e+001, -7.8358000e+000, -8.6669300e+001, 5.7853200e+001};

C.4 Griewank F7 (non Rotated)

$$f = -179 + \frac{\sum_{d=1}^{10} (x_d - o_d)^2}{4000} - \prod_{d=1}^{10} \cos\left(\frac{x_d - o_d}{\sqrt{d}}\right) \tag{28}$$

The search space is $[-600, 600]^{10}$. The solution point is the offset $O = (o_1, \ldots, o_{10})$, where $f = -180$.

Offset (C code)

static double offset_5[30] =
{ -2.7626840e+002, -1.1911000e+001, -5.7878840e+002, -2.8764860e+002, -8.4385800e+001, -2.2867530e+002, -4.5815160e+002, -2.0221450e+002, -1.0586420e+002, -9.6489800e+001};

C.5 Ackley F8 (non Rotated)

$$f = -120 + e + 20e^{-0.2\sqrt{\frac{1}{D}\sum_{d=1}^{10}(x_d - o_d)^2}} - e^{\frac{1}{D}\sum_{d=1}^{10}\cos(2\pi(x_d - o_d))} \tag{29}$$

The search space is $[-32, 32]^{10}$ .The solution point is the offset $O = (o_1, \ldots, o_{10})$, where $f = -140$.

Offset (C code)

static double offset_6[30] =
{ -1.6823000e+001, 1.4976900e+001, 6.1690000e+000, 9.5566000e+000, 1.9541700e+001, -1.7190000e+001, -1.8824800e+001, 8.5110000e-001, -1.5116200e+001, 1.0793400e+001};

[10] The Rosenbrock function is indeed multimodal as soon the dimension is greater than three [27].

C.6 Rastrigin F9

$$f = -230 + \sum_{d=1}^{30} \left((x_d - o_d)^2 - 10 \cos \left(2\pi \left(x_d - o_d \right) \right) \right)$$

The search space is $[-5, 5]^{10}$. The solution point is the offset $O = (o_1, \ldots, o_{10})$, where $f = -330$.

Offset (C code)

```
static double offset_3[30] =
{ 1.9005000e+000, -1.5644000e+000, -9.7880000e-001, -2.2536000e+000,
2.4990000e+000, -3.2853000e+000, 9.7590000e-001, -3.6661000e+000,
9.8500000e-002, -3.2465000e+000};
```

C.7 Tripod

The function to be minimised is ([9])

$$\begin{aligned} f = & \frac{1-sign(x_2)}{2} \left(|x_1| + |x_2 + 50| \right) \\ & + \frac{1+sign(x_2)}{2} \frac{1-sign(x_1)}{2} \left(1 + |x_1 + 50| + |x_2 - 50| \right) \\ & + \frac{1+sign(x_1)}{2} \left(2 + |x_1 - 50| + |x_2 - 50| \right) \end{aligned} \tag{30}$$

with

$$\begin{aligned} sign(x) &= -1 \quad \text{if} \quad x \le 0 \\ &= 1 \quad \text{else} \end{aligned}$$

The search space is $[-100, 100]^2$. The solution point is $(0, -50)$, where $f = 0$.

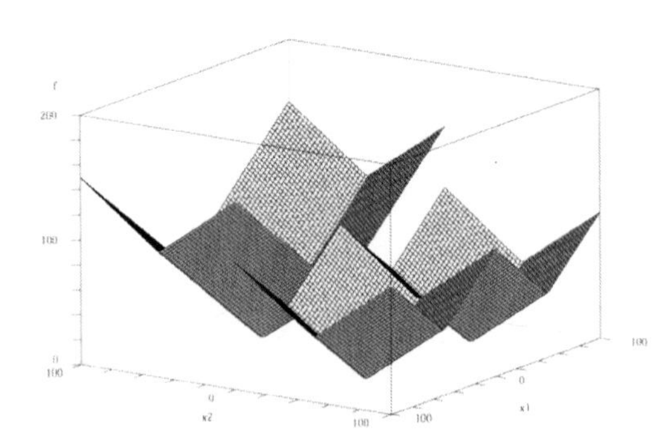

Fig. 7. Tripod function. Although the dimension is only two, finding the minimum is not that easy

C.8 Pressure Vessel

We give only an outline of this problem. For more details, see[26, 4, 19]. There are four variables

$$\begin{aligned} &x_1 \in [1.125, 12.5] \text{ granularity } 0.0625\\ &x_2 \in [0.625, 12.5] \text{ granularity } 0.0625\\ &x_3 \in]0, 240]\\ &x_4 \in]0, 240] \end{aligned}$$

and three constraints

$$\begin{aligned} g_1 &:= 0.0193x_3 - x_1 \leq 0\\ g_2 &:= 0.00954x_3 - x_2 \leq 0\\ g_3 &:= 750 \times 1728 - \pi x_3^2 \left(x_4 + \tfrac{4}{3}x_3\right) \leq 0 \end{aligned}$$

The function to be minimised is

$$f = 0.6224x_1x_3x_4 + 1.7781x_2x_3^2 + x_1^2\left(3.1611x + 19.84x_3\right) \tag{31}$$

The analytical solution is $(1.125, 0.625, 58.2901554, 43.6926562)$, which gives the fitness value 7,197.72893. To take the constraints into account, a penalty method is used.

C.9 Compression Spring

For more details, see[26, 4, 19]. There are three variables

$$\begin{aligned} &x_1 \in \{1, \ldots, 70\} \text{ granularity } 1\\ &x_2 \in \quad [0.6, 3]\\ &x_3 \in [0.207, 0.5] \text{ granularity } 0.001 \end{aligned}$$

and five constraints

$$\begin{aligned} g_1 &:= \frac{8C_f F_{max} x_2}{\pi x_3^3} - S \leq 0\\ g_2 &:= l_f - l_{max} \leq 0\\ g_3 &:= \sigma_p - \sigma_{pm} \leq 0\\ g_4 &:= \sigma_p - \frac{F_p}{K} \leq 0\\ g_5 &:= \sigma_w - \frac{F_{max} - F_p}{K} \leq 0 \end{aligned}$$

with

$$\begin{aligned} C_f &= 1 + 0.75\frac{x_3}{x_2 - x_3} + 0.615\frac{x_3}{x_2}\\ F_{max} &= 1000\\ S &= 189000\\ l_f &= \frac{F_{max}}{K} + 1.05\left(x_1 + 2\right)x_3\\ l_{max} &= 14\\ \sigma_p &= \frac{F_p}{K}\\ \sigma_{pm} &= 6\\ F_p &= 300\\ K &= 11.5 \times 10^6 \frac{x_3^4}{8x_1x_2^3}\\ \sigma_w &= 1.25 \end{aligned}$$

and the function to be minimised is

$$f = \pi^2 \frac{x_2 x_3^2 (x_1 + 1)}{4} \tag{32}$$

The best known solution is $(7, 1.386599591, 0.292)$, which gives the fitness value 2.6254214578. To take the constraints into account, a penalty method is used.

C.10 Gear Train

For more details, see[26, 19]. The function to be minimised is

$$f(x) = \left(\frac{1}{6.931} - \frac{x_1 x_2}{x_3 x_4} \right)^2 \tag{33}$$

The search space is $\{12, 13, \ldots, 60\}^4$. There are several solutions, depending on the required precision. For this chapter, we used 1×10^{-13}. So, a possible solution is $f(19, 16, 43, 49) = 2.7 \times 10^{-12}$.

C.11 Cellular Phone

This problem arises in a real application in the telecommunications domain. However, here, it is over simplified: all constraints has been removed, except of course the ones given by the search space itself, and there is no time varying variables. We have a square flat domain $[0, 100]^2$, in which we want to put M stations. Each station m_k has two coordinates $(m_{k,1}, m_{k,2})$. These are the $2M$ variables of the problem. We consider each "integer" point of the domain, i.e. $(i, j), i \in \{0, 1, \ldots, 100\}, j \in \{0, 1, \ldots, 100\}$. On each lattice point, the field induced by the station m_k is given by

$$f_{i,j,m_k} = \frac{1}{(i - m_{k,1})^2 + (j - m_{k,2})^2 + 1} \tag{34}$$

and we want to have at least one field that is not too weak. Finally, the function to be minimised is

$$f = \frac{1}{\sum_{i=1}^{100} \sum_{j=1}^{100} \max_k (f_{i,j,m_k})} \tag{35}$$

In this chapter, we set $M = 10$. Therefore the dimension of the problem is 20. The objective value is 0.005530517. This is not the true minimum, but enough from an engineering point of view. Of course, in reality we do not know the objective value. From figure 8 we can see a solution found by SPSO over five runs of at most 20000 fitness evaluations. The represented solution is

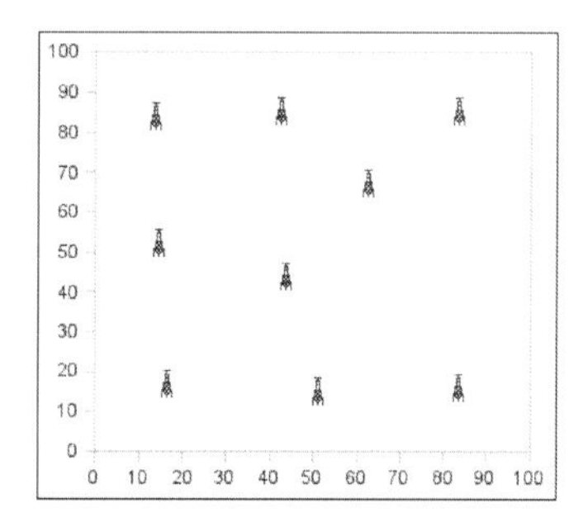

Fig. 8. Cellular phone problem. A possible (approximate) solution for 10 stations, found by SPSO after 20000 fitness evaluations

$$
\begin{array}{c}
(84.8772524383, 48.0531002672)\\
(63.0652824840, 66.9082795241)\\
(84.0251981196, 85.2946222198)\\
(42.9532153205, 85.0583215663)\\
(13.9838548728, 84.0053734843)\\
(44.0541093162, 43.8945288153)\\
(84.0621036253, 16.0606612136)\\
(16.7185336955, 16.9123022615)\\
(14.9363969042, 51.9481381014)\\
(51.6936885883, 14.9308551889)
\end{array}
$$

and has been found after 6102 evaluations.

Actually, for this simplified problem, more efficient methods do exist (Delaunay's tesselation, for example), but those can not be used as soon as we introduce a third dimension and more constraints.

D Results

Table 3. Results with the PSO variants discussed in this chapter. In each cell, we have the mean best value over 100 runs (including the values for Rosenbrock, although it is questionable as shown in E), the success rate if not null, and the mean number of fitness evaluations if the success rate is 100%. The rows in gray correspond to our small benchmark with moderately multimodal and low dimensional functions. Results for Standard PSO 2007, and the "Nearer is Better" approach

	FE_{max}	**Standard**	**NisB**	**NisB**
	Precision		**global**	***K*=3**
Sphere F1	300000	10^{-6}	10^{-6}	10^{-6}
	10^{-6}	100%	100%	100%
		13861	2618	7965
Schwefel F2	100000	10^{-5}	10^{-5}	10^{-5}
	10^{-5}	100%	100%	100%
		9682	3427	6815
Rosenbrock F6	100000	*4.16*	*72.71*	*429.9*
	10^{-2}	70%	57%	
Griewank F7	100000	0.055	0.25	0.16
	10^{-2}	4%	1%	2%
Ackley F8	100000	0.096	11.9	2.35
	10^{-4}	93%	0%	11%
Rastrigin F9	300000	52.18	279.82	194.35
	10^{-2}			
Tripod	10000	0.51	0.88	0.84
	10^{-4}	51%	38%	39%
Pressure vessel	50000	26.76	1295.3	20.78
	10^{-5}	86%	0%	0%
Compression spring	20000	0.027	74.51	3.57
	10^{-10}	33%	0%	0%
Gear train	20000	1×10^{-8}	6×10^{-7}	9×10^{-8}
	10^{-13}	3%	0%	0%
Cellular phone	20000	4.4×10^{-6}	4.27×10^{-5}	1.8×10^{-5}
	10^{-9}	26%	0%	0%

Table 4. Results for different options of Balanced PSO

	H2 D1	H2 D0	R2	D0 V1 R2 P2 H2
Sphere F1	10^{-6}	10^{-6}	10^{-6}	10^{-6}
	100%	100%	100%	100%
	19293	14534	12783	13747
Schwefel F2	10^{-5}	10^{-5}	10^{-5}	10^{-5}
	100%	100%	100%	100%
	12053	10187	9464	10003
Rosenbrock F6	6.28	4.57	1.66	1.30
	66%	65%	67%	71%
Griewank F7	0.07	0.05	0.05	0.046
	2%	6%	6%	6%
Ackley F8	9×10^{-5}	0.01	0.035	0.06
	100%	99%	97%	95%
	6377			
Rastrigin F9	7.96	53.87	80.92	82.33
Tripod	0.65	0.61	0.19	0.20
	46%	50%	69%	73%
Pressure vessel	27.86	13.47	11.11	7.5
	7%	91%	91%	94%
Compression spring	0.066	0.026	0.03	0.03
	7%	34%	36%	42%
Gear train	5×10^{-9}	10^{-9}	3.4×10^{-10}	3.7×10^{-10}
	3%	9%	24%	22%
Cellular phone	2.2×10^{-6}	5.1×10^{-6}	1.7×10^{-5}	1.7×10^{-5}
	34%	17%	0%	3%

E When the Mean Best Is Meaningless

Let us run our Modular PSO on the Rosenbrock function, and see what happens when the number of runs increases (table 5). It is clear that the mean best value does not converge at all, while the success rate does. The mean best value is therefore is not an acceptable performance criterion. Let us try to explain this phenomenon.

E.1 Distribution of the Errors for Rosenbrock 2D

We run the algorithm 5000 times, with 5000 fitness evaluations for each run, i.e. just enough to have a non zero success rate. Each time, we save the best value found. We can then estimate the shape of the distribution of these 5000 values, seen as occurrences of a random variable (see Figure 9).Contrary to what is sometimes claimed, this distribution is far from normal (Gaussian). Indeed, the main peak is very sharp, and there are some very high values. Even if these are

Table 5. For Rosenbrock, the mean best value is highly dependent on the number of runs (50000 fitness evaluations for each run). The success rate is more stable

Runs	Success rate	Mean best value
100	16%	10.12
500	15%	12.36
1000	14,7%	15579.3
2000	14%	50885.18

rare, it implies that the mean value is not really representative of the performance of the algorithm. It would be better to consider the value on which the highest peak (the mode) lies. For SPSO, it is about 7 (the correct value is 0), and the mean is 25101.4 (there are a few very bad runs). The small peak (around 10, as the correctt value is 4) corresponds to a local optimum. Sometimes (but rarely) the swarm gets quickly trapped in it. Actually, as soon as there are local optima such a distribution will indeed have some peaks, at least for a small number of fitness evaluations.

As we can see from figure9, the models used there are quite accurate. Here, we used the union of a power law (on the left of the main peak), and a Cauchy law (on the right). One should note that finding these models may not be easy though. The formulae are:

$$\begin{aligned} frequency &= \alpha \frac{m^k}{class^{k+1}} && \text{if } class \leq m \\ &= \frac{1}{\pi} \frac{\gamma}{(class-m)^2+\gamma^2} && \text{else} \end{aligned} \tag{36}$$

with $\gamma = 1.294$, $m = 7$, and $k = 6.5$. Note that a second power law for the right part of the curve (instead of the Cauchy) would not be suitable: although it could be better for class values smaller than, say, 15, it would "forget" the important fact that the probability of high values is far from zero. In fact, even the Cauchy model is over-optimistic, as we can see from the magnified version (classes 40-70) of figure 9, but at least the probability is not virtually equal to zero, as happens with the power law model.

E.2 Mean Best vs. Success Rate as Criterion

A run is said to be *successful* if the final value is smaller than a "small" ε , and *bad* if the final value is greater than a "big" M. For one run, let p_M be the probability of that run being bad. Then, the probability, over N runs, that at least one of the runs is bad is

$$p_{M,N} = 1 - (1 - p_M)^N$$

This probability increases quickly with the number of runs. Now, let f_i be the final value of the run i. The estimate of the mean best value is usually given by

$$\mu_N = \frac{\sum_{i=1}^{N} f_i}{N}$$

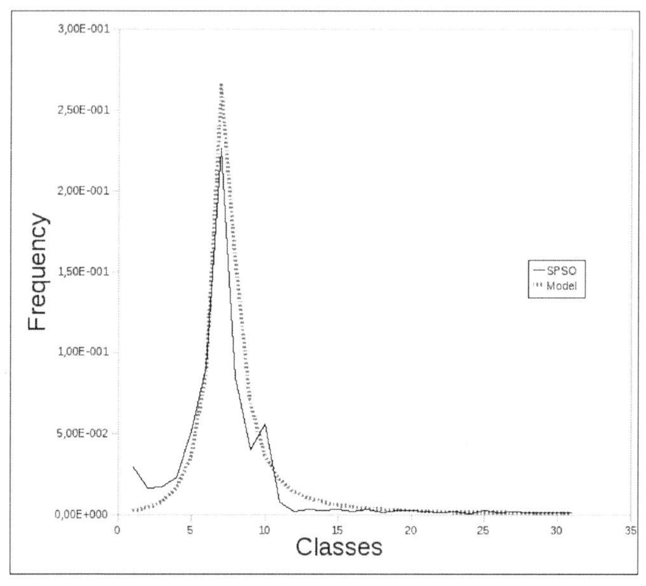

(a) Global shape

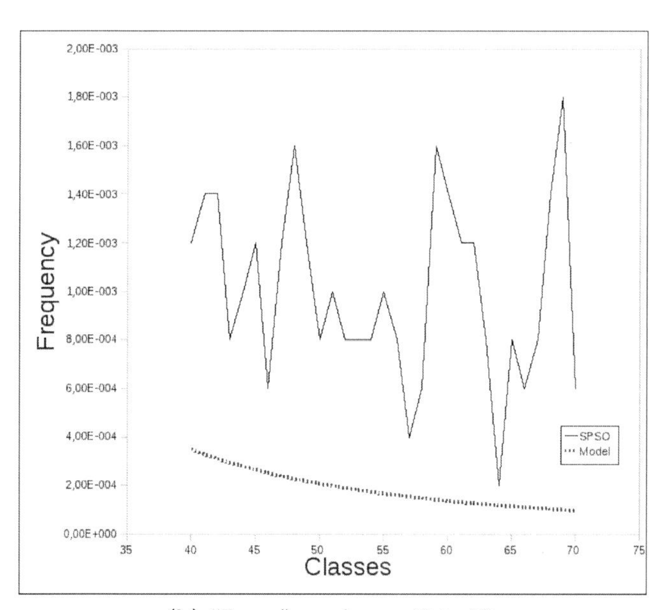

(b) "Zoom" on classes 40 to 70

Fig. 9. Rosenbrock. Distribution of the best value over 5000 runs. On the "zoom", we can see that the Cauchy model, although optimistic, gives a better idea of the distribution than the power law model for class values greater than 40

Let us say the success rate is ς . It means we have ςN successful runs. Let us consider another sequence of N runs, exactly the same, except that k runs are "replaced" by bad ones. Let m be the maximum of the corresponding f_i in the first sequence of N runs. The probability of this event is

$$p_{M,N,1} = p_M^k \left(1 - p_M\right)^{N-k}$$

For the new success rate ς' , we have

$$\varsigma \geq \varsigma' \geq \varsigma - \frac{k}{N}$$

For the new estimate μ'_N of the mean best, we have

$$\mu'_N > \mu_N + k\frac{M - m}{N}$$

We immediately see that there is a problem when a "big value" M is possible with a non negligible probability: when the number of runs N increases the success rate may slightly decrease, but then the mean dramatically increases. Let us suppose that, for a given problem and a given algorithm, the distribution of the errors follows a Cauchy law. Then we have

$$p_M = 0.5 - \frac{1}{\pi} \arctan\left(\frac{M}{\gamma}\right)$$

With the parameters of the model of figure 9, we have, for example, $p_{5000} = 8.3 \times 10^{-5}$. Over $N = 30$ runs, the probability of having at least one bad run (fitness value greater than $M = 5000$) is low, just 2.5×10^{-3}. Let us say we find an estimate of the mean to be m. Over $N = 1000$ runs, the probability is 0.08, which is quite high. It may easily happen. In such a case, even if for all the other runs the best value is about m, the new estimate is about $(4999m + 5000)/1000$, which may be very different from m. If we look at table 5, this simplified explanation shows that for Rosenbrock a Cauchy law based model is indeed optimistic.

In other words, if the number of runs is too small, you may never have a bad one, and therefore, you may wrongly estimate the mean best, even when it exists. Note that in certain cases the mean may not even exist (for example, in case of a Cauchy law), and therefore any estimate of a mean best is wrong. That is why it is important to estimate the mean for different N values (but of course with the same number of fitness evaluations). If it seems not stable, forget this criterion, and just consider the success rate, or, as seen above, the mode. As there are a lot of papers in which the probable existence of the mean is not checked, it is worth insisting on it: if there is no mean, giving an "estimate" of it is not technically correct. Worse, comparing two algorithms based on such an "estimate" is simply wrong.

What Makes Particle Swarm Optimization a Very Interesting and Powerful Algorithm?

J.L. Fernández-Martínez[1,2,3] and E. García-Gonzalo[3]

[1] Energy Resources Department. Stanford University, Palo Alto, USA
[2] Department of Civil and Environmental Engineering. University of California Berkeley-Lawrence Berkeley Lab., Berkeley, USA
[3] Department of Mathematics. University of Oviedo, Oviedo, Spain
jlfm@uniovi.es, espe@uniovi.es

Abstract. Particle swarm optimization (PSO) is an evolutionary computational technique used for optimization motivated by the social behavior of individuals in large groups in nature. Different approaches have been used to understand how this algorithm works and trying to improve its convergence properties for different kind of problems. These approaches go from heuristic to mathematical analysis, passing through numerical experimentation. Although the scientific community has been able to solve a big variety of engineering problems, the tuning of the PSO parameters still remains one of its major drawbacks.

This chapter reviews the methodology developed within our research group over the last three years, which is based in adopting a completely different approach than those followed by most of the researchers in this field. By trying to avoid heuristics we proved that PSO can be physically interpreted as a particular discretization of a stochastic damped mass-spring system. Knowledge of this analogy has been crucial in deriving the PSO continuous model and to deduce a family of PSO members with different properties with regard to their exploitation/exploration balance: the generalized PSO (GPSO), the CC-PSO (centered PSO), CP-PSO (centered-progressive PSO), PP-PSO (progressive-progressive PSO) and RR-PSO (regressive-regressive PSO). Using the theory of stochastic differential and difference equations, we fully characterize the stability behavior of these algorithms. For well posed problems, a sufficient condition to achieve convergence is to select the PSO parameters close to the upper limit of second order stability. This result is also confirmed by numerical experimentation for different benchmark functions having an increasing degree of numerical difficulties. We also address how the discrete GPSO version (stability regions and trajectories) approaches the continuous PSO model as the time step decreases to zero. Finally, in the context of inverse problems, we address the question of how to select the appropriate PSO version: CP-PSO is the most explorative version and should be selected when we want to perform sampling of the posterior distribution of the inverse model parameters. Conversely, CC-PSO and GPSO provide higher convergence rates. Based on the analysis shown in this chapter, we can affirm that the PSO optimizers are not heuristic algorithms since there exist mathematical results that can be used to explain their consistency/convergence.

Keywords: particle swarm, PSO continuous model, GPSO, CC-GPSO, CP-GPSO, stochastic stability analysis, convergence.

B.K. Panigrahi, Y. Shi, and M.-H. Lim (Eds.): Handbook of Swarm Intelligence, ALO 8, pp. 37–65.
springerlink.com

1 Introduction

Particle Swarm Optimization [1] is a global optimization algorithm that it is based on a sociological model to analyze the behavior of individuals in groups in nature. One of the main features of this algorithm is its apparent simplicity when applied to solve optimization problems:

1. Individuals, or particles, are represented by vectors whose length is the number of degrees of freedom of the optimization problem.
2. The part of the model space where we look for solutions is called the search space and is the only prior knowledge we require to solve any optimization problem. To start, a population of particles is initialized with random positions $(\mathbf{x}_i^0)$ and velocities $(\mathbf{v}_i^0)$. Usually the positions of the particles try to intelligently sampling a prismatic volume on the model space. The velocities are the perturbations of the model parameters needed to find the global minimum, supposing that it does exist and is unique. Initially they are set to zero or they might be randomized with values not greater than a certain percentage of the search space in each direction.
3. A misfit (in the case of an inverse problem) or cost function is evaluated for each particle of the swarm in each iteration. In inverse problems this stage is called the forward problem and typically involves the numerical solution of a set of partial or ordinary differential equations, integral equations or algebraic systems. Thus, the use of global optimization techniques to solve inverse problems is much restricted by the speed needed to solve the so-called forward problem and by the number of parameters used to describe the model space (curse of dimensionality).
4. As time advances, the positions and velocities of each particle are updated as a function of its misfit and the misfit of its neighbors. At time-step $k+1$, the algorithm updates positions $\left(\mathbf{x}_i^{k+1}\right)$ and velocities $\left(\mathbf{v}_i^{k+1}\right)$ of the individuals as follows:

$$\begin{aligned}\mathbf{v}_i^{k+1} &= \omega\mathbf{v}_i^k + \phi_1(\mathbf{g}^k - \mathbf{x}_i^k) + \phi_2(\mathbf{l}_i^k - \mathbf{x}_i^k),\\ \mathbf{x}_i^{k+1} &= \mathbf{x}_i^k + \mathbf{v}_i^{k+1}\end{aligned}$$

   with

$$\phi_1 = r_1 a_g \ \phi_2 = r_2 a_l \ r_1, r_2 \to U(0,1) \ \omega, a_l, a_g \in \mathbb{R}.$$

   $\mathbf{l}_i^k$ is the best position found so far by $i-$th particle and $\mathbf{g}^k$ is the global best position on the whole swarm or in a neighborhood if a local topology is used. ω, a_l, a_g are called the inertia and the local and global acceleration constants and constitute the parameters we have to tune for the PSO to achieve convergence. r_1, r_2 are uniform random numbers used to generate the stochastic global and local accelerations, ϕ_1 and ϕ_2. Due to the stochastic effect introduced by these numbers PSO trajectories should be considered as stochastic processes.

A great effort has been deployed to provide PSO convergence results through the stability analysis of the trajectories [2]-[14]. These studies were aimed at understanding theoretically why PSO algorithm works and why under certain conditions it might fail to find a good solution. From these studies and numerical experiments three different promising parameter sets of inertia (ω), and local and global acceleration constants (a_l and a_g) were proposed [4]-[6]. The presence of instabilities in the PSO is a major issue, as shown by the theoretical study performed by Clerc and Kennedy [5]. They introduced a constriction factor to avoid them. First, the inertia parameter was introduced to win stability, nevertheless the problem was not completely solved. Also, a common practice in PSO algorithm is to clamp the particle velocities trying to get stability and avoid hitting the boundaries of the search space. As we will show in the section devoted to the GPSO analysis the numerical constriction factor is the time discretization step, Δt.

The reason why stability and convergence are related is the following: in PSO each particle of the swarm has two points of attraction, the global best, $\mathbf{g}^k$, and each particle previous best position, $\mathbf{l}_i^k$. The algorithm can be interpreted as a two discrete gradient method with random effects introduced in the global and local acceleration constants, by uniform random numbers r_1, r_2. The following stochastic vectorial difference equation is involved for each particle in the swarm:

$$\begin{cases} \mathbf{x}_i^{k+1} + (\phi - \omega - 1)\mathbf{x}_i^k + \omega \mathbf{x}_i^{k-1} = \phi \mathbf{o}_i^k = \phi_1 \mathbf{g}^k + \phi_2 \mathbf{l}_i^k, \\ \mathbf{x}_i^0 = \mathbf{x}_{i0}, \\ \mathbf{x}_i^1 = \varphi\left(\mathbf{x}_i^0, \mathbf{v}_{i0}\right), \end{cases}, \tag{1}$$

where $\phi = \phi_1 + \phi_2$, and $\mathbf{x}_i^0$, $\mathbf{v}_{i0}$ are the initial positions and velocities [10]. Particle trajectories oscillate around the point:

$$\mathbf{o}_i^k = \frac{a_g \mathbf{g}^k + a_l \mathbf{l}_i^k}{a_g + a_l}.$$

The stability of the mean trajectories (or first order stability) depends only on how the parameters ω, and $\overline{\phi} = \frac{a_g + a_l}{2}$ are chosen. The first order stability region turns to be:

$$S_D = \left\{ \left(\omega, \overline{\phi}\right) : |\omega| < 1,\ 0 < \overline{\phi} < 2\left(\omega + 1\right) \right\}. \tag{2}$$

Thus, choosing the $\left(\omega, \overline{\phi}\right)$ parameters on S_D (2) makes the particle trajectories oscillating around $\mathbf{o}_i^k$, stabilizing their position with iterations. A sufficient condition for PSO to find the global minimum of the cost function, called $\min c$, is that $\mathbf{o}_i^k$ should approach $\min c$ with iterations, and the particles get attracted towards $\mathbf{o}_i^k$. As we will show later in this chapter, the attraction potential of the oscillation center is related to the stability of the variance of the trajectories (second order moments).

Most of the work done on the PSO trajectories analysis assumed that the center was stagnated, and/or the attractors had a deterministic character. In fact, these approaches are equivalent to analyze the stability of the homogeneous trajectories in difference equation(1). A complete study of this subject can be consulted in [10]. Four different zones of stability arise for the first order

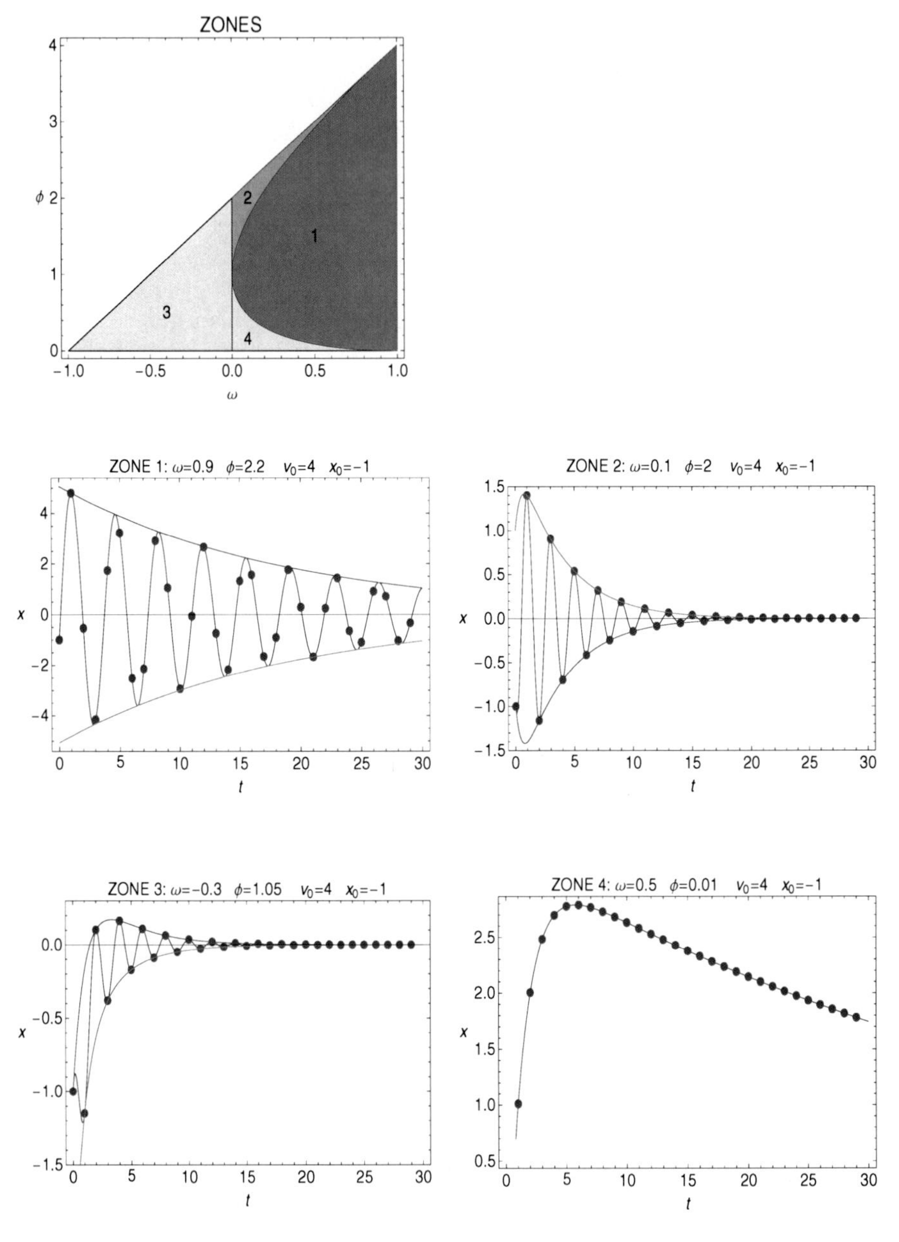

Fig. 1. Homogeneous trajectories (without taking into account the effect of the center of attraction) for particles in different zones of the PSO deterministic stability region: ZONE 1: Complex zone. ZONE 2: Real symmetrical oscillating zone. ZONE 3: Real asymmetrical oscillating zone. ZONE 4: Real non-oscillating zone. The points represent the PSO values. Continuous line is the solution of the PSO-discrete model at real times. Envelopes curves for trajectories are also shown.

trajectories: the complex zone, the real symmetrical oscillating zone, the real asymmetrical zone and the real non-oscillating zone (figure 1).

Two major axes of advancement to understand the PSO convergence were:

1. The use of **PSO physical analogies** ([15], [16] and [10]), which allowed a study of the PSO dynamics from a physical point of view. These interdisciplinary points of view brought important advances since the PSO algorithm was first proposed based on a sociological analogy. Particularly, the mechanical analogy as stated in [10]-[14] is very important in this theoretical development.
2. The **stochastic analysis of the PSO dynamics**. A few attempts to understand the behavior of the PSO in the presence of stochasticity have been made ([17]-[20], [9], [11]-[13]). Nevertheless it is important to note that in most of these approaches stagnation was adopted as a simplification hypothesis.

The results presented here, developed through these last three years of research on PSO, combine a simple mathematical analysis and a comparison to numerical experiments. The roadmap of this contribution is the following:

1. First, based on a mechanical analogy, we introduce the PSO continuous model and we briefly explain how to perform the stochastic stability analysis of its first and second order moments. We show that its first and second order moments are governed by two systems of first order linear differential equations, where the cost function enters as a force term through the mean trajectory of the center of attraction and via the covariance functions between the center of attraction, the trajectories, and their derivatives. We also show that the oscillation center dynamics are very similar for different type of benchmark functions.
2. Based on this mechanical analysis, we derive a family of PSO algorithms and we deduce their first and second stability regions. In the GPSO case we show how these trajectories (first and second order moments) approach their continuous counterparts as the time step, Δt, goes towards zero. This is done for both the homogeneous and transient cases. We also show that using these models we are able to match acceptably well the results observed in real runs.
3. Finally, we assess how to choose the right PSO member and how to tune the PSO parameters: inertia, local and global accelerations. A main conclusion of this analysis is that there are no "magic" PSO tuning points highly dependent on the kind of cost function we want to optimize, but regions where the all the PSO versions have a higher probability to reach a good solution. These regions are close to the second order stability region on the intersection to the corresponding median lines of the first order stability region where no temporal correlation between trajectories exist. These results allow us to propose the cloud-PSO algorithm where each particle in the swarm has different inertia (damping) and acceleration (rigidity) constants. This feature allows the algorithm to better control the velocities update and to find

the sets of parameters that are better suited for each optimization/inverse problem.

2 How Physics Can Help Heuristics: The Continuous PSO Model

Difference equation (1) presented earlier is the result of applying a centered discretization in acceleration

$$\mathbf{x}_i''(t) \simeq \frac{\mathbf{x}_i(t+\Delta t) - 2\mathbf{x}_i(t) + \mathbf{x}_i(t-\Delta t)}{\Delta t^2}, \tag{3}$$

and a regressive schema in velocity

$$\mathbf{x}_i'(t) \simeq \frac{\mathbf{x}_i(t) - \mathbf{x}_i(t-\Delta t)}{\Delta t}. \tag{4}$$

in time $t = k \in \mathbb{N}$ to the following system of stochastic differential equations:

$$\begin{cases} \mathbf{x}_i''(t) + (1-\omega)\,\mathbf{x}_i'(t) + \phi \mathbf{x}_i(t) = \phi_1 \mathbf{g}(t) + \phi_2 \mathbf{l}_i(t), \ t \in \mathbb{R}, \\ \mathbf{x}_i(\mathbf{0}) = \mathbf{x}_{i0}, \\ \mathbf{x}_i'(\mathbf{0}) = \mathbf{v}_{i0}, \end{cases} \tag{5}$$

adopting a unit discretization time step, $\Delta t = 1$. $\mathbf{v}_{i0}$, $\mathbf{x}_{i0}$ are the initial velocity and position of particle i. Model (5) has been addressed as the PSO continuous model and it has been derived from a mechanical analogy [10]: a damped mass-spring system unit mass, damping factor, $1-\omega$, and stochastic stiffness constant, ϕ. In this model, particle coordinates interact via the cost function terms $\mathbf{g}(t)$ and $\mathbf{l}_i(t)$.

Using the spring-mass analogy Fernández Martínez and García Gonzalo [11] derived the generalization of PSO for any iteration time and step discretization step, the so-called GPSO:

$$\begin{aligned} \mathbf{v}_i(t+\Delta t) &= (1-(1-\omega)\,\Delta t)\,\mathbf{v}_i(t) + \phi_1 \Delta t\,(\mathbf{g}(t) - \mathbf{x}_i(t)) + \phi_2 \Delta t\,(\mathbf{l}_i(t) - \mathbf{x}_i(t)), \\ \mathbf{x}_i(t+\Delta t) &= \mathbf{x}_i(t) + \Delta t.\mathbf{v}_i(t+\Delta t), \ t, \ \Delta t \in \mathbb{R} \\ \mathbf{x}_i(0) &= \mathbf{x}_{i0}, \ \mathbf{v}_i(0) = \mathbf{v}_{i0}. \end{aligned} \tag{6}$$

The introduction of parameter Δt makes model (6) physically correct with respect to the physical units. Also, the systematic study of PSO and GPSO trajectories and their comparison to their continuous counterparts in terms of attenuation, oscillation and center attraction [10], [11], was initially used to select promising PSO parameter areas, and to clarify the success of some parameter sets proposed in the literature [4]-[6]. The first and second stability regions for GPSO with $\Delta t = 1$ are the same found in earlier research work, nevertheless, the GPSO result is more general since it describes how the first and second order stability regions tend to the continuous counterparts as Δt goes to zero.

2.1 Stochatic Analysis of the PSO Continuous Model

The PSO continuous model (5) is a second order stochastic differential equation with randomness effects introduced by the rigidity parameter, ϕ, and by the forcing term, $\phi_1 \mathbf{g}(t) + \phi_2 \mathbf{l}_i(t)$. In this section we apply the well established theory of stochastic processes [21] to analyze the stability of random vibrations associated to the PSO mass-spring system [22]. Knowledge gained from the analysis of the linear continuous PSO model is very important to properly understand the GPSO dynamics, and to separately account for the role of the PSO parameters and that of the cost function on the first and second order trajectories [13].

The major assumption made in this model is that $\mathbf{g}(t)$ and $\mathbf{l}_i(t)$ are stochastic processes that have a certain similarity to the trajectory particle trajectory, $\mathbf{x}(t)$. For simplicity these attractors will be called $g(t)$ and $l(t)$ when they referred to any particle coordinate.

Interpreted in the mean square sense[1], the mean of the stochastic process $x(t)$, $\mu(t) = E(x(t))$, fulfills the following ordinary differential equation:

$$\begin{aligned} &\mu''(t) + (1-\omega)\,\mu'(t) + \overline{\phi}\mu(t) = E(\phi o(t)) = \frac{a_g E(g(t)) + a_l E(l(t))}{2}, \quad t \in \mathbb{R}, \\ &\mu(0) = E(x(0)), \\ &\mu'(0) = E(x'(0)). \end{aligned} \tag{7}$$

To derive (7) we suppose that ϕ is independent from $x(t)$, and also $\phi_1, g(t)$ and $\phi_2, l(t)$ are independent.These facts were observed in all the experimental simulations we have performed with benchmark functions using its discrete approximation: the GPSO algorithm.

The solution $\mu(t)$ of (7) can be written:

$$\mu(t) = \mu_h(t) + \mu_p(t), \tag{8}$$

where $\mu_h(t)$ is the general solution of the corresponding homogeneous differential equation, and

$$\mu_p(t) = \frac{a_g E(g(t)) + a_l E(l(t))}{a_g + a_l} = E(o(t)).$$

This physically means that $E(o(t))$ is the oscillation center for the mean trajectory $\mu(t)$.

Expressions for $\mu_h(t)$ have been presented in [10], as a function of the eigenvalues of the characteristic equation associated to (7). The region of first order stability are:

$$S_C = \{(\omega, \overline{\phi}) : \ \omega < 1, \overline{\phi} > 0\}. \tag{9}$$

Parabola

$$\overline{\phi} = \frac{(1-\omega)^2}{4}, \tag{10}$$

separates the regions where the first order eigenvalues are complex or real, defining three different zones for the continuous PSO mean trajectories [10].

---

[1] Mean square convergence allows to interchange the derivative and the mean operators.

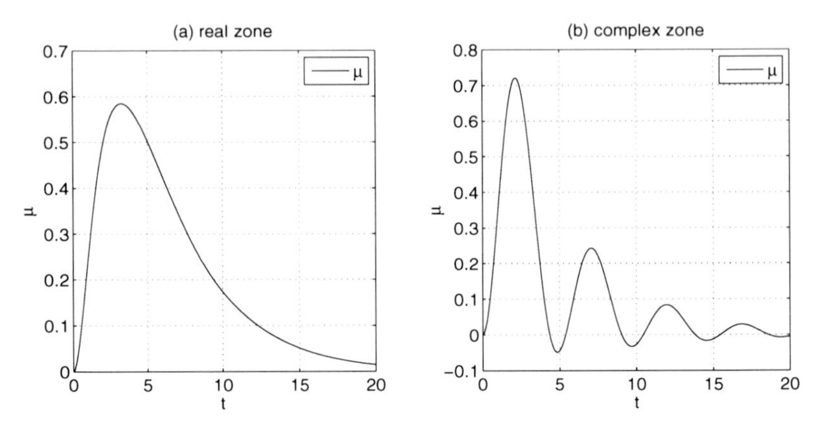

Fig. 2. Non homogeneous trajectories for two different points locagted on the first order stability region: a) $(\omega, \overline{\phi}) = (0.6, 1.7)$ belonging to the complex zone, and b) $(\omega, \overline{\phi}) = (-1, 0.9)$ located on the real zone. In this case the oscillation center folloes a decreasing exponential behavior $E(o(t)) = e^{-0.25t}$.

Figure 2 illustrates, for the case of a decreasing exponential behavior $E(o(t)) = e^{-0.25t}$, the mean trajectory for $(\omega, \overline{\phi}) = (0.6, 1.7)$ belonging to the complex zone of first order stability region, and also for $(\omega, \overline{\phi}) = (-1, 0.9)$ located on the real zone. As a main conclusion the shape of $\mu(t)$ depends on both, the position of the $(\omega, \overline{\phi})$ point on the first order stability region, and on the mean trajectory of the oscillation center, named $E(o(t))$.

Proceeding in the same way we can deduce the system of differential equations for the second order moments:

$$\frac{d}{dt}\begin{pmatrix} \mathrm{Var}(x(t)) \\ \mathrm{Cov}(x(t), x'(t)) \\ \mathrm{Var}(x'(t)) \end{pmatrix} = A_\sigma \begin{pmatrix} \mathrm{Var}(x(t)) \\ \mathrm{Cov}(x(t), x'(t)) \\ \mathrm{Var}(x'(t)) \end{pmatrix} + \mathbf{b}_\sigma(t), \tag{11}$$

where

$$A_\sigma = \begin{pmatrix} 0 & 2 & 0 \\ -\overline{\phi} & \omega - 1 & 1 \\ 0 & -2\overline{\phi} & 2(\omega - 1) \end{pmatrix},$$

and

$$\mathbf{b}_\sigma(t) = \begin{pmatrix} 0 \\ \overline{\phi}\,\mathrm{Cov}(x(t), o(t)) \\ 2\overline{\phi}\,\mathrm{Cov}(x'(t), o(t)) \end{pmatrix}.$$

Thus, $\mathbf{b}_\sigma(t)$ includes the statistical similarity functions between processes $x(t)$, $x'(t)$ and the oscillation center $o(t)$. The second order trajectories will depend on the type of eigenvalues of matrix A_σ and on the similarity functions $E(x(t)\,o(t))$ and $E(x'(t)\,o(t))$. The relationship (11) is known in the literature devoted to stochastic processes and random vibrations as the Lyapunov equation. The region of second order stability, that is, the part of the $(\omega, \overline{\phi})$ plane where the eigenvalues of A_σ are on the unit circle, coincides with S_C (9). Also the line separating the zones where the eigenvalues of matrix A_σ are real or complex is the parabola (10).

Solutions of the first order system

$$\frac{d\sigma(t)}{dt} = A_\sigma \sigma(t) + \mathbf{b}_\sigma(t),$$
$$\sigma(0) = \sigma_0,$$

can be written as follows:

$$\sigma(t) = \sigma_h(t) + \sigma_p(t) = e^{A_\sigma t}\sigma_0 + \int_0^t e^{A_\sigma(t-\tau)}\mathbf{b}_\sigma(\tau)\, d\tau.$$

More precisely the homogeneous part of $\sigma(t)$, $\sigma_h(t)$, can be expressed as a function of the eigenvalues of matrix A_σ:

1. $\lambda_1, \lambda_2, \lambda_3 \in \mathbb{R}: \ \sigma_h(t) = \mathbf{C}_1 e^{\lambda_1 t} + \mathbf{C}_2 e^{\lambda_2 t} + \mathbf{C}_3 e^{\lambda_3 t}$ in the real eigenvalue region.
2. $\lambda_1 = \omega - 1 \in \mathbb{R}, \lambda_2, \lambda_3 \in \mathbb{C}: \ \sigma_h(t) = (\mathbf{C}_1 + \mathbf{C}_2 \cos(\beta t + \mathbf{C}_3)) e^{(\omega-1)t}$ in the complex eigenvalue region, where β is imaginary part of the complex eigenvalues, λ_2, λ_3 and the real part of λ_2, λ_3 is $\omega - 1$.
3. $\lambda_1 = \lambda_2 = \lambda_3 = \omega - 1: \sigma_h(t) = \left(\mathbf{C}_1 + \mathbf{C}_2 t + \mathbf{C}_3 t^2\right) e^{(\omega-1)t}$ in the limit parabola (10) between both regions.

Figure 3 shows the homogeneous solution, $\sigma_h(t) = e^{A_\sigma t}\sigma_0$, for two points located on the real and complex zones of the second order stability region, S_c. Using the same methodology it is possible to account for the interaction between any two coordinates, i, j, of any two particles in the swarm ([13]).

2.2 The Oscillation Center Dynamics

The previous analysis has shown that the cost function enters the first and second order moments dynamics through two different forcing terms including respectively the mean trajectory of the center of attraction, $\overline{o}(t)$, and the covariance functions between $x(t)$, $x'(t)$ and $o(t)$. The question is if we can attach any regular behavior to these functions. Before any numerical experiment we can expect the following results:

1. $E(x(t)) \to E(o(t))$. The velocity of convergence should depend on the exploratory/exploitative character of the $(w, \overline{\phi})$ point that we had adopted.
2. $\mathrm{Var}(o(t))$ goes to zero with time, that is., the oscillation center tends to stagnate with iterations. Also, similarity between $\mathrm{Var}(o(t))$ and $\mathrm{Cov}(x(t), o(t))$ should increase with time as $E(x(t)) \to E(o(t))$ .Obviously the presence of local minima will also influence the shape of these two curves.
3. $\mathrm{Cov}(x(t), o(t))$ tends to $\mathrm{Var}(o(t))$ with time.

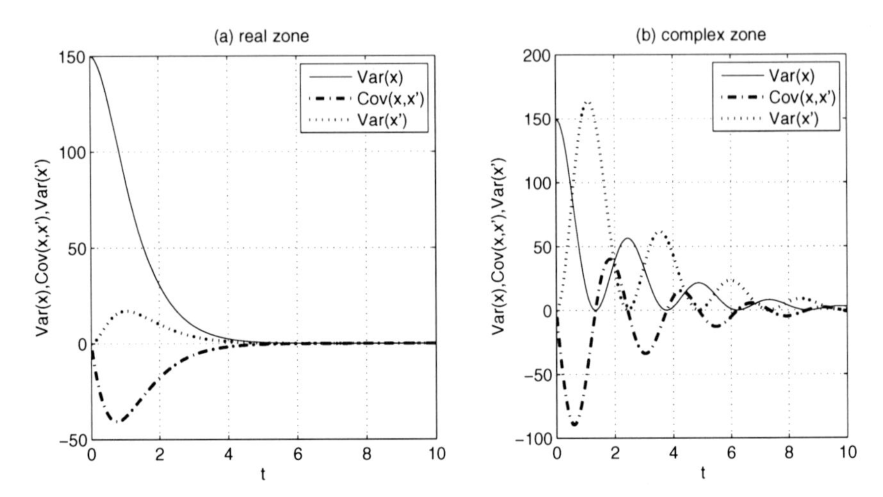

Fig. 3. Homogeneous trajectories for two points located on the real and complex zones of the second order stability region, S_c.

4. Taking into account that

$$\mathrm{Cov}\left(x'\left(t\right),o\left(t\right)\right)=\lim_{\Delta t\to 0}\frac{1}{\Delta t}\left(\mathrm{Cov}\left(x(t),o\left(t\right)\right)-\mathrm{Cov}\left(x(t-\Delta t),o\left(t\right)\right)\right),$$

it is possible to predict that this function might exhibit a greater variability. When the PSO parameters are selected in zones of greater variability this function typically alternates its sign with time

Figure 4 shows in the PSO case the univariate and bivariate dynamics of the oscillation center for a well known family of benchmark functions for $\left(\omega,\overline{\phi}\right)=(0.6,0.035)$ located on the first order real zone. To produce these results we have performed 3000 different runs using a swarm of ten particles evolving during 100 iterations. This allows us to have at our disposal a large statistical sample in each case to infer experimentally these functions. The results are also similar on other zones of the first order stability, the biggest difference being observed in $\mathrm{Cov}\left(x'\left(t\right),o\left(t\right)\right)$. In conclusion, the experimental functions describing the first and second order moments of the univariate and bivariate distributions of the oscillation center are quite similar for different benchmark functions and can be fitted using functions of exponential type. The shape of these functions depends on several factors, such as the initial conditions for the first and second order differential systems; on the search space type and on its dimension (number of degrees of freedom of the optimization problem); on the selected PSO parameters; and finally, on the type of benchmark function used for optimization and its numerical complexities. Nevertheless the results seemed to be very consistent for a wide range of functions. This circumstance might explain why the PSO algorithm performs acceptably well for a wide range of benchmark functions, depending its success more on the selection of the PSO parameters itself than on the type of cost function.

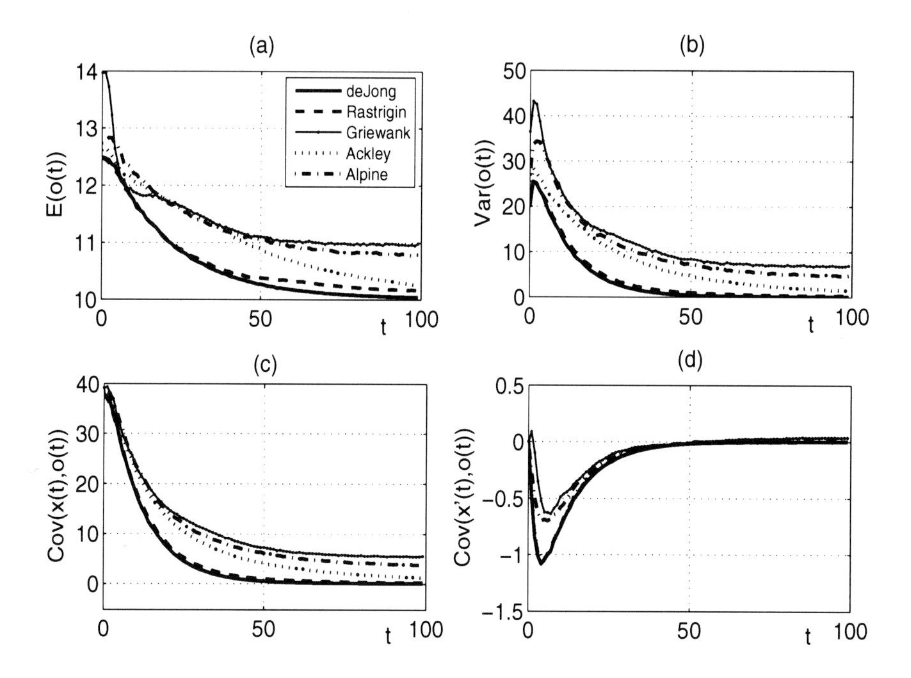

Fig. 4. The PSO univariate and bivariate dynamics of the oscillation center for a different benchmark functions for $(\omega, \overline{\phi}) = (0.6, 0.035)$, in the first order real zone.

3 Other Family Members

In this section we summarize the major achievements shown in [12]. Two other PSO family members have been recently introduced known as the PP-PSO and the RR-PSO [14]. They correspond respectively to a progressive-progressive and regressive-regressive finite differences schemes. These algorithms have some peculiarities that make them different from other family members, but their analysis is not presented in this chapter.

Let us consider a PSO continuous model where the oscillation center may be delayed a time $t_0 = k\Delta t$ with respect to the trajectory $\mathbf{x}_i(t)$:

$$\mathbf{x}_i''(t) + (1-\omega)\,\mathbf{x}_i'(t) + \phi\mathbf{x}_i(t) = \phi_1\mathbf{g}\,(t-t_0) + \phi_2\mathbf{l}_i\,(t-t_0)\,,\ t, t_0 \in \mathbb{R}. \quad (12)$$

This delay parameter is aimed at increasing exploration and/or at simplifying the final algorithm design.

Let us adopt a centered discretization in acceleration (3) and a β-discretization in velocity, where $\beta \in [0,1]$ is a real parameter:

$$x'(t) \simeq \frac{(\beta-1)x(t-\Delta t) + (1-2\beta)x(t) + \beta x(t+\Delta t)}{\Delta t}. \quad (13)$$

Then, for $\beta = 0$ we obtain the centered-regressive version (CR-GPSO) or just GPSO. For $\beta = 1$ the discretization in velocity is progressive and we obtain the CP-GPSO. Finally, for $\beta = 0.5$ the velocity discretization is centered and we obtain the CC-GPSO. Also, other versions can be obtained taking different values of $\beta \in [0,1]$ (β-GPSO). All these algorithms can be written in terms of the absolute position and velocity $(x(t), v(t))$ as follows:

$$\begin{pmatrix} x(t+\Delta t) \\ v(t+\Delta t) \end{pmatrix} = M_\beta \begin{pmatrix} x(t) \\ v(t) \end{pmatrix} + b_\beta,$$

where

$$M_\beta = \begin{pmatrix} 1+(\beta-1)\Delta t^2\phi & \Delta t(1+(\beta-1)(1-w)\Delta t) \\ \Delta t\phi\dfrac{(1-\beta)\beta\Delta t^2\phi-1}{1+(1-w)\beta\Delta t} & (1-\beta\Delta t^2\phi)\dfrac{1+(1-w)(\beta-1)\Delta t}{1+(1-w)\beta\Delta t} \end{pmatrix},$$

and

$$b_\beta = \begin{pmatrix} \Delta t^2(1-\beta)\left(\phi_1 g(t-t_0)+\phi_2 l(t-t_0)\right) \\ \Delta t\frac{\phi_1(1-\beta)(1-\beta\Delta t^2\phi)g(t-t_0)+\beta\phi_1 g(t+\Delta t-t_0)+\phi_2(1-\beta)(1-\beta\Delta t^2\phi)l(t-t_0)+\phi_2\beta l(t+\Delta t-t_0)}{1+(1-w)\beta\Delta t} \end{pmatrix}.$$

The value adopted for the delay parameter t_0 is zero for GPSO and CC-GPSO, and $t_0 = \Delta t$ for the CP-GPSO case. Other values of $t_0 = k\Delta t$ could be taken, but the ones proposed here simplify the algorithm designs.

Finally for the β-GPSO the following stochastic second order difference equation is obtained:

$$x(t+\Delta t) - A_\beta x(t) - B_\beta x(t-\Delta t) = C_\beta(t), \tag{14}$$

where

$$A_\beta = \frac{2-\phi\Delta t^2-(1-\omega)(1-2\beta)\Delta t}{1+(1-\omega)\beta\Delta t,}, \tag{15}$$

$$B_\beta = -\frac{1+(1-\omega)(\beta-1)\Delta t}{1+(1-\omega)\beta\Delta t,}, \tag{16}$$

$$C_\beta(t) = \frac{\phi_1 g(t-t_0)+\phi_2 l(t-t_0)}{1+(1-\omega)\beta\Delta t}\Delta t^2. \tag{17}$$

The first order stability region of the β-PSO depends on β:

1. $0 \le \beta < 0.5$

$$S^1_{\beta-pso} = \left\{(\omega,\overline{\phi}) : 1-\frac{2}{(1-2\beta)\Delta t} < \omega < 1,\ 0 < \overline{\phi} < \frac{2(2+(1-\omega)(2\beta-1)\Delta t)}{\Delta t^2}\right\}. \tag{18}$$

The straight line, $\overline{\phi} = \frac{2(2+(1-\omega)(2\beta-1)\Delta t)}{\Delta t^2}$, always passes through the point $(\omega,\overline{\phi}) = (1, \frac{4}{\Delta t^2})$ for any β.

2. $\beta = 0.5$

$$S^1_{\beta-pso} = \left\{(\omega,\overline{\phi}) : \omega < 1,\ 0 < \overline{\phi} < \frac{4}{\Delta t^2}\right\}.$$

3. $0.5 < \beta \leq 1$, this region is composed of two different disjoint zones, and is:

$$S^1_{\beta-pso} = \left\{ \left(\omega, \overline{\phi}\right) : D^1_{\beta_1} \cup D^1_{\beta_2} \right\}.$$

$$D^1_{\beta_1} = \left\{ \left(\omega, \overline{\phi}\right) : -\infty < \omega < 1,\ 0 < \overline{\phi} < \frac{2(2 + (1-\omega)(2\beta - 1)\Delta t)}{\Delta t^2} \right\},$$

$$D^1_{\beta_2} = \left\{ 1 + \frac{2}{(2\beta - 1)\,\Delta t} < \omega < +\infty,\ \frac{2(2 + (1-\omega)(2\beta - 1)\Delta t)}{\Delta t^2} < \overline{\phi} < 0 \right\}.$$

4. The borderline separating the real and complex roots in the first stability regions are ellipses for $\beta \in (0,1)$, and only on the limits $\beta = 0$ (GPSO) and $\beta = 1$ (CP-GPSO) these lines become parabolas. β-PSO is the particular case when the time step is $\Delta t = 1$.

The second order stability region follows a similar behavior than the first order stability region, is embedded in the first order region, and has as limit of second order stability the line:

$$\phi_{\beta-pso}\left(\beta, \omega, \alpha, \Delta t\right) = \frac{12}{\Delta t} \frac{(1-\omega)\left((\omega - 1)\,(2\beta - 1)\,\Delta t - 2\right)}{(4\,(-1 + (\omega - 1)\,(1+\beta)\,\Delta t) + (2-\alpha)\,\alpha\,(2 + (1-\omega)\,(2\beta - 1)\,\Delta t)},$$

where $\alpha = \frac{a_g}{\overline{\phi}} = \frac{2a_g}{a_g + a_l}$ is the ratio between the global acceleration and the total mean acceleration $\overline{\phi} = \frac{a_g + a_l}{2}$, and varies in the interval $[0, 2]$. Low values of α imply for the same value of $\overline{\phi}$ that the local acceleration is bigger than the global one, and thus, the algorithm gets more explorative.

Figures 5 and 6 show the first and second order stability regions and their associated spectral radii for PSO, CC-PSO, CP-PSO and β-GPSO with $\beta = 0.75$, for the case $a_g = a_l$. The spectral radii are related to the attenuation of the first and second order trajectories. In the PSO case, the first order spectral radius is zero in $\left(\omega, \overline{\phi}\right) = (0, 1)$, that is, the mean trajectories are instantly attenuated. It can be noted that the first order stability zone only depends on $\left(w, \overline{\phi}\right)$, while the second order stability region depends on (w, a_g, a_l). Also, the second order stability region is embedded in the first order region, and depends symmetrically on α, reaching its maximum size when $\alpha = 1$ $(a_l = a_g)$.

Since all the PSO members come from adopting a different finite difference scheme in velocity of the same continuous model (12), their first order stability regions were found to be isomorphic [12], that is, there exist a correspondence between their first order homogeneous discrete trajectories. This is an important result since the trajectories study should be done only once for a member of the family. The change of variables to make a β-PSO version with parameters a_g, $a_l, w, \Delta t$, correspond to a standard PSO ($\Delta t = 1$) with parameters b_g, b_l, and γ is:

$$b_g = \frac{\Delta t^2}{1 + (1-w)\,\beta\Delta t} a_g,\ b_l = \frac{\Delta t^2}{1 + (1-w)\,\beta\Delta t} a_l,\ \gamma = \frac{1 + (1-w)\,(\beta - 1)\,\Delta t}{1 + (1-w)\,\beta\Delta t}. \tag{19}$$

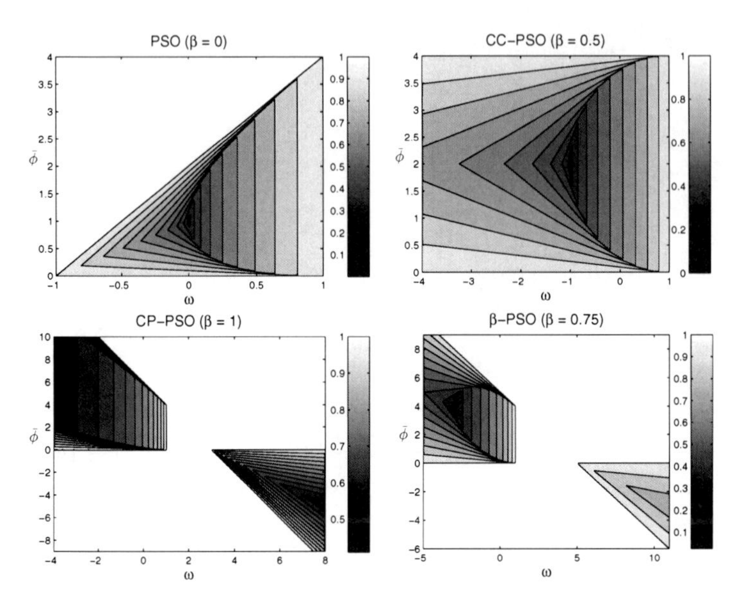

Fig. 5. First order stability regions for different PSO family members as a function of β : PSO $(\beta = 0)$, CC-PSO $(\beta = 0.5)$, CP-PSO $(\beta = 1)$, and β-PSO with $\beta = 0.75$.

It is important to note that in the CC-PSO and CP-PSO cases, the regions of first stability become unbounded, while in the PSO case this zone is limited by the straight line $\overline{\phi} = 2(\omega + 1)$, which is in fact due to the unit time step adopted, since when Δt goes to zero this line goes to infinity. For the different PSO versions there is a correspondence between the different zones of first order trajectories: complex zone, real symmetrical zigzagging zone, real asymmetrical oscillating zone and real non-oscillating zone. In the CC-PSO case the real asymmetric zigzagging zone becomes unbounded, while in the CP-PSO case these four zones become unbounded. Also, in the CP-PSO case the corresponding points to the PSO points located in the neighborhood of $(\omega, \overline{\phi}) = (0, 1)$ (where the PSO first order spectral radius is zero), lie in $(\omega, \overline{\phi}) \to (-\infty, +\infty) \cup (\omega, \overline{\phi}) \to (+\infty, -\infty)$. That means that the region of very fast trajectories attenuation for the PSO and CC-PSO cases is not present for the CP-PSO algorithm. Thus the CP-PSO version is the one that has the greater exploratory capabilities.

These correspondences make the algorithms linearly isomorphic, but when applied to real optimization problems they perform very different, mainly due to two main reasons:

1. The introduction of a delay parameter increases the exploration and makes the algorithms different in the way they update the force terms. This is the case for the CP-PSO with delay one.
2. The different way these algorithms update the positions and velocities of the particles.

A detailed analysis can be found in [12].

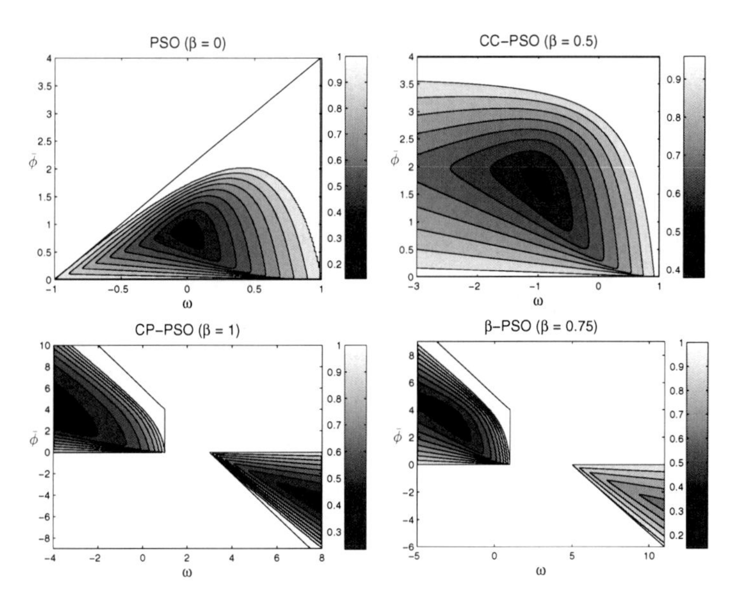

Fig. 6. Second order stability regions for different PSO family members as a function of β.

4 Methodology for Stochastic Analysis of the Discrete Trajectories

Let us consider the discrete trajectories generated by any PSO family member (GPSO, CC-PSO or CP-PSO) as stochastic processes fulfilling the following vector difference equation

$$\mathbf{x}(t+\Delta t) - A\mathbf{x}(t) - B\mathbf{x}(t-\Delta t) = C\,(t)\,. \tag{20}$$

A, B, are random variables, independent of the trajectories $x(t)$ and $x(t-\Delta t)$, and $C\,(t)$ is a stochastic process that depends on the center of attraction (see for instance relations 15, 16 and 17) :

$$\mathbf{o}\,(t;t_0) = \frac{a_g \mathbf{g}\,(t-t_0) + a_l \mathbf{l}\,(t-t_0)}{a_g + a_l}.$$

In this case, for design purposes we allow the center of attraction $\mathbf{o}\,(t;t_0)$ to be delayed a time $t_0 = k\Delta t$ with respect to the trajectory $\mathbf{x}(t)$.

Let $x\,(t)$ be the trajectory of any particle coordinate in the swarm. The first order moment vector

$$\mu\,(t+\Delta t) = \begin{pmatrix} E\,(x\,(t+\Delta t)) \\ E\,(x\,(t)) \end{pmatrix},$$

fulfills the following first order affine dynamical system:

$$\mu(t+\Delta t) = A_\mu . \mu(t) + \mathbf{b}_\mu(t),$$

where

$$A_\mu = \begin{pmatrix} E(A) & E(B) \\ 1 & 0 \end{pmatrix},$$

and

$$\mathbf{b}_\mu(t) = \begin{pmatrix} E(C(t)) \\ 0 \end{pmatrix}.$$

Stability of the first order moments is related to the eigenvalues of the iteration matrix A_μ, and the theoretical asymptotic velocity of convergence[2] of $x(t)$ towards the mean trajectory of the oscillation center, $E(o(t))$, depends on the spectral radius of first order iteration matrix, A_μ [11]-[14].

Similarly, the non-centered second order moments vector

$$\mathbf{r}_2(t) = \begin{pmatrix} E\left(x^2(t)\right), \\ E\left(x(t)x(t-\Delta t)\right), \\ E\left(x^2(t-\Delta t)\right) \end{pmatrix},$$

fulfills the following second order affine dynamical system:

$$\mathbf{r}_2(t+\Delta t) = A_\sigma \mathbf{r}_2(t) + \mathbf{b}_\mathbf{r}(t), \tag{21}$$

where

$$A_\sigma = \begin{pmatrix} E\left(A^2\right) & 2E(AB) & E\left(B^2\right) \\ E(A) & E(B) & 0 \\ 1 & 0 & 0 \end{pmatrix},$$

and

$$\mathbf{b}_\mathbf{r}(t) = \begin{pmatrix} E\left(C^2(t)\right) + 2E\left(AC(t)\,x(t)\right) + 2E\left(BC(t)\,x(t-\Delta t)\right) \\ E\left(C(t)\,x(t)\right) \\ 0 \end{pmatrix}.$$

The term $E\left(C^2(t)\right)$ is related to the oscillation center variability. Terms $E\left(AC(t)\,x(t)\right)$, $E\left(BC(t)\,x(t-\Delta t)\right)$ and $E\left(C(t)\,x(t)\right)$ introduce in the second order affine dynamical system the correlation functions between the oscillation center, $o(t)$, and the trajectories, $x(t)$ and $x(t-\Delta t)$. Similar considerations for stability and asymptotic velocity of convergence of the second order moments can be done, considering in this case the eigenvalues of iteration matrix A_σ. The centered moments follow a similar relationship but with a different force term.

Finally, interactions between any two coordinates, $x_i(t)$ and $x_j(t)$ of any two particles in the swarm can be modeled in the same way and it is possible to show that their respective non-centered second order moments vector have the same stability condition than matrix A_σ [13].

---

[2] The asympthotic velocity of convergence is $-\ln(\rho)$, where ρ is the spectral radius of the iteration matrix, that is, the supremum among the absolute values of the elements in its spectrum (eigenvalues).

4.1 The GPSO Second Order Trajectories

This methodology can be applied to the study of the GPSO second order trajectories. The analysis shown here serves to clarify the role of the inertia, and the local and global accelerations on the second order moments, and to explain how the cost function enters into the second order dynamical system.

In the GPSO case the constants in the second order difference equation are:

$$A\left(\omega, \phi, \Delta t\right) = 2 - \left(1 - \omega\right) \Delta t - \phi \Delta t^2, \tag{22}$$

$$B\left(\omega, \Delta t\right) = \left(1 - \omega\right) \Delta t - 1, \tag{23}$$

and

$$C\left(t\right) = (\phi_1 \mathbf{g}\left(t\right) + \phi_2 \mathbf{l}\left(t\right))\Delta t^2, \tag{24}$$

gathers the influence of the global and local best attractors.

First order trajectories have been systematically studied in [10] and in [11]. Let us call $\mu\left(t + \Delta t\right) = \begin{pmatrix} E\left(x\left(t + \Delta t\right)\right) \\ E\left(x\left(t\right)\right) \end{pmatrix}$ the vector containing the mean trajectories of $x(t+\Delta t)$ and $x(t)$. The following first order linear system is involved:

$$\mu\left(t + \Delta t\right) = \begin{pmatrix} E\left(A\right) & B \\ 1 & 0 \end{pmatrix} \mu\left(t\right) + \mathbf{d}_{\mu}\left(t\right), \tag{25}$$

where

$$\mathbf{d}_{\mu}\left(t\right) = \begin{pmatrix} \Delta t^2 E\left(\phi_1 g\left(t\right) + \phi_2 l\left(t\right)\right) \\ 0 \end{pmatrix} \overset{ind.}{=} \begin{pmatrix} \overline{\phi} \Delta t^2 E\left(o\left(t\right)\right) \\ 0 \end{pmatrix},$$

and A, B are given respectively by (22) and (23). To deduce (25) we also have supposed the independence between ϕ and $x\left(t\right)$, ϕ_1 from $g\left(t\right)$, and ϕ_2 and $l\left(t\right)$. The first order stability region turns to be (18) with $\beta = 0$.

The non-centered second order vector

$$\mathbf{r}_2\left(t\right) = \begin{pmatrix} E\left(x^2(t)\right) \\ E\left(x(t)x(t - \Delta t)\right) \\ E\left(x^2(t - \Delta t)\right) \end{pmatrix},$$

follows the following second order affine dynamical system:

$$\mathbf{r}_2\left(t + \Delta t\right) = A_{\sigma} \mathbf{r}_2\left(t\right) + \mathbf{d}_{\mathbf{r}}\left(t\right), \tag{26}$$

where

$$A_{\sigma} = \begin{pmatrix} E\left(A^2\right) & 2BE\left(A\right) & B^2 \\ E\left(A\right) & B & 0 \\ 1 & 0 & 0 \end{pmatrix},$$

and

$$\mathbf{d_r}(t) = \begin{pmatrix} E\left(C^2(t)\right) + 2BE\left(C(t)\,x(t-\Delta t)\right) + 2E\left(AC(t)\,x(t)\right) \\ \overline{\phi}\Delta t^2 E\left(o(t)\,x(t)\right) \\ 0 \end{pmatrix}.$$

The term $E\left(C^2(t)\right)$ includes the variabilities due to $\phi_1, \phi_2, l(t)$ and $g(t)$, while the other two terms include the correlations between $x(t)$ and $x(t-\Delta t)$ and the attractors $l(t)$ and $g(t)$.

The analysis of the GPSO second order trajectories (variance and covariance) when $\mathbf{d_c}(t) = \mathbf{0}$, allow us to separate in the GPSO second order trajectories the contribution of the GPSO parameters (homogeneous parts) from those of the forcing terms (influence of the cost function). Analysis of the homogeneous first order trajectories has been also presented in [10].

Second order homogeneous trajectories are classified as a function of ω, a_g, a_l and the time step Δt, depending on the eigenvalues of the iteration matrix A_σ. Solutions of the dynamical system

$$\mathbf{r}_2(t+\Delta t) = A_\sigma(\omega, a_g, a_l, \Delta t)\,.\mathbf{r}_2(t),$$

can be expanded in terms of the eigenvalues, $(\lambda_1, \lambda_2, \lambda_3)$, and the eigenvectors, $B_v = \{\mathbf{v}_1, \mathbf{v}_2, \mathbf{v}_3\}$, of A_σ:

$$\mathbf{r}_2(n\Delta t) = P \begin{pmatrix} \lambda_1^n & 0 & 0 \\ 0 & \lambda_2^n & 0 \\ 0 & 0 & \lambda_3^n \end{pmatrix} P^{-1}\mathbf{r}_2(0) = \sum_{i=1}^{3} c_i \lambda_i^n \mathbf{v}_i,$$

where $\mathbf{c} = (c_1, c_2, c_3)$ are the coordinates of initial conditions $\mathbf{r}_2(0)$ expressed on basis B_v. This imply that variance and covariance trajectories will depend on the kind of A_σ eigenvalues.

The analysis is similar to that shown for the PSO continuous model:

1. When the eigenvalues are real and positive the trajectories are monotonous decreasing.
2. When the eigenvalues are complex, the second order trajectories are oscillatory. Also, negative real eigenvalues or complex eigenvalues with a negative real part originates zigzagging on the second order trajectories. A main difference to the PSO continuous model is that the zigzagging phenomenon does not exist.

Figure 7 shows for the PSO case $(\Delta t = 1)$ and $\alpha = 1$ $(a_g = a_l)$ .the different stability zones of the second order moments, together with the frequency of the oscillation.

The following zones are clearly differentiated in this graph:

1. The real region, which is composed of two different parts: the zigzagging and the exponential decreasing zones. In the zigzagging zone the frequency of the oscillation is $\frac{1}{2}$, and is associated to the presence of negative real eigenvalues.
2. The complex region which is composed also of two different non-overlapping zones which are separated by the line of temporal uncorrelation between

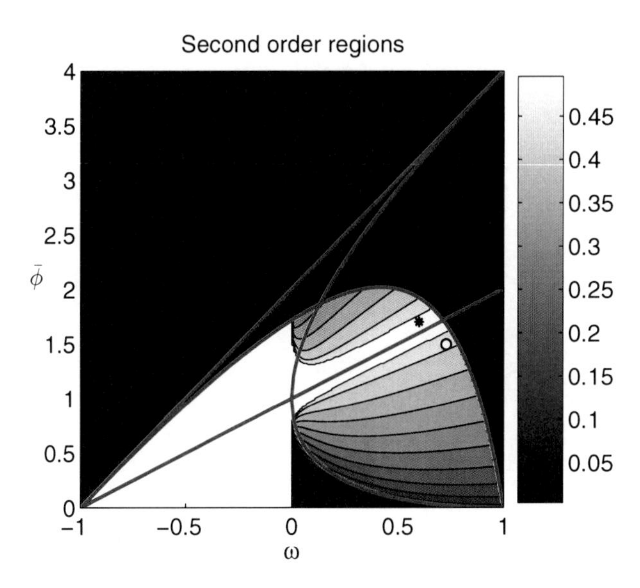

Fig. 7. PSO second order trajectories stable regions. Frequency of oscillation as a function of $(w, \overline{\phi})$ for $\alpha = 1$ $(a_g = a_l)$ and $\Delta t = 1$. Trelea (asterisk) and Clerc and Kennedy (circle) points are also located on these plots

trajectories (median of the first order stability triangle). The frequency in the complex zone increases from $\overline{\phi} = 0$ and from the upper border of second order stability, towards the limit with the zigzagging real zone.

In this plot we also show the location of two performance parameter sets found in the literature [5], [6]. It can be observed that Clerc-Kennedy point lies on the complex zone while the Trelea point lies on the zigzagging zone. Carslile and Dozier [4] point also lies on the complex zone of A_σ, but in this case $\alpha = 0.64$ $(a_l \simeq 2a_g)$. Comparison between these three points can be established analyzing the spectral radius and the oscillation frequency of their respective second order trajectories. Trelea's point has the higher frequency $\left(\frac{1}{2}\right)$ and the lowest spectral radius (0.889). Carlisle and Dozier's and Clerc and Kennedy's points have the same first order trajectories and approximately the same frequency of oscillation for the second order trajectories $(0.463 - 0.462)$, but the Carlisle and Dozier's point has a higher second order spectral radius (0.975) than Clerk and Kennedy's point (0.943). This feature implies a higher exploratory capacity for the Carlisle and Dozier's point. Finally, figure 8 shows the homogenous variance and covariance trajectories for these three points. The temporal covariance between trajectories, $\mathrm{Cov}(x(t), x(t - \Delta t))$, for the Trelea's point goes to zero faster than those of the Clerc and Kennedy's and Carlisle and Dozier's points. In these two last points the covariance function is almost all the time positive, that is, trajectories $x(t - \Delta t)$ and $x(t)$ have a certain degree of similarity, while in Trelea's point the temporal covariance is negative. This feature can be explained because over the median line of the first order stability triangle the temporal

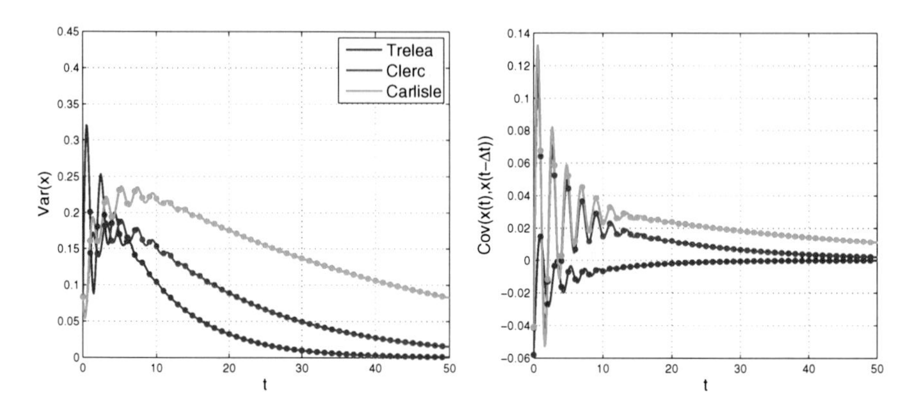

Fig. 8. Variance and covariance trajectories for some popular parameter sets.

covariance has a negative sign. The amplitude of the variance increases with $\overline{\phi}$, reaching its maximum value in the highest point of the second order stability region, $(w, \overline{\phi}) = (0.42,\ 2.02)$. In this point the variance stabilizes in the value of 0.497. Above the line of second order stability the variance grows with iterations (it is not bounded). Close to the median line of the first order stability triangle the temporal covariance goes to zero very fast. Its minimum negative value is also reached close to the highest point of the second order stability region, where the covariance stabilizes to -0.209.

5 Does GPSO Converge towards the Continuous PSO?

Although this can be regarded as an anecdotal question from the computational point of view, in this section we will show that GPSO approaches the PSO continuous model when the time step Δt decreases towards zero. We will also show that the corresponding GPSO first and second order stability regions tend to those of the continuous PSO. This is done first with the homogeneous systems and later with the transitory solutions, making comparisons with real simulations. This final comparison serves to numerically quantify how these linear models (continuous and discrete) account for the variability observed in real runs. We show that mean trajectories are always very well matched, while the second order moments are only approximately predicted on regions of high variability. Based on the results, it is conjectured that nonlinear dependencies between the local and global attractors and the trajectories are more important on zones of high variability. In conclusion it is possible to affirm that PSO algorithm does not have a heuristic character since we are able to state some theoretical results that describe very accurately its behavior.

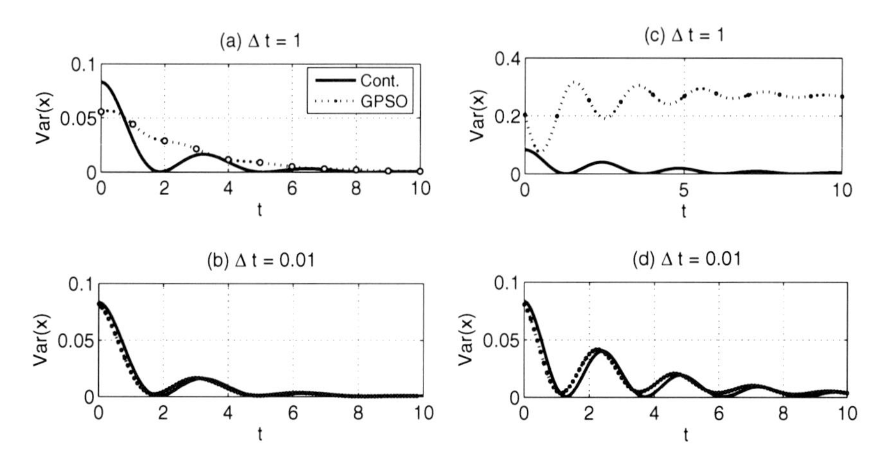

Fig. 9. Homogeneous second order trajectories in two different points: a) and b) $(\omega, \overline{\phi}) = (0.7, 1.7)$; c) and d) $(\omega, \overline{\phi}) = (0.5, 1)$.

5.1 The Second Order Homogeneous Trajectories

Let us first consider the case where both dynamic systems (continuous and discrete) become homogeneous, i.e., the effect of the center of attraction $o(t)$ is not taken into account. Figure 9 shows the continuous and discrete (GPSO) variance trajectories for two different $(\omega, \overline{\phi})$ points on the second order stability region. Figure 9-a shows this comparison in $(\omega, \overline{\phi}) = (0.7, 1.7)$ located in the complex zone of the second order stability regions, for both, the GPSO and the continuous PSO. It can be observed that both trajectories are of the same kind, and the GPSO variance trajectory approaches the PSO continuous variance as time step goes to zero (figures 9-b). Figure 9-c shows the same analysis in the point $(\omega, \overline{\phi}) = (0.5, 1)$ for $\Delta t = 1$. These dynamics are different, because for $\Delta t = 1$ this point is located on the complex region for the continuous PSO and in the real zigzagging zone for the GPSO. As time step decreases the second order boundary hyperbola goes to infinity and the two dynamics approach each other, almost overlapping when $\Delta t = 0.01$ (figures 9-b and 9-d). Also when Δt decreases, the first and second GPSO zigzagging zones tend to disappear and the GPSO sampling becomes denser.

5.2 The Transient Trajectories

The same analysis has been done with a moving center of attraction using the 1D Griewank function. The analysis shown here was also performed for other benchmark functions and the results obtained were very similar. Covariance functions between the oscillation center and the trajectories, needed in $\mathbf{b}_r(t)$ and in $\mathbf{d}_r(t)$ to solve numerically the continuous and GPSO dynamical systems, have been empirically calculated as shown in section 3, devoted to the analysis of

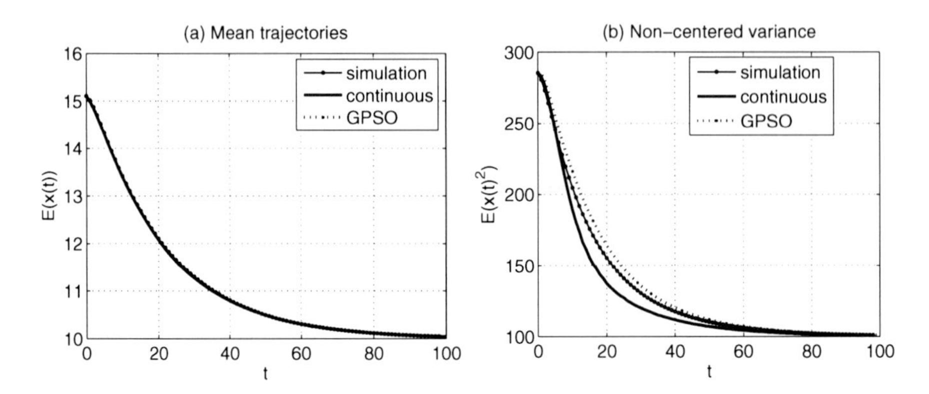

Fig. 10. First and second order transient trajectories on the point $(\omega, \overline{\phi}) = (-3,\ 3.5)$ with $\Delta t = 0.1$. Compared to real runs.

the center dynamics. Obviously the results shown here will be affected by the accuracy of the empirical identification of the center dynamics.

Figure 10 shows the mean, $E\left(x\left(t\right)\right)$, and the non centered trajectories, $E\left(x^2\left(t\right)\right)$, for the continuous PSO, GPSO with $\Delta t = 0.1$,compared to the corresponding empirical counterparts, for the point $(\omega, \overline{\phi}) = (-3,\ 3.5)$ located on the real first and second order stability zones. It can be observed that mean trajectories do coincide (figure 10-a). For the second order trajectories, figure 10-b shows that the simulated trajectories are imbedded between those of continuous PSO and GPSO.

In areas of higher variability this match worsens for the second order moments but the first order trajectories are perfectly fitted (figure 11). This feature is

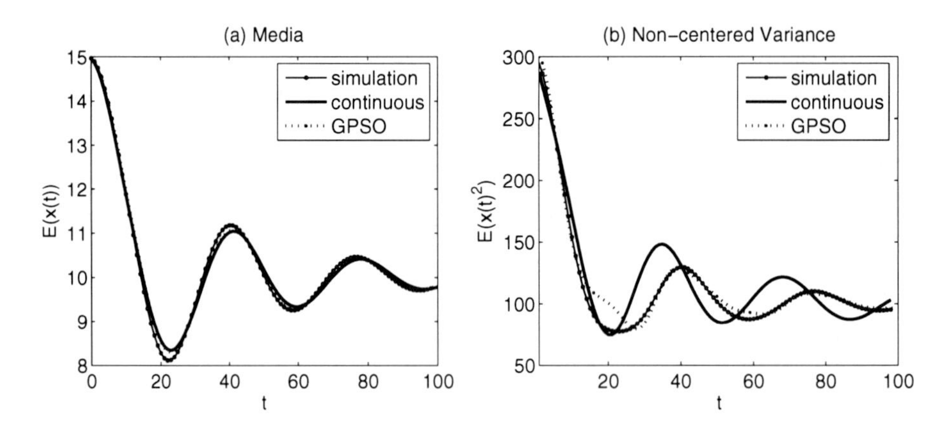

Fig. 11. First and second order transient trajectories for a point $(\omega, \overline{\phi})$ located on the first and second order complex zone with $\Delta t = 0.1$. Comparison to real runs.

related to the difficulty of performing a correct identification of the covariances terms involved in our linear models.

As a main conclusion, the GPSO dynamic approaches the continuous PSO dynamic when the time step Δt goes to zero, and both models are able to describe most of the variability observed in real runs. This is quite an impressive result if we bear in mind that mathematical models are always a simplification of reality. In this case our major assumption in the second order moments is to model the interaction to the cost function through the covariances functions between the attractors, $l(t)$, $g(t)$, and the trajectories, $x(t)$.

6 Tuning of the PSO Parameters

When applying the PSO family to real problems several questions arise to the modeler:

1. Which family member has to be used? Will all of them provide similar results?
2. Once the family member has been selected, how to tune the PSO parameters (inertia and local and global accelerations)? Are there some magic tuning points highly dependent on the kind of problem we would like to solve?. Or, is it possible to recommend "robust" and "any purpose" sets of PSO parameters?

These questions have been analyzed in [13] by means of numerical experiments using benchmark functions. Basically the answer is as follows:

1. The performance of each algorithm will depend on the degree of numerical difficulties of the cost function and the number of dimensions. In general terms CC-PSO and GPSO are the most exploitative versions while CP-PSO is the most explorative one. This is due to how the algorithm updates the position and velocity. PSO can be considered as an integrator, since first the velocity is updated and then the position. CC-PSO performs numerical differentiation since the position is calculated first and then the velocity using two consecutive attractors positions. It seems to be the most exact algorithm since it has the lowest approximation error with respect to the continuous model. The CP-PSO updates positions and velocities at the same time and thus, taking into account Heisenberg's uncertainty principle, will be the most explorative version. When numerical difficulties increase, exploration might be needed and the CP-PSO could eventually also provide very good results, even better than the other more exploitative versions that can be trapped in a local minima or in a flat area. For most of the benchmark functions we have optimized, the CC-PSO and the PSO were the best performing versions [12]. For instance, figure 12 shows how these three algorithms perform for the Rosenbrock and Griewank in ten dimensions. The convergence curves are the median curves obtained after 500 different simulations, with a swarm of 100 particles and 200 iterations in each simulation. The $(w, \overline{\phi})$ point adopted

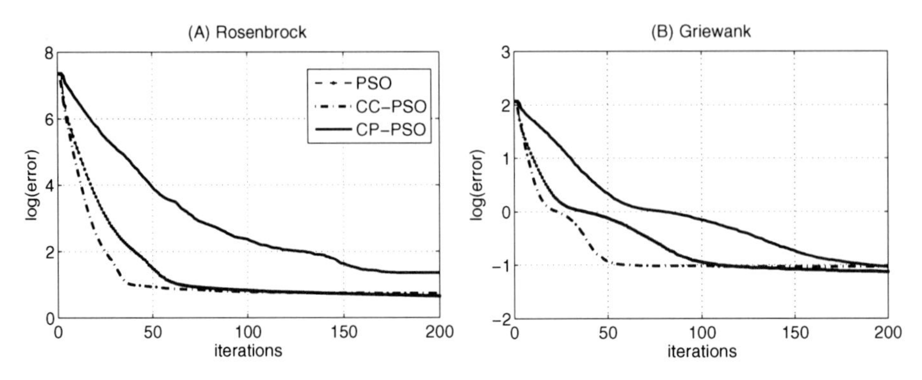

Fig. 12. Comparison (in $\log_{10}$ scale) of different PSO versions in the corresponding points of Trelea point: a) Rosenbrock, b) Griewank.

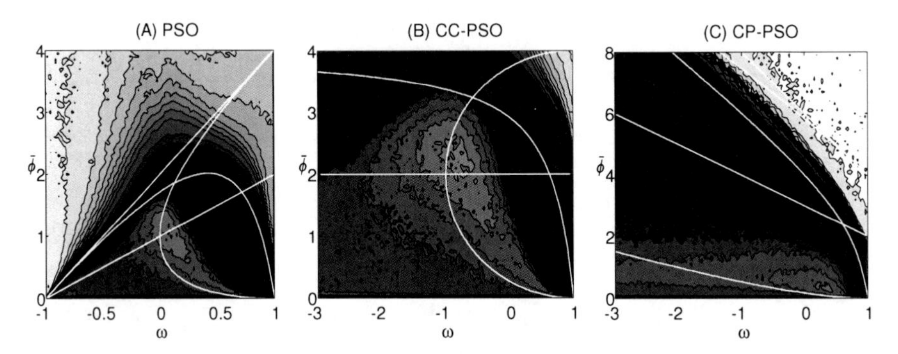

Fig. 13. Benchmark in ten dimensions. Median error contourplot (in $\log_{10}$ scale) for the Griewank function: a) PSO, b) CC-PSO, c) CP-PSO.

for the simulations was the Trelea's point [6], and the corresponding points for CC-PSO and CP-PSO using the relationships (19).

2. The selection of the PSO parameters in each case follows a clear logic. For different benchmark functions, the parameter sets with a high probability of success are close to the upper border of second order stability and the median line of temporal uncorrelation. This property means that an important exploration is needed to achieve good results. Also, when the number of dimensions increases, the good parameter sets move towards the limit of stability $w = 1$, close to the limit of second order stability [12]. This means that when dimensions increase a positive temporal covariance between the trajectories is needed to track the global minimum. For instance, figure 13 shows the best parameters sets for the PSO, CC-PSO and CP-PSO using the Griewank in ten dimensions. It can be observed that parameter sets with smallest misfits (in logarithmic scale) follow the pattern on which we have commented above. To produce these median error plots we have performed 50 simulations, with a swarm of 100 particles and 500 iterations, in each $(\omega, \overline{\phi})$ point of a grid that includes the first order stability regions of the PSO family members.

These results are very important in practice because it shows two main features common to all the PSO members:

1. All the PSO members perform fairly well in a very broad area of the PSO parameter space (ω, a_g, a_l).
2. These regions are fairly the same for benchmark functions with different numerical pathologies (having the global minimum in a flat valley or surrounded by other local minima). This means that with the same cloud it is possible to optimize a wide range of cost functions.

Based on this idea we have designed the so called cloud-PSO algorithm where each particle in the swarm has different inertia (damping) and acceleration (rigidity) constants [23]. This work has been recently expanded to the other family versions [14]. This feature allows the PSO algorithm to control the velocity update and to find the sets of parameters that are better suited for each optimization/inverse problem. Some of the particles will have a more exploratory behavior while other will show a higher exploitative character. Once the cloud has been generated, the particles are randomly selected depending on the iterations and the algorithm keep track of the best parameters used to achieve the global best solution. The criteria for choosing the cloud points it is not very rigid and points close to the second order convergence border work well, especially those inside the first order convergence complex zone. In this algorithm design the Δt parameter arises as a natural numerical constriction factor to achieve stability. The first and second order stability regions increases their size and tend to the continuous stability region when this parameter approaches to zero. When this parameter is less than one the exploration around the global best solution is increased. Conversely when Δt is greater than one, the exploration in the whole search space is increased, helping to avoid entrapment in local minima. This feature has inspired us to create the lime and sand algorithm that combines values of different values of Δt depending on the iterations [11].

7 How to Select the PSO Member: Application to Inverse Problems

Our main interest in global algorithms is the solution of different kind of inverse problems in science and technology [29]. Some examples in Environmental Geophysics and Reservoir Optimization can be consulted for example in [26], [27], [28], [24] and [25].

Most of the inverse problems can be written in discrete form as:

$$\mathbf{d} = \mathbf{F}(\mathbf{m})$$

where $\mathbf{d} \in \mathbf{R}^s$ is the observed data, $\mathbf{m} \in \mathbf{R}^n$ is the vector containing the model parameters, and $\mathbf{F} : \mathbf{R}^n \to \mathbf{R}^s$ is the physical model, that typically involves the solution of a set partial differential equations, integral equations or algebraic system. Given a particular observed data set $\mathbf{d}$, the inverse problem is then solved as an optimization problem, that is, finding the model that minimizes the data prediction error expressed in a certain norm $\|\mathbf{d} - \mathbf{F}(\mathbf{m})\|_p$.

The above optimization problem turns out to be ill-posed because the forward model $\mathbf{F}$ is a simplification of reality (numerical approximations included); the data are noisy and discrete in number, that is, there is not a sufficient number of data to uniquely determine one solution. These three points cause an inverse problem to be very different from any other kind of optimization problem since physics and data are involved on the cost function. In addition, the topography of the prediction error function usually corresponds to functions having the global minimum located in a very flat and elongated valley or surrounded by many local minima, as the Rosenbrock and Griewank functions. The type of the numerical difficulty found depends mainly on the forward functional $\mathbf{F}$, that is, the problem physics. The effect of data noise is to increase the presence of local minima and/or the size of the valley topography. Combinations of both pathologies are also possible in real problems.

Local optimization methods are not able to discriminate among the multiple choices consistent with the end criteria and may land quite unpredictably at any point on that area. These pathologies are treated through regularization techniques and the use of "good" prior information and/or initial guesses. Global optimization methods, such as genetic algorithms, simulated annealing, particle swarm, differential evolution, etc, are very interesting because instead of solving the inverse problem as an optimization problem, they are able to sample the region of the model space containing the models that fit the observed data within a given tolerance, that is, they are able to provide an idea of the posterior distribution of the inverse model parameters. To perform this task they do not need in principle any prior model to stabilize the inversion and are able to avoid the strong dependence of the solution upon noisy data.

Particle swarm optimization and its variants are interesting global methods since they are able to quickly approximate the posterior distribution of the model parameters. To correctly perform this task a good balance between exploration and exploitation, and the CP-PSO version seems to be better than the GPSO and CC-PSO. Conversely when only a good model (the candidate to global minimum) is needed and no uncertainty analysis is performed, the CC-PSO and GPSO versions have better convergence rates to locate this solution. These two facts can be taken into account to select the appropriate PSO version when facing a real problem.

8 Conclusions

Particle swarm optimization is a stochastic search algorithm whose consistency can be analyzed through the stability theory of stochastic differential and difference equations. The damped mass-spring analogy known as the PSO continuous model allowed us to derive a whole family of Particle Swarm optimizers. PSO is one of the most performant algorithms of this family in terms of rate of convergence. Other familiy members such as CP-PSO have better exploration capabilities. Numerical experiments with different benchmark functions have shown that most performing points for GPSO, CC-PSO and CP-PSO are close to the upper limit of second order stability. The time step parameter arises as a natural constriction factor to achieve stability and increasing around the global best solution, or to avoid entrapment in local minima (when this parameter is increased). Knowledge of these facts has been used to design a family of particle-cloud with variable time step. The cloud design avoids two main drawbacks of the PSO algorithm: the tuning of the PSO parameters and the clamping of the particle velocities. Considering all the theoretical developments and numerical results provided in this chapter we can affirm that the consistency/convergence of the Particle Swarm algorithms can be characterized through stochastic stability analysis.

Acknowledgments

This work benefited from a one-year sabbatical grant at the University of California Berkeley (Department of Civil and Environmental Engineering), given by the University of Oviedo (Spain) and by the "Secretaría de Estado de Universidades y de Investigación" of the Spanish Ministry of Science and Innovation. We also acknowledge the financial support for the present academic year coming from the University of California Berkeley, the Lawrence Berkeley National Laboratory (Earth Science Division) and the Energy Resources Engineering Department at Stanford University, that is allowing us to apply this methodology to several geophysical and reservoir inverse problems (Stanford Center for Reservoir Forecasting and Smart Field Consortiums). This has greatly helped us to grasp a better understanding of the theoretical results presented in this chapter. We finally acknowledge Laura Nikravesh (UC Berkeley) for style corrections.

References

1. Kennedy, J., Eberhart, R.: Particle swarm optimization. In: Proceedings of the IEEE International Conference on Neural Networks (ICNN 1995), Perth,WA, Australia, vol. 4, pp. 1942–1948 (November-December, 1995)
2. Ozcan, E., Mohan, C.K.: Analysis of a simple particle swarm optimization system. In: Intelligent Engineering Systems Through Artificial Neural Networks, vol. 8, pp. 253–258. ASME Press, St. Louis (1998)
3. Ozcan, E., Mohan, C.K.: Particle swarm optimization: surfing the waves. In: Proceedings of the IEEE Congress on Evolutionary Computation (CEC 1999), July 1999, vol. 3, pp. 1939–1944. IEEE Service Center, Washington (1999)
4. Carlisle, A., Dozier, G.: 1001, An off-the-shelf PSO. In: Proceedings of The Workshop On particle Swarm Optimization, Indianapolis, USA (2001)
5. Clerc, M., Kennedy, J.: The particle swarm—explosion, stability, and convergence in a multidimensional complex space. IEEE Transactions on Evolutionary Computation 6(1), 58–73 (2002)
6. Trelea, I.C.: The particle swarm optimization algorithm: convergence analysis and parameter selection. Information Processing Letters 85(6), 317–325 (2003)

7. Zheng, Y.-L., Ma, L.-H., Zhang, L.-Y., Qian, J.-X.: On the convergence analysis and parameter selection in particle swarm optimisation. In: Proceedings of the 2nd International Conference on Machine Learning and Cybernetics (ICMLC 2003), Xi'an, China, vol. 3, pp. 1802–1807 (November 2003)
8. van den Bergh, F., Engelbrecht, A.P.: A study of particle swarm optimization particle trajectories. Information Sciences 176(8), 937–971 (2006)
9. Poli, R.: Dynamics and Stability of the Sampling Distribution of Particle Swarm Optimisers via Moment Analysis. Journal of Artificial Evolution and Applications, Article ID 761459, 10 (2008), doi:10.1155/2008/761459
10. Fernández Martínez, J.L., García Gonzalo, E., Fernández Alvarez, J.P.: Theoretical analysis of particle swarm trajectories through a mechanical analogy. International Journal of Computational Intelligence Research 4(2), 93–104 (2008)
11. Fernández Martínez, J.L., García Gonzalo, E.: The generalized PSO: a new door for PSO evolution. Journal of Artificial Evolution and Applications Article ID 861275, 15 (2008), doi:10.1155/2008/861275
12. Fernández Martínez, J.L., García Gonzalo, E.: The PSO family: deduction, stochastic analysis and comparison. Special issue on PSO. Swarm Intelligence 3, 245–273 (2009), doi:10.1007/s11721-009-0034-8
13. Fernández Martínez, J. L., García Gonzalo, E.: Stochastic stability analysis of the linear continuous and discrete PSO models. Technical Report, Department of Mathematics, University of Oviedo, Spain (June 2009), Submitted to IEEE Transactions on Evolutionary Computation (2009)
14. García-Gonzalo, E., Fernández-Martínez, J. L.: The PP-GPSO and RR-GPSO. Technical Report. Department of Mathematics. University of Oviedo, Spain (December 2009b), Submitted to IEEE Transactions on Evolutionary Computation (2009)
15. Brandstätter, B., Baumgartner, U.: Particle swarm optimization: mass-spring system analogon. IEEE Transactions on Magnetics 38(2), 997–1000 (2002)
16. Mikki, S.M., Kishk, A.A.: Physical theory for particle swarm optimisation. Progress in Electromagnetics Research 75, 171–207 (2007)
17. Clerc, M.: Stagnation analysis in particle swarm optimisation or what happens when nothing happens, Tech. Rep. CSM-460, Department of Computer Science, University of Essex (August 2006)
18. Kadirkamanathan, V., Selvarajah, K., Fleming, P.J.: Stability analysis of the particle dynamics in particle swarm optimizer. IEEE Transactions on Evolutionary Computation 10(3), 245–255 (2006)
19. Jiang, M., Luo, Y.P., Yang, S.Y.: Stochastic convergence analysis and parameter selection of the standard particle swarm optimization algorithm. Information Processing Letters 102, 8–16 (2007)
20. Poli, R., Broomhead, D.: Exact analysis of the sampling distribution for the canonical particle swarm optimiser and its convergence during stagnation. In: Proceedings of the 9th Genetic and Evolutionary Computation Conference (GECCO 2007), pp. 134–141. ACM Press, London (2007)
21. Sun, J.Q.: Stochastic Dynamics and Control. In: Luo, A.C.J., Zaslavsky, G. (eds.) Monographs series on Nonlinear Science and Complexity, p. 410. Elsevier, Amsterdam (2006)
22. Lutes, L.D., Sarkani, S.: Random vibrations. In: Analysis of Structural and Mechanical Systems, p. 638. Elsevier, Amsterdam (2004)

23. García Gonzalo, E., Fernández Martínez, J.L.: Design of a simple and powerful Particle Swarm optimizer. In: Proceedings of the International Conference on Computational and Mathematical Methods in Science and Engineering, CMMSE 2009, Gijón, Spain (2009)
24. Fernández-Martínez, J.L., García-Gonzalo, E., Álvarez, J.P.F., Kuzma, H.A., Menéndez-Pérez, C.O.: PSO: A Powerful Algorithm to Solve Geophysical Inverse Problems. Application to a 1D-DC Resistivity Case. Jounal of Applied Geophysics (2010), doi:10.1016/j.jappgeo.2010.02.001
25. Fernández Martínez, J.L., García Gonzalo, E., Naudet, V.: Particle Swarm Optimization applied to the solving and appraisal of the Streaming Potential inverse problem. Geophysics (2010) special issue in Hydrogeophysics (accepted for publication)
26. Fernández Martínez, J.L., Fernández Álvarez, J.P., García Gonzalo, E., Menéndez Pérez, C.O., Kuzma, H.A.: Particle Swarm Optimization (PSO): a simple and powerful algorithm family for geophysical inversion: SEG Annual Meeting. SEG Expanded Abstracts 27, 3568 (2008)
27. Fernández Martínez, J.L., Kuzma, H., García Gonzalo, E., Fernández Díaz, J.M., Fernández Alvarez, J.P., Menéndez Pérez, C.O.: Application of global optimization algorithms to a salt water intrusion problem. In: Symposium on the Application of Geophysics to Engineering and Environmental Problems, vol. 22, pp. 252–260 (2009a)
28. Fernández Martínez, J.L., Ciaurri, D.E., Mukerji, T., Gonzalo, E.G.: Application of Particle Swarm optimization to Reservoir Modeling and Inversion. In: Fernández Martínez, J.L. (ed.) International Association of Mathematical Geology (IAMG 2009), Stanford University (August 2009b)
29. Fernández Martínez, J.L., García Gonzalo, E., Fernández Muñiz, Z., Mukerji, T.: How to design a powerful family of Particle Swarm Optimizers for inverse modeling. New Trends on Bio-inspired Computation. Transactions of the Institute of Measurement and Control (2010c)

Developing Niching Algorithms in Particle Swarm Optimization

Xiaodong Li

School of Computer Science and Information Technology, RMIT University,
Melbourne, Australia
xiaodong.li@rmit.edu.au

Abstract. Niching as an important technique for multimodal optimization has been used widely in the Evolutionary Computation research community. This chapter aims to provide a survey of some recent efforts in developing state-of-the-art PSO niching algorithms. The chapter first discusses some common issues and difficulties faced when using niching methods, then describe several existing PSO niching algorithms and how they combat these problems by taking advantages of the unique characteristics of PSO. This chapter will also describe a recently proposed *lbest* ring topology based niching PSO. Our experimental results suggest that this *lbest* niching PSO compares favourably against some existing PSO niching algorithms.

1 Introduction

Stochastic optimization algorithms such as Evolutionary Algorithms (EAs) and more recently Particle Swarm Optimization (PSO) algorithms have shown to be effective and robust optimization methods for solving difficult optimization problems. The original and many existing forms of EAs and PSOs are usually designed for locating a single global solution. These algorithms typically converge to one final solution because of the global selection scheme used. However, many real-world problems are "multimodal" by nature, that is, multiple satisfactory solutions exist. For such an optimization problem, it might be desirable to locate all global optima and/or some local optima that are also considered as being satisfactory. Numerous techniques have been developed in the past for locating multiple optima (global or local). These techniques are commonly referred to as "niching" methods. A niching method can be incorporated into a standard EA to promote and maintain the formation of multiple stable subpopulations within a single population, with an aim to locate multiple optimal or suboptimal solutions. Niching methods are of great value even when the objective is to locate a single global optimum. Since a niching EA searches for multiple optima in parallel, the probability of getting trapped on a local optimum is reduced.

Niching methods have also been incorporated into PSO algorithms to enhance their ability to handle multimodal optimization problems. This chapter aims to

B.K. Panigrahi, Y. Shi, and M.-H. Lim (Eds.): Handbook of Swarm Intelligence, ALO 8, pp. 67–88.
springerlink.com

provide a survey of several state-of-the-art PSO niching algorithms. The chapter will begin with a brief background on niching methods in general, and then identify some difficulties faced by existing niching methods. The chapter will then go on to describe the development of several PSO niching algorithms and how they are designed to resolve some of these issues by taking advantages of the inherent characteristics of PSO. In particular, the chapter will describe in detail a recently proposed *lbest* ring topology based niching PSO. Experimental results on this *lbest* niching PSO will be compared against an existing PSO niching algorithm, and their strengthes and weaknesses will be examined. Finally the chapter concludes by summing up the important lessons learnt on developing competent PSO niching methods, and possible future research directions.

2 Niching Methods

Just like Evolutionary Algorithms themselves, the notion of niching is inspired by nature. In natural ecosystems, individual species must compete to survive by taking on different roles. Different species evolve to fill different *niches* (or subspaces) in the environment that can support different types of life. Instead of evolving a single population of individuals indifferently, natural ecosystems evolve different species (or subpopulations) to fill different niches. The terms species and niche are sometimes interchangeable. Niching methods were introduced to EAs to allow maintenance of a population of diverse individuals so that multiple optima within a single population can be located [25]. One of the early niching methods was developed by De Jong in a scheme called *crowding*. In *crowding*, an offspring is compared to a small random sample taken from the current population, and the most similar individual in the sample is replaced. A parameter CF (*crowding factor*) is commonly used to determine the size of the sample. The most widely used niching method is probably *fitness sharing*. The sharing concept was originally introduced by Holland [16], and then adopted by Goldberg and Richardson [14] as a mechanism to divide the population into different subpopulations according to the similarity of the individuals in the population. Fitness sharing was inspired by the *sharing* concept observed in nature, where an individual has only limited resources that must be shared with other individuals occupying the same niche in the environment. A sharing function is often used to degrade an individual's fitness based on the presence of other neighbouring individuals. Although *fitness sharing* has proven to be a useful niching method, it has been shown that there is no easy task to set a proper value for the sharing radius parameter σ_{share} in the sharing function without prior knowledge of the problems [13].

Apart from the above, many more niching methods have been developed over the years, including *derating* [1], *deterministic crowding* [24], *restricted tournament selection* [15], *parallelization* [2], *clustering* [37], and *speciation* [30, 20]. Niching methods have also been developed for PSO, such as *NichePSO* [31], *SPSO* [26], and *VPSO* [33].

2.1 Difficulties Facing Niching Methods

Most of these niching methods, however, have difficulties which need to be overcome before they can be applied successfully to real-world multimodal problems. Some identified issues include the following:

- Reliance on prior knowledge of some niching parameters, which must be set with some optimal values so that the optimization algorithm can perform well. A common use of a niching parameter is to tell how far apart two closest optima are. A classic example is the sharing parameter σ_{share} in fitness sharing [14]. Other uses of niching parameters include *crowding factor* in *crowding method* [12], the window size w in *restricted tournament selection* [15], or the number of clusters in *k-means clustering methods* [37, 17].
- Difficulty in maintaining found solutions in a run. Some found solutions might be lost in successive generations. For example, the original De Jong's crowding has been shown unable to maintain all found peaks during a run [24]. A good niching algorithm should be able to form and maintain stable subpopulations over the run.
- In traditional niching EAs, it was observed that crossover between two fit individuals from different niches could produce far less fit offspring than the parents [25]. How can we minimize such detrimental crossover operations across different niches?
- Some existing niching methods are designed only for locating all global optima, while ignoring local optima. Examples include the sequential niche GA (SNGA) [1], clearing [30], SCGA [20], NichePSO [31], and SPSO [21, 26]. However, it might be desirable to obtain both global and local optima in a single run.
- Most niching methods are evaluated on test functions of only 2 or 3 dimensions. How well these niching algorithms perform on high dimensional problems remain unclear.
- Higher computational complexity. Most of the niching algorithms use global information calculated from the entire population, therefore require at least $\mathcal{O}(N^2)$ computational complexity (where N is the population size). Many niching algorithms suffer from this problem.
- Most existing niching methods are evaluated using static functions. When functions can vary over time, ie., the multimodal fitness landscape may change over time, most existing niching methods are unable to cope with the dynamically changing environments.

Problems with Niching Parameters

Most existing niching methods, however, suffer from a serious problem - their performance is subjected heavily to some niching parameters, which are often difficult to set by a user. For example the sharing parameter σ_{share} in *fitness sharing* [14], the species distance σ_s in *species conserving GA* (SCGA) [20], the distance measure σ_{clear} in *clearing* [30], and the species radius r_s in the

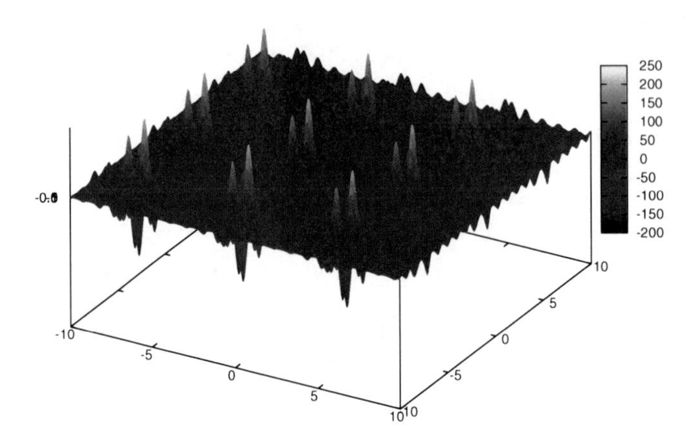

Fig. 1. Inverted Shubert 2D function.

speciation-based PSO (SPSO) [26]. Sometimes niching parameters can be under different disguises, such as the *crowding factor* in *crowding* [12], the window size w in *restricted tournament selection* [15], or the number of clusters in *k-means clustering methods* [37, 17]. The performance of these EAs depend very much on how these parameters are specified. Unfortunately, in many real-world problems such prior knowledge are often unavailable. Some recent works by Bird and Li [4, 5] attempted to reduce the sensitivity of the SPSO to the niche radius parameter values. However, either this parameter still remains (though made more robust), or several new parameters are introduced. It would be desirable if a user can be completely freed from specifying any niching parameters.

Fig. 1 shows an example of a function fitness landscape that has 9 pairs of global optima and numerous local optima. Within each pair, two global optima are very close to each other but optima from different pairs are further away. A niching algorithm relying on a fixed niche radius value to determine a particle's membership in a niche would have a significant difficulty to work properly on such a landscape. To capture all peaks, a niching EA would have to set its niche radius extremely small so that the closest two peaks can be distinguished. However, doing so would form too many small niches, with possibly too few individuals in each niche. As a result, these niches tend to prematurely converge. On the other hand, if the niche radius is set too large, peaks with a distance between them smaller than this value will not be distinguished. In short, it is likely that there is no optimal value for the niche radius parameter. Dependency on a fixed niche radius is a major drawback for niching methods that rely on such a parameter. For example in [20], on the inverted Shubert 2D function (as shown in Fig. 1), SCGA had to be tuned with a radius value of 0.98 and a population size of 1000 in order to locate all 18 global peaks reliably [20].

For Shubert 3D, SCGA used a population size of 4000 in order to locate all 81 global peaks. As dimension increased to 4, SCGA was only able to identify groups of global peaks, but not individual global optima within each group. Another similar niching algorithm SPSO [26] suffers the same problem.

3 Particle Swarm Optimization

Particle Swarm Optimization (PSO) is a Swarm Intelligence technique originally developed from studies of social behaviours of animals or insects, e.g., bird flocking or fish schooling [18]. In a canonical PSO, the velocity of each particle is modified iteratively by its *personal best* position (i.e., the position giving the best fitness value so far), and the position of best particle from the entire swarm. As a result, each particle searches around a region defined by its personal best position and the position of the population best. Let's use $\mathbf{v}_i$ to denote the velocity of the i-th particle in the swarm, $\mathbf{x}_i$ its position, $\mathbf{p}_i$ the best position it has found so far, and $\mathbf{p}_g$ the best position found from the entire swarm (so called global best). $\mathbf{v}_i$ and $\mathbf{x}_i$ of the i-th particle in the swarm are updated according to the following two equations [10]:

$$\mathbf{v}_i \leftarrow \chi(\mathbf{v}_i + \mathbf{R}_1[0,\varphi_1] \otimes (\mathbf{p}_i - \mathbf{x}_i) + \mathbf{R}_2[0,\varphi_2] \otimes (\mathbf{p}_g - \mathbf{x}_i)), \tag{1}$$

$$\mathbf{x}_i \leftarrow \mathbf{x}_i + \mathbf{v}_i, \tag{2}$$

where $\mathbf{R}_1[0,\varphi_1]$ and $\mathbf{R}_2[0,\varphi_2]$ are two separate functions each returning a vector comprising random values uniformly generated in the range $[0, \varphi_1]$ and $[0, \varphi_2]$ respectively. φ_1 and φ_2 are commonly set to $\frac{\varphi}{2}$ (where φ is a positive constant). The symbol $\otimes$ denotes point-wise vector multiplication. A constriction coefficient χ is used to prevent each particle from exploring too far away in the search space, since χ applies a dampening effect to the oscillation size of a particle over time. This Type 1" constricted PSO suggested by Clerc and Kennedy is often used with χ set to 0.7298, calculated according to $\chi = \frac{2}{\left|2-\varphi-\sqrt{\varphi^2-4\varphi}\right|}$, where $\varphi = \varphi_1 + \varphi_2 = 4.1$ [10].

3.1 PSO Niching Methods

This section describes several representative niching methods that have been developed in conjunction with PSO.

Stretching Method

In [27], Parsopoulos and Vrahitis introduced a method in which a potentially good solution is isolated once it is found, then the fitness landscape is 'stretched' to keep other particles away from this area of the search space [28], similar to the derating method used in SNGA [1]. The isolated particle is checked to see if it is a global optimum, and if it is below the desired accuracy, a small population is generated around this particle to allow a finer search in this area.

The main swarm continues its search in the rest of the search space for other potential global optima. With this modification, Parsopoulos and Vrahitis' PSO was able to locate all the global optima of some selected test functions successfully. However, the drawback is that this *stretching* method introduces several new parameters which are difficult to specify in the *stretching* function, as well as the risk of introducing false optima as a result of *stretching*.

NichePSO

Brits *et al.* proposed NichePSO [31], which further extended Parsopoulos and Vrahitis's model. In NichePSO, multiple subswarms are produced from a main swarm population to locate multiple optimal solutions in the search space. Subswarms can merge together, or absorb particles from the main swarm. NichePSO monitors the fitness of a particle by tracking its variance over a number of iterations. If there is little change in a particle's fitness over a number of iterations, a subswarm is created with the particles closest neighbor. The issue of specifying several user parameters still remains. The authors also proposed *nbest* PSO in [9], where a particle's neighbourhood best is defined as the average of the positions of all particles in its neghbourhood. By computing the Euclidean distances between particles, the neighbourhood of a particle can be defined by its k closest particles, where k is a user-specified parameter. Obviously the performance of *nbest* PSO depends on how this parameter is specified.

Speciation-Based PSO

The speciation-based PSO (SPSO) model was developed based on the notion of species [21]. The definition of species depends on a parameter r_s, which denotes the radius measured in Euclidean distance from the center of a species to its boundary. The center of a species, the so-called species seed, is always the best-fit individual in the species. All particles that fall within the r_s distance from the species seed are classified as the same species.

The procedure for determining species seeds, introduced by Pétrowski in [30] and also Li et al. in [20], is adopted here. By applying this algorithm at each iteration step, different species seeds can be identified for multiple species and these seeds' $\mathbf{p_i}$ can be used as the $\mathbf{p_g}$ (like a neighbourhood best in a *lbest* PSO) for different species accordingly. Algorithm 1 summarizes the steps for determining species seeds.

Algorithm 1 is performed at each iteration step. The algorithm takes as an input L_{sorted}, a list containing all particles sorted in decreasing order of their $\mathbf{x_i}$ fitness. The species seed set S is initially set to $\emptyset$. All particles' $\mathbf{x_i}$ are checked in turn (from best to least-fit) against the species seeds found so far. If a particle does not fall within the radius r_s of all the seeds of S, then this particle will become a new seed and be added to S. Fig. 2 provides an example to illustrate the working of this algorithm. In this case, applying the algorithm will identify s_1, s_2 and s_3 as the species seeds. Note also that if seeds have their radii overlapped (e.g., s_2 and s_3 here), the first identified seed (such as s_2) will dominate over

```
input  : L_sorted - a list of all particles sorted in their decreasing f(x_i) values
output: S - a list of all dominating particles identified as species seeds
begin
    S = ∅;
    while not reaching the end of L_sorted do
        Get best unprocessed p ∈ L_sorted;
        found ← FALSE;
        for all s ∈ S do
            if d(s,p) ≤ r_s then
                found ← TRUE;
                break;
        if not found then
            let S ← S ∪ {p};
end
```

Algorithm 1. The algorithm for determining species seeds according to all $f(\mathbf{x_i})$ values.

those seeds identified later from the list L_{sorted}. For example, s_2 dominates s_3 therefore p should belong to the species led by s_2.

Since a species seed is the best-fit particle's $\mathbf{x_i}$ within a species, other particles within the same species can be made to follow the species seed's $\mathbf{p_i}$ as the newly identified neighborhood best. This allows particles within the same species to be attracted to positions that make them even fitter. Because species are formed around different optima in parallel, making species seeds the new neighborhood bests provides the right guidance for particles in different species to locate multiple optima.

Since species seeds in S are sorted in the order of decreasing fitness, the more highly fit seeds also have a potentially larger influence than the less fit seeds. This also helps the algorithm to locate the global optima before local ones.

Once the species seeds have been identified from the population, we can then allocate each seed's $\mathbf{p_i}$ to be the $\mathbf{p_g}$ to all the particles in the same species at each iteration step. The speciation-based PSO (SPSO) accommodating the algorithm for determining species seeds described above can be summarized in Algorithm 2.

In SPSO, a niche radius must be specified in order to define the size of a niche (or species). Since this knowledge might not be always available a priori, it might be difficult to apply this algorithm to some real-world problems. To combat this problem, two extensions to SPSO aiming to improve the robustness to such a niching parameter were proposed in [4, 5]. In [4], population statistics were used to adaptively determine the niching parameters during a run (see also section 3.1), whereas in [5], a time-based convergence measure was used to directly enhance SPSO's robustness to the niche radius value. These extensions to SPSO made it more robust. Nevertheless, the need to specify the niche radius still remains.

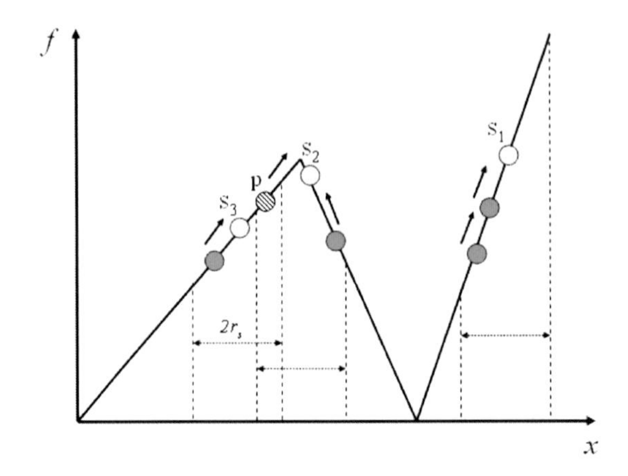

Fig. 2. An example of how to determine the species seeds from a population of particles. s_1, s_2 and s_3 are chosen as the species seeds. Note that p follows s_2.

```
//initialization;
for i=1 to popSize do
    randomly initialize i-th particle: vi, xi;
    pi ← xi
repeat
    for i=1 to popSize do
        evaluate f(xi);
        if f(xi) > f(pi) then
            pi ← xi
    Sort all particles according to their fitness values (from the best-fit to the
    least-fit);
    Call the speciation procedure in Algorithm 1 to identify species seeds;
    Assign each identified species seed's pi as the pg to all individuals identified
    in the same species;
    Adjust particle positions using PSO update equations (1) and (2);
    Check each species to see if the numParticles > pmax; If so, replace the
    excess particles with random particles into the search space;
until the termination condition is met ;
```

Algorithm 2. The species based PSO algorithm.

Adaptive Niching PSO

Instead of requiring a user to specify the niche radius r, the Adaptive Niching PSO (ANPSO) proposed in [4, 3] adaptively determines it from the population statistics at each iteration. More specifically, ANPSO sets r to the average distance between every particle and its closest neighbour (see Fig. 3), as follows:

$$r = \frac{\sum_{i=1}^{N} min_{j \neq i} ||x_i - x_j||}{N}. \tag{3}$$

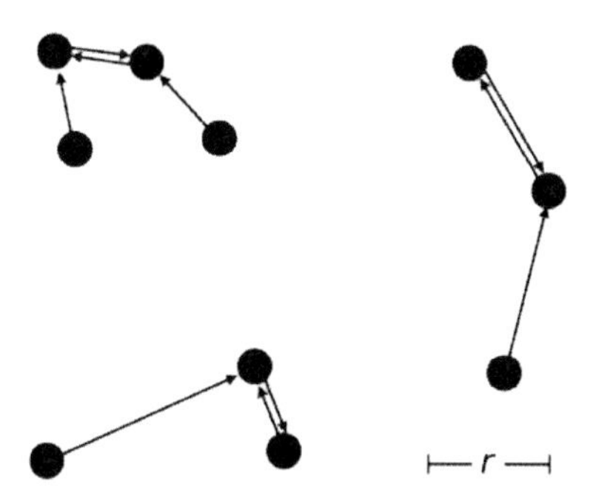

Fig. 3. Calculating the distance from each particle to the particle closest to it. r is calculated by averaging these distances.

An undirected graph g is then created containing a node for each particle. If ANPSO finds pairs of particles that are within r of each other for several iterations, a niche is formed. The remaining unconnected particles (ie., unniched) are mapped onto a von Neumann neighbourhood. At each iterations, particles can join or be removed from existing niches. Whereas the standard PSO updates are applied to particles in each niche, the *lbest* PSO according to the von Neumann neighbourhood topology is used to update particles that are classified as unniched. The unniched particles are useful especially to search more broadly around the problem space. ANPSO removes the need to specify niche radius r in advance, however, at the same time, it introduces two new parameters, the number of steps two particles must be close before forming a niche, and the maximum number of particle in each niche. Nevertheless, at least on some multimodal test functions, ANPSO's performance was shown to be less sensitive to these two parameters.

Fitness-Euclidean Distance Ratio Based PSO

A PSO based on Fitness-Euclidean distance Ratio (FER-PSO) was proposed in [22]. In FER-PSO, personal bests of the particles are used to form a *memory-swarm* to provide a stable network retaining the best points found so far by the population, while the current positions of particles act as parts of an *explorer-swarm* to explore broadly around the search space. Instead of using a single global best, each particle is attracted towards a fittest-and-closest neighbourhood point $\mathbf{p}_n$ which is identified via computing its FER (Fitness and Euclidean-distance Ratio) value:

$$FER_{(j,i)} = \alpha \cdot \frac{f(\mathbf{p}_j) - f(\mathbf{p}_i)}{||\mathbf{p}_j - \mathbf{p}_i||}, \quad (4)$$

where $\alpha = \frac{||s||}{f(\mathbf{p}_g) - f(\mathbf{p}_w)}$ is a scaling factor, to ensure that neither fitness nor Euclidean distance becomes too dominated over one another. $||s||$ is the size of the

```
input  : A list of all particles in the population
output: Neighbourhood best p_n based on the i-th particle's FER value

FER ← 0, tmp ← 0, euDist ← 0 ;
for j = 1 to Population Size do
    Calculate the Euclidean distance euDist from p_i to the j-th particle's
    personal best p_j;
    if (euDist not equal to 0) then
        Calculate FER according to equation (4) ;
        if (j equal to 1) then tmp ← FER;
        if (FER > tmp) then
            tmp ← FER ;
            p_n ← p_j ;
return p_n
```

Algorithm 3. The pseudocode of calculating $\mathbf{p}_n$ for the i-th particle under consideration, according to its FER value. To obtain $\mathbf{p}_n$ for all particles, this algorithm needs to be iterated over the population.

search space, which can be estimated by its diagonal distance $\sqrt{\sum_{k=1}^{Dim}(x_k^u - x_k^l)^2}$ (where x_k^u and x_k^l are the upper and lower bounds of the k-th dimension of the search space). $\mathbf{p}_w$ is the worst-fit particle in the current population.

FER-PSO is able to reliably locate all global optima, given that the population size is sufficiently large. One noticeable advantage is that FER-PSO does not require specification of niching parameters. Nevertheless, it introduces a parameter α which needs to be determined by the upper and lower bounds of the variables. Since the algorithm uses global information, the complexity of the algorithm is $\mathcal{O}(N^2)$ (where N is the population size).

Vector-Based PSO

In [34, 33], a vector-based PSO (VPSO) was developed by treating each particle as a vector and simply carrying out vector operations over them. For each particle, VPSO computes the dot product Δ of two differential vectors, $\mathbf{p_i} - \mathbf{x_i}$ and $\mathbf{p_i} - \mathbf{x_i}$. A niche is defined by a niche radius determined by the distance between $\mathbf{p_g}$ and the nearest particle with a negative dot product (ie., moving in an opposite direction). Niche identification is done in a sequential manner. Once a niche is determined, it is excluded from the population, and the process is repeated on the remaining population, until the entire population is grouped into various niches. In VPSO it is not required to specify the niche radius parameter. However, the distance calculations can be expensive since every particle has to be compared with all remaining particles in the population.

In a subsequent work [35], PVPSO which is a parallel version of VPSO was proposed. In PVPSO, different niches can be maintained in parallel. A special procedure was also introduced to merge niches if they are too close to each other (below a specified threshold ϵ).

Clustering-Based PSO

The use of clustering techniques for PSO was first proposed by Kennedy in [17], where the k-means clustering algorithm was used to identify the centers of different clusters of particles in the population, and then use these cluster centers to substitute the personal bests or neighborhood bests. However, Kennedy's clustering technique was used to help locate a single global optimum, rather than multiple optima, as niching normally does. Inspired by this work, a k-means clustering PSO (kPSO) for niching was proposed in [29]. In kPSO, k-means is repeatedly applied to the swarm population at a regular interval. Between each interval, PSO is executed in the normal manner. Particles in different clusters at an early stage of a run could end up in the same cluster as they converge towards the same local optimum. The parameter k is estimated by using the Bayesian information criterion (BIC) [36]. More specifically, k-means is repeatedly applied to the population with different k values (usually from 2 to $N/2$), and the resulting clustering that has the highest BIC value is chosen. By doing this, there is no need to specify k in kPSO. It was shown that the performance of kPSO was comparable to existing PSO niching algorithms such as SPSO and ANPSO on some multimodal test functions.

Niching PSOs for Dynamically Changing Multimodal Environments

Many real-world optimization problems are dynamic and require optimization algorithms capable of adapting to the changing optima over time. An environment that is both multimodal and dynamic presents additional challenges. In fully dynamic multimodal environments, optima may shift spatially, change both height and shape or come into or go out of existence. One useful approach in handling this is to divide the population into several subpopulations, with each subpopulation searches for a promising region of the search space simultaneously. This is the core idea of several recently proposed PSO niching algorithms for handling a dynamical multimodal landscape such as the Dynamic SPSO [26] and the multi-swarm PSO (MPSO) [8], and rSPSO [6]. Several additional issues must be addressed, including outdated memory, population re-diversification, change detections and response strategies. For further information, readers can refer to [7].

4 New Niching Methods Using a *lbest* PSO

In [23], a novel PSO niching method was developed using a simple ring neighbourhood topology, which belongs to the class so called *lbest* PSO models. This

PSO niching method makes use of the inherent characteristics of PSO and does not require any niching parameters. It can operate as a niching algorithm by using individual particles' local memories to form a stable network retaining the best positions found so far, while these particles explore the search space more broadly. Given a reasonably large population uniformly distributed in the search space, the ring topology based niching PSOs are able to form stable niches across different local neighbourhoods, eventually locating multiple global/local optima. This section describes several such ring topology based niching PSO variants in detail, and how PSO's inherent characteristics such as *memory-swarm* and *explorer-swarm* can be utilized to induce stable niching behaviours.

Generally speaking, two common approaches of choosing $\mathbf{p}_g$ in equation (1) are known as *gbest* and *lbest* methods. In a *gbest* PSO, the position of each particle in the search space is influenced by the best-fit particle in the entire population, whereas a *lbest* PSO only allows each particle to be influenced by the best-fit particle chosen from its neighborhood. The *lbest* PSO with a neighborhood size set to the population size is equivalent to a *gbest* PSO. Kennedy and Mendes [19] studied PSOs with various population topologies. One of common population topologies suggested was a ring topology, where each particle on the population array is only allowed to interact with its two immediate neighbours. Among all topologies studied, the ring topology was considered to be "*the slowest, most indirect communication pattern*", whereas the *gbest* PSO represents "*the most immediate communication possible*".

Clearly the ring topology is desirable for locating multiple optima, because ideally we would like to have individuals to search thoroughly in its local neighbourhood before propagating the information throughout the population. The consequence of any quicker than necessary propagation would result in the population converging onto a single optimum (like *gbest* PSO).

As we will demonstrate in the following sections, the ring topology is able to provide the right amount of communication needed for inducing stable niching behaviour.

4.1 Memory-Swarm vs. Explorer-Swarm

In PSO, interactions among particles play an important role in particles' behaviour. A distinct feature of PSO (which is different from many EAs) is that each particle carries a *memory* of its own, i.e., its personal best. We can never underestimate the significance of using *local memory*. As remarked by Clerc in [11], a swarm can be viewed as comprising of two sub-swarms according to their differences in functionality. The first group, *explorer-swarm*, is composed of particles moving around in large step sizes and more frequently, each strongly influenced by its velocity and its previous position (see equation (1) and (2)). The *explorer-swarm* is more effective in exploring more broadly the search space. The second group, *memory-swarm*, consists of personal bests of all particles. This *memory-swarm* is more stable than the *explorer-swarm* because personal bests

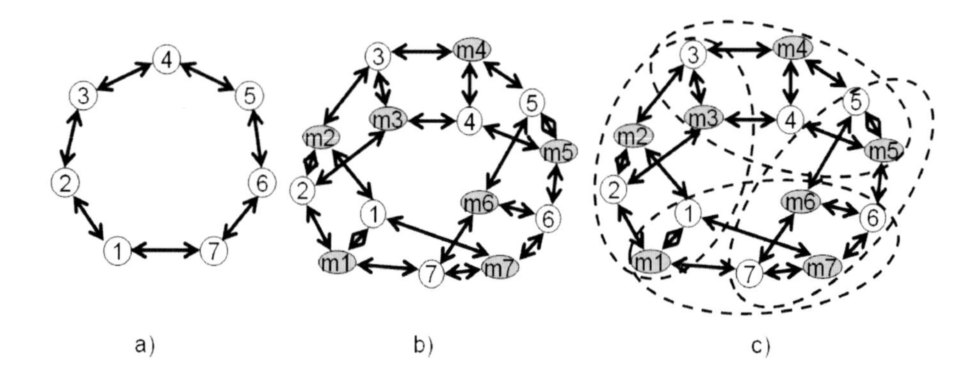

Fig. 4. a) The ring topology used in a conventional EA. Each member interacts only with its immediate left and right neighbours, with no *local memory* used; b) Graph of influence for a *lbest* PSO using the same ring topology (see also p.89 in [11]). Each particle possesses a *local memory*; c) The same as b) but also showing the overlapping subpopulations, each consisting of a particle and its two immediate neighbours, and their corresponding memories.

represent positions of only the best positions found so far by individual particles. The *memory-swarm* is more effective in retaining better positions found so far by the swarm as a whole.

Fig. 4 a) shows an example of a conventional EA using a ring topology with a population of 7 individuals. Fig. 4 b) shows a swarm of 7 particles using a ring topology, as illustrated by using a 'graph of influence' as suggested by Clerc [11]. The 'graph of influence' can be used to explicitly demonstrate the source and receiver of influence for each particle in a swarm. A particle that informs another particle is called 'informant'. Here the *explorer-swarm* consists of particles as marked from numbers 1 to 7, and the *memory-swarm* consists of particles as marked from *m1* to *m7*. Each particle has 3 informants, from two neighbouring particles' memories and its own memory. Each particle's memory also has 3 informants, from two neighbouring particles and the particle itself. In stark contrast, Fig. 4 a) shows that no local memories are used in a conventional EA using a ring topology.

The idea of memory-swarm and explorer-swarm inspired us to develop effective PSO niching algorithms. With an aim to locate and maintain multiple optima, the more stable personal best positions retained in the *memory-swarm* can be used as the 'anchor' points, providing the best positions found so far. Meanwhile, each of these positions can be further improved by the more exploratory particles in the *explorer-swarm*.

4.2 *lbest* PSO Using a Ring Topology

As shown in Fig. 4, in a *lbest* PSO with a ring topology, each particle interacts only with its immediate neighbours. An implementation of such a *lbest* PSO

```
Randomly generate an initial population
repeat
    for i = 1 to Population Size do
    |  if fit(x_i) > fit(p_i) then p_i = x_i;
    end
    for i = 1 to Population Size do
    |  p_{n,i} = neighbourhoodBest(p_{i-1}, p_i, p_{i+1});
    end
    for i = 1 to Population Size do
    |  Equation (1);
    |  Equation (2);
    end
until termination criterion is met ;
```

Algorithm 4. The pseudocode of a *lbest* PSO using a ring topology. Note that in equation (1), $\mathbf{p}_g$ should be replaced by the i-th particle's neighbourhood best $\mathbf{p}_{n,i}$.

using a ring topology is provided in Algorithm 4. Note that we can conveniently use population indices to identify the left and right neighbours of each particle. Here we assume a 'wrap-around' ring topology, i.e., the first particle is the neighbour of the last particle and vice versa. The *neighbourhoodBest*() function returns the best-fit personal best in the i-th neighbourhood, which is recorded as $\mathbf{p}_{n,i}$ (denoting the neighbourhood best for the i-th particle). This $\mathbf{p}_{n,i}$ is then used as the *local leader* when updating the i-th particle in Equation (1) and (2).

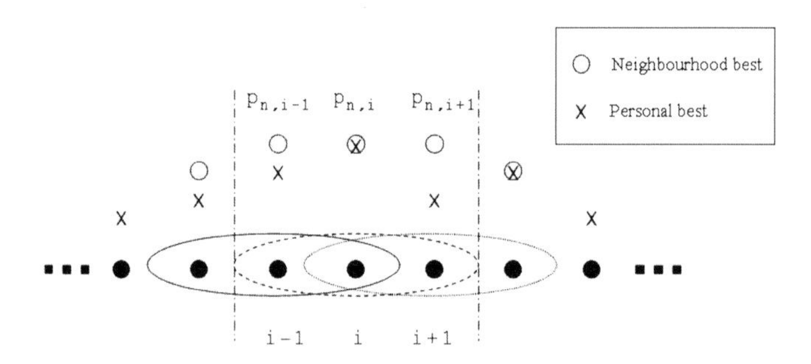

Fig. 5. A ring topology with each member interacting with its 2 immediate neighbours (left and right). Local neighbourhoods are overlapped with each other. The i-th particle's neighbourhood best $\mathbf{p}_{n,i}$ is the same as those of its 2 immediate neighbouring particles, but differs from those particles in the neighbourhoods further out.

Note that different particles residing on the ring can have different $\mathbf{p}_n$[1], and they do not necessarily converge into a single value over time. As illustrated in

[1] We use $\mathbf{p}_n$ to denote a non-specific 'neighbourhood best'.

Fig. 5, the ring topology not only provides a mechanism to slow down information propagation in the particle population, but also allows different neighbourhood bests to *coexist* (rather than becoming homogeneous) over time. This is because a particle's $\mathbf{p}_n$ can only be updated if there is a better personal best in its neighbourhood, but not by a better $\mathbf{p}_n$ of its neighbouring particle. Assuming that particles from the initial population are uniformly distributed across the search space, niches can naturally emerge as a result of the coexistence of multiple $\mathbf{p}_n$ positions being the local attraction points for the particles in the population. With a reasonable population size, such a *lbest* PSO is able to form stable niches around the identified neighbourhood bests $\mathbf{p}_n$.

Apart from its simplicity, the ring topology *lbest* PSO does not require any prior knowledge of (neither the need to specify) any niching parameters, e.g., the niche radius or the number of peaks, since niches emerge naturally from the initial population. The complexity of the algorithm is only $\mathcal{O}(N)$ (where N is the population size), as the calculation to obtain a neighbourhood best is only done locally from each particle's local neighbourhood.

4.3 Numerical Examples

To evaluate the niching ability of the above *lbest* PSO with a ring topology, we used 3 multimodal optimization test functions of different characteristics.[2] f_1 *Equal Maxima* has 5 evenly spaced global maxima, whereas f_2 *Uneven Maxima* has 5 global maxima unevenly spaced. f_3 *Inverted Shubert function* is the inverted Shubert function, as shown in Fig. 1, the inverted Shubert 2D function has 9 groups of global optima, with 2 very close global optima in each group. For n-dimensional Shubert function, there are $n \cdot 3^n$ global optima unevenly distributed. These global optima are divided into 3^n groups, with each group having n global optima being close to each other. For f_3 Shubert 3D, there are 81 global optima in 27 groups; whereas for f_3 Shubert 4D, there are 324 global optima in 81 groups. f_3 will pose a serious challenge to any niching algorithm relying on a fixed niche radius parameter.

The *lbest* PSO with a ring topology as described above has overlapping local neighbourhoods. To further restrain the influence from a few dominant $\mathbf{p}_n$ points, we could reduce the neighbourhood size or even completely remove the overlaps. In our experiments, we used the following ring topology based PSO variants:

- **r3pso**: a *lbest* PSO with a ring topology, each member interacts with its immediate member on its left and right (as in Fig. 5);
- **r2pso**: a *lbest* PSO with a ring topology, each member interacts with only its immediate member to its right;
- **r3pso-lhc**: the same as **r3pso**, but with no overlapping neigbourhoods. Basically multiple PSOs search in parallel, like local hill climbers.
- **r2pso-lhc**: the same as **r3pso-lhr**, but with each member interacts with only its next member on the population array.

[2] These 3 functions are also described in [23].

Table 1. Success rates.

fnc	r2pso	r3pso	r2pso-lhc	r3pso-lhc	SPSO
f_1	98%	**100%**	**100%**	100%	**100%**
f_2	**100%**	**100%**	**100%**	100%	**100%**
f_3(2D)	94%	**100%**	**100%**	98%	60%

For any particle with its $\mathbf{x}_i$ exceeding the boundary of the variable range, its position is reset to a value which is twice of the right (or left boundary) subtracting $\mathbf{x}_i$.

The performance of the above PSO variants were compared with SPSO [26], which is a typical niching algorithm requiring a user to pre-specify a niche radius parameter.

To compare the performance of niching algorithms, we first allow a user to specify a *level of accuracy* (typically $0 < \epsilon \leq 1$), i.e., how close the computed solutions to the expected optimal solutions are. If the distance from a computed solution to an expected optimum is below the specified ϵ, then we can consider the optimum is found. For all comparing niching algorithms in this paper, we used SPSO's procedure for identifying species seeds (as described in the previous section) to check if a niching algorithm has located all expected global optima. Note that this procedure was only used for the purpose of performance measurement, but not for optimization in the proposed PSO niching methods, with the only exception of SPSO itself.

All PSO niching algorithms' performance were measured in terms of *success rate*, i.e., the percentage of runs in which all global optima are successfully located, for a given number of evaluations in each run.

4.4 Results and Discussion

For the relatively simple f_1 and f_2, a population size of 50 was used. The PSO niching variants were run until all the known global optima were found, or a maximum of 100,000 evaluations was reached. For the more challenging f_3 2D and 3D, a population size of 500 was used. For f_3 3D, we allowed a maximum of 200,000 evaluations for each run. For f_3 4D, a population size of 1000 was used, and we allowed a maximum of 400,000 evaluations. All results were averaged over 50 runs. For all PSO niching methods (except SPSO) ϵ and r (niche radius) were used purely for the purpose of performance measurement. In order to measure more accurately each niching algorithm's ability in forming niches in the vicinities of all known global optima, for f_1 and f_2, both ϵ and r were set to 0.01. For f_3 2D, ϵ was set to 0.1, and for f_3 3D and 4D, ϵ was set to 0.2. For all f_3 2D, 3D and 4D, r was set to 0.5.

Table 1 presents the success rates on f_1, f_2 and f_3 2D. On f_1 nd f_2, almost all comparing PSOs achieved a 100% success rate. However, for the more challenging f_3 2D, SPSO did not perform very well, whereas the ring topology PSOs achieved success rates greater than 90%. Bear in mind that SPSO was tuned with the

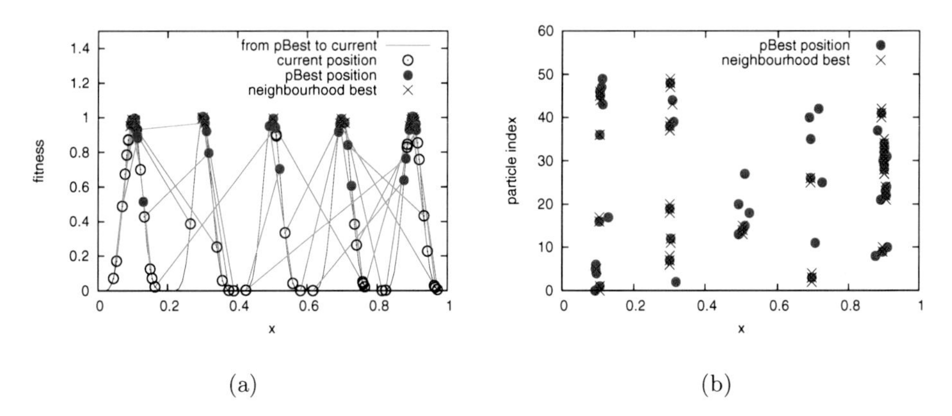

Fig. 6. a) Niches formed when using **r3pso** variant on the f_1 at iteration 15 (a population size of 50 was used); b) Particles' $\mathbf{p}_i$ and their $\mathbf{p}_n$ on the population array at iteration 15, corresponding to the run in a).

Table 2. Averaged peak ratios on f_{11} Inverted Shubert 3D and 4d over 50 runs.

fnc	ϵ	r	r2pso	r3pso	r2pso-lhc	r3pso-lhc	SPSO
f_3 (3D)	0.2	0.5	0.16	0.61	0.27	**0.66**	0.01
f_3 (4D)	0.2	0.5	0.00	**0.25**	0.00	0.14	0.00

optimal niche radius, whereas the ring topology based PSOs did not depend on any niching parameters, showing greater robustness.

Fig. 6 a) shows an example of running **r3pso** on f_1. At iteration 15, all 5 global peaks were located by the $\mathbf{p}_n$ points identified for individual particles on the population array. Although particles' $\mathbf{x}_i$ points (i.e., current positions) tended to be more exploratory oscillating around peaks, their $\mathbf{p}_n$ points converged stably on the tips of the peaks, even if we ran the model for a large number of iterations. Niches formed from neighbouring particles (as shown by their indices on the population array) are clearly visible in Fig. 6 b). It can be also observed that for each of the 5 peaks, **r3pso** formed multiple small niches centered around the $\mathbf{p}_n$ points.

Fig. 7 shows that **r3pso** was able to locate all 18 global peaks on f_3 the inverted Shubert 2D by iteration 75 in a single run. Multiple emerged niches are clearly visible.

For the more challenging f_3 Inverted Shubert 3D and 4D functions, since no run can find all peaks, hence we used *averaged peak ratio* (instead of *success rate*) as the performance measure. Peak ratio measures the percentage of global peaks found in a single run. Table 2 shows the averaged peak ratios on f_3 Inverted

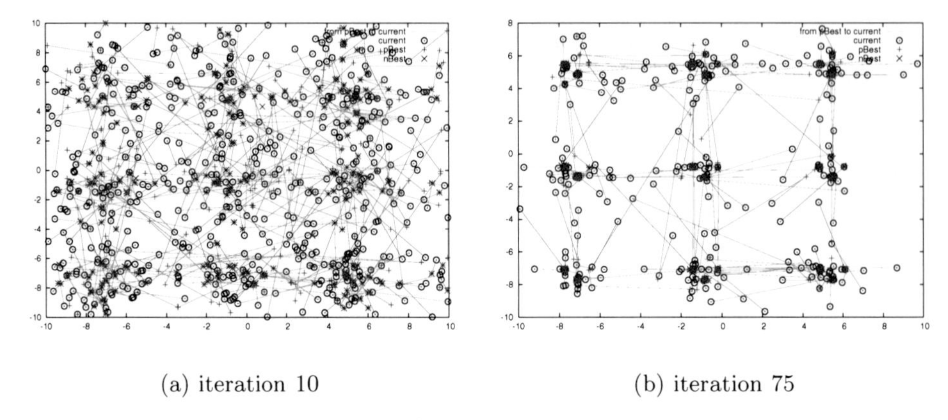

(a) iteration 10 (b) iteration 75

Fig. 7. The niching behaviour of the **r3pso** (with a population size of 500) on f_3 the inverted Shubert 2D function at iteration 10 and 75 of a run.

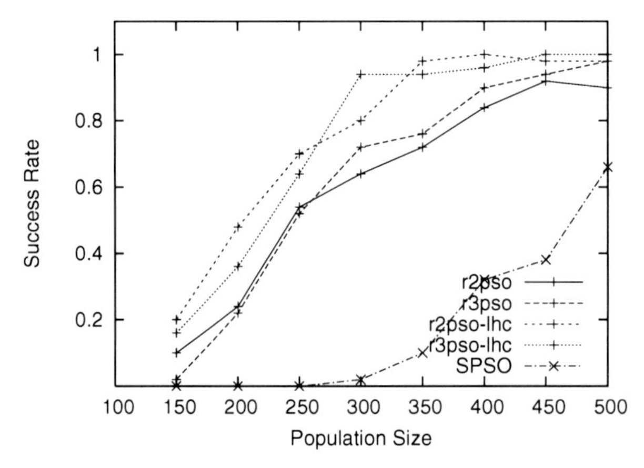

Fig. 8. Success rates for varying population sizes on f_3 the inverted Shubert 2D function.

Shubert 3D and 4D over 50 runs. As can be seen in Table 2, **r3pso** and **r3pso-lhc** are the best performers, whereas SPSO is the worst. **r2pso** and **r2pso-lhc** were able to find a few global peaks on f_3 3D, but failed to find almost any peaks on f_3 4D.

4.5 Effect of Varying Population Size

For the proposed ring topology based PSOs, the only parameter that needs to be specified is population size. Given a reasonably large population size, these PSOs are able to locate all global optima reliably. Fig. 8 shows that on f_3 2D, with a population size of 450 or above, the ring topology based PSOs achieved 90% or above success rates. In contrast, even with a population size of 500, SPSO only managed to achieve 60% success rate. Another similar niching algorithm, SCGA [20], which also required a user to specify a niche radius parameter, needed a population size of 1000 or above in order to locate all 18 global optima.

It is worth noting that the local hill-climber variants **r2pso-lhc** and **r3pso-lhc** performed better than **r2pso** and **r3pso** on f_3 2D. This indicates that when handling problems with a large number of global optima, it might be more effective to have multiple local hill climbers each optimizing independently than a niching algorithm working with a single large population.

5 Conclusions

Niching methods have been developed mostly in the context of EAs, and have been around for more than two decades. Recent advances in Swarm Intelligence and in particular PSO has made possible to design novel and competent niching methods for multimodal optimization. This chapter has presented a survey of several state-of-the-art PSO niching algorithms, and described how some of the challenging issues faced by classic niching methods can be addressed. Apart from the fact that existing niching methods developed in the early days of EAs can be easily incorporated into a PSO, more importantly, it has been shown here that the inherent characteristics of PSO can be utilized to design highly competitive niching algorithms. In particular, it is shown that in a *lbest* PSO, *local memory* and *slow communication topology* are the two key elements for its success as an effective niching algorithms. In fact it is foreseeable that other population based stochastic optimization methods characterized by these two key elements can be also used to induce stable niching behaviour.

In future, we will be interested in investigating how to increase the search capability of small niches so that the performance of these niches will scale well with increasing dimensions, since *lbest* PSO niching algorithms tend to generate multiple small niches. Ideally a function generator suitable for this kind of evaluation will need to offer controllable features such as the number global optima and local optima, which are independent from the number of dimensions. One recently proposed function generator for multimodal function optimization in [32] seems to be a promising tool for this purpose. We will be also interested in developing techniques to adapt or self-adapt the population size, as this is the only parameter that still needs to be supplied by a user. Anther interesting research topic will be to apply the *lbest* niching PSO to tracking multiple peaks in a dynamic environment [26].

References

1. Beasley, D., Bull, D.R., Martin, R.R.: A sequential niche technique for multimodal function optimization. Evolutionary Computation 1(2), 101–125 (1993), `citeseer.ist.psu.edu/beasley93sequential.html`

2. Bessaou, M., Pétrowski, A., Siarry, P.: Island model cooperating with speciation for multimodal optimization. In: Deb, K., Rudolph, G., Lutton, E., Merelo, J.J., Schoenauer, M., Schwefel, H.-P., Yao, X. (eds.) PPSN 2000. LNCS, vol. 1917, pp. 16–20. Springer, Heidelberg (2000), citeseer.ist.psu.edu/bessaou00island.html
3. Bird, S.: Adaptive techniques for enhancing the robustness and performance of speciated psos in multimodal environments, phd thesis. Ph.D. dissertation, RMIT University, Melbourne, Australia (2008)
4. Bird, S., Li, X.: Adaptively choosing niching parameters in a PSO. In: Cattolico, M. (ed.) Proceedings of Genetic and Evolutionary Computation Conference, GECCO 2006, Seattle, Washington, USA, July 8-12, pp. 3–10. ACM, New York (2006), http://doi.acm.org/10.1145/1143997.1143999
5. Bird, S., Li, X.: Enhancing the robustness of a speciation-based PSO. In: Yen, G.G. (ed.) Proceedings of the 2006 IEEE Congress on Evolutionary Computation, July 16-21, pp. 843–850. IEEE Press, Vancouver (2006), http://ieeexplore.ieee.org/servlet/opac?punumber=11108
6. Bird, S., Li, X.: Using regression to improve local convergence. In: Proceedings of the 2007 IEEE Congress on Evolutionary Computation, Singapore, pp. 1555–1562 (2007)
7. Blackwell, T., Branke, J., Li, X.: Particle swarms for dynamic optimization problems. In: Blum, C., Merkle, D. (eds.) Swarm Intelligence - Introduction and Applications, pp. 193–217. Springer, Heidelberg (2008)
8. Blackwell, T.M., Branke, J.: Multi-swarm optimization in dynamic environments. In: Raidl, G.R., Cagnoni, S., Branke, J., Corne, D.W., Drechsler, R., Jin, Y., Johnson, C.G., Machado, P., Marchiori, E., Rothlauf, F., Smith, G.D., Squillero, G. (eds.) EvoWorkshops 2004. LNCS, vol. 3005, pp. 489–500. Springer, Heidelberg (2004)
9. Brits, R., Negelbrecht, A., van den Bergh, F.: Solving systems of unconstrained equations using particle swarm optimizers. In: Proc. of the IEEE Conference on Systems, Man, Cybernetics, October 2002, pp. 102–107 (2002)
10. Clerc, M., Kennedy, J.: The particle swarm - explosion, stability, and convergence in a multidimensional complex space. IEEE Trans. on Evol. Comput. 6, 58–73 (February 2002)
11. Clerc, M.: Particle Swarm Optimization. ISTE Ltd., London (2006)
12. De Jong, K.A.: An analysis of the behavior of a class of genetic adaptive systems. Ph.D. dissertation, University of Michigan (1975)
13. Goldberg, D.E., Deb, K., Horn, J.: Massive multimodality, deception, and genetic algorithms. In: Männer, R., Manderick, B. (eds.) PPSN 2. Elsevier Science Publishers, B. V., Amsterdam (1992), citeseer.ist.psu.edu/goldberg92massive.html
14. Goldberg, D.E., Richardson, J.: Genetic algorithms with sharing for multimodal function optimization. In: Grefenstette, J. (ed.) Proc. of the Second International Conference on Genetic Algorithms, pp. 41–49 (1987)
15. Harik, G.R.: Finding multimodal solutions using restricted tournament selection. In: Eshelman, L. (ed.) Proc. of the Sixth International Conference on Genetic Algorithms, pp. 24–31. Morgan Kaufmann, San Francisco (1995), citeseer.ist.psu.edu/harik95finding.html
16. Holland, J.: Adaptation in Natural and Artificial Systems. University of Michigan Press, Ann Arbor (1975)

17. Kennedy, J.: Stereotyping: Improving particle swarm performance with cluster analysis. In: Proc. of IEEE Int. Conf. Evolutionary Computation, pp. 303–308 (2000)
18. Kennedy, J., Eberhart, R.: Swarm Intelligence. Morgan Kaufmann, San Francisco (2001)
19. Kennedy, J., Mendes, R.: Population structure and particle swarm performance. In: Proc. of the 2002 Congress on Evolutionary Computation, pp. 1671–1675 (2002)
20. Li, J.-P., Balazs, M.E., Parks, G.T., Clarkson, P.J.: A species conserving genetic algorithm for multimodal function optimization. Evol. Comput. 10(3), 207–234 (2002)
21. Li, X.: Adaptively choosing neighbourhood bests using species in a particle swarm optimizer for multimodal function optimization. In: Deb, K., et al. (eds.) GECCO 2004. LNCS, vol. 3102, pp. 105–116. Springer, Heidelberg (2004)
22. Li, X.: Multimodal function optimization based on fitness-euclidean distance ratio. In: Thierens, D. (ed.) Proc. of Genetic and Evolutionary Computation Conference 2007, pp. 78–85 (2007)
23. Li, X.: Niching without niching parameters: Particle swarm optimization using a ring topology. IEEE Trans. on Evol. Comput. 14(1), 150–169 (2010)
24. Mahfoud, S.W.: Crowding and preselection revisited. In: Männer, R., Manderick, B. (eds.) Parallel Problem Solving From Nature 2, pp. 27–36. North-Holland, Amsterdam (1992), `citeseer.ist.psu.edu/mahfoud92crowding.html`
25. Mahfoud, S.W.: Niching methods for genetic algorithms. Ph.D. dissertation, Urbana, IL, USA (1995), `http://citeseer.ist.psu.edu/mahfoud95niching.html`
26. Parrott, D., Li, X.: Locating and tracking multiple dynamic optima by a particle swarm model using speciation. IEEE Trans. on Evol. Comput. 10(4), 440–458 (2006)
27. Parsopoulos, K., Vrahatis, M.: Modification of the particle swarm optimizer for locating all the global minima. In: Kurkova, R.N.M.K.V., Steele, N. (eds.) Artificial Neural Networks and Genetic Algorithms, pp. 324–327. Springer, Heidelberg (2001)
28. Parsopoulos, K., Vrahatis, M.: On the computation of all global minimizers through particle swarm optimization. IEEE Trans. on Evol. Compu. 8(3), 211–224 (2004)
29. Passaro, A., Starita, A.: Particle swarm optimization for multimodal functions: a clustering approach. J. Artif. Evol. App. 2008, 1–15 (2008)
30. Pétrowski, A.: A clearing procedure as a niching method for genetic algorithms. In: Proc. of the 3rd IEEE International Conference on Evolutionary Computation, pp. 798–803 (1996)
31. Brits, A.E.R., van den Bergh, F.: A niching particle swarm optimizer. In: Proc. of the 4th Asia-Pacific Conference on Simulated Evolution and Learning 2002 (SEAL 2002), pp. 692–696 (2002)
32. Rönkkönen, J., Li, X., Kyrki, V., Lampinen, J.: A generator for multimodal test functions with multiple global optima. In: Li, X., Kirley, M., Zhang, M., Green, D., Ciesielski, V., Abbass, H.A., Michalewicz, Z., Hendtlass, T., Deb, K., Tan, K.C., Branke, J., Shi, Y. (eds.) SEAL 2008. LNCS, vol. 5361, pp. 239–248. Springer, Heidelberg (2008)
33. Schoeman, I.: Niching in particle swarm optimization, phd thesis. Ph.D. dissertation, University of Pretoria, Pretoria, South Africa (2009)

34. Schoeman, I., Engelbrecht, A.: Using vector operations to identify niches for particle swarm optimization. In: Proc. of the 2004 IEEE Conference on Cybernetics and Intelligent Systems, Singapore, pp. 361–366 (2004)
35. Schoeman, I., Engelbrecht, A.: A parallel vector-based particle swarm optimizer. In: Proc. of the 7th International Conference on Artificial Neural Networks and Genetic Algorithms (ICANNGA 2005), Coimbra, Portugal (2005)
36. Schwarz, G.: Estimating the dimension of a model. Annals of Statistics 6(2), 461–464 (1978)
37. Yin, X., Germay, N.: A fast genetic algorithm with sharing scheme using cluster analysis methods in multi-modal function optimization. In: The International Conference on Artificial Neural Networks and Genetic Algorithms, pp. 450–457 (1993)

Test Function Generators for Assessing the Performance of PSO Algorithms in Multimodal Optimization

Julio Barrera and Carlos A. Coello Coello

CINVESTAV-IPN (Evolutionary Computation Group), Computer Science Department, México
julio.barrera@gmail.com, ccoello@cs.cinvestav.mx

Summary. The particle swarm optimization (PSO) algorithm has gained increasing popularity in the last few years mainly because of its relative simplicity and its good overall performance, particularly in continuous optimization problems. As PSO is adopted in more types of application domains, it becomes more important to have well-established methodologies to assess its performance. For that purpose, several test problems have been proposed. In this chapter, we review several state-of-the-art test function generators that have been used for assessing the performance of PSO variants. As we will see, such test problems sometimes have regularities which can be easily exploited by PSO (or any other algorithm for that sake) resulting in an outstanding performance. In order to avoid such regularities, we describe here several basic design principles that should be followed when creating a test function generator for single-objective continuous optimization.

1 Introduction

Multimodal problems are those in which the search space has several local optima and possibly more than one global optimum. They constitute a type of optimization problem in which the particle swarm optimization (PSO) algorithm has been only scarcely applied [1]. Such multimodal problems are interesting not only because of the challenge that represents avoiding local optima or the localization of more than one global optimum at the same time, but because there exist several real-world problems presenting such features.

The most common multimodal test functions currently available in the specialized literature show regularities, such as symmetry with respect to one axis, uniform spacing among optima, exponential increase in the number of global optima with respect to the increase in the number of decision variables, among others. Such regularities can be exploited by an optimization algorithm such as PSO, decreasing their degree of difficulty [2]. To overcome these regularities and have a better testing environment to assess the performance of an optimization algorithm, some test functions generators have been developed [3] as well as methodologies to create new test functions by using a composition procedure or through the application of linear transformations on common test functions

B.K. Panigrahi, Y. Shi, and M.-H. Lim (Eds.): Handbook of Swarm Intelligence, ALO 8, pp. 89–117.
springerlink.com

[4]. Our particular interest are scalable test functions presenting several local optima, but only one global optimum.

In this chapter we present a brief introduction to particle swarm optimization, to linear transformations and to the composition of functions. We also provide guidelines to create a composition of functions in a simple way, using any sort of test functions at hand. Additionally, we review the state-of-the-art regarding test function generators, and conclude with some pointers towards promising directions to extend the test functions generators currently available in the specialized literature.

2 The Particle Swarm Optimization Algorithm

Kennedy and Eberhart introduced the PSO algorithm in the mid-1990s [5], and it quickly became popular as an optimizer mainly because of its ease of use and efficacy. In PSO, the position of a possible solution (a particle) is updated using the two rules shown in equations (2) and (1).

$$\mathbf{v}_{t+1} = \mathbf{v}_t + c_1 r_1(\mathbf{g} - \mathbf{x}_t) - c_2 r_2(\mathbf{p} - \mathbf{x}_t) \tag{1}$$

$$\mathbf{x}_{t+1} = \mathbf{x}_t + \mathbf{v}_{t+1} \tag{2}$$

where $\mathbf{x}_t$ and $\mathbf{v}_t$ are the current position and the current velocity of the particle, respectively, $\mathbf{p}$ is the position in which the particle has obtained its best (so far) fitness value, $\mathbf{g}$ is the position with the best (so far) fitness value obtained by the entire swarm, c_1 and c_2 are called the learning constants, r_1 and r_2 are random numbers in the range $[0, 1]$ generated using an uniform distribution. The initial tests of the PSO algorithm were made using the Schaffer F6 function and training a neural network. Further improvements to the PSO algorithm were later introduced by Eberhart and Shi [6], by adding an inertia weight constant to the update rule of the velocity, as shown in equation (3)

$$\mathbf{v}_{t+1} = \omega\mathbf{v}_t + c_1 r_1(\mathbf{g} - \mathbf{x}_t) - c_2 r_2(\mathbf{p} - \mathbf{x}_t) \tag{3}$$

The constant ω acts as a damping parameter, regulating the transition between the exploration and exploitation phases of the algorithm. In this case, the Schaffer F6 function was also adopted for the validation of the PSO algorithm. In some further work, Shi and Eberhart [7] presented a version in which the inertia weight value was linearly decreased. In that work the PSO algorithm was assessed using four test functions: the Sphere, Rosenbrock, Rastrigin, and Griewank. All of these test functions can be scaled to any number of variables.

An analysis of the PSO algorithm from the point of view of the dynamic systems was presented by Clerc and Kennedy [8]. This work introduced another modification to the velocity update rule, as expressed in equation (4).

$$\mathbf{v}_{t+1} = \chi\left[\mathbf{v}_t + c_1 r_1(\mathbf{g} - \mathbf{x}_t) - c_2 r_2(\mathbf{p} - \mathbf{x}_t)\right] \tag{4}$$

where the coefficient ω is computed according to equation (5).

$$\chi = \frac{2\kappa}{|2 - \phi - \sqrt{\phi^2 - 4\phi}|} \tag{5}$$

Here, κ is an arbitrary value in the range $[0, 1]$. It is common to use the value $\kappa = 1$. From equation (5) we can see that, in order to obtain a real value in the square root of the denominator, it is necessary that $\phi \leq 4$. In the case $\phi = 4$, with $\kappa = 1$, we obtain a value $\chi = 1$, and we get the original PSO update rules. In addition to the four test functions used by Shi and Eberhart [7], that can be scaled up to any number of variables, other four test functions were used, namely, De Jong F4, Schaffer F6, Foxholes (De Jong F5), and a Rosenbrock variant with only two variables. Of the last four test functions only the De Jong F4 test function can be used with any number of variables.

Other modifications to the PSO algorithm include the topology of the particles. Initially, the global best position $\mathbf{g}$ is computed by inspecting each of the particles in the swarm. In this case, all the particles can exchange information among them. Thus, the information of which particle has the best fitness value is transferred quickly and easily. In an attempt to slow down the information transfer (and favor diversity), the particles are aligned in a ring. Each particle has only two neighbors with whom they can share information. The position of a particle in the ring is not related to its position in the search space.

The first of these two models (in which all the particles are examined in order to determine the position of a particle) is called *global best* (or *gbest*). The second model (in which a ring topology is used) is called *local best* (or *lbest*). Eberhart and Kennedy introduced the *lbest* topology in their work [9]. Subsequent works from Kennedy [10] and Kennedy and Mendes [11] examined in more detail the effects of using different topologies in the PSO algorithm. The test functions used in this case include the Sphere, Rosenbrock, Ranstring, Griewank, and in [11], the Schaffer F6 test function is used as well. Along with the *gbest* and *lbest* topologies, the von Neumann topology [11] has also been popular, particularly when adopting sub-swarms [12, 13]. In the von Neumann topology, the particles

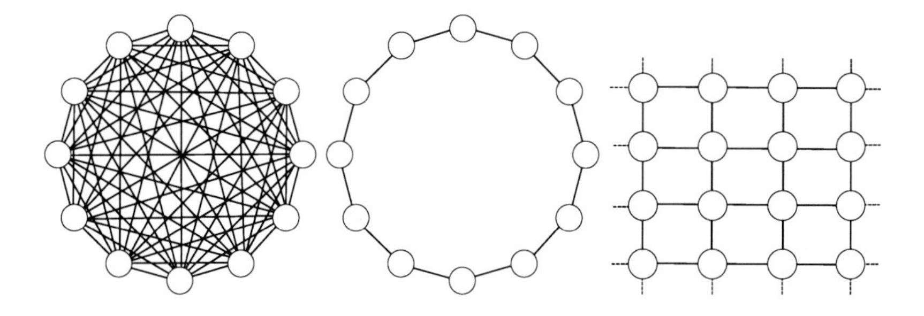

Fig. 1. Different topologies used with the PSO algorithm. From left to right: the *gbest* topology where all particles can share information among them, the *lbest* topology where each particle has only two neighbors and can only share information with them, and the von Neumann topology where the particles are arranged in a regular grid and each particle has four neighbors.

are arranged in a grid and each particle has four neighbors. Figure 1 illustrates the *gbest*, *lbest*, and von Neumann topologies.

Although other, more robust and elaborate, PSO variants have been proposed (see for example [14, 15]), most of them rely on the use of the constriction factor and the inertia weight model, along with the *gbest*, *lbest* or von Neumann topologies.

3 Linear Transformations and Homogeneous Coordinates

Most of the basic test functions have regularities that can be exploited by optimization algorithms. Examples of such regularities are that the position of the optimum is in the origin, or that it has equal values for all of its coordinates. These regularities can be overcome by using linear transformations. A translation transformation can displace the location of the optimum. Indeed, the position of the optimum can be not only displaced, but its coordinates can also have different values. Other transformations such as rotation and scaling can also be used to break regularities in the test functions.

A well-known drawback of using linear transformations is that such transformations are applied separately. It is, for example, common to first translate a point using vector operations, then multiply it by a scalar in order to apply a scaling transformation, and finally, multiply the vector by a matrix in order to rotate its position. This is partly due to that a translation cannot be expressed as a matrix for a given dimension D. If all linear transformations could be expressed by a matrix, they could all be combined using matrix multiplication, and then, by using a single vector-matrix operation, we could apply all the linear transformations to a point. This can be accomplished by using homogeneous coordinates.

Homogeneous coordinates are commonly used in computer graphics [16]. Although they are used for perspective and projection transformations, they are also useful to express a translation transformation as a matrix. To use homogeneous coordinates we only need to add an extra coordinate to the vector that represents the position. This extra coordinate is used only to help in the application of the linear transformations, and can be dropped after. For example, in a problem with two decision variables x and y, a point P of the search space is represented as $P = (x, y)$. The same point in homogeneous coordinates is expressed as $P' = (x, y, 1)$. A translation transformation that shifts an amount T_x the variable x, and an amount T_y the variable y is expressed as the matrix

$$\begin{bmatrix} 1 & 0 & 0 \\ 0 & 1 & 0 \\ T_x & T_y & 1 \end{bmatrix} \tag{6}$$

To apply the translation transformation, we multiply the vector representing the position P' by the matrix representing the translation transformation as follows

$$\begin{bmatrix} x & y & 1 \end{bmatrix} \begin{bmatrix} 1 & 0 & 0 \\ 0 & 1 & 0 \\ T_x & T_y & 1 \end{bmatrix} = \begin{bmatrix} (x + T_x) & (y + T_y) & 1 \end{bmatrix} \tag{7}$$

We obtain a translated point Q' with coordinates $(x+T_x, y+T_y, 1)$. Removing the last coordinate of Q' we obtain the point $Q = (x + T_x, y + T_y)$ which is the point $P = (x, y)$ translated by an amount T_x and T_y. The translation transformation can be expressed as a matrix for any number of variables. A generalization of the translation transformation matrix is shown in equation (8).

$$T = \begin{bmatrix} 1 & 0 & 0 & \cdots & 0 & 0 \\ 0 & 1 & 0 & \cdots & 0 & 0 \\ 0 & 0 & 1 & \cdots & 0 & 0 \\ \vdots & \vdots & \vdots & \ddots & \vdots & \vdots \\ 0 & 0 & 0 & \cdots & 1 & 0 \\ T_1 & T_2 & T_3 & \cdots & T_D & 1 \end{bmatrix} \tag{8}$$

A scaling transformation can be also represented as a matrix. Equation (9) shows a matrix that represents a scaling transformation for two variables. The values S_x and S_y in the diagonal of the matrix, represent the scaling factors for the variables x and y, respectively. Equation (10) shows the scaling transformation matrix in homogeneous coordinates for two variables.

$$S = \begin{bmatrix} S_x & 0 \\ 0 & S_y \end{bmatrix} \tag{9}$$

$$S = \begin{bmatrix} S_x & 0 & 0 \\ 0 & S_y & 0 \\ 0 & 0 & 1 \end{bmatrix} \tag{10}$$

The scaling transformation matrix can also be expressed for any number of variables. A scaling transformation matrix in D dimensions using homogeneous coordinates is represented in equation (11).

$$S = \begin{bmatrix} S_1 & 0 & 0 & \cdots & 0 & 0 \\ 0 & S_2 & 0 & \cdots & 0 & 0 \\ 0 & 0 & S_3 & \cdots & 0 & 0 \\ \vdots & \vdots & \vdots & \ddots & \vdots & \vdots \\ 0 & 0 & 0 & \cdots & S_D & 0 \\ 0 & 0 & 0 & \cdots & 0 & 1 \end{bmatrix} \tag{11}$$

In the case of a rotation transformation, a matrix can also represent it. However, the matrix that represents a rotation transformation has additional properties, since the matrix must be orthogonal. A rotation transformation can also be difficult to build. For example in three dimensions, and depending of the rotation axis, a rotation transformation matrix can be expressed in any of the following ways:

$$R_z = \begin{bmatrix} \cos\theta & \sin\theta & 0 \\ -\sin\theta & \cos\theta & 0 \\ 0 & 0 & 1 \end{bmatrix} \tag{12}$$

$$R_x = \begin{bmatrix} 1 & 0 & 0 \\ 0 & \cos\theta & \sin\theta \\ 0 & -\sin\theta & \cos\theta \end{bmatrix} \tag{13}$$

$$R_y = \begin{bmatrix} \cos\theta & 0 & \sin\theta \\ 0 & 1 & 0 \\ -\sin\theta & 0 & \cos\theta \end{bmatrix} \tag{14}$$

Rotation can be done around any axis, but the matrix, in each case, takes a different form. As we mentioned before, a matrix that represents a rotation transformation must be orthogonal. Salomon [2] describes a method to generate a matrix that represents the application of random rotation transformations for more that 2 variables. A brief description of this method is provided next:

1. A function that generates a rotation matrix is defined as follows: $Rot(A, i, j)$ returns an identity matrix A with a change in four elements, namely $a_{ii} = a_{jj} = rand_1$ and $a_{ij} = rand_2, a_{ji} = -rand_2$, where $rand_1$ and $rand_2$ are random numbers in the range $[-1, 1]$.
2. To create a square $n \times n$ orthogonal matrix, n rotation matrices are created $A_k = Rot(A, 2, k)$ for $k = 1, \dots, n$ and are multiplied to create A_1.
3. A second $n \times n$ matrix A_2 is created using the product of $A_m = Rot(A, m, n)$ matrices for $m = 3, \dots, n-1$.
4. The final matrix A_R is computed as the product $A_R = A_1 A_2$.

A rotation matrix in homogeneous coordinates follows the same form that the translation and scaling matrices. In order to represent a rotation transformation matrix we only need to add a row and a column and put a value of 1 in the lower-right element of the matrix as in equations (8) and (11).

Now, we can represent all the linear transformations as matrices, and we can combine any number of transformations of any type in a single matrix. This will be useful when we describe how to use linear transformations to generate new test functions in Section 6.

4 Function Composition

In this section, we present a brief review of the basic notions of function and function composition that we will use to generate test functions.

A function f is a rule that associates the elements of a set A called *domain* to the elements of a set B called *codomain*. This relation is commonly represented as $f : A \rightarrow B$, and it is usually said that f maps the elements $a \in A$ into elements $b \in B$. This is written as $a \mapsto b$ or $f(a) = b$. A function has the restriction that an element $a \in A$ is associated only with one element $b \in B$. Thus, an element $a \in A$ cannot be associated with two elements $b_1, b_2 \in B$, but an element $b \in B$ can be associated with two different elements $a_1, a_2 \in A$.

As an example, let's consider the function $f : R \rightarrow R$ that maps $x \in R$ to $y \in R$ using the rule $x \mapsto mx + b$ (which is usually written as $f(x) = mx + b$ or $y = mx + b$). This function maps a $x \in R$ to a $y \in R$ and it is easy to see that if we have two elements $x_1, x_2 \in R$ they map to different elements $y_1, y_2 \in R$. But, with the function $g : R \rightarrow R$, and the rule $x \mapsto x^2$, the elements $x, -x \in R$ are mapped to the same element $x^2 \in R$.

Function composition is the sequential application of two or more functions. That is, we apply the function f to a point x and then function g to $f(x)$. This is written as $g \circ f$. We use function composition so commonly that sometimes we do not realize it. Mathematical operations, such as an addition or a product are functions. When performing a multiplication along with an addition, a function composition is applied. The application of two or more transformations, as we did in the previous section, is also function composition. For example, the function $F(x) = (x + 3)^2 + 5$ is actually the composition of three functions: a function $f : R \rightarrow R$ such that $x \mapsto x + 3$, adds 3 to x, the function $g : R \rightarrow R$ with the rule $y \mapsto y^2$, which raises $(x + 3)$ to the second power. Then, the function $h : R \rightarrow R$, which does $z \mapsto z + 5$, adds 5 to $(x + 3)^2$. Thus, the final result of the composition of the three functions is $h \circ (g \circ f)(x) = F(x) = (x + 3)^2 + 5$.

Before using the transformations and function composition to generate test functions, we will provide a review of test functions that have been commonly adopted in the PSO literature on multimodal optimization.

5 Test Functions Commonly Adopted

This section describes the test functions that have been the most commonly adopted to assess performance of PSO-based algorithms. Details of each of them are also provided, such as the search range, the position of their known optima, and other relevant properties.

The Sphere test function is one of the most simple test functions available in the specialized literature. This test function can be scaled up to any number of variables. It belongs to a family of functions called *quadratic* functions and only has one optimum in the point $\mathbf{o} = (0, 0, \ldots, 0)$. The search range commonly used for the Sphere function is $[-100, 100]$ for each decision variable. Equation (15) shows the mathematical description of the Sphere function and Figure 2 shows its graphical representation with two variables.

$$f(\mathbf{x}) = \sum_{i=1}^{D} x_i^2 \tag{15}$$

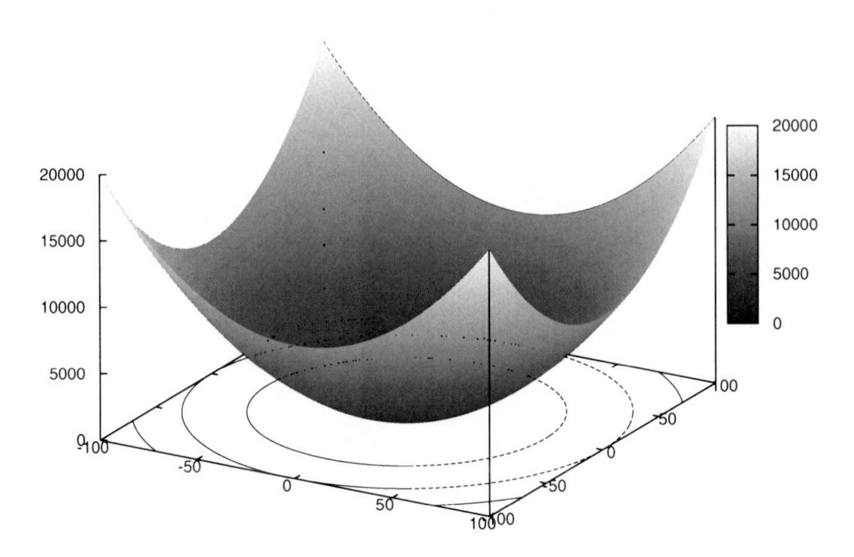

Fig. 2. Graphical representation of the Sphere test function in two dimensions.

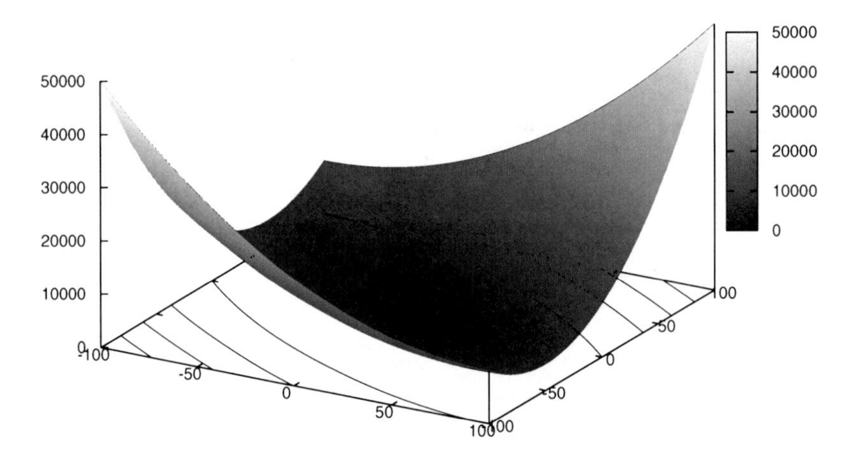

Fig. 3. Graphical representation of the first Schwefel test function in two dimensions.

The first Schwefel test function is also a quadratic function, and it is defined by equation (16). It also has only one optimum at the point $\mathbf{o} = (0, 0, \ldots, 0)$ and its search range is the same as that of the Sphere function (i.e., $[-100, 100]$ for each variable). The graphical representation of the first Schwefel function is shown in Figure 3.

$$f(\mathbf{x}) = \sum_{i=1}^{D} \left(\sum_{j=1}^{i} x_j \right)^2 \tag{16}$$

The Rosenbrock test function is defined by equation (17) and its graphical representation with two variables is shown in Figure 4. Although this picture shows an extense flat region, it only has one optimum located at the point $\mathbf{o} = (1, 1, \ldots, 1)$. It is also a quadratic function, and its search range is $[-30, 30]$ for each variable.

$$f(\mathbf{x}) = \sum_{i=1}^{D-1} \left[100(x_{i+1} - x_i^2) + (x_i - 1)^2 \right] \tag{17}$$

The second Schwefel test function includes a trigonometric function in the equation that defines it (see equation (18)). This provides the function with multiple local optima in the search range, which is, in this case, $[-500, 500]$ for each variable. However, it only has one optimum located at the point $\mathbf{o} = (420.96, 420.96, \ldots, 420.96)$. Its graphical representation, using two variables is shown in Figure 5.

$$f(\mathbf{x}) = -\sum_{i=1}^{D} x_i \sin(\sqrt{|x_i|}) \tag{18}$$

The generalized Rastrigin test function is also commonly adopted, and it is represented by equation (19). It includes a trigonometric function analogously

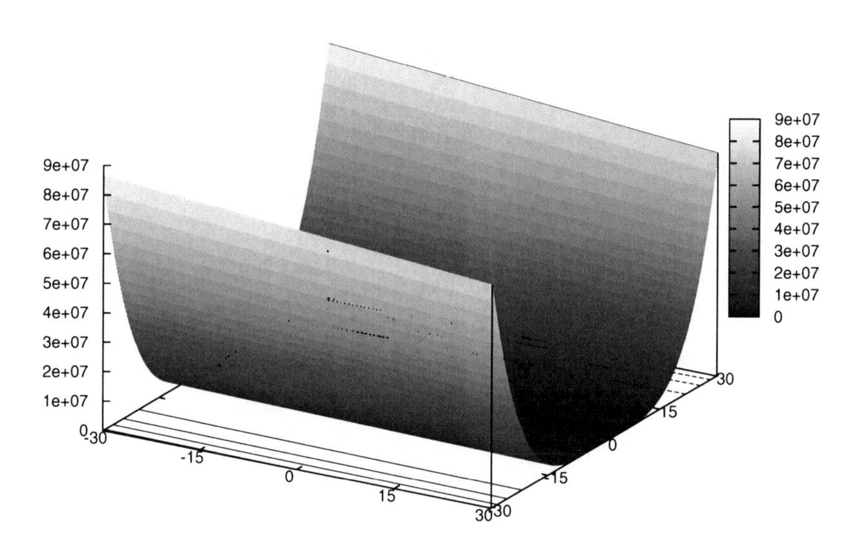

Fig. 4. Graphical representation of the Rosenbrock test function in two dimensions.

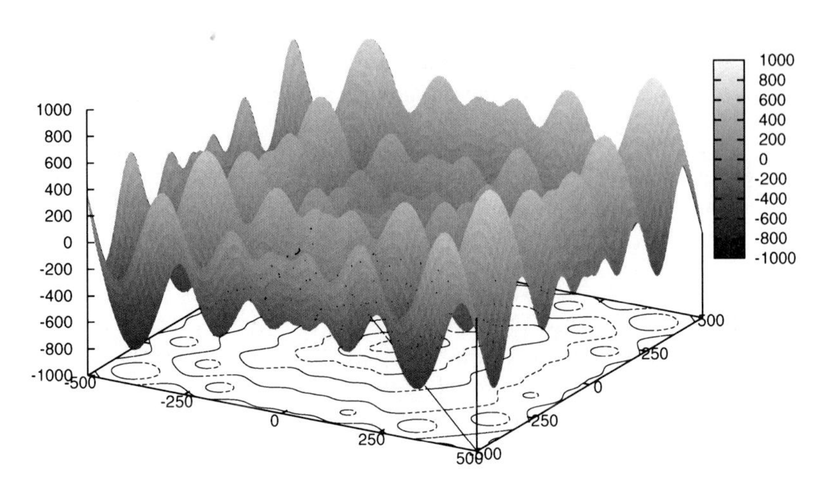

Fig. 5. Graphical representation of the second Schwefel test function in two dimensions.

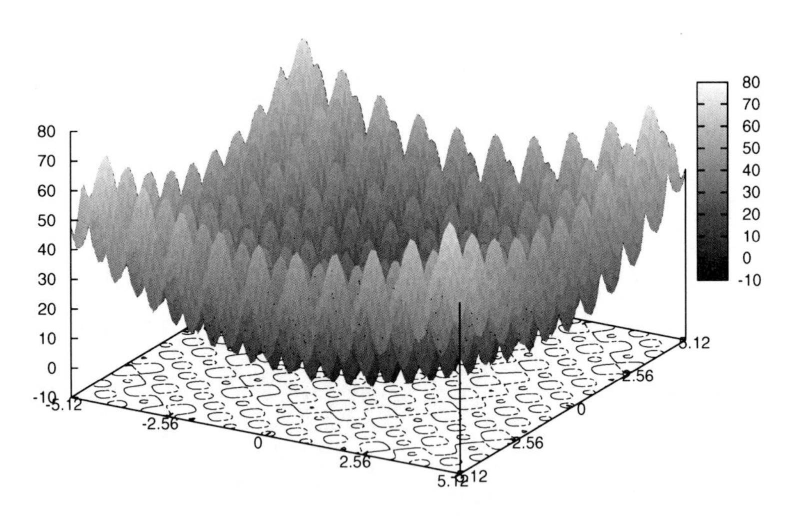

Fig. 6. Graphical representation of the Rastrigin test function in two dimensions.

to the second Schwefel function. The graphical representation of the Rastrigin function is shown in Figure 6. There, we can observe that it has several local optima arranged in a regular lattice, but it only has one global optimum located at the point $\mathbf{o} = (0, 0, \ldots, 0)$. The search range for the Rastrigin function is $[-5.12, 5.12]$ in each variable.

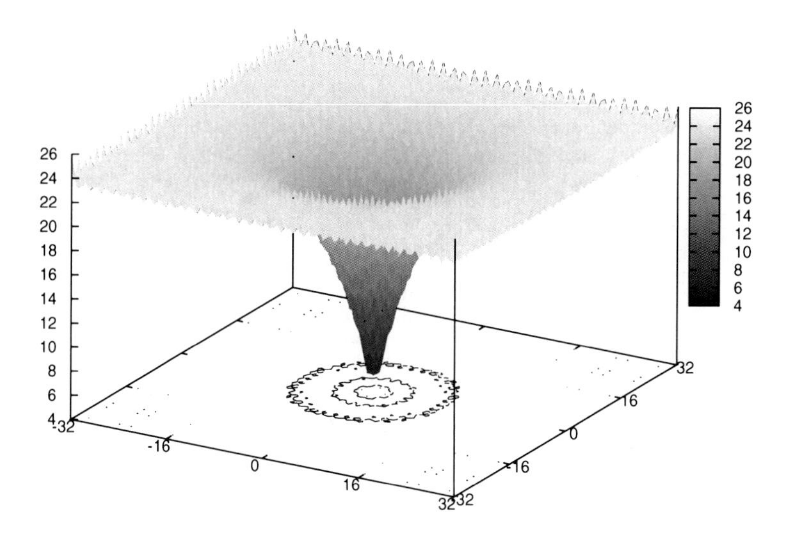

Fig. 7. Graphical representation of the Ackley test function in two dimensions.

$$f(\mathbf{x}) = 10 + \sum_{i=1}^{D} \left\{ x_i^2 - 10 \cos\left(2\pi x_i\right) \right\} \tag{19}$$

The Ackley test function is defined by equation (20) and its graphical representation is shown in Figure 7. As we can see, the Ackley test function has several local optima that, for the search range $[-32, 32]$, look more like noise, although they are located at regular intervals. The Ackley function only has one global optimum located at the point $\mathbf{o} = (0, 0, \ldots, 0)$.

$$\begin{aligned} f(\mathbf{x}) = & -20 \exp\left(-0.2\sqrt{\frac{1}{D}\sum_{i=1}^{D} x_i^2}\right) \\ & - \exp\left(\frac{1}{D}\sum_{i=1}^{D} \cos(2\pi x_i)\right) + 20 + e \end{aligned} \tag{20}$$

The Griewank test function is defined by equation (21). It also shows several local optima within the search region defined by $[-600, 600]$. Figure 8 shows its graphical representation for the case of two variables. It is similar to the Rastrigin function, but the number of local optima is larger in this case. It only has one global optimum located at the point $\mathbf{o} = (0, 0, \ldots, 0)$.

$$f(\mathbf{x}) = \sum_{i=1}^{D} \frac{x_i^2}{4000} - \prod_{i=1}^{D} \cos\left(\frac{x_i}{\sqrt{i}}\right) + 1 \tag{21}$$

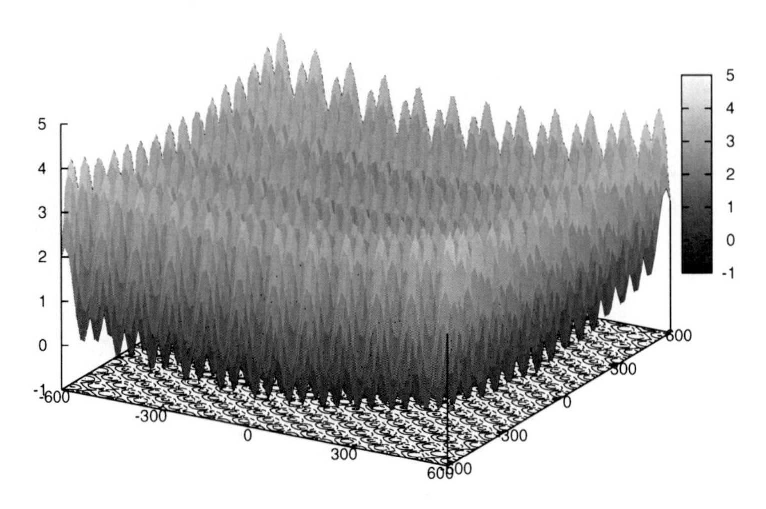

Fig. 8. Graphical representation of the Griewank test function in two dimensions.

The test function shown in this section correspond to those that are most commonly adopted in the specialized literature. They can be scaled up to any number of decision variables. There are also other commonly used test functions that were not included here because they are defined with only one or two decision variables, and are not scalable. We also did not include test functions such as Schubert's function, which has multiple global optima, since they are more suitable for methods dedicated to locate more than one optima in a single run. The interested reader is referred to the work of Bratton and Kennedy [17] for more test functions of this sort.

6 Generating Test Functions

This sections describes how to generate a new test function using a set of available test functions, by applying transformations and function composition.

6.1 Using Linear Transformations

We begin with the most simple function and a translation transformation. The Sphere test function is known to be a very simple function, since it only has one optimum centered in the origin. In order to change the position of the optimum of the Sphere function, we can apply a translation transformation. For example, if we use a translation transformation to move the location of the optimum of the Sphere function in two dimension from the origin to the point $(50, 50)$, we only need to multiply each point by a translation matrix before applying the Sphere function. That is, starting with a point $P = (x, y)$, we first add the dummy variable $w = 1$ and we get the point $P' = (x, y, 1)$. Then, we multiply this point by the matrix

$$\begin{bmatrix} 1 & 0 & 0 \\ 0 & 1 & 0 \\ -50 & -50 & 1 \end{bmatrix} \tag{22}$$

To obtain the translated point $Q' = (x - 50, y - 50, 1)$, we then drop the last coordinate of the point Q' in order to obtain the point $Q = (x - 50, y - 50)$. Finally, we apply the Sphere function to the point Q. Although this may look like a complicated operation to simply translate the optimum of the Sphere function, as we add more transformations, the procedure remains the same, and we only need to build a matrix representing all the transformations.

The effect of the translation applied to the Sphere function is shown in Figure 9. It is worth noting that the values used in the translation transformation are *negative*, and they translate the optimum to a *positive* position. This may be counterintuitive, but it has a reason: if we use positive values in the translation transformation $T_x = 50, T_y = 50$, the point where the optimum is located $\mathbf{o} = (0, 0)$ is translated to $\mathbf{o}' = (-50, -50)$. If we want the new optimum to be located at $P = (50, 50)$ after the translation, the values of T_x, T_y that we need to adopt for the translation transformation must be $T_x = -50, T_y = -50$. After applying the translation to the point $P = (50, 50)$, the outcome is the translated point $Q = (0, 0)$. So, the position of the optimum if we use as translation values $T_x = -50, T_y = -50$ will be located at the point $P = (50, 50)$. If we wish to translate the optimum to another point displaced by an amount x_1, y_1, we must use the values $T_x = -x_1, T_y = -y_1$ for the translation transformation.

It is worth mentioning that certain transformations do not have effect on the Sphere function. For example, a rotation does not have any effect on the complexity of this test function. In fact, rotation has no effect on any function with *radial symmetry*. To break the radial symmetry in the Sphere test function in two dimensions, we can apply a scaling transformation in only one of its variables. For example, to shrink its first coordinate, we can use the transformation matrix

$$S = \begin{bmatrix} 2.0 & 0 & 0 \\ 0 & 1 & 0 \\ 0 & 0 & 1 \end{bmatrix} \tag{23}$$

As in the case of the translation transformation in which we used the inverse value to translate the optimum, in this case, if we want to shrink by a 0.5 factor, the value that we must use in the transformation matrix is the inverse $1/0.5 = 2$. This also applies to the rotation transformation. The effect of the scaling transformation without a translation transformation is shown in Figure 10.

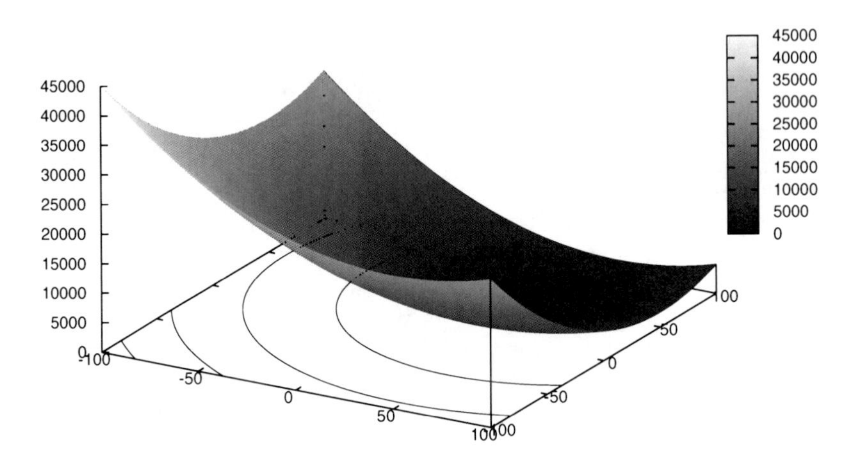

Fig. 9. The Sphere function with the optimum translated from the origin to the point (50, 50).

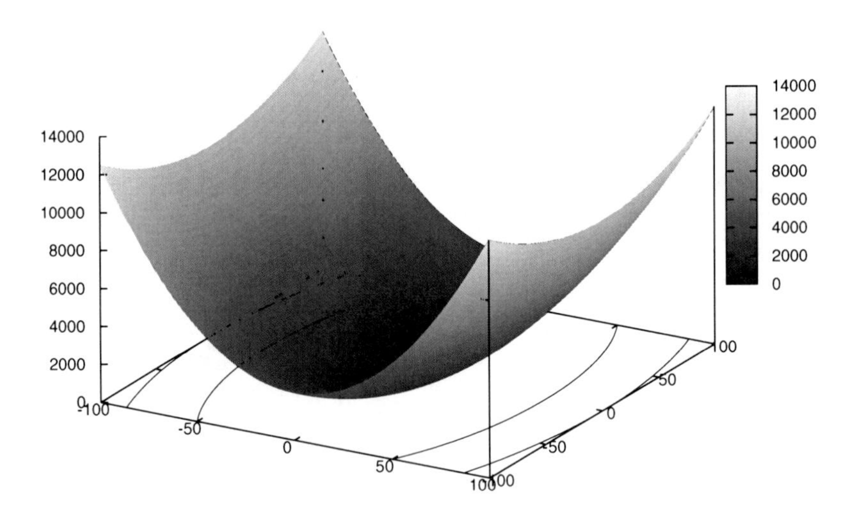

Fig. 10. The Sphere test function scaled.

Without its radial symmetry, a rotation transformation will have an effect in the Sphere test function. A translation of 30 degrees can be done using the following transformation matrix

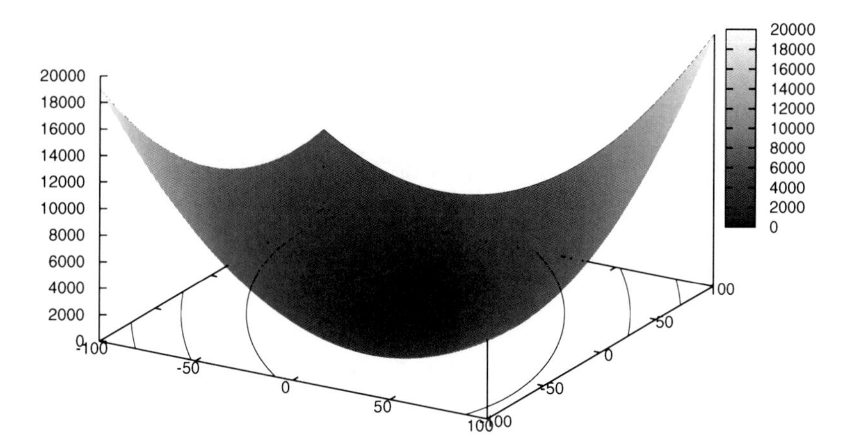

Fig. 11. The Sphere test function with both scaling and rotation transformations.

$$\begin{bmatrix} 0.866 & -0.5 & 0 \\ 0.5 & 0.866 & 0 \\ 0 & 0 & 1 \end{bmatrix} \tag{24}$$

The graphical representation of the Sphere test function with both scaling and rotation is shown in Figure 11.

So far, we have only applied two transformations at the same time to the Sphere function, but we can apply as many as we wish, without forgetting that it is necessary to break the radial symmetry before applying a rotation. As an example, we compute the matrix R that represents three transformations: a translation followed by a scaling and, finally a rotation. Using the same matrices as before, our transformation matrix is computed as follows:

$$R = \begin{bmatrix} 0.866 & -0.5 & 0 \\ 0.5 & 0.866 & 0 \\ 0 & 0 & 1 \end{bmatrix} \begin{bmatrix} 2 & 0 & 0 \\ 0 & 1 & 0 \\ 0 & 0 & 1 \end{bmatrix} \begin{bmatrix} 1 & 0 & 0 \\ 0 & 1 & 0 \\ -50 & -50 & 1 \end{bmatrix} \tag{25}$$

$$= \begin{bmatrix} 0.866 & 0.5 & 0 \\ -0.5 & 0.866 & 0 \\ 0 & 0 & 1 \end{bmatrix} \begin{bmatrix} 2 & 0 & 0 \\ 0 & 1 & 0 \\ -50 & -50 & 1 \end{bmatrix} \tag{26}$$

$$= \begin{bmatrix} 1.732 & -0.5 & 0 \\ 1.0 & 0.866 & 0 \\ -50 & -50 & 1 \end{bmatrix} \tag{27}$$

The graphical representation of the resulting Sphere test function after applying the transformations is shown in Figure 12

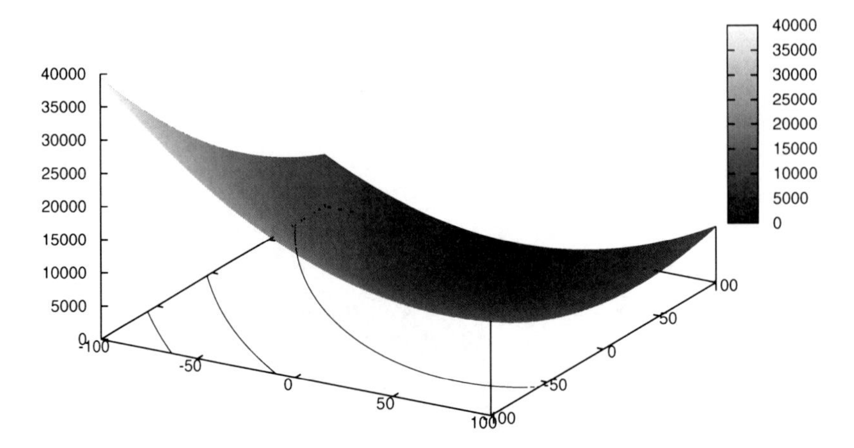

Fig. 12. The Sphere test function after applying three transformations.

The plots of the resulting Sphere test function with the same transformations applied in a different order are shown in Figure 13.

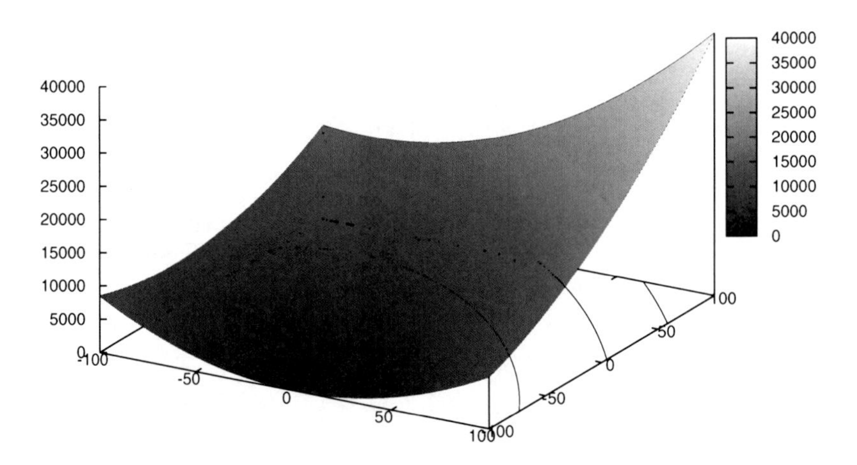

Fig. 13. The Sphere test function with three transformations applied in a different order.

Using linear transformations, we can eliminate some of the drawbacks of a test function, such as having an optimum in a position with repeated values, as well as having symmetry with respect to the axis, among others. However, we can also add more features to our test functions by using function composition.

6.2 Using Function Composition

In the previous subsection, we transformed the Sphere test function by changing the position of its optimum, by breaking its radial symmetry, and by adding a rotation with respect to its coordinate axis. However, the resulting Sphere test function still has only one optimum. The are two common options to do the composition of functions; one is to add the results of two or more functions. For example, if we have two Sphere functions f_1 and f_2, we generate the composite function F as follows

$$F(\mathbf{x}) = f_1(\mathbf{x}) + f_2(\mathbf{x}) \tag{28}$$

In general, if we have n functions, we can generate a composition function $F(x)$ by using the formula of equation (29).

$$F(\mathbf{x}) = \sum_{i=1}^{n} f_i(\mathbf{x}) \tag{29}$$

The second type of composition consists in computing the maximum or minimum of all our functions f_i. Again, in the case of two functions f_1 and f_2, the composite function F using the max function is:

$$F(\mathbf{x}) = \max\{f_1(\mathbf{x}), f_2(\mathbf{x})\} \tag{30}$$

and in general

$$F(\mathbf{x}) = \max_i\{f_i(\mathbf{x})\} \tag{31}$$

Both approaches have their own features and drawbacks that are explained using an example in which we generate two composite functions F_1 and F_2 using a sum and the min function, respectively, with two Sphere functions. If we use two Spheres without any transformation, no much complexity is added to the composite functions F_1 nor for F_2, so our Sphere function f_1 will be a standard Sphere function and f_2 will be a Sphere function with some of the transformations defined in the previous section.

We first use only translation on the f_2 function. The composite function F_1 must have two optima. However, an effect of the composition using addition is that the value of F_1 is different from zero in the search space. The f_1 function is only zero at the location of its optimum $P = (0,0)$. The value of F_1 in the point P is $F_1(P) = f_1(P) + f_2(P) = f_2(P)$, since f_2 is translated $f_2(P) \neq 0$. The same happens at the point Q where the optimum of the f_2 function is located. This has the consequence that the position and value of the optima of the composite function F_1 are different from those of the individual functions f_1 and f_2. A plot of F_1 is shown in Figure 14.

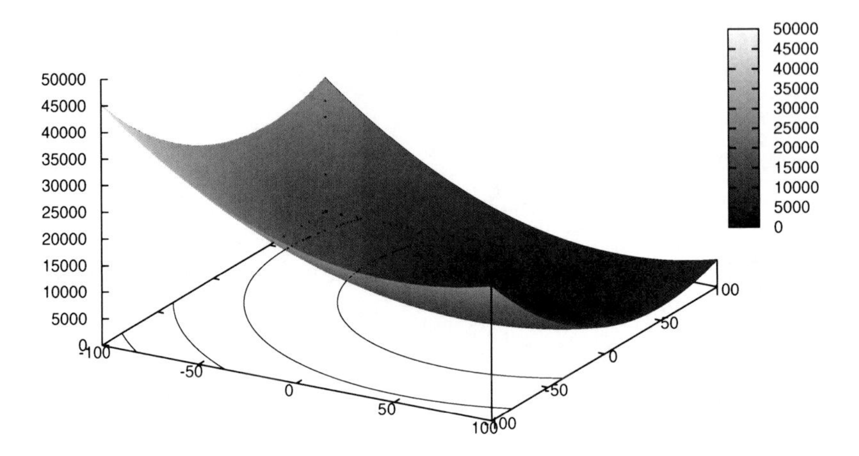

Fig. 14. Composite function F_1.

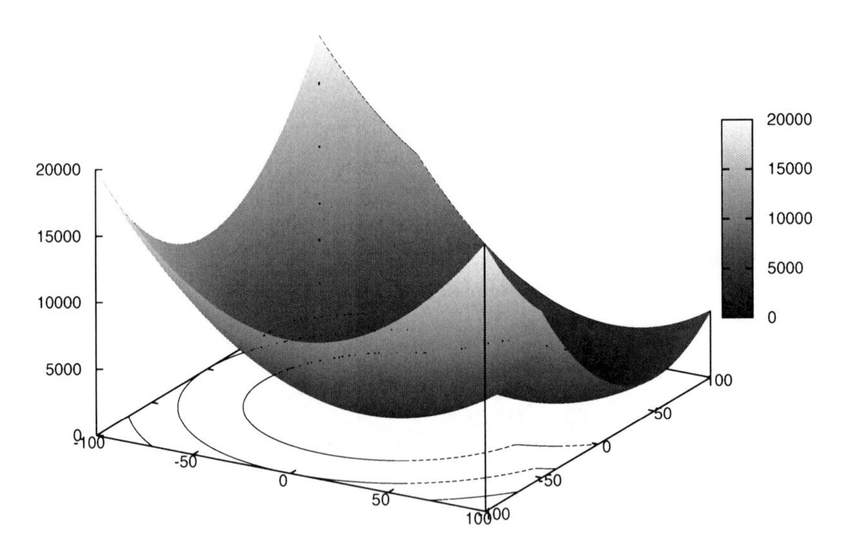

Fig. 15. Composite function F_2.

Now, we examine F_2. In this case, we use the min function, and thus, our composite function F_2 is

$$F_2(\mathbf{x}) = \min\{f_1(\mathbf{x}), f_2(\mathbf{x})\} \tag{32}$$

Contrary to the composite function F_1, the optima of the composite function F_2 is located in the same place of the optimum of f_1 and f_2, and the value of the optima is the same. However, if we use the min function to choose the minimal value of the two Sphere functions, it is possible that in some regions, F_2 is not differentiable, and has abrupt changes in its landscape. Figure 15 shows the plot of F_2.

The composite function F_2 has two optima with value 0 in the position of the optima of the f_1 and f_2 functions, but what if we want only one global optimum? The value of the optima can also be changed with a transformation, but in this case the transformation needs to be done after the application of the test function. Until now, all the transformations have taken place before the application of the function. Since we are dealing with single-objective functions not all transformations can be used. For example, a rotation has no meaning in one dimension. Following our example, we translate the resulting value after applying the Sphere function f_1 by adding a constant value of 5000 and by defining

$$f_1'(\mathbf{x}) = f_1(\mathbf{x}) + 5000 \tag{33}$$

Our composition functions are now defined as

$$F_1(\mathbf{x}) = f_1'(\mathbf{x}) + f_2(\mathbf{x}) \tag{34}$$

$$F_2(\mathbf{x}) = \min\{f_1'(\mathbf{x}), f_2(\mathbf{x})\} \tag{35}$$

The plot of the two composite functions is shown in Figure 17. The translation transformation applied to f_2 after passing the point to the Sphere function can also be represented as a matrix in homogeneous coordinates. In this case, it is represented by a 2×2 matrix, and a scaling transformation can be represented in homogeneous coordinates as well.

Following the example we can describe a general procedure to generate a new test function using linear transformations and function composition as follows:

1. We begin with a point P in the search space.
2. A point Q in homogeneous coordinates is computed using P.
3. The point Q is multiplied by a matrix representing a sequence of linear transformations to obtain Q'.
4. Using Q', we compute P', which gives us the point P transformed in the search space.
5. The value of the test function f_i is obtained passing the point P' as the argument to the test function, and $f(P')$ is computed.
6. Before computing the composite function F, we can additionally apply other linear transformations to $f_i(P')$ and obtain a $f_i'(P')$.
7. Finally, we compute the composite function F by adding the $f_i(P')$ values or by computing the max of them.

Although we only used the Sphere test function in our examples, any other test function can be adopted for the composition previously described. Some

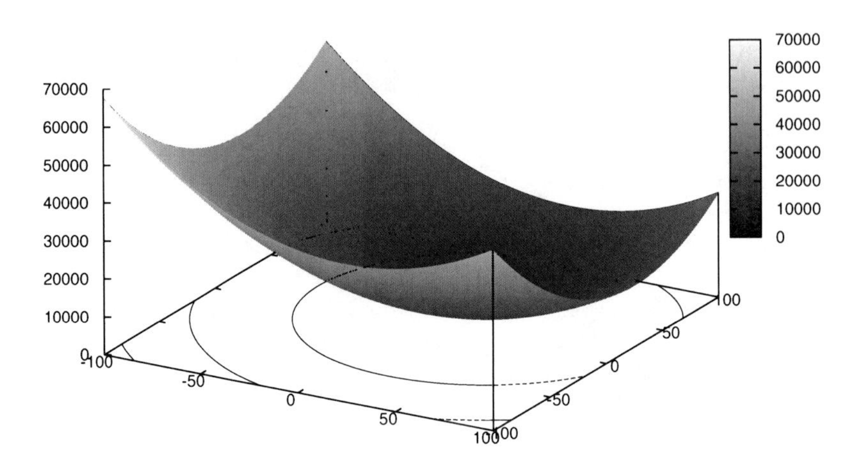

Fig. 16. Composite function F_1.

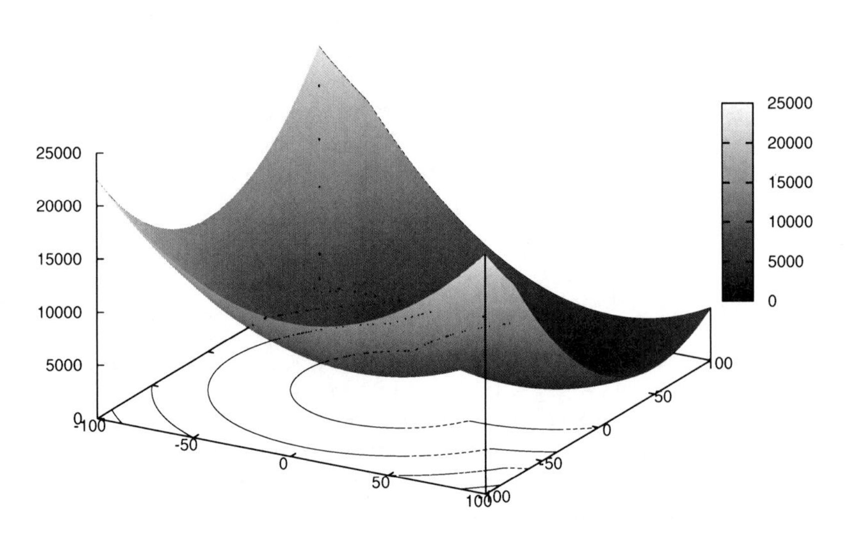

Fig. 17. Composite function F_2.

considerations must be taken when we are using different test functions in the composition. One of them is the search range, For example, if we wish to make a composite function using the Sphere and Ranstrigin tests functions, we need to consider that the range for the Sphere test function is usually $[-100, 100]$ and in the case of the Rastrigin test function the range is $[-5.12, 5.12]$. If we use the range of $[-100, 100]$ in the Rastrigin test function, we obtain the plot shown in

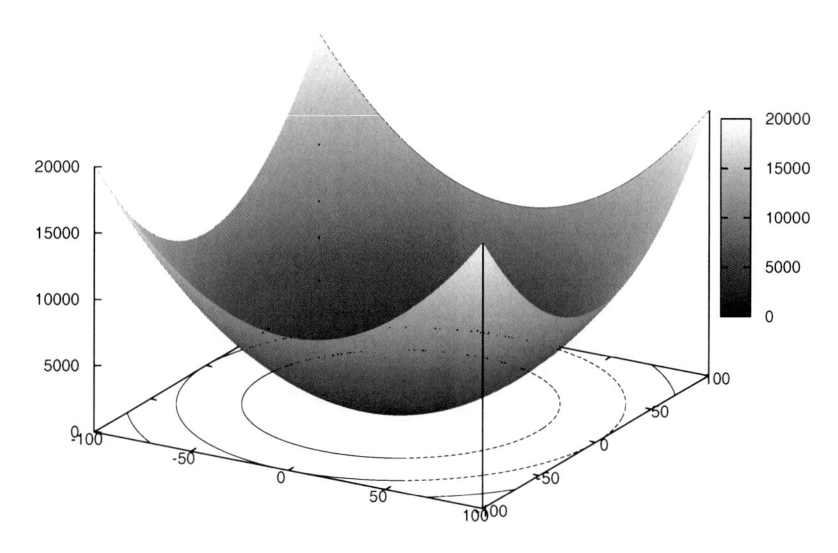

Fig. 18. The Rastrigin test function in the range $[-100, 100]$.

Figure 18. As we can observe, it looks like the Sphere function and does not show any of the original features of the Rastrigin function. Thus, it is necessary to apply a scaling transformation to it.

As a rule of thumb we first *normalize* by dividing by the length of the range of the function being scaled, and then we multiply by the length of the new range. In the case of the Rastrigin function, a scaling factor of $200/10.24 = 19.53$ is needed. In the transformation matrix we use the inverse of this value. If we scale the values of the search range, this does not mean that the values after applying the Rastrigin function are scaled. If we compare the plots of the Sphere and Rastrigin test functions (see Figures 2 and 6), the values that are computed using the Rastrigin test function are far smaller than those of the Sphere test function. It is then necessary to apply a scaling transformation after computing the values of the Rastrigin function. In general, the scaling factor depends on the maximum value of the functions involved. Such value may not be easy to compute. The plot of the composition of the Sphere and Ranstrigin test functions using scaling in the search range and after computing the Rastrigin value is shown in Figures 19 and 20.

Now we can generate new test functions from the common test functions using linear transformations and function composition. However, we should keep in mind the recommendations previously provided about the composition using the addition of two functions or the max or min functions, and be careful about the proper scaling of the search ranges and values.

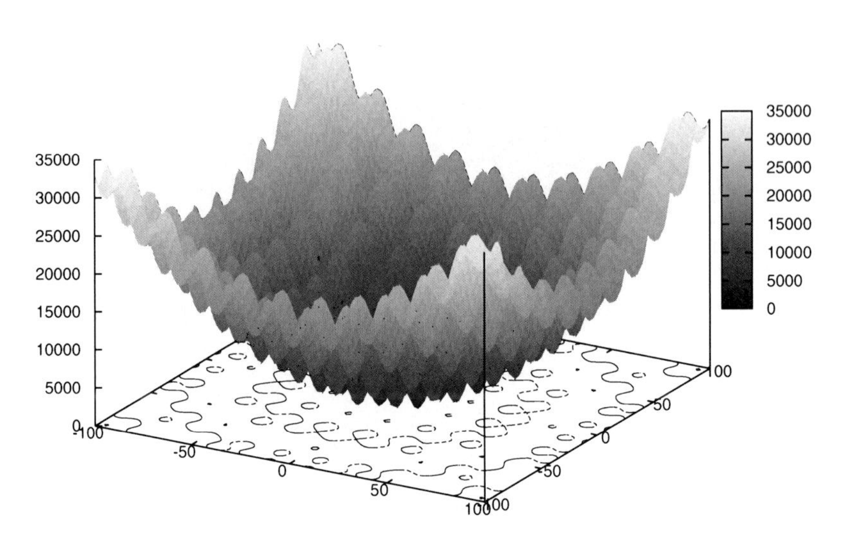

Fig. 19. The composition of the Rastrigin and the Sphere test functions using addition.

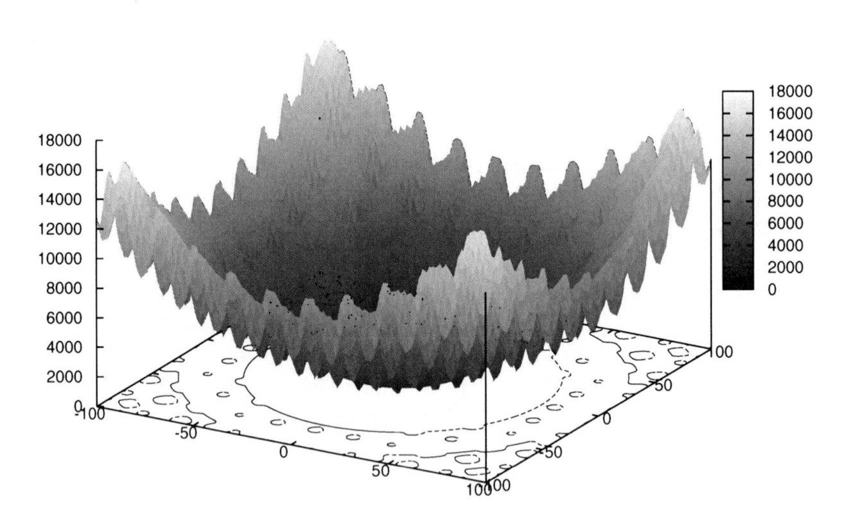

Fig. 20. The composition of the Rastrigin and the Sphere test functions using the min function.

7 Overview of Test Function Generators

In this section, we describe the state-of-the-art regarding test function generators involving the transformations and function composition procedures described in Sections 3 and 4.

We first examine the test function generator of Liang et al. [4]. In this work, the authors proposed a composition of functions using the formula of equation (36).

$$F(\mathbf{x}) = \sum_{i=1}^{n} \{w_i * [f_i'((\mathbf{x} - \mathbf{o}_i + \mathbf{o}_{old})/\lambda_i * M_i) + bias_i] + f_{bias}\} \tag{36}$$

If we examine equation (36), the argument of f_i' is a composition of linear transformations:

$$(\mathbf{x} - \mathbf{o}_i + \mathbf{o}_{old})/\lambda_i * M_i \tag{37}$$

The authors use a translation first to center the optimum of the function in the origin of the coordinate system with the $+\mathbf{o}_{old}$ factor. Then, we translate the optimum to a new position with the factor $-\mathbf{o}_i$. As we see in Section 3, it is not necessary to move first the optimum to the origin, since this displacement can be represented with a translation matrix T_i using homogeneous coordinates. Thus, equation (37) can be rewritten as follows

$$(\mathbf{x}T_i)/\lambda_i * M_i \tag{38}$$

Dividing by λ is also a linear transformation, and in this case, it is a *homogeneous* scaling transformation since they use the same scaling factor for each variable. This can also be represented by a scaling matrix S_i with a $1/\lambda_i$ scale factor in each element of the diagonal except for the last element (see Section 3). Again, equation (37) is rewritten as

$$\mathbf{x}T_iS_i * M_i \tag{39}$$

The $*$ operator in equation (37) represents a matrix product. Given M_i, which is a rotation transformation matrix, equation (37) can be expressed as a product of matrices representing linear transformations and the vector that represents a point in the search space, as expressed in equation (40).

$$\mathbf{x}T_iS_iM_i \tag{40}$$

Defining $R_i = T_iS_iM_i$, equation (36) can be written as

$$F(\mathbf{x}) = \sum_{i=1}^{n} \{w_i * [f_i'(\mathbf{x}R_i) + bias_i] + f_{bias}\} \tag{41}$$

Writing equation (36) in this form, it is easy to observe that after applying f_i' to the point $\mathbf{x}R_i$, a translation transformation is applied using the term $+bias_i$.

Analogously, a scaling transformation is applied using the factor w_i. However, these transformations are done after computing f_i'. If we represent the translation and scaling transformations with the matrices T_i' and S_i', respectively, equation (36) has the form

$$F(\mathbf{x}) = \sum_{i=1}^{n} \{f_i'(\mathbf{x}R_i)T_i'S_i' + f_{bias}\} \tag{42}$$

and we can define $R_i' = T_i'S_i'$, and

$$F(\mathbf{x}) = \sum_{i=1}^{n} [f_i'(\mathbf{x}R_i)R_i' + f_{bias}] \tag{43}$$

If f_{bias} is constant, it can be added as a second translation transformation T_i'', defining $R_i' = T_i'S_i'T_i''$. Therefore, equation (36) is reduced to

$$F(\mathbf{x}) = \sum_{i=1}^{n} f_i'(\mathbf{x}R_i)R_i' \tag{44}$$

After computing the transformation matrices R_i and R_i', the computation of the composite function $F(\mathbf{x})$ is easily done using matrix multiplication. The values used for the scaling transformations w_i and λ_i are related to the maximum value of the functions f_i and to the relative size of the search space, respectively, as indicated in Section 6.

In the work of Singh and Deb [18] a test function is proposed, providing the desired positions of the optima and a radius that defines the region in which the optimum is located. The function is described by equation (45).

$$f(\mathbf{x}) = \begin{cases} h_k \left[1 - \frac{d(x,k)}{r_k}^{\alpha_k}\right], & \text{if } d_{ik} \leq r_k \\ 0, & \text{otherwise} \end{cases} \tag{45}$$

This is more like a max function applied on a radius basis. It is not a coincidence that the function is similar to the equation to compute fitness sharing in genetic algorithms [19]. As we saw in Section 6, the use of a max or a min function allows the definition of regions in which the composite function is not differentiable. In this case, in the regions in which there is no intersection between the regions that contain the optima, the composite function f is constant with a value of 0. Another drawback is that depending on the value of α_k, the regions in which the optima lie can be very sharp.

In the work of Gallagher and Yuan [20], they use composition of exponential functions. The general equation for exponential functions is shown in equation (46).

$$g(x) = \left[\frac{1}{(2\pi)^{\frac{n}{2}} |\Sigma|^{\frac{1}{2}}} \exp\left(-\frac{1}{2}(x-\mu)\Sigma^{-1}(x-\mu)^T\right)\right]^{\frac{1}{n}} \tag{46}$$

where μ is a vector of means, that is, basically a translation of the center of the exponential. In this case, Σ is a square $n \times n$ covariance matrix, which corresponds to both scale and rotation transformations. They also suggest two types of composite functions:

$$F(x) = \sum_{i=1}^{m} w_i g_i(x) \tag{47}$$

and

$$G(x) = \max_i \{w_i g_i(x)\} \tag{48}$$

We discussed the features of each of these approaches in Section 6. It is easy to show that the baseline function in equation (46) is the exponential function

$$e^{x^2} \tag{49}$$

As they do not use homogeneous coordinates, the argument

$$(x - \mu)\Sigma^{-1}(x - \mu)^T \tag{50}$$

represents translation, rotation and scaling transformations. In homogeneous coordinates, the argument can be rewritten as

$$(xTRS)^2 \tag{51}$$

with T, R and S being translation, rotation and scaling transformations, respectively.

In a recent work of Rönkkönen et al. [3], a test function generator is proposed, in which three families of functions are defined: Cosine, Quadratic and Common families. They use rotation, translation and a non-homogeneous scaling using Bezier functions. They do not use homogeneous coordinates. The Cosine family is defined by equation (52).

$$f_{cos}(y) = \frac{\sum_{i=1}^{D} - \cos((G_i - 1)2\pi y_i) - \alpha \cos((G_i - 1)2\pi L_i y_i)}{2D} \tag{52}$$

The functions of the Quadratic family are computed using equation (53).

$$f_{quad}(x) = \min_i \left[(x - p_i)^T B_i^{-1}(x - p_i) - v_i\right] \tag{53}$$

As indicated before, the argument of these functions can be written as $(xTRS)^2$ which is the Sphere function. The Quadratic family coincides with the composition of the Sphere functions using the min function as described in Section 6. The Common family function consists of the Branin, Himmelblau, Shubert, Six-hump camelback, Vincent and the 1st & 3rd Deb's function which are functions in which the number of variables cannot be incremented. They do not use homogeneous coordinates either.

Finally, the work of Morrison and De Jong [21] is one of the few test functions generators designed to test algorithms for dynamic environments. They use function composition with the max function described by equation (54).

$$f(x, y) = \max_i \left[H_i - R_i \sqrt{(x - x_i)^2 + (y - y_i)^2} \right] \tag{54}$$

This is the Sphere function with the addition of a square root and the use of both a translation transformation H_i, and a scaling transformation R_i. They do not use rotation, since the scaling transformation is homogeneous and a rotation does not have any effect, as indicated in Section 6.

8 Guidelines to Build a Test Function Generator

So far we have discussed the basics of the test functions generators such as linear transformations and function composition, giving some advice on how to apply linear transformations to obtain a desired feature, and on the types of function compositions most commonly used. We have also reviewed some of the state-of-the-art test function generators available in the specialized literature. Using this knowledge, we can now offer some guidelines to create a test function generator.

The use of homogeneous coordinates is encouraged. Building a routine to transform a point in the search space into homogeneous coordinates is simple and the transformation itself does not consume a big amount of computer time. However, by using it, we are allowed to write all our linear transformations as a single matrix, and we can apply all of them to a point in a single operation. Returning a transformed point from homogeneous coordinates to the search space also requires a simple operation.

Using linear transformations allows to modify the properties of a test function without altering it at all. Thus, it is not necessary to build new test functions, since those already available can be used. A suitable combination of linear transformations can break the regularities of a test function, and there is no limit on the number of linear transformation that can be used. However, it is important to keep in mind that some regularities cannot be easily changed with linear transformations, as when having radial symmetry, or a regular spacing of the optima as in the Rastrigin function.

We are not limited to linear transformations, since function composition can be used to create new test functions. Like when using linear transformations, any available test function can be used in function composition. In previous sections we have described the features and drawbacks of the two approaches most commonly used for function composition. The use of any of them should be decided according to the features that are desired for the test functions to be generated.

Finally, the use of test function families can simplify the process of creating new test functions. For example, quadratic functions are easy to implement and can be manipulated using linear transformations to break regularities. Their use in function composition is also simple, since the search range and the function values do not change much when linear transformations are applied.

9 Conclusions

It is necessary to have test functions with properties that are more challenging for any optimization algorithm, particularly, those of metaheuristic nature (such as particle swarm optimization). We believe that this would lead to the development of more robust and effective algorithms.

Most of the test functions currently available for validating single-objective particle swarm optimization algorithms have regularities such as symmetry, uniform location of the optima, etc. These features can be exploited by metaheuristics such as PSO, and may turn out to be not as difficult to solve as originally intended.

Some of these drawbacks of the currently available test functions can be avoided by adopting transformations, but, as seen in this chapter, a more in-depth knowledge of the effect of such transformations is required before applying them, in order to avoid unexpected side-effects. Another alternative is to use function composition, but again, some previous knowledge about this procedure is required as well, in order to obtain the intended effects.

In this chapter, we have also reviewed the most representative test function generators reported in the specialized literature. Our review has shown that such generators are relatively complicated to implement, and that, in some cases, the test functions generated do not exhibit features that are sufficiently challenging for a metaheuristic. It is evidently necessary to produce new test function generators which are easier to implement, to configure and to use, and that, at the same time, produce test functions with more challenging features. The main intention of this chapter has been, precisely, to motivate the design of such a test function generator, adopting the transformations and function composition procedures described here.

Following the examples presented in this chapter, it can be clearly seen that the use of homogeneous coordinates simplifies the application of linear transformations. However, homogeneous coordinates are not used in any of the test function generators that we found in the specialized literature.

10 Future Work

The development of a truly simple and configurable test function generator is still an active topic of research. As we have seen in this chapter, the use of homogeneous coordinates can simplify the computations in a test function generator, and provides a more intuitive use of the linear transformations, hence the importance of incorporating homogeneous coordinates in test function generators.

We also believe that the generation of test functions must be focused on the features that we are interested on (e.g., non-uniform location of the local optima), rather than on the complexity of the test function itself (e.g., high nonlinearity in the objective function), since very scarce evidence exists regarding the actual features that turn out to be difficult for an algorithm such as PSO (or any other metaheuristic for that sake).

It is also desirable that the test function generators offer enough flexibility to allow a variety of combinations of features that we are interested in analyzing. Some of the most popular test functions in the current literature do not offer such flexibility. For example, in the Rastrigin function the optima are arranged in a regular lattice and it is impossible to break this property using linear transformations. The use of quadratic functions is therefore, more suitable, since it allows the addition of as many optima as needed, and each optima can be individually manipulated.

Acknowledgements

The first author acknowledges support from CONACyT through a postdoctoral fellowship at the Computer Science Department of CINVESTAV-IPN. The second author acknowledges support from CONACyT project no. 103570.

References

1. Engelbrecht, A.P.: Fundamentals of Computational Swarm Intelligence. John Wiley & Sons, Chichester (2006)
2. Salomon, R.: Re-evaluating genetic algorithm performance under coordinate rotation of benchmark functions. a survey of some theoretical and practical aspects of genetic algorithms. Biosystems 39(3), 263–278 (1996)
3. Rönkkönen, J., Li, X., Kyrki, V., Lampinen, J.: A generator for multimodal test functions with multiple global optima. In: Li, X., Kirley, M., Zhang, M., Green, D., Ciesielski, V., Abbass, H.A., Michalewicz, Z., Hendtlass, T., Deb, K., Tan, K.C., Branke, J., Shi, Y. (eds.) SEAL 2008. LNCS, vol. 5361, pp. 239–248. Springer, Heidelberg (2008)
4. Liang, J., Suganthan, P., Deb, K.: Novel composition test functions for numerical global optimization. In: Proceedings of the 2005 IEEE Swarm Intelligence Symposium (SIS 2005), pp. 68–75 (June 2005)
5. Kennedy, J., Eberhart, R.: Particle swarm optimization. In: Proceedings of the IEEE International Conference on Neural Networks, vol. 4, pp. 1942–1948 (December 1995)
6. Shi, Y., Eberhart, R.: A modified particle swarm optimizer. In: Proceedings of the 1998 IEEE International Conference on Evolutionary Computation. IEEE World Congress on Computational Intelligence, pp. 69–73 (May 1998)
7. Shi, Y., Eberhart, R.: Empirical study of particle swarm optimization. In: Proceedings of the 1999 Congress on Evolutionary Computation (CEC 1999), vol. 3, p. 1950 (1999)
8. Clerc, M., Kennedy, J.: The particle swarm - explosion, stability, and convergence in a multidimensional complex space. IEEE Transactions on Evolutionary Computation 6(1), 58–73 (2002)
9. Eberhart, R., Kennedy, J.: A new optimizer using particle swarm theory. In: Proceedings of the Sixth International Symposium on Micro Machine and Human Science (MHS 1995), pp. 39–43 (October 1995)
10. Kennedy, J.: Small worlds and mega-minds: effects of neighborhood topology on particle swarm performance. In: Proceedings of the 1999 Congress on Evolutionary Computation (CEC 1999), vol. 3, p. 1938 (1999)

11. Kennedy, J., Mendes, R.: Population structure and particle swarm performance. In: Proceedings of the 2002 Congress on Evolutionary Computation (CEC 2002), Washington, DC, USA, pp. 1671–1676. IEEE Computer Society, Los Alamitos (2002)
12. Passaro, A., Starita, A.: Particle swarm optimization for multimodal functions: a clustering approach. Journal Artificial Evolution and Applications 8(2), 1–15 (2008)
13. Bird, S., Li, X.: Enhancing the robustness of a speciation-based pso. In: IEEE Congress on Evolutionary Computation (CEC 2006), pp. 843–850 (2006)
14. Liang, J., Qin, A., Suganthan, P., Baskar, S.: Comprehensive learning particle swarm optimizer for global optimization of multimodal functions. IEEE Transactions on Evolutionary Computation 10(3), 281–295 (2006)
15. de Oca, M.M., Stützle, T., Birattari, M., Dorigo, M.: Frankenstein's PSO: A Composite Particle Swarm Optimization Algorithm. IEEE Transactions on Evolutionary Computation 13(5), 1120–1132 (2009)
16. Harrington, S.: Computer Graphics: A Programming Approach, 2nd edn. Mcgraw-Hill College Press, New York (1987)
17. Bratton, D., Kennedy, J.: Defining a standard for particle swarm optimization. In: Proceedings of the 2007 IEEE Swarm Intelligence Symposium (SIS 2007), pp. 120–127 (April 2007)
18. Singh, G., Deb, K.: Comparison of multi-modal optimization algorithms based on evolutionary algorithms. In: Proceedings of the 8th annual conference on Genetic and evolutionary computation (GECCO 2006), New York, NY, USA, pp. 1305–1312. ACM, New York (2006)
19. Deb, K., Goldberg, D.E.: An Investigation of Niche and Species Formation in Genetic Function Optimization. In: Schaffer, J.D. (ed.) Proceedings of the Third International Conference on Genetic Algorithms, San Mateo, California, George Mason University, pp. 42–50. Morgan Kaufmann Publishers, San Francisco (June 1989)
20. Gallagher, M., Yuan, B.: A general-purpose tunable landscape generator. IEEE Transactions on Evolutionary Computation 10(5), 590–603 (2006)
21. Morrison, R., Jong, K.D.: A test problem generator for non-stationary environments. In: Proceedings of the 1999 Congress Evolutionary Computation (CEC 1999), vol. 3, p. 2053 (1999)

Linkage Sensitive Particle Swarm Optimization

Deepak Devicharan and Chilukuri K. Mohan

Department of Electrical Eng. and Computer Science
4-206 Center for Science and Technology
Syracuse University, Syracuse, NY 13244-4100, USA
Deepak.Devicharan@oclaro.com, mohan@syr.edu

Abstract. In many optimization problems, the information necessary to search for the optimum is found in the linkages or inter-relationships between problem components or dimensions. Exploiting the information in linkages between problem components can help to improve the quality of the solution obtained and to reduce the computational effort required. Traditional particle swarm optimization (PSO) does not exploit the linkage information inherent in the problem. We develop a variant of particle swarm optimization that uses these linkages by performing more frequent simultaneous updates on strongly linked components. Prior to applying the linkage-sensitive variant of PSO to any optimization problem, it is necessary to obtain the nature of linkages between components specific to the problem. For some problems, the linkages are known beforehand or can be set by inspection. In most cases, however, this is not possible and the problem-specific linkages have to be learned from the data available for the problem under consideration. We show, using experiments conducted on several test problems, that the quality of the solutions obtained is improved by exploiting information held in the inter-dimensional linkages.

Keywords: Swarm Optimization, Component Linkage, Linkage Learning.

1 Introduction

In many problems, it is not sufficient to separately optimize each dimension or component of the solution representation. *Linkages,* i.e., interrelationships, exist between certain pairs of the dimensions, related to a phenomenon called *epistasis* [11, 12]. In such problems, finding an optimal solution requires the use of problem-specific linkage information; this may require the application of an algorithm to determine linkage relationships from available data.

Most optimization algorithms are "linkage-blind," whereas some algorithms have implicit linkage assumptions built into them, which may not be justifiable. For instance, traditional crossover operators in genetic algorithms break up problem representation without any regard to the problem-specific dependence between the components [23]. The canonical genetic algorithm (with one-point crossover) implicitly assumes strong

B.K. Panigrahi, Y. Shi, and M.-H. Lim (Eds.): Handbook of Swarm Intelligence, ALO 8, pp. 119–132.
springerlink.com

linkages between the adjacent components, whereas uniform crossover assumes equal linkages between all the dimensions of the problem [23]. As a result, classical crossover operators in genetic algorithms ignore and disrupt problem-specific linkages between the genes.

This chapter addresses the use of linkage information (and linkage learning algorithms in the context of the Particle Swarm Optimization [9] algorithm, using the following notation to describe the commonly used variant, i.e., the *gPSO* algorithm.

> f : Function to be minimized;
> $v(j)$: Current velocity vector of particle j;
> $p(j)$: Current position of particle j;
> $pbest(j)$: Best position encountered by particle j in all previous iterations;

and

> $gbest$: Best particle encountered by any particle so far.

Algorithm gPSO:

For each particle j,
 Randomly initialize $v(j)$ as well as $p(j)$ (and $pbest(j)=p(j)$, initially);
Compute $gbest$ to be the $p(j)$ with the least f among the current set of particles;
While computational limits are not exceeded, do {
 For each particle j, do {

 if $(f(p(j)) < f(pbest(j)))$
 then {$pbest(j) := p(j)$; if $f(p(j))<f(gbest)$ then $gbest := p(j)$;}

 $v(j) := w\, v(j) + c\, r_1\, (pbest(j)-p(j)) + d\, r_2\, (gbest - p(j))$ ....(1)
 where c and d are predetermined constants, and r_1 as well as r_2 are randomly chosen from the uniform distribution over [0,1];

 $p(j) := p(j) + v(j)$ ....(2)
 where w is an inertial parameter chosen from [0,1].
 }}

In the update equations (1) and (2), c and d represent the learning rate parameters which control the rate of convergence towards the optimum, and the inertia weight w controls the impact of the previous velocity on the velocity at iteration $(t+1)$. A larger value of w is used for exploring the global search space, whereas a smaller value of w will help in enhancing a local search. Variations of *gPSO* exist in which the value of w is adapted depending on progress through the optimization algorithm.

Traditional PSO does not take into account the linkages or correlations between the particle dimensions. Each dimension is optimized independent of every other dimension. However certain problems are characterized by the fact that the dimensions show strong linkages between themselves. This chapter shows that for this class of problems, search for the optimal solution is improved by incorporating such problem specific linkage information in the search process. We build on experience gained in addressing the problem of linkages in the genetic algorithm literature where linkage is considered to be the property that alleles for two genes need to be co-adapted in order to obtain good quality solutions [24]. In [23] a linkage-sensitive crossover operator was developed. In [7] crossover operators that are sensitive to linkages were developed and demonstrated to perform well on certain *GA-deceptive* problems.

When *a priori* linkage information is unavailable, we must infer the nature of the linkages from available data. In the context of genetic algorithms, problem-specific linkage information was considered to study correlations and to fine-tune a mutation operator [22, 23]. To this end, we describe in this chapter a method of learning and representing the linkage information, using a linkage-learning algorithm. We then develop a linkage-sensitive variant of the particle swarm optimization. We also examine different methods of specifying the linkage information to verify that linkage information improves performance in terms of the number of iterations taken to reach a optimum as well as the fitness of the discovered optimum.

In Section 2, we describe the development of a method to capture and represent linkage-information specific to each problem. We describe how the structure of the problems may influence the linkage structure and describe our method for evolving the linkage matrix. In Section 3, we explain how to utilize linkage-specific information in search algorithms and how to modify linkage-blind search procedures such as PSO in order to utilize problem-specific information that is provided in the form of a linkage matrix. Next, results are presented showing the performance of our algorithms on several common benchmark problems as well as on specific linkage problems designed to illustrate how the use of linkage information greatly improves search results. This chapter summarizes some of the results in [6]; an abbreviated version of this work also appears in [7].

2 Determining Linkages

In this section, we address the problem of defining problem specific linkages, and the development of an algorithm to learn these linkages.

In many cases, it is not possible to accurately specify relative linkage strengths from prior knowledge. In such cases linkage information must be learned by examining fitness values associated with some sample positions. In [22] and [23], certain algorithms were developed for learning linkages to be used crossover-specific in the context of linkage-sensitive genetic algorithms. Since they used concepts of inheritance which are not directly applicable in case of particle swarm optimization, we designed a new algorithm for learning linkages with particle swarms. This is the motivation for the linkage-learning algorithm described below, applied prior to the search using particle swarm optimization.

One possible method for learning linkages is outlined in *LiLA,* the linkage learning algorithm given below. LiLA samples the problem search space with a large number of particles. From each sampled particle, we consider the change in fitness resulting from a small increment or decrement (ε) in the particle position. The linkage matrix (with all entries initialized to 0) is updated based on the fitness changes resulting from the four perturbations. If the two components do not have significant linkage, a relatively small net change in fitness will result by perturbing them together. Entries corresponding to pairs of strongly linked elements will be more substantially updated in the linkage matrix. The Linkage Learning algorithm is run for a finite number of iterations prior to executing the main optimization procedure. The linkage matrix is not modified once the linkages are learned. The resulting linkage matrix is symmetric: component i is as strongly linked to j as component j is to i.

Notation: Let p^k_a denote the result of perturbing particle position p at dimension k by the amount a. In other words, $p^k_a[k] = p[k]+a$, and for all $j \neq k$, $p^k_a[j] = p[j]$.

Linkage Learning Algorithm (LiLA):

Initialize every element of the Linkage matrix L to 0;
Let ε be a small positive constant;
Sample the problem space with n particles;
For each of the n particles, do:
  For each dimension j in the particle, do:
    For $k=1$ to $j-1$, do:
      Increment $L[j,k]$ by the net change in fitness due to the four possible perturbations obtained by incrementing each of the components i and j by ±ε, i.e., $L[j,k] := L[j,k]+4f(p) - f((p^j_\varepsilon)^k_\varepsilon) - f((p^j_{-\varepsilon})^k_\varepsilon) - f((p^j_\varepsilon)^k_{-\varepsilon}) - f((p^j_{-\varepsilon})^k_{-\varepsilon})$
End;
For each j, For each $k>j$, let $L[j,k] = L[k,j]$.

When the explicit form of the function to be optimized is available, strong linkages can be presumed to exist between dimensions grouped within sub-expressions. For example, consider minimizing the objective function f given below, where x is a four-element vector.

$$f(x) = (x[0]*x[1])^n + x[2]+x[3] \quad \text{......(3)}$$

A simultaneous change to $x[0]$ and $x[1]$ can cause a larger net change to fitness than if these changes were applied to $x[0]$ and $x[2]$ instead. This suggests that stronger linkages exist between $x[0]$, $x[1]$ than between $x[0]$, $x[2]$. Depending on the value of the exponent n, it is clear that simultaneous changes to $x[0]$ and $x[1]$ would have a larger effect on net fitness than simultaneous changes to $x[0]$ and $x[3]$. If we use 20% perturbations in initial values of the particle dimensions and an index of size3, the total change in fitness obtained by modifying $x[0]$ and $x[3]$ simultaneously is only 82% of the total change in fitness obtained by modifying $x[0]$ and $x[1]$. The effect becomes more severe with increasing index. Similarly if weakly linked particles $x[2]$ and $x[3]$ are updated simultaneously, the net change is 27% of the change obtained updating $x[0]$ and $x[1]$.

LiLA extrapolates this concept to problems where the explicit form of the function is not available, but it is possible to evaluate the function at multiple points. Over several iterations, it is expected that the linkages (as in the above example) will be captured in the L matrix. The process can be applied to any problem characterized by grouping of dimensions under specific mathematical operators. Note that the nature of linkages might change depending on the range over which the function is optimized. Hence the linkage matrix obtained using LiLA must be over the same range as the optimization problem itself.

The figure below shows an L matrix learned using LiLA from data for a ten-dimensional function $(tan(x_0\, x_1\, x_2\, x_3)/200 + x_4\, x_5\, x_6 + x_7\, x_8\, x_9)$. The matrix was computed using 100 sampled positions; for clarity, only non-zero values are shown in the table below.

	x_0	x_1	x_2	x_3	x_4	x_5	x_6	x_7	x_8	x_9
x_0		0.01	0.40	6.21						
x_1	0.01		4.06	0.66						
x_2	0.40	4.06		0.85						
x_3	6.21	0.66	0.85							
x_4						0.04	0.02			
x_5					0.04					
x_6					0.02					
x_7									0.05	0.04
x_8								0.05		
x_9								0.04		

Three distinct clusters of strongly linked particle dimensions characterize the problem. For example, the first four variables form a strongly linked cluster; this is successfully learnt, and *L[0,1]*, *L[0,2]*, *L[0,3]*, *L[1,2]*, *L[1,3]*, and *L[2,3]* show non-zero numerical linkages. Similar reasoning can be extended to other clusters as well. Not all expected non-zero values emerged after learning with only 100 sampled points, but further learning alleviated this problem. However, some spurious linkage values emerge when the sample size increases substantially, due to random sampling effects.

3 Linkage-Sensitive Particle Swarm Optimization

Plain particle swarm optimization (*gPSO)* is linkage-blind. Updates are made to each and every particle position without regard to any linkage information. In order to exploit the linkage information in a problem captured by the linkage matrix, we develop a variant of traditional PSO that is sensitive to linkages, referred to as *Linkage-sensitive Particle Swarm Optimization (LiPSO)* algorithm.

The following key differences are noted between *gPSO* and *LiPSO*:

1. LiPSO does not maintain separate velocity information for each particle.
2. LiPSO assumes the availability of a Linkage matrix L, where *L[i,j]* denotes the linkage strength between the ith and j^{th} components. The L matrix is not assumed to be normalized; if LiLA is used, the values in L depend on the particles sampled.

3. Historical information about the previous best position *(pBest)* for each particle is maintained and used to modify particle positions, as in *gPSO*. But LiPSO does not attract a particle towards its own pBest.
4. In *gPSO*, every particle component is updated in each iteration. But in LiPSO, only a certain subset *S* (of components) is updated in each iteration. Even weakly linked components may be updated, although with low probability. Procedurally, one component *i* is selected randomly to be the first element for update in the subset *S*, and then a second element, *j*, is chosen to be included in *S* with a probability of selection proportional to the strength of the linkage *L[i,j]*.
5. In *gPSO*, every particle experiences an attracting force towards the same *gBest*. In *LiPSO*, however, the attractor for each particle is chosen separately, depending on the components chosen for update based on linkage strengths. The attractor would be that particle which gives maximum change in fitness per unit change in perturbations in the subset *S*. If *x* is the particle whose position is to be updated using the subset of components *S*, then the attractor is chosen to maximize the function Ω, defined below:

$$\Omega(S,x,y) = [f(pbest(y)) - f(x)] / \| proj(pbest(y),S) - proj(x,S) \| \quad (4)$$

where the fitness function *f* is to be maximized,
pbest(y) is the best position of the particle *y* (discovered so far),
$\|.\|$ is the Euclidean norm, and
proj(z,S) is the projection of *z* onto the positions in subset *S*,
e.g., *proj*((0.1, 0.3, 0.4, 0.8)), {1,4} = [0.1, 0.8].

The function Ω(S,x,y)considers changes that result only from perturbations to dimensions in subset *S*. The expression in the numerator, $[f(pbest(y)) - f(x)]$, favors attractors of higher fitness. The expression in the denominator, $\| proj(pbest(y),S) - proj(x,S) \|$, measures the "distance" between *x* and *y* when attention is restricted only to the components in *S*. Hence $\Omega(S,x,y)$ is relatively large when the potential attractor *y* is near *x*. Maximizing Ω ensures that the attractor is the *pBest* that provides the most fitness improvement modulated by the distance to the attractor; this approach was inspired by the fitness distance ratio PSO update algorithm *(FDR-PSO)*[12]. Points of higher fitness and greater proximity to *x* have a greater potential to attract *x*, but unlike the *FDR-PSO*, only one attractor is chosen at each update and only the subspace spanned by the components in *S* is relevant to the distance calculations.

Example: Consider the example in equation (3) with *n*=3,
S=[1,2],
pbest(*y*) = {1, 1, 2, 10},
current particle *x* = {1.2, 0.8, 1, 1}.
Then *proj(pbest(y),S)* = {1, 1} and *proj(x,S)* = {1.2, 0.8}.
The Euclidean distance between the above projections is
$d = \text{sqrt}\{ (1.2-1)^2 + (0.8-1)^2 \} \approx 0.28$

Since the original function f was to be minimized, we transform it to a maximization problem (e.g., maximizing the function $10-f(x)$). If the resulting fitness values for x and pbest(y) are 2.88 and 3.0, respectively (by transformation from equation (3)), the value of $\Omega(S,x,y) = (3-2.88)/0.28 = 0.428$.

Linkage-Sensitive Particle Swarm Optimization (LiPSO):

While computational limits are not exceeded, do
    For each particle k in the swarm, do
        Randomly select position i;
        Gather a subset S of indices (i.e., component numbers), so that the probability that S contains j is proportional to $L(i,j)$;
        Find particle y that maximizes Ω(S,x,y) defined in equation (4) above;
        For each component j in S, perform the update:

$$x_j := (1-c)* x_j + c*pbest(y_j) \qquad (5)$$

        if $(f(p(k)) > f(pbest(k))$ then $pbest(k) := p(k)$;
    End.

In the update equation, the learning rate parameter c affects the magnitude of the particle position update. We have empirically set the value of c, though adaptation algorithms may be used to determine an optimum value of c. Note that a subset of components are updated in each iteration, depending on the size of S, which can be empirically determined for each problem depending on dimensionality. The motivation is that in problems where linkages are important, updates need be applied only to a subset of dimensions which affect the fitness more strongly compared to others.

4 Results

In this section, results are given comparing traditional PSO (*gPSO*) with Linkage Sensitive PSO (*LiPSO*). More extensive results are contained in [6], including comparisons with simulated annealing and Fitness Distance Ratio - Particle Swarm Optimization [21] (*FDRPSO*).

First, we tested the algorithms on certain well-known benchmark problems used widely by the particle swarm research community. These benchmark problems were specified in detail for the Special Session on Real World Optimization at the Congress of Evolutionary Computing 2005 [26]. Strong linkages between particle components, however, do not characterize these problems. This can be verified by examining at the linkage matrices evolved for these problems. In cases where the evolved linkage matrices showed certain linkages, it turned out that nearly every particle dimension was linked to every other. In no cases were there strong intra cluster linkages with relatively weak inter cluster linkages. Consequently, problem-specific linkage information did not result in any improvement in performance.

We defined a different problem set characterized by the existence of clusters of components with strong linkages, and relatively weak linkages between the clusters. These *linkage strong* functions, shown below, include algebraic, trigonometric, logarithmic (with a constrained search space) and exponential operators. Tests were run with 100 particles. In case of *gPSO*, the values of constants *c* and *d* as described in equation (1) were pegged at 0.2 to equally weight the attraction towards to previous and global best estimates. Similarly the value of *c* in *LiPSO* as described in equation (5) was also 0.2. In case of *LiPSO*, the subset size for the best performance varied with the problem and depends on the relative size linkage clusters. The maximum value of *S* was fixed at half the number of dimensions.

4.1 Ten-Dimensional Functions

One set of functions is in 10-dimensional space, while another set is in 30-dimensional space. In these examples, the linkages are visible in the groupings of particle dimensions under specific mathematical operations. However, given the nature of the mathematical operations and the numerical weighting applied to each grouping; it might not be straightforward to specify numerical linkage strengths. We examined functions containing common mathematical functions and operators, to verify that the linkage learning algorithm and related linkage sensitive PSO do indeed yield a performance improvement. We explored the effect of particle dimensionality on the performance to verify that the performance improvement persists as the dimensionality increases.

The functions were not derived with any specific benchmark in mind, except that they are all characterized by the existence of linkages, as mentioned above.

Function	Form
F1	`x[0]*x[1]*x[2]*x[3] + x[4]*x[5]*x[6] +x[7]*x[8]*x[9]`
F2	`pow((x[0]*x[1]),2) + pow((x[2]*x[3]),2) + pow((x[4]*x[5]),2) + pow((x[6]*x[7]),2) + pow((x[8]*x[9]),2)`
F3	`pow((x[0]+x[1]+x[2]+x[3]),3)+pow((x[4]+x[5]+x[6]+x[7]),2)+pow((x[8]+x[9]),4)`
F4	`sin(x[0]+x[1]+x[2]) + cos(x[3]+x[4]+x[5]) + sin(x[6]+x[7]+x[8]) + x[9]`
F5	`cos(x[0]*x[1]) + pow(x[2]*x[3],4) + pow(x[4]*x[5],4) + cos(x[6]*x[7]*x[8]*x[9])`
F6	`0.005*tan(x[0]*x[1]*x[2]*x[3]) + x[4]*x[5]*x[6] + x[7]*x[8]*x[9]`
F7	`cos(x[0]*x[1]) + pow(x[2]*x[3],4) + pow(x[4]*x[5],4) + cos(x[6]*x[7]*x[8]*x[9])`
F8	`0.01*exp(x[0]+x[1]) + 0.01*exp(x[2]+x[3]) + 0.01*exp(x[4]+x[5]) + 0.01*exp(x[6]+x[7]) +x[8]+x[9]`
F9	`pow((x[0]*x[1]*x[2]),1) + x[3]+ pow((x[4]*x[5]*x[6]),2) + pow((x[7]*x[8]*x[9]),2)`
F10	`exp(x[0]*x[1]*x[2]*x[3]*x[4]) + exp(x[5]*x[6]*x[7]*x[8]*x[9])`

Results on attempts to minimize the 10-dimensional functions are shown below. The entries in the table are median (13th among 25 runs) for each algorithm after 1000 generations (swarms). Values below 1.0E-7 are abbreviated as *0*.

Function	gPSO	LiPSO
F1	5.12E-4	0
F2	5.22E-4	4.76E-3
F3	7.95E-5	6.57E-5
F4	9.67E-5	0
F5	9,91E-6	0
F6	2.57E-4	0
F7	1.34E-4	0
F8	7.00E-5	0
F9	8.00E-7	0
F10	0	0

From the above table, we infer that *LiPSO* outperforms *gPSO* in all cases except F2.

4.2 Thirty-Dimensional Functions

Results on the 30-dimensional functions are shown below. The entries in the table are median (13th best fitness achieved among 25 runs) for each algorithm, after 1000 generations (iterations). Values below 1.0E-7 are abbreviated as *0*.

Functions	Form
F1	g1 + g2 + g3, where g1 = x[0]*x[1]*x[2]...x[9], g2 = x[10]*x[11]*x[12]...x[19], g3 = x[20]*x[21]*x[22]...x[29]
F2	g1 + g2 + g3 + g4 + g5 + g6,where g1 = x[0]*x[1]*x[2]...x[4], g2 = x[5]*x[6]*x[7]...x[9], g3 = x[10]*x[11]*x[12]...x[14], g4 = x[15]*x[16]*x[17]...x[19], g5 = x[20]*x[21]*x[22]...x[23], g6 = x[24]*x[25]*x[26]...x[29]
F3	g1 + g2 + g3 +x[29] ,where g1 = x[0]*x[1]*x[2]...x[7], g2 = x[8]*x[9]*x[10]...x[14], g3 = x[15]*x[16]*x[17]...x[21], g4 = x[22]*x[23]*x[24]...x[28]
F4	sin(g1) + cos(g2) + sin(g3) + x[29], where g1 = x[0]*x[1]*x[2]...x[9], g2 = x[10]*x[11]*x[12]...x[19], g3 = x[20]*x[21]*x[22]...x[29]
F5	cos(g1) + pow(g2) + pow(g3) + cos(g4) + cos(g5) , where g1 = x[0]*x[1]*x[2]...x[4], g2 = x[5]*x[6]*x[7]...x[8], g3 = x[9]*x[10]*x[11]...x[13], g4 = x[14]*x[16]*x[17]...x[24], g5 = x[25]*x[26]*x[27]...x[29]

F6	0.005*g1 + g2 + g3, where g1 = x[0]*x[1]*x[2]…x[9], g2 = x[10]*x[11]*x[12]…x[19], g3 = x[20]*x[21]*x[22]…x[29]
F7	0.00001*g1 + 0.00001*g2 + 0.00001*g3 + 0.00001*g4 + g5, where g1 = x[0]*x[1]*x[2]…x[4], g2 = x[5]*x[6]*x[7]…x[9], g3 = x[10]*x[11]*x[12]…x[14], g4 = x[15]*x[16]*x[17]…x[19], g5 = x[20]*x[21]*x[22]…x[29]
F8	cos(g1) + pow(g2,4) + pow(g3,2) + cos(g4), where g1 = x[0]*x[1]*x[2]…x[7], g2 = x[8]*x[9]*x[10]…x[14], g3 = x[15]*x[16]*x[17]…x[21], g4 = x[22]*x[23]*x[24]…x[29]
F9	pow(g1,2) + x[8] + pow(g2,2) + pow(g3,2) + g4, where g1 = x[0]*x[1]*x[2]…x[8], g2 = x[9]*x[10]*x[11]…x[17], g3 = x[18]*x[19]*x[20]…x[26], g4 = x[27]*x[28]*x[29]
F10	exp(g1) + exp(g2/100) + exp(g3/100)+exp(g4/100) + exp(g5/100) + exp(g6/100) + exp(g7/100) + exp(g8/100) + exp(g9/100) + exp(g10/100), where g1 = x[0]*x[1]*x[2], g2 = x[3]*x[4]*x[5], g3 = x[6]*x[7]*x[8], g4 = x[9]*x[10]*x[11], g5 = x[12]*x[13]*x[14], g6 = x[15]*x[16]*x[17], g7 = x[18]*x[19]*x[20], g8 = x[21]*x[22]*x[23], g9 = x[24]*x[25]*x[26], g10 = x[27]*x[28]*x[29]

Function	gPSO	LiPSO
F1	6.29E-4	0
F2	5.21E+4	2.26E+0
F3	5.68E-3	0
F4	8.13E-6	5.64E-3
F5	2.72E-4	0
F6	1.19E-4	0
F7	6.67E-6	0
F8	5.61E+0	0
F9	5.81E-1	0
F10	4.54E-4	2.56E+0

From the above table, we infer that *LiPSO* outperformed *gPSO* in 8 out of 10 cases.

4.3 Complexity Analysis

Criteria for evaluating algorithm complexity were prescribed in the Special Session on Real Parameter Optimization at the Congress of Evolutionary Computing, 2005 [25].

The time to compute a test function is evaluated and recorded. The test function includes mathematical operators of algebraic, logarithmic, and exponential types. This time is averaged over n = 1000000 evaluations of this function and is referred to as T0.

Next, the time to compute the high conditioned elliptical function (function 3 of the benchmark function set) is recorded over n = 2000000 evaluations and termed as T1. This is repeated each time the dimensionality of the problem is changed.

Finally, the time it takes for the algorithm to compute the optimal value of the highly conditioned elliptical function, over n = 2000000 iterations is recorded as T2.

The algorithm complexity is estimated by (Ť2-T1)/T0, where Ť2 is the mean of T2 obtained over 5 independent calculations. The results are shown below. The algorithms were all implemented in C and tested on an Intel Xeon Dual processor machine with the CPU working at 3.20 GHz.

Time Complexity Results (T2-T1)/T0:

Dimensionality	LiPSO	gPSO
D = 10	644	694
D = 30	1320	1040
D = 50	2021	1314

The results show that the superior performance of *LiPSO* comes at the expense of increased time complexity, although the additional time required is no more than 55%. The reasons for the additional time requirements are:

- The need for a separate linkage learning step prior to optimization.
- The method of choosing an attractor. In *gPSO,* each particle in the swarm has the same global attractor and its previous best is kept in memory for every iteration. In contrast, for *LiPSO*, the attractor for each particle is different and needs a search step in every iteration. As the swarm size increases this step consumes more computational time.
- We need to calculate the *g* function detailed in Section 3, in addition to calculating the fitness function. This also increases the time required to complete one iteration of the algorithm.

4.4 Linkage Learning Error

To verify that the linkage information does really help, the linkage learning algorithm described in section 2 was evaluated against other methods of capturing linkage information. The basic linkage sensitive PSO body was useed in these tests, with linkage matrices specified using as described below.

1. **Hand Tailored Linkage:** *A priori* problem-specific information is used to determine entries in the linkage matrix, prior to the optimization. Values in the hand-tailored linkage matrix are set to 1 or 0, depending on whether a linkage is obvious from the explicit function description.
2. **LiLA:** In this variant of linkage sensitive PSO, the Linkage Learning Algorithm is used to generate the linkage matrix before the optimization commences.

3. **Fixed Linkage PSO:** All entries in the linkage matrix are set to 1, as a control case.
4. **Random Linkage PSO:** Every element of the L matrix was set to a random value between 0 and 1 prior to the optimization.
5. **ALINX3 based Linkage Learning:** We adapted Salman's ALinX3 algorithm [18, 1] for genetic algorithms to PSO. The original ALinX3 was designed specifically with discrete optimization problems in mind. The adaptation compared to LiLA was specifically designed for continuous parameter optimizations as in case of PSO. The algorithm updates the linkage matrix based on net fitness changes resulting from jointly perturbing pairs of particle dimensions.

In each case, the performance of the PSO variant with the Adaptive Linkage Learning Algorithm performed better in terms of final best fitness achieved. Setting the linkage matrix elements to equal fixed values gave the worst performance in terms of best fitness.

We propose the following error metric to evaluate performance of the linkage learning algorithm: $E = \Sigma (I - L)^2$
where I denotes the ideal (hand-tailored) linkage matrix for a particular problem, and L denotes the linkage matrix obtained by the linkage learning procedure. Here, both I and L are normalized before calculating the metric. The largest matrix L was first normalized to 1 and in a second step, the elements of matrices I and L are divided by the squared sum of all elements of the respective matrix.

This metric measures the difference between the hand-tailored and learned matrices. Results for the average linkage matrix learning error are shown below.

N=10	1	2	3	4	5	6	7	8	9	10
Random	1.923	2.170	1.721	1.813	2.137	1.977	1.932	2.165	1.912	1.406
Fixed	1.800	1.800	1.500	1.640	1.640	1.520	1.640	1.800	1.640	1.2
ALINX3	1.386	1.844	1.297	2.106	1.781	1.542	1.777	1.991	1.595	0.289
LiLA	0.493	0.030	0.862	0.563	1.360	1.010	1.335	0.417	0.302	0.064

N=30	1	2	3	4	5	6	7	8	9	10
Random	1.608	2.095	1.861	1.724	1.901	1.542	1.957	1.831	1.840	2.376
Fixed	1.400	1.736	1.562	1.436	1.596	1.400	1.622	1.509	1.529	1.867
ALINX3	1.500	0.226	1.903	2.092	1.932	1.662	1.747	2.202	1.937	2.245
LiLA	0.952	0.202	0.883	1.025	1.839	0.913	1.058	1.915	1.904	0.041

The above results show that LiLA resulted in the least error in 17 out of the 20 functions.

5 Conclusions

This chapter describes a linkage-based approach to Particle Swarm Optimization. We developed a new algorithm, LiLA, to learn problem specific linkage information, adapted to particle swarm optimization. The information is captured in the form of a Linkage Matrix prior to the optimization step.

We developed a new variant of particle swarm optimization, LiPSO, which uses problem specific linkage information to improve search performance both in terms of the quality of the final best solution achieved as well as the number of algorithm iterations it takes to reach this solution.

We tested the effect of incorporating linkage information and found that in problems with such strong inter-dimensional linkages, the linkage sensitive LiPSO consistently outperformed the linkage-blind *gPSO*. Tests run by specifying different forms of linkage matrices showed that the learning the linkage matrix improves performance, as opposed to merely specifying linkage information without regard to the problem specific linkages.

Linkage information can also be useful when applied to other search algorithms. A linkage sensitive variant of simulated annealing and an algorithm hybridizing particle swarm optimization and simulated annealing were developed and tested on the benchmark problems as well as the linkage strong problems; the results are presented in [6]. The performance of these algorithms was intermediate between linkage sensitive particle swarm optimization and their respective linkage blind variants.

References

[1] Bergh, F.: An Analysis of Particle Swarm Optimizers. Ph.D. Dissertation, University of Pretoria, Pretoria, Republic of South Africa (2001)

[2] Beyer, H.G.: The Theory of Evolution Strategies. Springer, Heidelberg (2001)

[3] Binos, T.: Evolving Neural Network Architecture and Weights Using an Evolutionary Algorithm. Master's chapter, Department of Computer Science, RMIT

[4] Clerc, M.: The Swarm and the Queen: Towards a Deterministic and Adaptive Particle Swarm Optimization. In: Proc. Congress on Evolutionary Computation, Washington, DC, pp. 1927–1930 (1999)

[5] Dasgupta, D., Michalewicz, Z.: Evolutionary Algorithms in Engineering Applications. Springer, Heidelberg (1997)

[6] Devicharan, D.: Particle Swarm Optimization with Adaptive Linkage Learning. M.S. Thesis, Dept. of EECS, Syracuse University (2006)

[7] Devicharan, D., Mohan, C.K.: Particle, Swarm Optimization with Adaptive Linkage Learning. In: IEEE Congress on Evolutionary Computing (June 2004)

[8] DeJong, K.A., Potter, M.A., Spears, W.M.: Using Problem Generators to Explore the Effects of Epistasis. In: Proc. Seventh International Conference on Genetic Algorithms, pp. 338–345. Morgan Kaufmann, San Francisco (1997)

[9] Eberhart, R.C., Kennedy, J.: A New Optimizer using Particle Swarm Theory. In: Proc. the Sixth International Symposium on Micro Machine and Human Science, Nagoya, Japan, pp. 39–43 (1995)

[10] Eberhart, R.C., Shi, Y.: Comparison Between Genetic Algorithms and Particle Swarm Optimization. In: Proc. 7th International Conference on Evolutionary Programming, San Diego, California, USA (1998)

[11] Goldberg, D.E., Lingle, R.: Alleles, loci and the traveling salesman problem. In: Proc. An International Conference on Genetic Algorithms, pp. 10–19. Morgan Kaufmann, San Francisco

[12] Harik, G.: Learning gene linkage to efficiently solve problems of bounded difficulty using genetic algorithms. ILLIGAL Report No. 97005

[13] Holland, J.H.: Adaptation in Natural and Artificial Systems. University of Michigan Press (1975)
[14] Holland, J.H.: Hidden Order: How Adaptation Builds Complexity. Addison-Wesley, Reading (1995)
[15] Kennedy, J., Eberhart, R.: Swarm Intelligence. Morgan Kaufmann Publications, San Francisco (2001)
[16] Kennedy, J.: The Particle Swarm: Social Adaptation of Knowledge. In: Proc. International Conference on Evolutionary Computation, Indianapolis, Ind., pp. 303–308 (1997)
[17] Kennedy, J., Eberhart, R.C.: Particle Swarm Optimization. In: Proc. IEEE International Conference on Neural Networks, pp. 1942–1948 (1995)
[18] Krishnamurthy, E.V., Sen, S.K.: Numerical Algorithms-Computations in Science and Engineering. Affiliated East-West Press (1997)
[19] Michalewicz, Z.: Genetic Algorithms + Data Structures = Evolution Programs, 3rd edn. Springer, Heidelberg (1999)
[20] Mohan, C.K., Al-kazemi, B.: Discrete Particle Swarm Optimization. In: Proc. Workshop on Particle Swarm Optimization, Indianapolis, Ind., Purdue School of Engineering and Technology, IUPUI (2001)
[21] Peram, T., Veeramachaneni, K., Mohan, C.K.: Fitness-Distance Ratio-Based Particle Swarm Optimization. In: Proc. IEEE Swarm Intelligence Symposium, Indianapolis (IN) (April 2003)
[22] Salman, A.: Linkage Crossover Operator for Genetic Algorithms. Ph.D Dissertation, Department of Electrical Engineering and Computer Science, Syracuse University (December 1999)
[23] Salman, A., Mehrotra, K., Mohan, C.K.: Adaptive Linkage Crossover. Evolutionary Computation 8(3), 341–370 (2000)
[24] Singh, A., Goldberg, D.E., Chen, Y.P.: Modified Linkage Learning Genetic Algorithm for Difficult Non-Stationary Problems. In: Proc. Genetic and Evolutionary Computation Conference (2002)
[25] Suganthan, P.N., Hansen, N., Liang, J.J., Deb, K., Chen, Y.-P., Auger, A., Tiwari, S.: Problem Definitions and Evaluation Criteria for the CEC 2005 Special Session on Real-Parameter Optimization, Technical Report, Nanyang Technological University, Singapore, AND KanGAL Report #2005005, IIT Kanpur, India (May 2005)
[26] Suganthan, P.N.: Particle Swarm Optimizer with Neighborhood Operator. In: Proc. Congress on Evolutionary Computation, Piscataway, N.J., pp. 1958–1962. IEEE Service Center, Piscataway (1999)

Parallel Particle Swarm Optimization Algorithm Based on Graphic Processing Units

Ying Tan and You Zhou

Key Laboratory of Machine Perception (MOE), Peking University
Department of Machine Intelligence, School of Electronics Engineering and Computer Science, Peking University, Beijing 100871, P.R. China
ytan@pku.edu.cn, zhouyoupku@pku.edu.cn

Summary. A novel parallel approach to implement particle swarm optimization(PSO) algorithm on graphic processing units(GPU) in a personal computer is proposed in this chapter. By using the general-purpose computing ability of GPU and under the software platform of compute unified device architecture(CUDA) which is developed by NVIDIA, the PSO algorithm can be executed in parallel on the GPU. The process of fitness evaluation, as well as the updating of the velocity and the position of all the particles in the swarm are parallelized and described in details in the context of this chapter. Experiments are conducted by running the PSO both on the GPU and the CPU, respectively, to optimize several benchmark test functions. The running time of the PSO based on GPU(GPU-PSO, for short) is greatly shortened compared to that of the PSO based on CPU(CPU-PSO, for short). A ***40***× speedup can be obtained by our implemented GPU-PSO on a display card of NVIDIA Geforce 9800GT, with the same optimizing performance. Compared to the CPU-PSO, the GPU-PSO has special speed advantages on large-scale population and high-dimensional problems which are often happened and widely used in a lot of real-world optimization applications.

Keywords: Particle Swarm Optimization, Graphic Processing Units, Parallelization, CUDA.

List of Abbreviations and Symbols

PSO	Particle Swarm Optimization
GPU	Graphic Processing Units
CUDA	Compute Unified Device Architecture
CPU-PSO	the PSO based on CPU
GPU-PSO	the PSO based on GPU

1 Introduction

Particle swarm optimization(PSO), developed by Eberhart and Kennedy in 1995, is a stochastic global optimization technique inspired by social behavior of bird flocking or fish schooling [1]. In the PSO, each particle in the swarm adjusts its position in the search space based on the best position it has found so far as well as the position of the known best-fit particle of the entire swarm, and finally converges to the global best point of the whole search space.

B.K. Panigrahi, Y. Shi, and M.-H. Lim (Eds.): Handbook of Swarm Intelligence, ALO 8, pp. 133–154.
springerlink.com

Compared to other swarm based algorithms such as genetic algorithm and ant colony algorithm, PSO has the advantage of easy implementation, while maintaining strong abilities of convergence and global search. In recent years, PSO has been used increasingly as an effective technique for solving complex and difficult optimization problems in practice. PSO has been successfully applied to problems such as function optimization, artificial neural network training, fuzzy system control, blind source separation, machine learning and so on.

In spite of those advantages, it takes PSO algorithm a long time to find solutions for large scale problems, such as problems with large dimensions and problems which need a large swarm population for searching in the solution space. The main reason for this is that the optimizing process of PSO requires a large number of fitness evaluations, which are usually done in a sequential way on CPU, so the computation task can be very heavy, thus the running speed of PSO may be quite slow.

In recent years, Graphics Processing Unit(GPU) which has traditionally been a graphics-centric workshop, has shifted its attention to the non-graphics and general-purpose computing applications. Because of its parallel computing mechanism and fast float-point operation, GPU has shown great advantages in scientific computing fields, and many successful applications have been achieved.

In order to perform general-purpose computing on GPU more easily and conveniently, some software platforms have been developed, such as BrookGPU (Stanford University) [8], CUDA(Compute Unified Device Architecture, NVIDIA Corporation) [10]. These platforms have greatly simplified the GPU-based programming.

In this chapter, we present a novel method to run PSO on GPU in parallel, based on CUDA, which is a new but powerful platform for programming on GPU. With a good optimization performance, the PSO implemented on GPU can solve problems with large-scale population and high dimension, speed up its running dramatically and provide users with a feasible solution for complex optimizing problems in reasonable time. As GPU chips can be found in any ordinary PC nowadays, more and more people will be able to solve large-scale problems in real-world applications by applying this parallel algorithm.

This chapter is organized as follows. In Section 2, the related work is presented in details. In Section 3, we briefly introduce the backgrounds of GPU based computing. Algorithmic implementations of our proposed approach are elaborated in Section 4. A number of experiments are done on four benchmark functions and analysis of results are reported in Section 5. Finally, we conclude the chapter and give our future works in Section 6.

2 Related Work

For the sake of simplicity, we call the particle swarm optimization algorithm presented by Eberhart and Kennedy in 1995 as original particle swarm optimization (original PSO). After the original PSO was proposed, many researchers have taken great effort to improve the performance of it, and many variants have also been developed [2, 3, 4, 5, 6, 7]. In this section, we will briefly review the original PSO as well as the so-called standard PSO which is proposed by Bratton and Kennedy in 2007.

2.1 Original Particle Swarm Optimization

In original PSO, each solution of the optimization problem is called a particle in the search space. The search of the problem space is done by a swarm with a specific number of particles.

Assume that the swarm size is N and the problem dimension is D. Each particle i $(i = 1, 2, ...N)$ in the swarm has the following properties: a current position X_i, a current velocity V_i, a personal best position $\tilde{P}_i$. And there is a global best position $\hat{P}$, which has been found in the search space since the start of the evolution. During each of the iteration, the position and velocity of every particle are updated according to $\tilde{P}_i$ and $\hat{P}$. This process in original PSO can be formulated as follows:

$$V_{id}(t+1) = wV_{id}(t) + c_1 r_1(\tilde{P}_{id}(t) - X_{id}(t)) + c_2 r_2(\hat{P}_d(t) - X_{id}(t)) \quad (1)$$

$$X_{id}(t+1) = X_{id}(t) + V_{id}(t) \quad (2)$$

where $i = 1, 2, ...N$, $d = 1, 2, ...D$. In (1) and (2), the learning factors c_1 and c_2 are nonnegative constants, r_1 and r_2 are random numbers uniformly distributed in the interval $[0, 1]$, $V_{id} \in [-V_{max}, V_{max}]$, where V_{max} is a designated maximum velocity which is a constant preset according to the objective optimization function. If the velocity on one dimension exceeds the maximum, it will be set to V_{max}. This parameter controls the convergence rate of the PSO and can prevent it from growing too fast. The parameter w is the inertia weight, which is a constant in the interval $[0, 1]$ used to balance the global and local search abilities.

2.2 Standard Particle Swarm Optimization

In these two decades, many researchers have taken great effort to improve the performance of original PSO by exploring the concepts, issues, and applications of the algorithm. In spite of this attention, there has as yet been no standard definition representing exactly what is involved in modern implementations of the technique.

In 2007, Daniel Bratton and James Kennedy designed a standard Particle Swarm Optimization (standard PSO) which is a straightforward extension of the original algorithm while taking into account more recent developments that can be expected to improve performance on standard measures [9]. This standard algorithm is intended for use both as a baseline for performance testing of improvements to the technique, as well as to represent PSO to the wider optimization community.

Standard PSO is different from original PSO mainly in two aspects, namely:

Swarm Communication Topology

Original PSO uses a global topology as shown in Fig. 1(a). In this topology, the global best particle, which is responsible for the velocity updating of all the particles, is chosen from the whole swarm population, while in standard PSO there is no global best, instead, each particle relies on a local best particle for velocity updating, which is chosen from its left and right neighbors as well as itself. We call this a local topology, which is shown in Fig. 1(b).

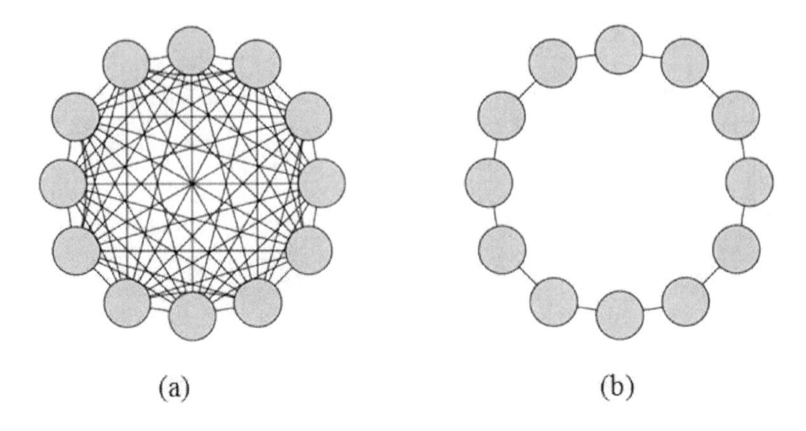

Fig. 1. Original PSO and Standard PSO Topologies

Inertia Weight and Constriction

In original PSO, an inertia weight parameter was designed to adjust the influence of the previous particle velocities on the optimization process. By adjusting the value of w, the swarm has a greater tendency to eventually constrict itself down to the area containing the best fitness and explore that area in detail. Similar to the parameter w, standard PSO introduced a new parameter χ known as the constriction factor, which is derived from the existing constants in the velocity update equation:

$$\chi = \frac{2}{|2 - \varphi - \sqrt{\varphi^2 - 4\varphi}|}, \qquad \varphi = c_1 + c_2$$

and the velocity updating formula in standard PSO is

$$V_{id}(t+1) = \chi(V_{id}(t) + c_1 r_1(\tilde{P}_{id}(t) - X_{id}(t)) + c_2 r_2(\hat{P}_{id}(t) - X_{id}(t))) \quad (3)$$

where $\hat{P}$ is no longer the global best but the local best position.

Statistical tests have shown that standard PSO can find better solutions than original PSO, while retaining the simplicity of original PSO. The introduction of standard PSO can give researchers a common grounding to work from. Standard PSO can be used as a means of comparison for future developments and improvements of PSO, and thus prevent unnecessary effort being expended on "reinventing the wheel" on rigorously tested enhancements that are being used at the forefront of the field.

3 GPU-Based Computing

GPU was at first designed especially for the purpose of image and graphic processing on computers, where compute-intensive and highly parallel computing is required. Compared to CPU, GPU shows many advantages [10].

- GPU computes faster than CPU. GPU devotes more transistors to data processing rather than data caching and flow control, which enables it to do much more float-point operations per second than CPU.
- GPU is more suitable for data-parallel computations. It is especially well-suited to solve problems that can be expressed as data-parallel computations with high arithmetic intensity - the ratio of arithmetic operations to memory operations.

3.1 Programming Model for GPU

The programming model for GPU is illustrated by Fig. 2. A shader program operates on a single input element stored in the input registers, and after all the operations are finished, it writes the execution results into the output registers. This process is done in parallel by applying the same operations to all the data, during which the data stored on other memory patterns such as texture and temp registers can also be retrieved.

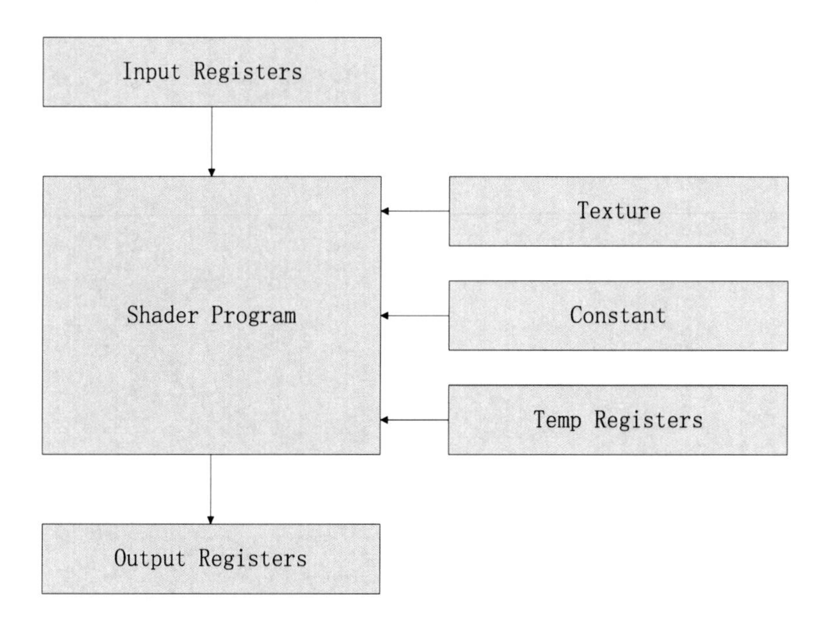

Fig. 2. Programming Model for GPU

3.2 Applications of GPU Based Computing

GPU has already been successfully used in some computation fields. R. Yang and G.Welch use GPU to perform real-time image segmentation and image morphology operations, and a performance increase of over 30% has been achieved [11]. Kenneth E.Hoff et.al present a method for rapid computation of generalized discrete Voronoi diagrams in two and three dimensions using graphics hardware [12]. Luo et.al use graphic hardware to speed up the computation of artificial neural network [13]. GPU is entering the main stream of computing [14].

3.3 Compute Unified Device Architecture (CUDA)

NVIDIA CUDA technology is a C language environment that enables programmers and developers to write software to solve complex computational problems by tapping into the many-core parallel processing power of GPUs. It is a new hardware and software architecture for issuing and managing computations on the GPU as a data-parallel computing device without the need of mapping them to a graphics API. Two core concepts in programming through CUDA are *thread batching* and *memory model* [15].

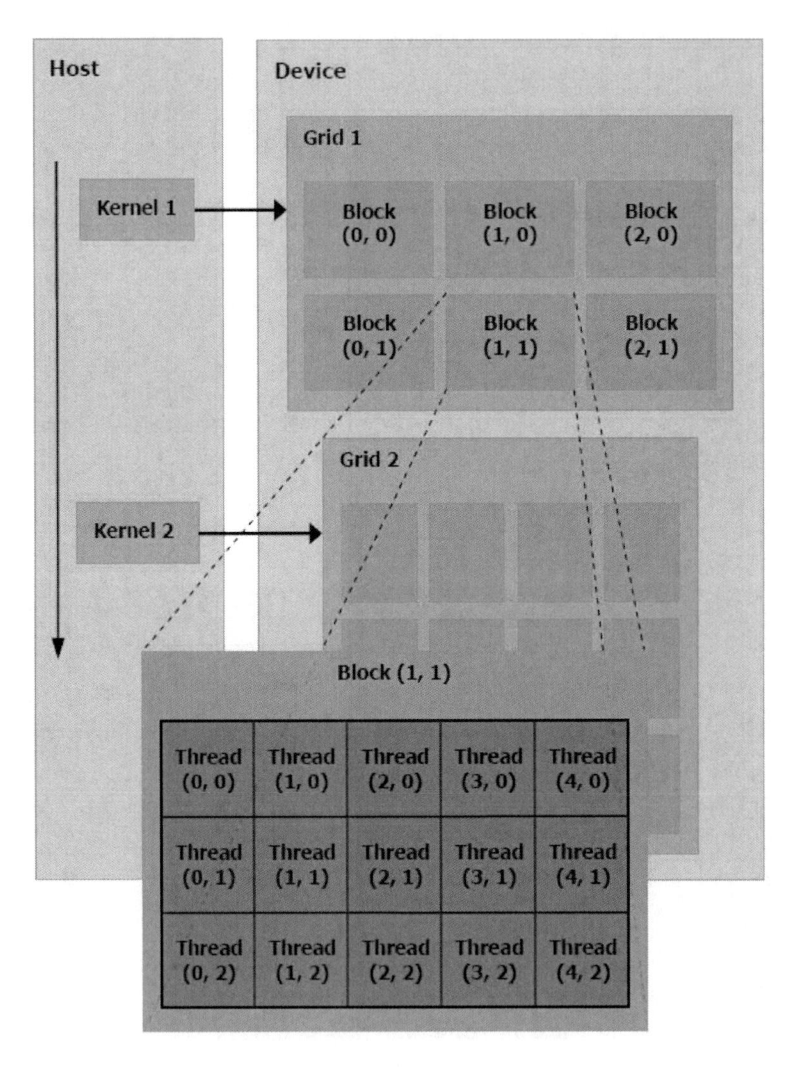

Fig. 3. Thread Batching of a Kernel in CUDA

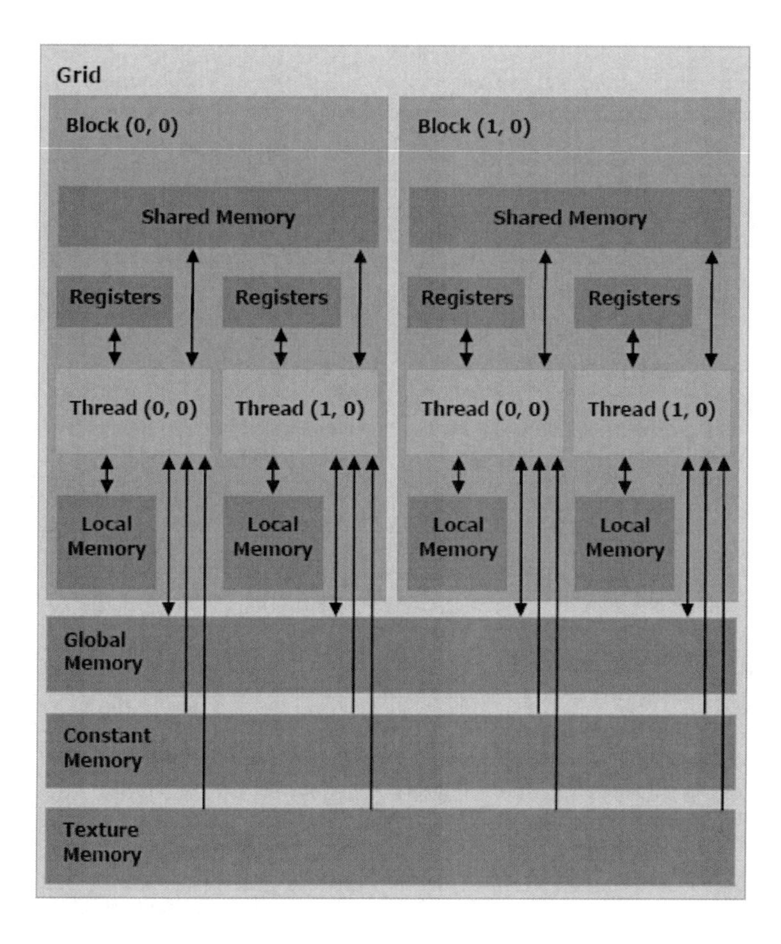

Fig. 4. Memory Model of CUDA

Thread Batching

When programming through CUDA, the GPU is viewed as a compute device capable of executing a very high number of threads in parallel. A function in CUDA is called as a *Kernel*, which is executed by a batch of threads. The batch of threads is organized as a grid of thread blocks. A thread block is a batch of threads that can cooperate together by efficiently sharing data through some fast shared memory and synchronizing their execution to coordinate memory accesses.

Each thread is identified by its thread ID. To help with complex addressing based on the thread ID, an application can also specify a block as a 2 or 3-dimensional array of arbitrary size and identify each thread using a 2 or 3-component index instead. For a two dimensional block of size $D_x \times D_y$, the ID of the thread with the index (x,y) is $y * D_x + x$. The thread batching mechanism can be illustrated by Fig. 3.

Memory Model

The memory model of CUDA is tightly related to its thread bathing mechanism. There are several kinds of memory spaces on the device:

- Read-write per-thread registers
- Read-write per-thread local memory
- Read-write per-block shared memory
- Read-write per-grid global memory
- Read-only per-grid constant memory
- Read-only per-grid texture memory

The memory model can be illustrated by Fig. 4. Registers and local memory can only be accessed by threads, the shared memory is only accessible within a block, and global memory is available to all the threads in a grid. In this paper, we mainly use the shared memory and global memory for our implementation.

Some applications have already been developed based on the platform of CUDA, for example, matrix multiplication, parallel prefix sum of large arrays, image denoising, sobel edge detection filter and so on. For details about programming through CUDA, interested readers can visit the website of NVIDIA CUDA ZONE.

4 Implementation of PSO on GPU

Standard PSO is as simple as original PSO, but with a better performance. So we would implement standard PSO rather than original PSO on GPU. The PSO algorithm mentioned in the remaining parts of this chapter will be actually standard PSO. Our purpose is to accelerate the running speed of PSO on GPU (GPU-PSO), during the search for the global best in an optimization problem. Meanwhile, performances of GPU-PSO should not be deteriorated. Furthermore, by making full use of the parallel computing ability of GPU, we expect GPU-PSO can solve optimization problems of high dimension and those with large swarm size.

4.1 Data Organization

In our method, position and velocity information of all the particles is stored on the global memory of GPU chips. As the global memory only allows the allocation of one dimensional arrays, so only one-dimensional arrays can be used here for storing data, which includes the position, velocity and fitness values of all the particles.

Given the dimension of the problem is D, and the swarm population is N, an array X of length $D*N$ is used here to represent this swarm by storing all the position values. But the array should be logically seen as a two-dimensional array Y. An element with the index of (i, j) in Y corresponds to the element in X with the index of $(i * N + j)$. This address mapping method can be formulized as:

$$Y(i, j) = X(i * N + j) \tag{4}$$

where the element $Y(i, j)$ stores the corresponding value for the i-th dimension of the j-th particle in the swarm.

4.2 Variable Definition and Initialization

Suppose the fitness value function of the problem is $f(X)$ in the domain $[-r, r]$. The variable definitions are given as follows: (notice that every array is one-dimensional)

- Particle position array: X
- Particle velocity array: Y
- Personal best position: $\tilde{P}$
- Local best position: $\hat{P}$
- Fitness value of particles: F
- Personal best fitness value: PF
- Local best fitness value: LF

where the sizes of $X, Y, \tilde{P}$ and $\hat{P}$ are $D * N$; the sizes of F,PF and LF are N.

4.3 Random Number Generation

During the process of optimization, PSO needs lots of random numbers for velocity updating. The absence of high precision integer arithmetic in current generation GPUs makes random numbers generating on GPU very tricky though it is still possible [16]. In order to focus on the implementation of PSO on GPU, we would rather generate random numbers on CPU and transfer them to GPU. However the data transportation between GPU and CPU is quite time consuming. If we generate random numbers on CPU and transfer them to GPU during each iteration of PSO, it will greatly slow down the algorithm's running speed due to mountains of data to be transferred. So data transportation between CPU and GPU should be avoided as much as possible.

We solve this problem in this way: M $(M >> D{*}N)$ random numbers are generated on CPU before the running of PSO algorithm. Then they are transferred to GPU once for ado and stored in an array R on the global memory, serving as a random number "pool". Whenever the velocity updating process is undergone, we just pass two random integer numbers P_1,$P_2 \in [0,M\text{-}D{*}N]$ from CPU to GPU, then $2{*}D{*}N$ numbers can be drawn from array R starting at positions P_1 and P_2, respectively, instead of transferring $2 * D * N$ numbers from CPU to GPU. The running speed can be obviously improved by using this technique.

4.4 Algorithmic Flow for GPU-PSO

The algorithmic flow for GPU-PSO is illustrated by Algorithm 1. Here $Iter$ is the maximum number of iterations that GPU-PSO runs, which serves as the stop condition for the optimizing process of functions.

4.5 Methods of Parallelization

The difference between a CPU function and a GPU *kernel* is that the operations of a kernel should be parallelized. So we must design the methods of parallelization carefully for all the sub-processes of GPU-PSO algorithm.

Algorithm 1. Algorithmic Flow for GPU-PSO

Initialize the positions and velocities of all particles.
Transfer data from CPU to GPU.

// sub-processes in "for" are done in parallel
for i=1 to $Iter$ **do**
  Compute the fitness values of all particles
  Update $\tilde{P}$ of all particles
  Update $\hat{P}$ of all particles
  Update velocities and positions of all particles
end for
Transfer data back to CPU and output.

Compute Fitness Values of Particles

The computation of fitness values is the most important task during the whole search process, where high density of arithmetical computation is required. It should be carefully designed for parallelization so as to improve the overall performance (attention is mainly paid to the running speed) of GPU-PSO.

The algorithm for computing fitness values of all the particles is shown in Algorithm 2. From Algorithm 2, we can see that the iteration is only applied to dimension index $i = 1, 2..., D$, while on CPU, it should also be applied to the particle index $j = 1, 2, ...N$. The reason is that the arithmetical operation on all the N data of all the particles in dimension i is done in parallel (synchronously) on GPU.

Algorithm 2. Compute Fitness Values

Initialize, set the *'block size'* and *'grid size'*, with the number of threads in a grid equaling to the number of particles(N).

for each dimension i **do**
  Map all threads to the N position values one-to-one
  Load N data from global to shared memory
  Apply arithmetical operations to all N data in parallel
  Store the result of dimension i with $f(X_i)$
end for

Combine $f(X_i)$ ($i = 1, 2...D$) to get the final fitness values $f(X)$ of all particles, store them in array F.

Mapping all the threads to the N data in a one-dimensional array should follow two steps:

- Set the block size to $S_1 \times S_2$ and grid size $T_1 \times T_2$. So the total number of threads in the grid is $S_1 * S_2 * T_1 * T_2$. It must be guaranteed that $S_1 * S_2 * T_1 * T_2$=N, only in this case can all the data of N particles be loaded and processed synchronously.

- Assuming that the thread with the index (T_x, T_y) in the block whose index is (B_x, B_y), is mapped to the I-th datum in a one-dimensional array, then the relationship between the indexes and I is:

$$I = (B_y * T_2 + B_x) * S_1 * S_2 + T_y * S_2 + T_x \tag{5}$$

In this way, all the threads in a kernel are mapped to N data one to one. Then applying an operation to one thread will cause all the N threads to do exactly the same operation synchronously. This is the core mechanism for explaining why GPU can accelerate the computing speed greatly.

Update $\tilde{P}$ and $\hat{P}$

After the fitness values are updated, each particle may arrive at a better position $\tilde{P}$ than ever before and a new local best position $\hat{P}$ may be found. So $\tilde{P}$ and $\hat{P}$ (refer to equation. 3) must be updated according to the current status of the particle swarm. The updating process of $\tilde{P}$ (PF at the same time) can be achieved by Algorithm 3.

Algorithm 3. Update $\tilde{P}$

Map all the threads to N particles one-to-one.
Transfer all the N data from global to shared memory.

//Do operations to thread i ($i = 1, ..., N$) in parallel
if $F(i)$ is better than $PF(i)$ **then**
  $PF(i)$= $F(i)$
  for each dimension d **do**
    Store the position $X(d * N + i)$ to $\tilde{P}(d * N + i)$
  end for
end if

The updating of $\hat{P}$ (LF at the same time) is similar to that of $\tilde{P}$. Compare a particle's previous $\hat{P}$ to the current $\tilde{P}$ of the right neighbor, left neighbor and its own, respectively, then choose the best one as the new $\hat{P}$ for that particle.

Update Velocity and Position

After the personal best and local best positions of all the particles have been updated, the velocities and positions should also be updated according to Equation. 3 and Equation. 2, respectively, by making use of the new information provided by $\tilde{P}$ and $\hat{P}$. This process is conducted dimension by dimension. On the same dimension d ($d = 1, 2, ..., D$), the velocity values of all the particles are updated in parallel, using the same technique mentioned in the previous algorithms. What should be paid special attention to is that two random integers P_1 and P_2 should be provided for fetching random numbers from array R, namely the random number "pool".

5 Experimental Results and Discussion

The experimental platform for this chapter is based on Intel Core 2 Duo 2.20GHz CPU, 3.0GRAM, NVIDIA GeForce 9800GT, and Windows XP. Performance comparisons between GPU-PSO and CPU-PSO is made based on four classical benchmark test functions as shown in Table 1.

Table 1. Benchmark Test Functions

NO.	Name	Equation	Bounds
f_1	Sphere	$\sum_{i=1}^{D} x_i^2$	$(-100, 100)^D$
f_2	Rastrigin	$\sum_{i=1}^{D} [x_i^2 - 10 * \cos(2\pi x_i) + 10]$	$(-10, 10)^D$
f_3	Griewangk	$\frac{1}{4000} \sum_{i=1}^{D} x_i^2 - \prod_{i=1}^{D} \cos(x_i/\sqrt{i}) + 1$	$(-600, 600)^D$
f_4	Rosenbrock	$\sum_{i=1}^{D-1} (100(x_{i+1} - x_i^2)^2 + (x_i - 1)^2)$	$(-10, 10)^D$

The variables of f_1,f_2 and f_3 are independent but variables of f_4 are dependent, namely there are related variables such as the i-th and $(i+1)$-th variable. The optimal solution of all the four functions are 0.

PSO is run both on GPU and CPU, and we call them as GPU-PSO and CPU-PSO, respectively. Now we define *speedup* as the times that GPU-PSO runs faster than CPU-PSO.

$$\gamma = \frac{T_{CPU}}{T_{GPU}} \tag{6}$$

where γ is *speedup*, T_{CPU} and T_{GPU} is the time used by CPU-PSO and GPU-PSO to optimize a function during a specific number of iterations, respectively.

In the following paragraphs, *Iter* is the number of iterations that SPSO runs, *D* is the dimension, *N* is the swarm population, namely the number of particles in the swarm; *CPU-Time* and *GPU-Time* are the average time that CPU-PSO and GPU-PSO consumed in the 20 runs, respectively, with second as the unit of time. *CPU-Optima* and *GPU-Optima* stand for the mean final optimized function values of running PSO for 20 times on CPU and GPU, respectively.

The experimental results and analysis are given as follows.

5.1 Running Time and Speedup versus Swarm Population

We run both GPU-PSO and CPU-PSO on f_1,f_2, f_3 and f_4 for 20 times independently, and the results are shown in Table 2, 3, 4, 5. (D=50, $Iter$=2000 for f_1 and f_3, $Iter$=5000 for f_2 and f_4).

After analyzing the data given in Table 2, 3, 4 and 5, we can make the following conclusions.

Table 2. Results of CPU-PSO and GPU-PSO on f_1 (D=50)

N	CPU-Time	GPU-Time	Speedup	CPU-Optima	GPU-Optima
256	3.11	1.17	**2.7**	7.68E-8	5.54E-8
512	6.15	1.25	**4.9**	5.92E-8	4.67E-8
768	9.22	1.46	**6.3**	4.49E-8	3.97E-8
1024	12.4	1.62	**7.7**	3.89E-8	3.94E-8
1280	15.97	1.89	**8.5**	4.17E-8	3.59E-8

Table 3. Results of CPU-PSO and GPU-PSO on f_2 (D=50)

N	CPU-Time	GPU-Time	Speedup	CPU-Optima	GPU-Optima
256	16.91	3.01	**5.6**	132.10	133.99
512	34.46	3.28	**10.5**	113.64	111.43
768	51.53	3.75	**13.7**	118.62	109.72
1024	70.27	4.12	**17.1**	110.78	106.94
1280	88.61	4.78	**18.5**	108.72	104.11

Table 4. Results of CPU-PSO and GPU-PSO on f_3 (D=50)

N	CPU-Time	GPU-Time	Speedup	CPU-Optima	GPU-Optima
256	5.94	1.22	**4.9**	1.52E-09	3.87E-08
512	11.86	1.32	**9.0**	1.21E-09	1.18E-08
768	18.05	1.53	**11.8**	9.04E-10	5.96E-09
1024	24.58	1.73	**14.2**	6.87E-10	0
1280	30.85	1.91	**16.2**	6.97E-10	0

Table 5. Results of CPU-PSO and GPU-PSO on f_4 (D=50)

N	CPU-Time	GPU-Time	Speedup	CPU-Optima	GPU-Optima
256	9.17	3.02	**3.0**	26.09	23.40
512	18.03	3.25	**5.5**	16.21	14.21
768	27.75	3.79	**7.3**	11.62	16.20
1024	37.81	4.20	**9.0**	12.14	10.15
1280	46.62	4.89	**9.5**	6.82	8.88

Optimizing Performance Comparison

GPU-PSO uses a random number "pool" for updating velocity instead of instant random number generation. To some extend it may affect the results. However, seen from the tables above, on all of the four functions, GPU-PSO and CPU-PSO can find optimal solutions of almost the same magnitude, namely the precision of the result given by GPU-PSO is almost the same with CPU-PSO (even better in some cases). So we can say that GPU-PSO is reliable in optimizing functions.

Population Size Setup

When the swarm size grows (from 512 to 1280), the optima of a function found by both CPU-PSO and GPU-PSO are almost of the same magnitude, no obvious precision improvement can be seen. So we can say that a large swarm population is not always necessary. But exceptions may still exist when large swarm population is required in order to obtain better optimizing results, especially in real-world optimization problems.

Running Time

Compared with f_1, the fitness evaluation of f_3 is much more complex. f_1 contains only square arithmetic, while f_3 contains square, cosine and square root arithmetic as well. Similarly we can say that f_2 is much more complex than f_4.

Fig. 5 and Fig. 6 depict how running time of both CPU-PSO and GPU-PSO changes when the swarm size(N)grows. On all of the 4 functions, the running time of GPU-PSO as well as CPU-PSO is proportional to the swarm size(N), namely the time increases

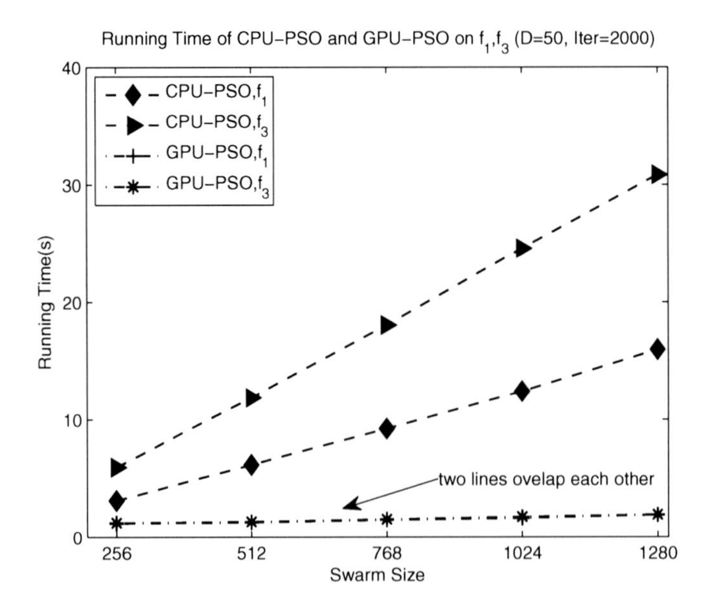

Fig. 5. Running Time and Swarm Population(f_1 and f_3)

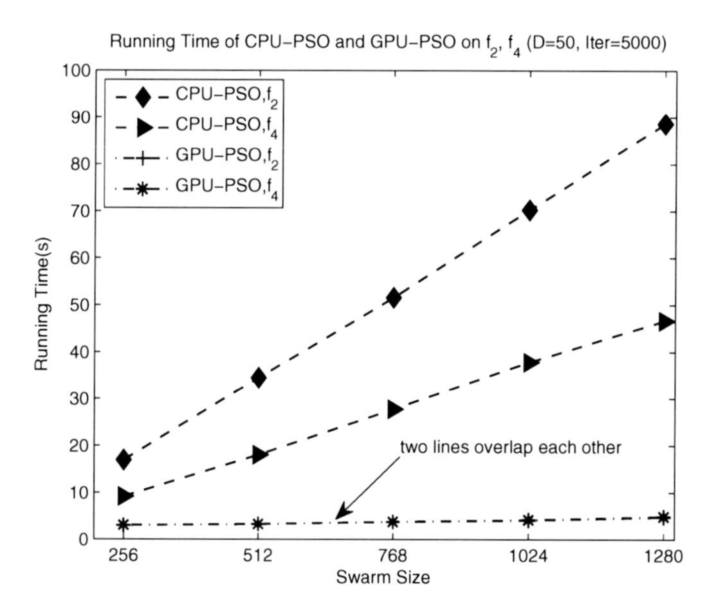

Fig. 6. Running Time and Swarm Population(f_2 and f_4)

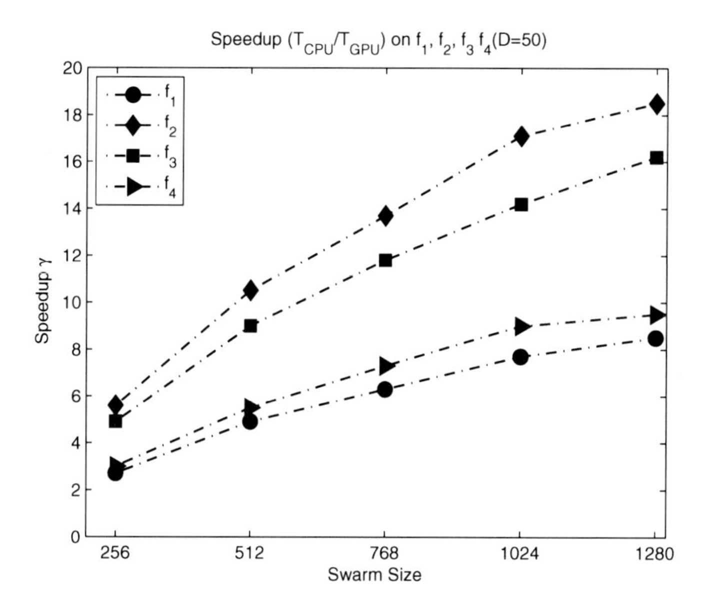

Fig. 7. Speedup and Swarm Population

linearly with N, while keeping the other parameters constant. And when N, D and $Iter$ are fixed, it takes much more time for CPU-PSO to optimize the function with more complex arithmetic than the one with less complex arithmetic.

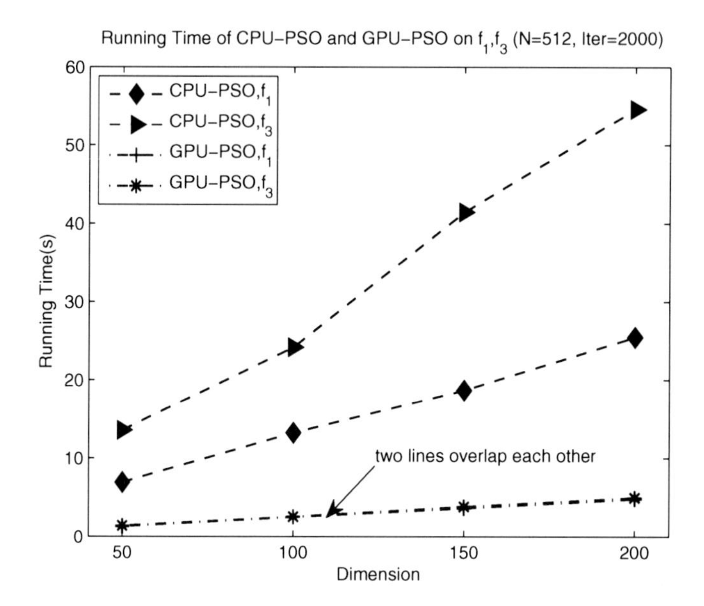

Fig. 8. Running Time and Dimension(f_1 and f_3)

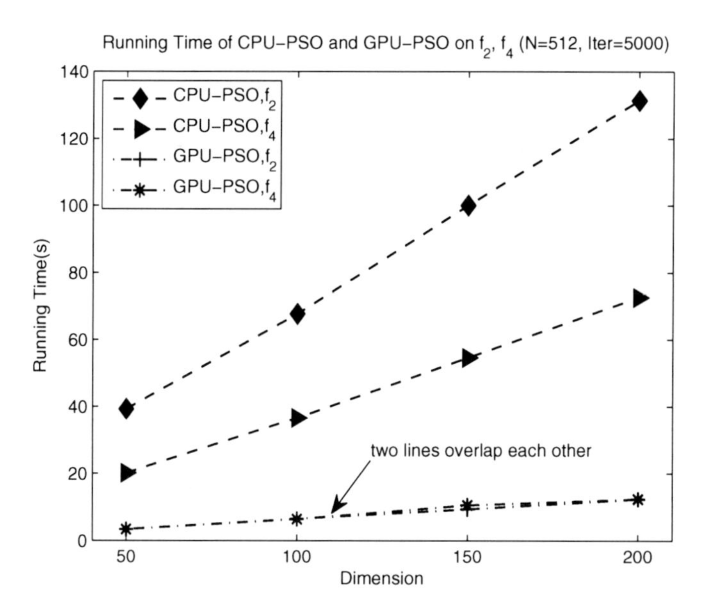

Fig. 9. Running Time and Dimension(f_2 and f_4)

Speedup

As seen from Fig. 7, For a certain function, as the swarm size grows, the *speedup* also increases, but it is limited to a specific constant. Furthermore, the line of a function with more complex arithmetic lies above the line of those with less complex arithmetic (f_3 above f_1, and f_2 above f_4), that is to say, optimizing a function with more complex arithmetic by GPU-PSO can reach a relatively higher speedup.

5.2 Running Time and Speedup versus Dimension

Now we fix the swarm size to 50, and vary the dimension of the functions(D). Analysis about the relationship between running time (as well as *Speedup*) and D is given here. We run both GPU-PSO and CPU-PSO on f_1,f_2 and f_3 for 20 times independently, and the results are shown in Table 6, 7, 8, 9(N=512, $Iter$=2000 for f_1 and f_3, $Iter$=5000 for f_2 and f_4).

From Table 6, 7, 8 and 9, we can make the following conclusions.

Running Time

As seen from Fig. 8 and Fig. 9, the running time of both GPU-PSO and CPU-PSO increases linearly with the dimension, keeping the other parameters constant. When D, N and $Iter$ are fixed, functions with more complex arithmetic(f_2 and f_3) need more time than the function with much less complex arithmetic (f_4 and f_1) to be optimized by CPU-PSO, while the time is almost the same when optimized by GPU-PSO, just as mentioned in Section 5.1.

Speedup

It can be seen from Fig. 10 that the *speedup* remains almost the same when the dimension grows. The reason is, in GPU-PSO, the parallelization can only be applied to swarm size(N), but not to the dimension. Still, the function with more complex arithmetic has a relatively higher speedup under the same conditions.

5.3 Other Characteristics of GPU-PSO

Maximum Speedup

In some applications, large swarm size is needed during the optimizing process. In this case, GPU-PSO can greatly benefit the optimization by improving the running speed dramatically. Now we will carry through an experiment to find out how much speedup GPU-PSO can reach. We run GPU-PSO and CPU-PSO on f_2, respectively. Set D=50, $Iter$=5000, and both GPU-PSO and CPU-PSO are run only once (as the time needed for each run is almost the same). The results are shown in Table 10.

As shown in Table 10, when optimizing f_2, GPU-PSO can reach a *speedup* of almost ***40***, when the swarm population size is 16384. And on the functions more complex than f_2, the speedup may be even greater.

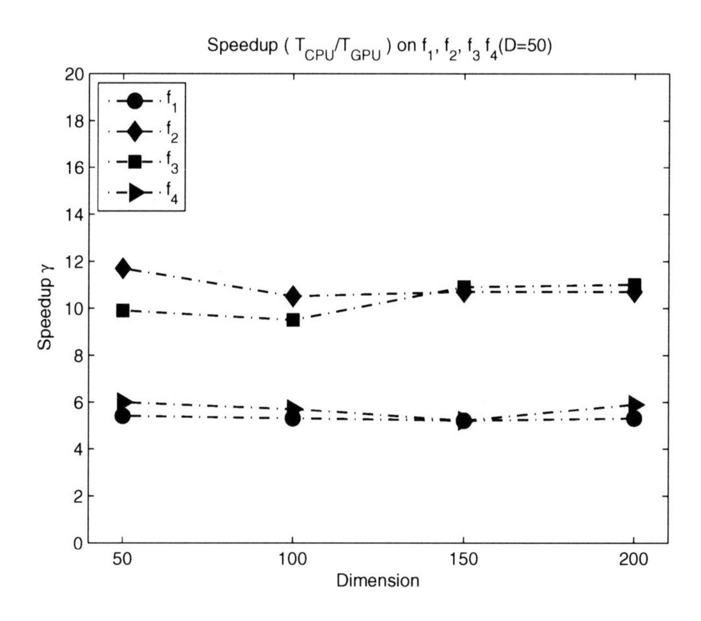

Fig. 10. Speedup and Dimension

Table 10. GPU-PSO and CPU-PSO on f_2 (D=50)

N	CPU-Time	GPU-Time	CPU-Optima	GPU-Optima	**Speedup**
8192	648.1	25.9	101.49	96.52	**25.02**
16384	2050.3	51.7	83.58	87.56	**39.7**

High Dimension Application

In some real world applications such as face recognition and fingerprint recognition, the problem dimension may be very high. Running PSO on CPU to optimize high dimensional problems can be quite slow, but the speed can be greatly accelerated if running it on GPU. Now we run both GPU-PSO and CPU-PSO on f_2 once(N=512, $Iter$=5000). The results are given in Table 11.

Table 11. GPU-PSO and CPU-PSO on f_3 (N=512)

D	CPU-Time	GPU-Time	**Speedup**
1000	934.5	87.9	**10.6**
2000	2199.7	128.2	**17.2**

From Table 11, we can find out that even when the dimension is as large as 2000, GPU-PSO can run more than 6.5 times faster than GPU-PSO.

5.4 Comparison with Related Works

The PSO algorithm was also implemented on GPU in an other way by making use of the texture-rendering of GPU [17]. We call it as *Texture-PSO* for short. This method used the textures on GPU chips to store particle information, and the fitness evaluation, velocity and position updating were done by means of texture rendering. But Texture-PSO has the following disadvantages which make it almost useless when doing real-world optimizations.

- The dimension must be set to a specific number, and it can not be changed unless redesigning the data structures.
- When the swarm population is small, for example smaller than 512, the running time of the Texture-PSO may even be longer than corresponding PSO that runs on CPU.
- The functions which can be optimized by Texture-PSO must have completely independent variables as a result of the architecture of GPU textures. So functions like f_4 can not be optimized by the Texture-PSO.

Instead of using the textures on GPU, we use the global memory to implement GPU-PSO in this paper. Global memory is more like memory on CPU than textures do. So GPU-PSO has overcome all the three disadvantages mentioned above:

- The dimension serves as a changeable parameter and it can be set to any reasonable numbers. High dimensional problems are also solvable by using our GPU-PSO.
- When the swarm population is small, for example smaller than 512, a remarkable speedup can also be achieved.
- The functions with dependent variables such as f_4 can also be optimized by GPU-PSO.

6 Conclusions

In this paper, a novel way to implement PSO on GPU(GPU-PSO) is presented, based on the software platform of CUDA from NVIDIA Corporation. GPU-PSO has the following features:

- The running time of GPU-PSO is greatly shortened over CPU-PSO, while maintaining similar optimizing performance. Under certain conditions, the speedup can be as large as nearly *40*$\times$ on f_2. On other functions with more complex arithmetic, a larger speedup can be expected. On GPU chips that have much more multiprocessors than the Geforce 9800GT used in this paper, the GPU-PSO is expected to run tens of times faster than CPU-PSO.
- The running time and swarm population size take a linear relationship. This is also true for running time and dimension. And it takes almost the same time for GPU-PSO to optimize functions with different arithmetic complexity, while CPU-PSO takes much more time to optimize functions with more complex arithmetic, with the same swarm population, dimension and number of iterations. Furthermore, function with more complex arithmetic has a higher speedup.

- The swarm population can be very large, and the larger the population is, the faster GPU-PSO runs than CPU-PSO. So GPU-PSO can especially benefit optimization with large swarm size.
- High dimensional problems and functions with dependent variables can also be optimized by GPU-PSO, and noticeable speedup can be reached.
- Because in current common PC, there are GPU chips in the display card, more researchers can make use of our parallel GPU-PSO to solve their practical problems in a quick way.

Because of these features of GPU-PSO, it can be applied to a large scope of practical optimization problems.

Our future research will focus on implementing genetic algorithm and other swarm intelligence algorithms in terms of similar methods presented in the paper. We will also try to put our GPU-PSO onto real-world applications.

Acknowledgment

This work was supported by National Natural Science Foundation of China (NSFC), under Grant No. 60875080 and 60673020, and also in part supported by the National High Technology Research and Development Program of China (863 Program), with Grant No. 2007AA01Z453.

References

1. Kennedy, J., Eberhart, R.: Particle Swarm Optimization. In: IEEE International Conference on Neural Networks, Perth, WA, Australia, pp. 1942–1948 (November 1995)
2. Shi, Y.H., Eberhart, R.: Parameter selection in particle swarm optimization. In: Porto, V.W., Waagen, D. (eds.) EP 1998. LNCS, vol. 1447, pp. 591–600. Springer, Heidelberg (1998)
3. Clerc, M., Kennedy, J.: The Particle Swarm-Explosion, Stability, and Convergence in a Multidimensional Complex Space. IEEE Trans. on Evolutionary Computation 6(1), 58–73 (2002)
4. van den Bergh, F., Engelbrecht, A.P.: A cooperative approach to particle swarm optimization. IEEE Trans. on Evolutionary Computation 8, 225–239 (2004)
5. Tan, Y., Xiao, Z.M.: Clonal Particle Swarm Optimization and Its Applications. In: IEEE Congress on Evolutionary Computation, pp. 2303–2309 (August 2007)
6. Zhang, J.Q., Xiao, Z.M., Tan, Y., He, X.G.: Hybrid Particle Swarm Optimizer with Advance and Retreat Strategy and Clonal Mechanism for Global Numerical Optimization. In: IEEE Congress on Evolutionary Computation, Hong Kong, China, pp. 2059–2066 (June 2008)
7. Tan, Y., Zhang, J.Q.: Magnifier Particle Swarm Optimisation. In: Chiong, R. (ed.) Nature-Inspired Algorithms for Optimisation, SCI, vol. 193, pp. 279–298. Springer, Heidelberg (2009)
8. Buck, I., et al.: Brook for GPUs: Stream Computing on Graphics Hardware. ACM, 777–786 (2004)
9. Bratton, D., Kennedy, J.: Defining a Standard for Particle Swarm Optimization. In: IEEE Swarm Intelligence Symposium, pp. 120–127 (April 2007)
10. NVIDIA, NVIDIA CUDA Programming Guide1.1, Introduction to CUDA, ch. 1 (2007)

11. Yang, R., Welch, G.: Fast Image Segmentation and Smoothing Using Commodity Graphics Hardware. Journal of Graphics Tools, special issue on Hardware-Accelerated Rendering Techniques, 91–100 (2003)
12. Kenneth, E.H., John, K., Lin, M., et al.: Fast Computation of Generalized Voronoi Diagrams Using Graphics Hardware. In: Proceedings of SIGGRAPH, pp. 277–286 (1999)
13. Luo, Z.W., Liu, H.Z., Wu, X.C.: Artificial Neural Network Computation on Graphic Process Unit. In: Proceedings of International Joint Conference on Neural Networks, Montreal, Canada (2005)
14. Macedonia, M.: The GPU Enters Computing's Mainstream. Entertainment Computing, 106–108 (October 2003)
15. NVIDIA, NVIDIA CUDA Programming Guide1.1, ch. 2: Programming Model (2007)
16. Langdon, W.B.: A Fast High Quality Pseudo Random Number Generator for Graphics Processing Units. In: IEEE World Congress on Evolutionary Computation, pp. 459–465 (2008)
17. Li, J.M., et al.: A parallel particle swarm optimization algorithm based on fine grained model with GPU accelerating. Journal of Harbin Institute of Technology, 2162–2166 (December 2006) (in Chinese)

Velocity Adaptation in Particle Swarm Optimization

Sabine Helwig[1], Frank Neumann[2], and Rolf Wanka[1]

[1] Department of Computer Science, University of Erlangen-Nuremberg, Germany
{sabine.helwig,rwanka}@informatik.uni-erlangen.de
[2] Max-Planck-Institut für Informatik, Saarbrücken, Germany
fne@mpi-inf.mpg.de

Summary. Swarm Intelligence methods have been shown to produce good results in various problem domains. A well-known method belonging to this kind of algorithms is particle swarm optimization (PSO). In this chapter, we examine how adaptation mechanisms can be used in PSO algorithms to better deal with continuous optimization problems. In case of bound-constrained optimization problems, one has to cope with the situation that particles may leave the feasible search space. To deal with such situations, different bound handling methods were proposed in the literature, and it was observed that the success of PSO algorithms highly depends on the chosen bound handling method. We consider how velocity adaptation mechanisms can be used to cope with bounded search spaces. Using this approach we show that the bound handling method becomes less important for PSO algorithms and that using velocity adaptation leads to better results for a wide range of benchmark functions.

1 Introduction

Stochastic search methods have been widely applied for different discrete and continuous optimization problems. In the case of discrete optimization, approaches such as (iterated) local search [1, 19], ant colony optimization [10] or genetic algorithms [15, 12] often produce good results for wide classes of problems. Continuous optimization problems may be tackled by other well-known approaches such as evolution strategies [30], differential evolution [26], or particle swarm optimization [21, 22].

Particle swarm optimization is a population-based stochastic search algorithm which has become very popular for solving continuous optimization problems in recent years. The algorithm is inspired by the interactions observed in social groups such bird flocks, fish schools, or human societies. Continuous optimization problems are often defined on a bounded subspace of $\mathbb{R}^n$. Search points that do not belong to this area are considered as infeasible. One question is how to deal with particles that reach the infeasible region. Various methodologies were proposed in the literature, e.g., [5, 2]. Such methods often handle infeasible particles by guiding them towards the feasible region of the search space. Recently, it was observed that many particles leave the feasible region of the

B.K. Panigrahi, Y. Shi, and M.-H. Lim (Eds.): Handbook of Swarm Intelligence, ALO 8, pp. 155–173.
springerlink.com

underlying search space at the beginning of the optimization process [18]. These results highlight the importance of the bound handling strategy for the success of a particle swarm optimizer.

Our goal is to show how to avoid that many particles leave the feasible region of the parameter space of a bounded continuous optimization problem. We consider the model examined in [18] and point out that the problem of constructing many infeasible solutions occurs due to high velocities of the particles. Based on this observation it seems to be natural to place upper bounds on the velocities of the particles. To deal with the fact that sometimes high velocities are needed for an efficient search process, we propose to adjust the length of a particle's velocity vector in a similar way to step size adaptation in evolution strategies [28, 31]. This has the effect that if the algorithm is not able to construct feasible solutions by high velocities, the velocities become smaller such that it is more likely to produce feasible solutions within the bounded search space. We examine the proposed velocity adaptation for particle swarm optimization and compare it to a standard particle swarm optimizer [3]. The effects of using different bound handling methods are investigated for the two approaches. When using no bounds or constant bounds on the velocities, the use of a suitable bound handling strategy is crucial for the success of the algorithm, especially for high-dimensional optimization problems. We show that our approach is almost invariant with respect to the bound handling method which means that we get more robust PSO algorithms by using velocity adaptation. Further experimental studies show that the use of velocity adaptation in PSO algorithms leads to significantly better results for a wide range of benchmark functions compared to a standard PSO approach. The chapter is based on results that were presented in [16]. Additional experiments were performed to analyze the introduction of local velocity lengths, and the specific velocity scaling strategy of PSO with velocity adaptation.

The outline of the chapter is as follows. In Section 2, we describe particle swarm optimization for bounded search spaces. Section 3 considers the effect of velocities for staying within the boundary constraints. Based on these investigations, we introduce a particle swarm optimization algorithm using velocity adaptation in Section 4. Experimental investigations of our approach that point out the benefits of velocity adaption are reported in Section 5. Finally, we finish with some concluding remarks.

2 Particle Swarm Optimization

In this section, firstly the PSO algorithm is presented. Afterwards, bound-constrained optimization problems are formally defined, and PSO strategies to cope with these contraints are outlined. Finally, some existing adaptive particle swarm optimizers are shortly described.

Particle swarm optimization (PSO) [21, 22] is a population-based stochastic search algorithm for global optimization. The goal is to minimize or to maximize an objective function $f : S \subseteq \mathbb{R}^n \to \mathbb{R}$ that is defined over an n-dimensional *search space* S. Without loss of generality, f is a minimization problem. PSO is inspired by the social interaction of individuals living together in groups. In order to optimize the given problem, a population of individuals, also denoted as *particle swarm*, moves through the n-dimensional search space S. Each particle i has a *position* $\mathbf{x}_{i,t}$, a *velocity* $\mathbf{v}_{i,t}$, and a *fitness value* $f(\mathbf{x}_{i,t})$, where t is the iteration counter. Based on the social metaphor, the particles' movement is guided by own experiences, and by the experiences of other group members. Therefore, each particle i remembers the best position it has visited so far as its *private guide* $\mathbf{p}_{i,t}$. Each particle i is able to communicate with a subset of all particles, which is denoted as its *neighborhood*. In each iteration, it determines the best private guide of all its neighbors, its *local guide* $\mathbf{l}_{i,t}$. The iteration step is given by:

$$\mathbf{v}_{i,t} = \omega \cdot \mathbf{v}_{i,t-1} + c_1 \cdot \mathbf{r}_1 \odot (\mathbf{p}_{i,t-1} - \mathbf{x}_{i,t-1}) + c_2 \cdot \mathbf{r}_2 \odot (\mathbf{l}_{i,t-1} - \mathbf{x}_{i,t-1}) \quad (1)$$

$$\mathbf{x}_{i,t} = \mathbf{x}_{i,t-1} + \mathbf{v}_{i,t} \quad (2)$$

where

- the *inertia weight* ω and the so-called *acceleration coefficients* c_1 and c_2 are user-defined parameters,
- $\odot$ denotes component-wise vector multiplication,
- and $\mathbf{r}_1$ and $\mathbf{r}_2$ are vectors of random real numbers chosen uniformly at random in $[0, 1]$, and independently drawn every time they occur.

After updating position and velocity of each particle, the private guides are considered. The private guide of a particle is updated to the particle's current position if the new position is strictly better than the private guide. If the new particle position and the private guide are of equal fitness, the private guide is updated with probability 1/2.

2.1 PSO for Bound-Constrained Problems

In this chapter, we consider optimization problems with boundary constraints. This means that the search space $S = [lb_1, ub_1] \times [lb_2, ub_2] \times \ldots \times [lb_n, ub_n]$ consists of n real-valued parameters $(x_1, \ldots, x_n)$ where each parameter x_i, $1 \leq i \leq n$, is bounded with respect to some interval $[lb_i, ub_i]$. Without loss of generality, $S = [-r, r]^n$, i. e., $lb_i = -r$ and $ub_i = r$, $1 \leq i \leq n$. There exist a lot of strategies to handle boundary constraints in the literature, e.g., [5, 2, 35, 29, 3, 18], among them:

- *Infinity:* Particles are allowed to move to an infeasible solution. In this case, its position and velocity remain unchanged, and the evaluation step is skipped [3].
- *Absorb:* Invalid particles are set on the nearest boundary position. The respective velocity components are set to zero [6].

- *Random:* Invalid components of a particle's position vector are set to a random value. Afterwards, the velocity is adjusted to $\mathbf{v}_{i,t+1} = \mathbf{x}_{i,t+1} - \mathbf{x}_{i,t}$ [35].

Previous studies analyzed the initial swarm behavior in high-dimensional bound-constrained search spaces [17, 18]. It was proven that particles are expected to be initialized very close to the search space boundary [17]. Furthermore, it was shown that, using three different velocity initialization strategies, at least all particles which have a better neighbor than themselves leave the search space in the first iteration with overwhelming probability w. r. t. the search space dimensionality. Hence, when solving high-dimensional problems, the behavior of a particle swarm optimizer strongly depends on the chosen bound handling mechanism, at least at the beginning of the optimization process. Experimental studies confirmed these results: Significant performance differences were observed on the investigated testproblems if the bound handling strategy was varied [18].

2.2 Adaptive Particle Swarm Optimization

Adaptive particle swarm optimizers try to adapt the algorithmic PSO parameters and/or the swarm behavior to the characteristics of the underlying optimization problem by considering, for instance, the particles' behavior or success. In the so-called *Tribes* algorithm of Clerc [4, 6], the number of particles and their social network is adapted to the search process based on the performance of the particle swarm. Other approaches combine concepts from the field of evolutionary computation with particle swarm optimization, in order to adapt the algorithmic parameters to the problem at hand. Examples are the so-called *Evolutionary PSO algorithm* of Miranda and Fonseca [24, 25], and the *Efficient Multi-Objective PSO* of Toscano-Pulido et al. [27]. Furthermore, the neighborhood graph can be adjusted to the search progress. In the *Hierarchical Particle Swarm Optimizer* of Janson and Middendorf [20], good particles are moved up in a hierachically structured neighborhood graph to increase their impact on other swarm members.

In this chapter, PSO with velocity adaptation is discussed. The adaptation mechanism is based on the 1/5 rule of Rechenberg [28]: Particle velocities are adapted to the search process by analyzing the success of the particle swarm.

The lengths of the particle velocity vectors have strong impact on particle swarm behavior and particle swarm performance. The use of a constant maximum particle velocity, also denoted as *velocity clamping*, can improve the solution quality produced by a PSO algorithm [11]. In order to increase the swarm's convergence rate, Fan [13] suggests to deterministically decrease the maximum particle velocities during the optimization from a pre-defined value V_{max} to zero. Cui et al. [9, 8] propose the use of a stochastic velocity threshold. Although inspired by evolutionary programming, the particles' success is not considered when computing the velocity threshold. Takahama and Sakei [33] adapt the particles' velocities based on the feasibility of the particles. In their subswarm approach, the maximum particle velocity of the worst subswarm is modified such that it approaches the one of the best-performing subswarm. Fourie and Groenwold [14] adapt the maximum particle velocities based on the particles' success. Whenever the best found solution was not improved during a specified number of iterations, the maximum particle velocity is decreased by multiplying each component with a constant β, where $0 < \beta < 1$. The main differences to our approach are that Fourie and Groenwold do never increase particle velocities, that they use a componentwise maximum particle velocity, and that only the improvement of the best found solution is considered in their adaptation procedure.

3 Theoretical Observations

In the following, the effect of the velocity values on particle behavior is analyzed, under the assumption of a bound-constrained optimization problem. The investigation is independent of the function to be optimized. To examine the effect of maximum velocities we assume that the particles of a PSO algorithm are drawn uniformly at random from the underlying bounded search space. Additionally, we assume that the velocities are drawn from a fixed interval uniformly at random. As we do not take any properties of the optimization problem into account, we do not present rigorous results for later stages of the optimization process. Instead, we complement the theoretical studies carried out in this section with experimental investigations that confirm our findings. The following theorem generalizes previous results [18] to velocities that are drawn uniformly at random from $[-\frac{r}{s}, \frac{r}{s}]^n$, where $s \geq 1$ is a parameter that determines the maximum velocity with respect to a particular direction.

Theorem 1. *Let $\mathbf{x}_{i,t-1}$ and $\mathbf{v}_{i,t}$ be independently drawn from a uniform distribution over $[-r, r]^n$ and $[-\frac{r}{s}, \frac{r}{s}]^n$ with $s \geq 1$, respectively, for an arbitrary particle i at time step t. Then, the probability that particle i leaves the search space in iteration t is*

$$1 - \left(1 - \frac{1}{4s}\right)^n .$$

Proof. According to Eq. (2), particle i's position at time step t evaluates to $\mathbf{x}_{i,t} = \mathbf{x}_{i,t-1} + \mathbf{v}_{i,t}$. Hence, the d-th component of $\mathbf{x}_{i,t}$, $x_{i,t,d}$, is the sum of two independent uniformly distributed random variables $x_{i,t-1,d}$ and $v_{i,t,d}$. Their probability density functions are given by

$$f_{x_{i,t-1,d}}(z) = \begin{cases} \frac{1}{2r} & \text{for } -r \leq z \leq r \\ 0 & \text{otherwise} \end{cases}$$

$$f_{v_{i,t,d}}(z) = \begin{cases} \frac{s}{2r} & \text{for } -\frac{r}{s} \leq z \leq \frac{r}{s} \\ 0 & \text{otherwise} . \end{cases}$$

Hence, the density function $f_{x_{i,t,d}}$ of $x_{i,t,d}$ is trapezoidal and can be determined by convolution to

$$f_{x_{i,t,d}}(z) = \int_{-\infty}^{\infty} f_{x_{i,t-1,d}}(t) f_{v_{i,t,d}}(z-t)\mathrm{d}t$$

$$= \begin{cases} \frac{s}{4r^2} z + \frac{s}{4r} + \frac{1}{4r} & \text{for } -r - \frac{r}{s} \le z \le -r + \frac{r}{s} \\ \frac{1}{2r} & \text{for } -r + \frac{r}{s} < z < r - \frac{r}{s} \\ -\frac{s}{4r^2} z + \frac{s}{4r} + \frac{1}{4r} & \text{for } -\frac{r}{s} + r \le z \le r + \frac{r}{s} \\ 0 & \text{otherwise} . \end{cases}$$

Thus, the probability that particle i leaves the search space in dimension d is

$$\mathrm{Prob}(|x_{i,t,d}| > r) = \int_{-\infty}^{-r} f_{x_{i,t,d}}(z)\mathrm{d}z + \int_{r}^{\infty} f_{x_{i,t,d}}(z)\mathrm{d}z = \frac{1}{4s} .$$

As each component of $\mathbf{x}_{i,t}$ is computed independently, the probability $p(n,s)$ that particle i becomes infeasible in iteration t evaluates to

$$p(n,s) = \mathrm{Prob}(\mathbf{x}_{i,t} \notin [-r,r]^n) = 1 - \left(1 - \frac{1}{4s}\right)^n .$$

Note that $\lim_{n\to\infty}(1 - \left(1 - \frac{1}{4s}\right)^n) = 1 - e^{-\frac{n}{4s}}$ holds. Theorem 1 shows that the probability that a particle leaves the bounded search space crucially depends on the interval from which the velocities are chosen. For example, if s is a constant, the probability that a particle becomes infeasible rapidly approaches 1 when the dimensionality n of the optimization problem increases. On the other hand, $s = n$ implies that, with constant probability, particles do not leave the bounded search space.

This theoretical result stresses the importance of adapting velocities during the optimization process. Large velocities are important for search space exploration and to gain large improvements. However, the probability that particles leave the search space increases when velocities are allowed to grow large, which might deteriorate the swarm's performance. Reducing the particles' velocities also reduces the probability that particles become infeasible, down to a constant probability if velocities are, for instance, drawn uniformly at random in $[-\frac{r}{n}, \frac{r}{n}]^n$. Hence, small velocities are necessary to cope with situations where large improvements are not possible. These considerations lead us to the PSO with velocity adaptation presented in the subsequent section.

4 PSO with Velocity Adaptation

The investigations carried out in the previous section have pointed out the influence of velocities on the probability that particles leave the bounded search space. It has been shown how the probability of leaving the bounded search space depends on the interval from which the velocity is chosen. The probability of leaving the bounded search space decreases when the velocities decrease. However, small velocities lead to stagnation which implies that the algorithm achieves less progress. Therefore, we propose to use velocity adaptation in particle swarm optimization such that a large progress can be achieved by high

Algorithm 1. PSO with velocity adaptation

Require: Objective function $f : \mathcal{S} \subseteq \mathbb{R}^n \to \mathbb{R}$, $SuccessProbability \in \mathbb{R}_{[0,1]}$, Initial velocity length l
1: Initialize particle positions, velocities, and private guides
2: Initialize neighborhood graph
3: $SuccessCounter \leftarrow 0$
4: $VelocityLength \leftarrow l$
5: Scale each initial velocity (up or down) so that its length is exactly $VelocityLength$
6: **for** $t := 1, \ldots, maxIteration$ **do**
7:   **for** $i := 0, \ldots, m-1$ **do**
8:     Update velocity of particle i according to standard PSO equation (Eq. (1))
9:     Scale velocity (up or down) so that its length is exactly $VelocityLength$
10:     Update position of particle i according to standard PSO equation (Eq. (2))
11:   **end for**
12:   **for** $i := 0, \ldots, m-1$ **do**
13:     Update private guide of particle i
14:     **if** (success($\mathbf{x}_{i,t}, \mathbf{p}_{i,t-1}$)) **then**
15:       $SuccessCounter \leftarrow SuccessCounter + 1$
16:     **end if**
17:   **end for**
18:   **if** $t \mod n = 0$ **then**
19:     $SuccessRate \leftarrow \frac{SuccessCounter}{n \cdot m}$
20:     **if** $SuccessRate > SuccessProbability$ **then**
21:       $VelocityLength \leftarrow 2 \cdot VelocityLength$
22:     else
23:       $VelocityLength \leftarrow \frac{VelocityLength}{2}$
24:     **end if**
25:     $SuccessCounter \leftarrow 0$
26:   **end if**
27: **end for**

velocities. On the other hand, low velocities should be attained if the algorithm observes to have difficulties for finding better solutions. Our scheme for velocity adaptation is similar to the use of step size adaptation known in the field of evolution strategies. The idea is to increase velocities in the case that the particles have found improvements during the last iterations. In the case that no progress could be achieved, we reduce the velocities. The amount of progress that has been made during the last iterations is measured by the number of successes. A success occurred if the private guide of a particle is updated to its current position. This happens if the new particle position is strictly better than its private guide. If the new particle position and the private guide are of equal fitness, the private guide is updated with probability 1/2.

Definition 1 (Success). *A particle i is called* successful *in iteration t if its private guide $\mathbf{p}_{i,t}$ is updated to its current position $\mathbf{x}_{i,t}$.*

Our particle swarm optimizer using velocity adaptation is shown in Algorithm 1. It distinguishes itself from standard PSO approaches by using the described velocity adaptation mechanism. The algorithm uses a counter *SuccessCounter* which counts the number of successes according to Definition 1. In total n iterations are considered to decide whether velocities are increased or decreased. The success rate of the algorithm for a number of n iterations is given by $SuccessCounter/(n \cdot m)$, where m is the number of particles. If the observed success rate is higher than a given threshold (called *SuccessProbability*) the velocities are increased by a factor of 2 otherwise the velocities are scaled down by a factor of 1/2. In most of our experimental studies we will use a success probability of 0.2 which is motivated by the 1/5-rule used in evolution strategies [28].

5 Experimental Results

In the subsequent experimental analysis, PSO with velocity adaptation is compared to a standard particle swarm optimizer [3]. For both approaches, the following parameter setting was used: $c_1 = c_2 = 1.496172$, $\omega = 0.72984$ [3]. As proposed by Kennedy and Mendes [23], the swarm is connected via the so-called *von Neumann* topology, a two-dimensional grid with wrap-around edges. The population size was set to $m = 49$ to arrange the particles in a 7×7 grid. In the non-adaptive PSO, velocity clamping was applied, i.e., each velocity component is restricted to $[-r \dots r]$ [11], where $[-r \dots r]^n$ is the parameter space of the optimization problem to be solved. As bound handling methods, Absorb, Random, and Infinity were used.

Experiment 1 compares PSO with velocity adaptation to a standard particle swarm optimizer. Therefore, six widely-used benchmarks, Sphere, Rosenbrock, Rastrigin, Griewank, Schwefel 2.6, and Ackley (function descriptions see, e.g., [3]), and the CEC 2005 benchmarks f1–f14 [32], were used as 100- and 500-dimensional problems. From the CEC 2005 benchmarks, all noisy functions and all functions without search space boundaries were excluded. Additionally, f8 was omitted because all swarms were equally unsuccessful in preliminary experiments, and f12 was skipped due to its high computation time. For all benchmarks, particles were initialized uniformly at random in the whole search space. Historically, individuals are often initialized in a subspace of the feasible region in order to avoid that the performance of algorithms with center bias is overestimated. E.g., the initialization space of the Sphere function may be set to $[50 \dots 100]^n$ whereas the parameter space is $[-100 \dots 100]^n$ [3]. However, we do not use these asymmetric initialization ranges due to the following two reasons: First, we also investigate several CEC 2005 benchmarks which do not have their global optimum at the center of the search space. Hence, center bias can be identified considering all results. Second, we explicitely do not want to design an algorithm which has to explore outside the initialization space as such behavior is not needed for real applications. Velocities were initialized with half-diff initialization [7]: Let $\mathbf{x}_{i,0}$ be the initial position of particle i, and $\mathbf{y}_i$ a vector drawn uniformly at random in S. The initial velocity $\mathbf{v}_{i,0}$ of particle i is then set to $\frac{1}{2}(\mathbf{y}_i - \mathbf{x}_{i,0})$. If not mentioned otherwise, the success probability is set to $SuccessProbability = 0.2$ in PSO with velocity adaptation, in accordance to the 1/5 rule of Rechenberg [28]. The initial velocity length was set to $l = r$, or more generally, to half of the average search space range:

$$l = \frac{\sum_{d=1}^{n} \frac{ub_d - lb_d}{2}}{n}$$

Experiments 2–5 are dedicated to the investigation of different parameter settings for PSO with velocity adaptation, and were only run on 100-dimensional problems.

In our experimental study, each run was terminated after 300,000 function evaluations, and each configuration was repeated 50 times. The performance of two algorithms A and B was compared by using the one-sided Wilcoxon rank sum test [34] with null-hypothesis $H_0 : F_A(z) = F_B(z)$, and the one-sided alternative $H_1 : F_A(z) < F_B(z)$, where $F_X(z)$ is the distribution of the results of algorithm X. The significance level was set to $\alpha = 0.01$. Additionally, we show the obtained average objective values and corresponding standard errors as well as convergence plots for selected functions where appropriate.

5.1 Experiment 1: Comparison with a Standard PSO

PSO with velocity adaptation was compared to a standard particle swarm optimizer on all 100- and 500-dimensional problems. The obtained average values and corresponding standard errors are shown in Table 1. The results of the one-sided Wilcoxon rank sum tests are shown in Table 2. The experiment was conducted to investigate the following two questions: First, which algorithm, the adaptive PSO or the standard PSO, performs better on the chosen testbed? Second, is PSO with velocity adaptation more invariant with respect to the bound handling mechanism than a standard particle swarm optimizer?

Table 1 shows that the adaptive PSO (denoted as *Absorb-A*, *Random-A*, and *Infinity-A*, depending on the bound handling method) provides superior average results than the used standard PSO (*Absorb-S*, *Random-S*, *Infinity-S*) on most testfunctions. The only exceptions for the 100-dimensional functions are Schwefel, f9, and f10. When solving the 500-dimensional test suite, exceptions are again f9 and f10, but the relative difference in the average objective values is diminished. In some 500-dimensional cases, *Random-A* was not able to provide satisfactory results for the optimization problem under consideration. It was observed earlier that *Random* can distract particles from the boundary [17] which might be a reason for the bad performance of *Random-A* on some functions. Note that in high-dimensional spaces, most of the volume is located near the boundary (see, e.g., Theorem 3.1. in [17]), and a search algorithm should therefore be able to explore boundary regions. Table 2 lists the results of the one-sided Wilcoxon rank sum test applied on the outcome of this experiment. The tables show that the adaptive variants significantly outperform their non-adaptive counterparts on much more testfunctions (highlighted by bold numbers) than vice versa (see italic numbers).

Table 1. Experimental results on the 100- and 500-dimensional benchmarks. Average objective values and standard errors are shown. The best objective values are presented together with the function name.

100-dimensional benchmarks

	Sphere (0)	Rosenbrock (0)	Ackley (0)	Griewank (0)
Absorb-S	6.069e-06±1.175e-07	191.06±8.785	1.3959±0.11837	2.765e-03±7.761e-04
Random-S	6.078e-06±1.2964e-07	195.45±8.2053	1.7332±0.10218	3.751e-03±7.749e-04
Infinity-S	6.208e-06±1.4284e-07	221.06±7.0927	1.5847±0.1094	7.252e-03±2.529e-03
Absorb-A	1.047e-06±9.3267e-09	114.03±4.7795	3.7094e-06±1.3119e-08	2.709e-03±8.757e-04
Random-A	1.059e-06±1.0115e-08	120.75±4.5423	3.6963e-06±1.5878e-08	1.479e-03±6.171e-04
Infinity-A	1.044e-06±9.9384e-09	107.08±3.6154	3.7032e-06±1.7861e-08	5.928e-04±3.418e-04

	Rastrigin (0)	Schwefel (≈ -41898.3)	f1 (-450)	f2 (-450)
Absorb-S	282.2±4.504	-27841±242.12	-450±0	25223±1166.3
Random-S	239.56±4.6447	-24826±207.29	-450±0	23035±973.35
Infinity-S	276.34±5.6791	-23705±234.82	-450±0	9951.7±395.9
Absorb-A	93.91±2.3929	-24430±180.2	-450±0	286.74±27.998
Random-A	87.716±2.1884	-22341±170.7	-450±0	1685.6±733.36
Infinity-A	93.499±2.3445	-22837±187.56	-450±0	269.45±33.939

	f3 (-450)	f5 (-310)	f6 (390)	f9 (-330)
Absorb-S	39163000±1884400	27545±501.09	594.24±8.27	29.387±8.8699
Random-S	78163000±5844400	30534±497.03	590.71±8.142	48.163±6.6587
Infinity-S	17457000±727450	33148±485.31	593.54±6.4731	226.72±12.379
Absorb-A	8120300±222840	23100±333.67	511.6±5.3632	147.44±10.078
Random-A	1.1002e+08±12771000	35603±639.93	530.09±6.3345	160.73±6.9788
Infinity-A	7234600±154480	26836±355.33	517.21±5.576	174.25±10.714

	f10 (-330)	f11 (-460)	f13 (-130)	f14 (-300)
Absorb-S	30.313±7.7389	217.56±0.81253	-69.427±1.9555	-253.47±0.058467
Random-S	98.299±8.9825	221.94±0.88144	-68.36±1.5856	-253.7±0.068529
Infinity-S	261.59±13.889	224.14±0.94733	-69.666±1.6095	-253.54±0.053858
Absorb-A	85.767±9.1288	185.36±1.1865	-112.95±0.45879	-254.01±0.070051
Random-A	149.52±10.355	195.1±1.2822	-113.28±0.36477	-254.13±0.060984
Infinity-A	195.87±10.924	185.43±1.2738	-113.12±0.48073	-254.15±0.085889

500-dimensional benchmarks

	Sphere (0)	Rosenbrock (0)	Ackley (0)	Griewank (0)
Absorb-S	1669.3±236.6	397590±32299	9.1571±0.13512	15.692±1.9949
Random-S	1523.8±136.19	877140±157910	9.4522±0.16661	15.428±1.1851
Infinity-S	780890±5098.7	2.3589e+09±22756000	20.057±0.018442	7023.7±47.232
Absorb-A	0.54291±0.15378	2188±45.632	1.712±0.030091	0.19666±0.050727
Random-A	0.8328±0.35029	2232.3±53.622	1.7225±0.028023	0.12122±0.029484
Infinity-A	0.4738±0.095705	2210.7±36.524	1.762±0.025693	0.09308±0.017517

	Rastrigin (0)	Schwefel (≈ -209491.5)	f1 (-450)	f2 (-450)
Absorb-S	1917.1±29.846	-122640±847.19	3260.3±296.88	2369600±32679
Random-S	1642±25.489	-106940±647.94	31185±1302.3	2556700±40344
Infinity-S	6704.4±21.475	-11628±276.33	2174600±7890.1	97764000±5238200
Absorb-A	422.85±13.534	-99575±981.1	-430.82±4.9622	1071200±18936
Random-A	375.63±10.86	-88646±985.12	201560±4265.1	2999200±123610
Infinity-A	408.62±11.456	-93069±973.41	-414.15±20.668	866340±10705

	f3 (-450)	f5 (-310)	f6 (390)	f9 (-330)
Absorb-S	1.6426e+09±28943000	-310±0	6.1811e+08±86998000	3288.4±40.28
Random-S	3.7348e+09±1.2208e+08	-310±0	1.5943e+09±1.5081e+08	3218.5±29.644
Infinity-S	1.1268e+11±1.2292e+09	-310±0	2.0502e+12±1.4785e+10	10220±23.611
Absorb-A	4.795e+08±5962900	-310±0	705350±83435	3306.4±18.958
Random-A	6.3402e+09±1.2971e+08	-310±0	4.5585e+09±4.7332e+08	3417±18.45
Infinity-A	4.2855e+08±6008000	-310±0	309230±58497	3352.9±20.115

	f10 (-330)	f11 (-460)	f13 (-130)	f14 (-300)
Absorb-S	4959.3±74.314	909.67±2.2174	10650±1854.9	-54.763±0.050616
Random-S	5560.7±74.102	918.3±1.844	3594.6±227.73	-54.835±0.062736
Infinity-S	16972±48.023	1025.1±1.5096	45526000±1002400	-52.774±0.05418
Absorb-A	7881.6±46.142	751±3.701	404.46±10.313	-60.306±0.1963
Random-A	7975.7±64.224	849.35±5.8459	409.14±11.209	-59.999±0.15094
Infinity-A	7803.3±39.094	756.8±3.822	406.86±11.149	-60.105±0.14006

Table 2. Summary of one-sided Wilcoxon rank sum test with significance level 0.01 for all benchmarks. For each algorithmic combination (A, B), this matrix shows how often A performed significantly better than B. *Example:* Entry 6 in the first row shows that Absorb-S significantly outperformed Infinity-S on 6 benchmarks.

100-dimensional benchmarks						
	1	2	3	4	5	6
Absorb-S (1)	0	5	6	*3*	5	3
Random-S (2)	1	0	5	1	*4*	3
Infinity-S (3)	2	2	0	0	3	*1*
Absorb-A (4)	**12**	12	14	0	7	3
Random-A (5)	10	**10**	12	0	0	1
Infinity-A (6)	11	12	**14**	1	4	0
Total number of benchmarks: 16						

500-dimensional benchmarks						
	1	2	3	4	5	6
Absorb-S (1)	0	8	15	*2*	7	3
Random-S (2)	2	0	15	3	*6*	3
Infinity-S (3)	0	0	0	0	0	*0*
Absorb-A (4)	**12**	12	15	0	7	1
Random-A (5)	8	**8**	15	1	0	1
Infinity-A (6)	12	12	**15**	3	7	0
Total number of benchmarks: 16						

The second question was whether the adaptive PSO is more invariant to the bound handling strategy compared to the standard PSO. We already mentioned an exception: *Random-A* was sometimes significantly outperformed by *Absorb-A* and *Infinity-A*, especially in the 500-dimensional case. Hence, our adaptive mechanism sometimes cannot prevent that particles are not able to converge to boundary regions when using *Random* bound handling. However, a comparison of the average objective values presented in Table 1 shows that invariance with respect to the bound handling strategy was nevertheless improved: The obtained avarage objective values of the three adaptive variants are mostly more similar to one another than the ones of the non-adaptive PSO. This observation is confirmed by the results of the Wilcoxon rank sum test presented in Table 2: Especially in the 500-dimensional case, *Absorb-S*, *Random-S*, and *Infinity-S* showed significant performance differences among each other. *Infinity-S* was outperformed by the other two variants in 15 of 16 functions. Although performance differences can be observed among the adaptive PSOs as well, they occur less often. *Absorb-A* and *Infinity-A* provided solutions of similar quality on most 500-dimensional problems. We conclude that we partly achieved the second goal, but there is still room for improvement. Experiment 5 will show that initializing the velocity length to a smaller value can enhance invariance with respect to the bound handling strategy but also deteriorate performance.

Summarized, PSO with velocity adaptation yielded superior results than a standard PSO on most of the investigated 100- and 500-dimensional benchmarks. Invariance with respect to the bound handling mechanism was enhanced.

5.2 Experiment 2: Success Probability

The adaptation mechanism of our adaptive PSO is similar to the well-known *1/5 rule of Rechenberg* [28]: Whenever the success rate in the past n iterations exceeds 1/5, the velocity is doubled, otherwise it is halved. Of course, other values besides 1/5 can be chosen as success probablity threshold. We additionally investigated the following settings: 0.01, 0.1, 0.5, and 0.8. In Figures 1 and 2,

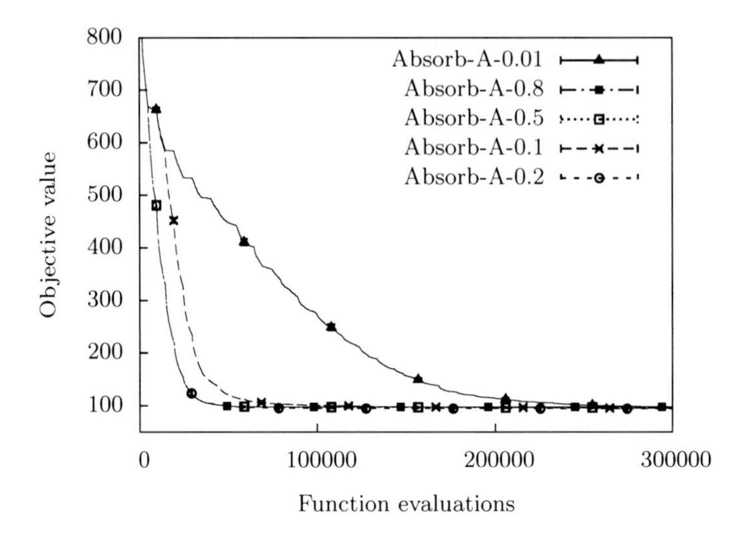

Fig. 1. Representative run (benchmark Rastrigin) of the adaptive PSO with different success probabilities (Experiment 2). In the plot, average objective values are shown. Vertical bars (very small) depict the standard error.

two representative runs are shown. Setting the success probability to 0.5 or 0.8 performed bad on most benchmarks, and can therefore not be recommended. The performance of 0.1 and 0.2 was quite similar with slight advantage for 0.2 considering average objective values and the result of the Wilcoxon rank sum tests for all functions. The behavior of choosing 0.01 differs from the other settings: Convergence is very slow, often too slow to obtain satisfactory results using 300,000 function evaluations (which already is quite much). When applying such a low success probability, we expect velocities to grow large. This can lead to a rather random behavior of the whole swarm and prevent convergence, as shown in Figure 1 and observed on approximately 7 functions. On the other hand, the same behavior is advantageous for other problems, see Figure 2: In these cases, using high success probabilities led to premature convergence, as velocities are decreased further and further if the swarm is not successful. However, the swarms using a threshold success probability of 0.01 kept exploring and finally found solutions of much better quality on approximately 6 of the investigated 16 benchmark functions. Summarized, both very low and very high success probability thresholds cannot be recommended. Using a very low success probability sometimes leads to exceptionally good results, but also often deteriorates particle swarm performance. However, values of 0.1 or 0.2 delivered solutions of good quality on most investigated problems.

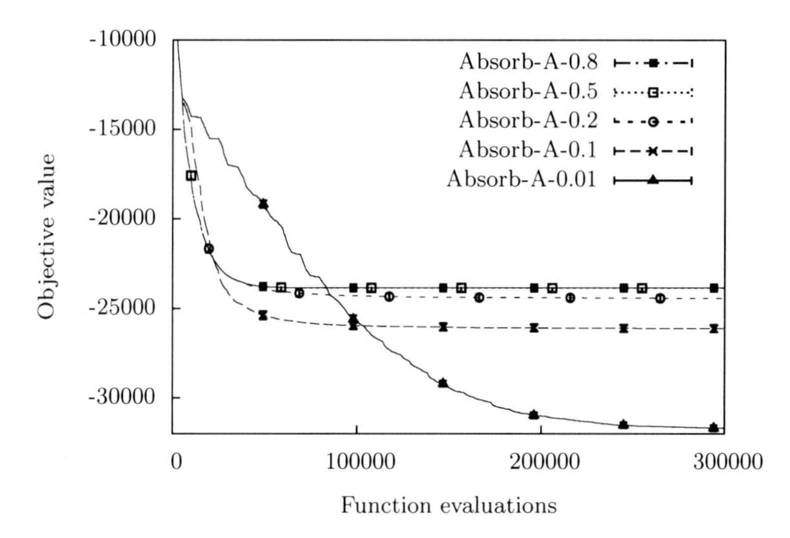

Fig. 2. Representative run (benchmark Schwefel) of the adaptive PSO with different success probabilities (Experiment 2). In the plot, average objective values and standard errors (vertical bars) are shown.

5.3 Experiment 3: Local Velocity Lengths

The adaptive PSO as presented in Algorithm 1 uses a global velocity length: The velocities of all particles are scaled to the same size. Alternatively, each particle could have its own value for the velocity length which is determined by its personal success. We tested this alternative procedure by using a resticted testbed of 6 benchmark functions with different characteristics: Rosenbrock, Rastrigin, Schwefel, f2, f9, and f10. The result of the Wilcoxon rank sum test is depicted in Table 3 (left), and show that the use of a global velocity length can be slightly preferred: Comparing only the variants which used the same bound handling procedure, the global variant significantly outperformed the local one on two functions (highlighted by bold numbers) whereas the local variant never provided significantly superior results than the respective global procedure (highlighted by italic numbers).

5.4 Experiment 4: Scaling Strategy

In our adaptive PSO, velocities are scaled such that their lengths exactly match the current velocity size. Traditional particle swarm optimizers sometimes use a constant maximum velocity length per component (e.g., [11]), which is often set to a fraction of the respective search space range.

Hence, it seems intuitive to only use an adaptive *maximum* velocity size instead of scaling velocity vectors up and down to a certain length. Using an

Table 3. Summary of one-sided Wilcoxon rank sum test with significance level 0.01 for Experiment 3 (left) and Experiment 4 (right). For each algorithmic combination (A, B), this matrix shows how often A performed significantly better than B.

Experiment 3

	1	2	3	4	5	6
Absorb-A (1)	0	3	2	**2**	4	3
Random-A (2)	0	0	1	0	**2**	2
Infinity-A (3)	0	1	0	2	4	**2**
Absorb-A-local (4)	*0*	1	1	0	2	2
Random-A-local (5)	0	*0*	0	0	0	1
Infinity-A-local (6)	0	1	*0*	0	1	0
Total number of benchmarks: 6						

Experiment 4

	1	2	3	4	5	6
Absorb-A (1)	0	7	3	**10**	13	12
Random-A (2)	0	0	1	9	**11**	9
Infinity-A (3)	1	4	0	11	12	**12**
Absorb-scale2 (4)	*0*	3	1	0	5	2
Random-scale2 (5)	0	*0*	0	0	0	0
Infinity-scale2 (6)	0	2	*0*	1	3	0
Total number of benchmarks: 16						

Table 4. Experimental results for Experiment 4 (100-dimensional benchmarks). Average objective values and standard errors are shown. The best objective values are presented together with the function name.

Experiment 4

	Sphere (0)	Rosenbrock (0)	Ackley (0)	Griewank (0)
Absorb-A	1.047e-06±9.3267e-09	114.03±4.7795	3.7094e-06±1.3119e-08	2.709e-03±8.757e-04
Random-A	1.059e-06±1.0115e-08	120.75±4.5423	3.6963e-06±1.5878e-08	1.479e-03±6.171e-04
Infinity-A	1.044e-06±9.9384e-09	107.08±3.6154	3.7032e-06±1.7861e-08	5.928e-04±3.418e-04
Absorb-scale2	1.072e-06±1.0606e-08	152.28±6.966	1.0596±0.13068	5.049e-03±2.16e-03
Random-scale2	1.086e-06±8.7608e-09	151.39±6.7687	1.1619±0.13762	3.056e-03±1.127e-03
Infinity-scale2	1.084e-06±1.0717e-08	158.41±6.968	1.1297±0.12853	3.681e-03±1.486e-03
	Rastrigin (0)	Schwefel (≈ -41898.3)	f1 (-450)	f2 (-450)
Absorb-A	93.91±2.3929	-24430±180.2	-450±0	286.74±27.998
Random-A	87.716±2.1884	-22341±170.7	-450±0	1685.6±733.36
Infinity-A	93.499±2.3445	-22837±187.56	-450±0	269.45±33.939
Absorb-scale2	91.153±2.5799	-24151±184.06	-450±0	659.12±64.551
Random-scale2	88.533±1.9747	-22877±207.74	-450±0	5027.4±2098.2
Infinity-scale2	90.976±2.459	-22682±241.05	-450±0	690.42±50.825
	f3 (-450)	f5 (-310)	f6 (390)	f9 (-330)
Absorb-A	8120300±222840	23100±333.67	511.6±5.3632	147.44±10.078
Random-A	1.1002e+08±12771000	35603±639.93	530.09±6.3345	160.73±6.9788
Infinity-A	7234600±154480	26836±355.33	517.21±5.576	174.25±10.714
Absorb-scale2	8569800±275270	27467±486.28	548.19±6.4877	242.39±7.1574
Random-scale2	1.5664e+08±15359000	38356±578.8	554.78±7.998	260.6±7.366
Infinity-scale2	7570700±228490	29642±422.59	554.55±6.0166	254.21±7.2082
	f10 (-330)	f11 (-460)	f13 (-130)	f14 (-300)
Absorb-A	85.767±9.1288	185.36±1.1865	-112.95±0.45879	-254.01±0.070051
Random-A	149.52±10.355	195.1±1.2822	-113.28±0.36477	-254.13±0.060984
Infinity-A	195.87±10.924	185.43±1.2738	-113.12±0.48073	-254.15±0.085889
Absorb-scale2	494.42±14.795	215.14±1.0666	-96.972±1.351	-253.24±0.07717
Random-scale2	553.59±12.038	218.34±1.2762	-95.976±1.6981	-253.32±0.067686
Infinity-scale2	538.2±14.136	214.85±1.4639	-98.821±1.0518	-253.27±0.06428

adaptive maximum velocity means that velocities which exceed the current velocity size are shortened to the desired length. However, velocities are not scaled up. This setting, denoted as *scale2*, was tested and compared with our original adaptive PSO as presented in Algorithm 1. The results of the one-sided Wilcoxon rank sum test are shown in Table 3 (right). Using a maximum velocity length instead of the proposed up- and downscaling mechanism led to significantly worse solution quality on many of the tested benchmark problems (highlighted by bold numbers in Table 3). The performance difference is also clearly visible when considering the obtained average objective values for both settings, shown in Table 4. Figure 3 shows the convergence plot for f10 as an example. We conclude that the proposed up- and down-scaling mechanism is an important feature of our adaptive PSO to control the velocity sizes as proposed by the adaptive strategy.

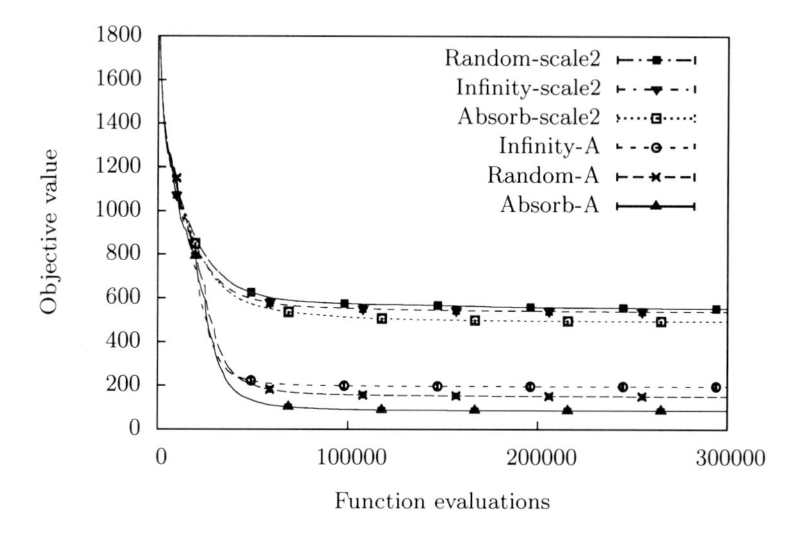

Fig. 3. Convergence plot of different scaling strategies (Experiment 4) on benchmark function f10.

5.5 Experiment 5: Initialization of Velocity Length l

The initial velocity length determines the exploration behavior of the particle swarm at the beginning of the optimization process. Choosing very small initial velocities can result in premature convergence on local optima. Two different settings were tested: $l = r$ and $l = r/\sqrt{n}$, where $[-r \dots r]^n$ is the parameter space of the problem to be solved. The second setting is denoted as *init2*. Initializing the velocity length to $l = r$ provided better results on most benchmark functions. In the other cases, results of similar quality were achieved. Especially for Schwefel, f9 and f10, great performance losses could be recognized when initializing the velocity length to $l = r/\sqrt{n}$, as shown in Figure 4. However, as indicated by the theoretical study in Section 3, choosing $l = r/\sqrt{n}$ (which corresponds to the order r/n per dimension) results in a PSO algorithm which is more invariant with respect to the bound handling method: Table 5 shows that *Absorb-init2*, *Random-init2*, and *Infinity-init2* less often significantly differed in performance among each other than *Absorb-A*, *Random-A*, and *Infinity-A*. However, the

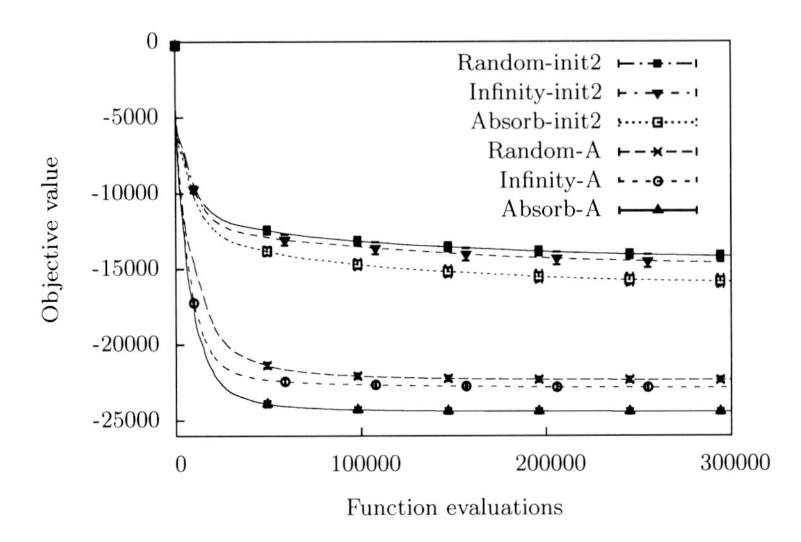

Fig. 4. Convergence plot for using different initial velocity lengths (Experiment 5) on benchmark function Schwefel.

difference is rather small so that initializing velocities to $l = r$ is to be preferred due to the desired explorative behavior of a search algorithm at the beginning of the optimization process. Considering Table 5, it is clear that the *Absorb-A*, *Random-A*, and *Infinity-A* variants more often significantly outperform the corresponding *init2* algorithms than vice versa. In the three cases in which the *init2* variants performed significantly better than the adaptive PSO with $l = r$ (Sphere, Ackley, and Griewank), both settings provided a satisfactory average solution quality very close to the global optimum.

Table 5. Summary of one-sided Wilcoxon rank sum tests with significance level 0.01. For each algorithmic combination (A, B), this matrix shows how often A performed significantly better than B.

Experiment 5

	1	2	3	4	5	6
Absorb-A (1)	0	7	3	**8**	9	7
Random-A (2)	0	0	1	6	**9**	6
Infinity-A (3)	1	4	0	8	9	**9**
Absorb-init2 (4)	*3*	6	3	0	4	0
Random-init2 (5)	3	*3*	3	0	0	0
Infinity-init2 (6)	3	6	*3*	0	3	0
Total number of benchmarks: 16						

6 Conclusion

Particle swarm optimization algorithms have found many applications in solving continuous optimization problems. Often such problems have boundary constraints and it has been observed that PSO algorithms have difficulties when dealing with such problems as many particles leave the feasible region of the search space due to high velocities. Because of this, different bound handling methods have been proposed in the literature and it has been observed that these methods have a large impact on the success of particle swarm optimization algorithms.

In this chapter, we have examined how velocity adaptation can help PSO algorithm to deal with boundary constraints. First, we explained the effect of velocity values from a theoretical point of view. Based on these observations, we examined a particle swarm optimization algorithm which uses velocity adaptation to deal with bound-constrained search spaces. This algorithm deals with boundary constraints by decreasing the velocities when no progress can be achieved. Our experimental analysis shows that the influence of the used bound handling method becomes less important when using velocity adaptation. Furthermore, the use of velocity adaptation leads to significantly better results for many benchmark functions in comparison to a standard particle swarm optimizer.

References

1. Aarts, E., Lenstra, J.K.: Local Search in Combinatorial Optimization. John Wiley & Sons, Chichester (2003)
2. Alvarez-Benitez, J.E., Everson, R.M., Fieldsend, J.E.: A MOPSO Algorithm Based Exclusively on Pareto Dominance Concepts. Evolutionlary Multi-Criterion Optimization, 459–473 (2005)
3. Bratton, D., Kennedy, J.: Defining a Standard for Particle Swarm Optimization. In: Proc. IEEE Swarm Intelligence Symp., pp. 120–127 (2007)
4. Clerc, M.: Tribes, a Parameter Free Particle Swarm Optimizer, `http://clerc.maurice.free.fr/pso/` (last checked: 22.12.2008)(2003)
5. Clerc, M.: Confinements and Biases in Particle Swarm Optimization (2006), `http://clerc.maurice.free.fr/pso/`
6. Clerc, M.: Particle Swarm Optimization. ISTE Ltd. (2006)
7. Clerc, M., et al.: Standard PSO 2007 (2007),`http://www.particleswarm.info` (standard_pso_2007.c)
8. Cui, Z., Cai, X., Zeng, J.: Stochastic velocity threshold inspired by evolutionary programming. In: Proceedings of the 2009 World Congress on Nature & Biologically Inspired Computing, pp. 626–631 (2009)
9. Cui, Z., Zeng, J., Sun, G.: Adaptive velocity threshold particle swarm optimization. In: Rough Sets and Knowledge Technology. Springer, Heidelberg (2006)
10. Dorigo, M., Stützle, T.: Ant Colony Optimization. MIT Press, Cambridge (2004)
11. Eberhart, R.C., Shi, Y.: Comparing Inertia Weights and Constriction Factors in Particle Swarm Optimization. In: Proceedings of the 2000 Congress on Evolutionary Computation, pp. 84–88 (2000)
12. Eiben, A.E., Smith, J.E.: Introduction to Evolutionary Computing, 2nd edn. Springer, Heidelberg (2007)

13. Fan, H.: A modification to particle swarm optimization algorithm. Engineering Computations 19(8), 970–989 (2002)
14. Fourie, P.C., Groenwold, A.A.: The particle swarm optimization algorithm in size and shape optimization. Structural and Multidisciplinary Optimization 23(4), 259–267 (2002)
15. Goldberg, D.E.: Genetic Algorithms in Search Optimization and Machine Learning. Addison-Wesley, Reading (1989)
16. Helwig, S., Neumann, F., Wanka, R.: Particle swarm optimization with velocity adaptation. In: Proceedings of the International Conference on Adaptive and Intelligent Systems (ICAIS 2009), pp. 146–151. IEEE, Los Alamitos (2009)
17. Helwig, S., Wanka, R.: Particle Swarm Optimization in High-Dimensional Bounded Search Spaces. In: Proc. IEEE Swarm Intelligence Symp., pp. 198–205 (2007)
18. Helwig, S., Wanka, R.: Theoretical Analysis of Initial Particle Swarm Behavior. In: Rudolph, G., Jansen, T., Lucas, S., Poloni, C., Beume, N. (eds.) PPSN 2008. LNCS, vol. 5199, pp. 889–898. Springer, Heidelberg (2008)
19. Hoos, H.H., Stützle, T.: Stochastic Local Search: Foundations & Applications. Elsevier / Morgan Kaufmann (2004)
20. Janson, S., Middendorf, M.: A hierarchical particle swarm optimizer and its adaptive variant. IEEE Transactions on Systems, Man, and Cybernetics, Part B 35(6), 1272–1282 (2005)
21. Kennedy, J., Eberhart, R.C.: Particle Swarm Optimization. In: Proc. of the IEEE Int. Conf. on Neural Networks, vol. 4, pp. 1942–1948 (1995)
22. Kennedy, J., Eberhart, R.C.: Swarm Intelligence. Morgan Kaufmann, San Francisco (2001)
23. Kennedy, J., Mendes, R.: Population Structure and Particle Swarm Performance. In: Proceedings of the IEEE Congress on Evol. Computation, pp. 1671–1676 (2002)
24. Miranda, V., Fonseca, N.: EPSO – Best-of-two-worlds Meta-heuristic Applied to Power System Problems. In: Proceedings of the 2002 IEEE Congress on Evolutionary Computation, vol. 2, pp. 1080–1085. IEEE Press, Los Alamitos (2002)
25. Miranda, V., Fonseca, N.: New evolutionary particle swarm algorithm (EPSO) applied to Voltage/Var control. In: Proceedings of the 14th Power Systems Computation Conference (2002)
26. Price, K.V., Storn, R.M., Lampinen, J.A.: Differential Evolution: A Practical Approach to Global Optimization. Springer, Heidelberg (2005)
27. Pulido, G.T., Coello, C.A.C., Santana-Quintero, L.V.: EMOPSO: A Multi-Objective Particle Swarm Optimizer with Emphasis on Efficiency. In: Obayashi, S., Deb, K., Poloni, C., Hiroyasu, T., Murata, T. (eds.) EMO 2007. LNCS, vol. 4403, pp. 272–285. Springer, Heidelberg (2007)
28. Rechenberg, I.: Evolutionsstrategie – Optimierung technischer Systeme nach Prinzipien der biologischen Evolution. Frommann-Holzboog, Verlag (1973)
29. Robinson, J., Rahmat-Samii, Y.: Particle Swarm Optimization in Electromagnetics. IEEE Trans. on Antennas and Propagation 52(2), 397–407 (2004)
30. Schwefel, H.-P.: Numerical optimization for computer models. John Wiley, Chichester (1981)
31. Schwefel, H.-P.P.: Evolution and Optimum Seeking: The Sixth Generation. John Wiley & Sons, Inc., New York (1993)
32. Suganthan, P.N., Hansen, N., Liang, J.J., Deb, K., Chen, Y., Auger, A., Tiwari, S.: Problem Definitions and Evaluation Criteria for the CEC 2005 Special Session on Real-Parameter Optimization. KanGAL Report 2005005. Nanyang Technological University, Singapore (2005)

33. Takahama, T., Sakai, S.: Solving constrained optimization problems by the ϵ constrained particle swarm optimizer with adaptive velocity limit control. In: Proceedings of the 2006 IEEE International Conference on Cybernetics and Intelligent Systems, pp. 1–7 (2006)
34. Wilcoxon, F.: Individual Comparisons by Ranking Methods. Biometrics Bulletin 1(6), 80–83 (1945)
35. Zhang, W.-J., Xie, X.-F., Bi, D.-C.: Handling Boundary Constraints for Numerical Optimization by Particle Swarm Flying in Periodic Search Space. In: Proc. of the IEEE Congress on Evol. Computation, vol. 2, pp. 2307–2311 (2004)

Integral-Controlled Particle Swarm Optimization

Zhihua Cui, Xingjuan Cai, Ying Tan, and Jianchao Zeng

Complex System and Computational Intelligence Laboratory,
Taiyuan University of Science and Technology,
No.66 Waliu Road, Wanbailin District, Taiyuan, Shanxi, China, 030024
cuizhihua@gmail.com, cai_xing_juan@sohu.com, zengjianchao@263.net

Summary. Particle swarm optimization (PSO) is a novel population-based stochastic optimization algorithm. However, it gets easily trapped into local optima when dealing with multi-modal high-dimensional problems. To overcome this shortcoming, two integral controllers are incorporated into the methodology of PSO, and the integral-controlled particle swarm optimization (ICPSO) is introduced. Due to the additional accelerator items, the behavior of ICPSO is more complex, and provides more chances to escaping from a local optimum than the standard version of PSO. However, many experimental results show the performance of ICPSO is not always well because of the particles' un-controlled movements. Therefore, a new variant, integral particle swarm optimization with dispersed accelerator information (IPSO-DAI) is designed to improve the computational efficiency. In IPSO-DAI, a predefined predicted velocity index is introduced to guide the moving direction. If the average velocity of one particle is superior to the index value, it will choice a convergent manner, otherwise, a divergent manner is employed. Furthermore, the choice of convergent manner or divergent manner for each particle is associated with its performance to fit different living experiences. Simulation results show IPSO-DAI is more effective than other three variants of PSO especially for multi-modal numerical problems. The IPSO-DAI is also applied to directing the orbits of discrete chaotic dynamical systems by adding small bounded perturbations, and achieves the best performance among four different variants of PSO.

1 Introduction

Particle swarm optimization (PSO) [1][2] is a novel population-based optimization technique firstly proposed in 1995. It simulates the animal social behaviors e.g. birds flocking, fish schooling and insects herding. Due to the simple concepts and fast convergent speed, it has been widely applied to many areas such as power system [3][4][5], structural damage identification [6], nonlinear system identification [7], ice-storage air conditioning system [8], financial classification problem [9], steel annealing process [10], portfolio optimization [11] , magnetoencephalography [12], layout optimization problem[13], wireless sensor networks[14], molecular docking[15] and multi-objective flexible job-shop scheduling problems[16].

In PSO methodology, each individual (called particle, in briefly) represents a potential solution without mass and volume, and flies within the search

B.K. Panigrahi, Y. Shi, and M.-H. Lim (Eds.): Handbook of Swarm Intelligence, ALO 8, pp. 175–199.
springerlink.com

space to seek the food (optimum). Suppose $\overrightarrow{x}_j(t) = (x_j^1(t), x_j^2(t), ..., x_j^D(t))$ (D denotes the dimension of problem space) is the position vector of jth particle at generation t, then it flies according to the following manner:

$$x_j^k(t+1) = x_j^k(t) + v_j^k(t+1), \quad k = 1, 2, ..., D \tag{1}$$

where the symbol $\overrightarrow{v}_j(t+1) = (v_j^1(t+1), v_j^2(t+1), ..., v_j^D(t+1))$ represents the velocity of particle j at time $t+1$, and is updated by:

$$v_j^k(t+1) = wv_j^k(t) + c_1 r_1(p_j^k(t) - x_j^k(t)) + c_2 r_2(p_g^k(t) - x_j^k(t)) \tag{2}$$

where $\overrightarrow{p}_j(t) = (p_j^1(t), p_j^2(t), ..., p_j^D(t))$ is the location with the most food resource particle j has been visited, $\overrightarrow{p}_g(t) = (p_g^1(t), p_g^2(t), ..., p_g^D(t))$ represents the best position found by the entire swarm. Inertia weight w is a positive number within 0 and 1. Cognitive learning factor c_1 and social learning c_2 are known as accelerator coefficients, r_1 and r_2 are two random numbers generated with uniform distribution within $(0, 1)$.

To keep the stability, a predefined velocity threshold v_{max} is used to limit the moving size of velocity vector such as:

$$|v_j^k(t+1)| \leq v_{max} \tag{3}$$

As a new swarm intelligence technique, it has attracted many researchers to focus their attention to this new area. J.Yisu et al. [17] proposed a PVPSO variant in which a distribution vector is used in the update of velocity, and can be adjusted automatically according to the distribution of particles in each dimension. To increase the convergence speed, Arumugam [18] proposed a new variant combining the crossover operator and root mean square strategy(RMS). S.H.Ling [19] designed a new mutation operator incorporating the properties of the wavelet theory to enhance the PSO performance so that it can explore the solution space more effectively to reach the solution. From the convergence viewpoint, Cui and Zeng designed a global convergence algorithm – stochastic particle swarm optimization [20], and it has been demonstrated that the mentioned algorithm converges with a probability of one towards the global optimum. Liu et al.[21][22][23] designed a turbulent particle swarm optimization by introducing a fuzzy logic controller to adjust the velocity threshold, simulation results show it is effective. Traditional PSO algorithms have been known to perform poorly especially on high dimensional problems [24], to combat this issue, Li and Yao [25] proposed a cooperative coevolving particle swarm optimization (CCPSO) algorithm incorporating random grouping and adaptive weighting, two techniques that have been shown to be effective for handling high dimensional problems. Huang and Mohan [26] proposed a micro-particle swarm optimizer to improve the performance for high dimensional optimization problems with a small population number. By introducing the fitness uniform selection strategy with a weak selection pressure and random walk strategy, Cui et al. [27] proposed a new variant used for high dimensional benchmarks with the dimension up to 3000. There are still many other modified methods [28][29] [30][31][32][33][34][35][36][37], due to the page limitation, the details of these modifications please refer to the corresponding references.

With z-translation, the structure of standard particle swarm optimization is viewed as a two-inputs, one-output feedback system. This phenomenon implies the controllers can be incorporated into the PSO structure to construct a new PSO model. As an exmaple, integral-controlled particle swarm optimization (ICPSO) [38] is designed by incorporating two integral controllers to increase the population diversity. However, due to the restriction of integral controllers, the performance is not well in some cases because of the small population diversity in the half past period. Therefore, a loose guide is added to make the trajectory is controlled partially for each particle to provide a larger diversity especially in the final stage.

The rest of this paper was organized as follows: section 2 provides a theoretical analysis for standard particle swarm optimization. In section 3, the details of the integral-controlled particle swarm optimization (ICPSO) is presented, the stability analysis and inertia weight selection strategy are also discussed. Furthermore, one extension, integral particle swarm optimization with dispersed accelerator information (IPSO-DAI) is presented in section 4.

2 Control Analysis for Standard Particle Swarm Optimization

To analyze the structure of the standard version of particle swarm optimization (SPSO), Eq.(1) is changed to $v_j^k(t+1) = x_j^k(t+1) - x_j^k(t)$, and is substituted to Eq.(2), then we have

$$\begin{aligned} x_j^k(t+1) = (w+1)x_j^k(t) - wx_j^k(t-1)) + \varphi_1(p_j^k(t) - x_j^k(t)) \\ +\varphi_2(p_g^k(t) - x_j^k(t)) \end{aligned} \tag{4}$$

where $\varphi_1 = c_1r_1$, $\varphi_2 = c_2r_2$, $\varphi = \varphi_1 + \varphi_2$.

Making z-transformation on Eq.(4), resulting

$$\begin{aligned} X_{jk}(z) = \frac{z}{z^2-(w+1)z+w}[\varphi_1(P_{jk}(z) - X_{jk}(z)) \\ +\varphi_2(P_{jk}(z) - X_{jk}(z))] \end{aligned} \tag{5}$$

Eq.(5) can be represented by Fig.1. Obviously, the behavior of SPSO is a two-order stochastic system, and can be influenced by adding some controllers (e.g. illustrated in Fig.2).

3 Integral-Controlled Particle Swarm Optimization

In this chapter, only the unconstrained minimization problems are considered:

$$\min \quad f(\overrightarrow{x}), \quad \overrightarrow{x} \in [Lower, Upper]^D \subseteq R^D \tag{6}$$

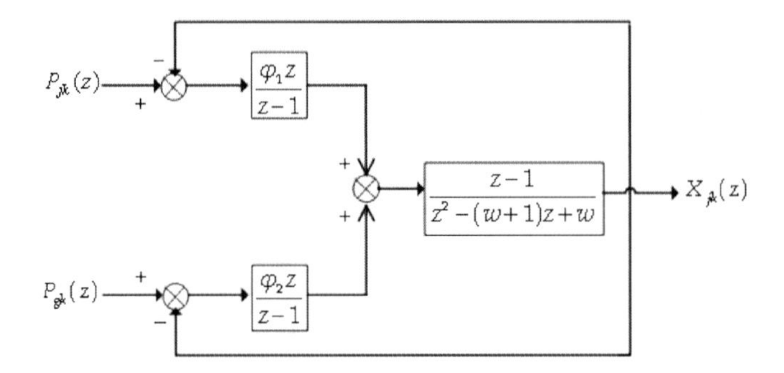

Fig. 1. System Structure of Standard PSO

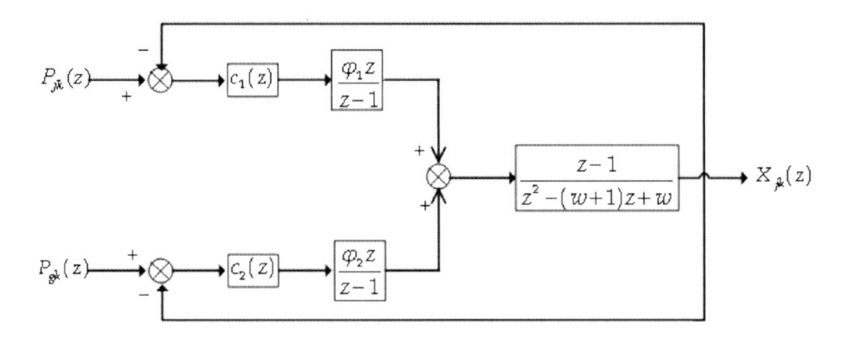

Fig. 2. System Structure of ICPSO

3.1 Update Equations

Integral-controlled particle swarm optimization (ICPSO) [38] is a new variant of particle swarm optimization by adding two integral controllers.

In control engineering, an integral controller is a feedback controller which drives the plant to be controlled with a weighted sum of the error (difference between the output and desired set-point) and the integral of that value. It is a special case of the common PID controller in which the derivative (D) of the error and the proportional (P) of the error are not used.

For an integral controller, there are two main advantages(properties):

(1)In an integral controller, steady state error should be zero.

(2)Integral control has a tendency to make a system slower.

From Fig.2, the input signals are $P_{jk}(z)$ and $P_{gk}(z)$, respectively, the output signal is $X_{jk}(z)$, and the error is the weighted sum of $P_{jk}(z)-X_{jk}(z)$ and $P_{gk}(z)-X_{jk}(z)$. Therefore, the steady state error with zero value (showed in property one) means that $P_{jk}(z) = P_{gk}(z) = X_{jk}(z)$, this is the necessary condition to guarantee the convergence of PSO.

The phenomenon illustrated in property two provides a complex behavior. As a stochastic system, slower convergent speed means more iterations are needed to reach the optimum point, therefore, it has more opportunities to escaping from a local optimum.

The z-transformation of integral controllers is $\frac{z}{z-1}$, and the corresponding updated equation of ICPSO is

$$X_{jk}(z) = \frac{z^2}{z^3 - (w+2)z^2 + (2w+1)z - w}[\varphi_1(P_{jk}(z) - X_{jk}(z)) + \varphi_2(P_{jk}(z) - X_{jk}(z))] \quad (7)$$

Applying inverse z-transformation, and combining the same items, the final updated equations of ICPSO are:

$$a_j^k(t+1) = wa_j^k(t) + c_1r_1(p_j^k(t) - x_j^k(t)) + c_2r_2(p_g^k(t) - x_j^k(t)) \quad (8)$$

$$v_j^k(t+1) = v_j^k + a_j^k(t+1) \quad (9)$$

$$x_j^k(t+1) = x_j^k + v_j^k(t+1) \quad (10)$$

Similarly with velocity threshold v_{max}, a predefined constant a_{max} is introduced to control the size of accelerator:

$$|a_j^k(t+1)| \leq a_{max} \quad (11)$$

Different from SPSO, there is additional accelerator item in ICPSO. This phenomenon means that ICPSO is a three-order stochastic system, and provides a more complex behavior than SPSO.

3.2 Stability Analysis

From Fig.2, the open-loop z-tranfer function for input $P_{jk}(z)$ is

$$G_{k1}(z) = \frac{\varphi_1 z^2}{(z-1)\cdot[z^2 - (1+w)z + w]} \quad (12)$$

The corresponding eigen equation is

$$1 + G_{k1}(z) = 0 \quad (13)$$

Combining the same items, we have

$$z^3 + (\varphi_1 - w - 2)z^2 + (2w+1)z - w = 0 \quad (14)$$

After applying the translation $z = \frac{y+1}{y-1}$, the above equation is changed to

$$\varphi_1 y^3 + \varphi_1 y^2 + (4 - 4w - \varphi_1)y + (4 + 4w - \varphi_1) = 0 \quad (15)$$

According to the Routh Theorem, the following system

$$b_0y^3 + b_1y^2 + b_2y + b_3 = 0 \tag{16}$$

is called stable if and only if the following conditions are satisfied:
(1) Coefficients b_0, b_1, b_2 and b_3 are all positive;
(2) $b_1b_2 - b_0b_3 > 0$.
For Eq.(15), the system is stable if and only if we have

$$b_0 > 0 \Rightarrow \varphi_1 > 0 \tag{17}$$

$$b_2 > 0 \Rightarrow 4 - 4w - \varphi_1 > 0 \tag{18}$$

$$b_3 > 0 \Rightarrow 4 + 4w - \varphi_1 > 0 \tag{19}$$

$$b_1b_2 - b_0b_3 > 0 \Rightarrow w < 0 \tag{20}$$

From Eq.(19) and Eq.(20), we have

$$\frac{1}{4}(\varphi_1 - 4) < w < 0 \tag{21}$$

With the same analysis, the stability condition with the input $P_{gk}(z)$ is

$$\frac{1}{4}(\varphi_2 - 4) < w < 0 \tag{22}$$

Therefore, the final stability condition for ICPSO is

$$\max\{\frac{1}{4}(\varphi_1 - 4), \frac{1}{4}(\varphi_2 - 4)\} < w < 0 \tag{23}$$

With this manner, the inertia weight is selected by

$$w = \max\{\frac{1}{4}(\varphi_1 - 4), \frac{1}{4}(\varphi_2 - 4)\} \cdot rand(0, 1) \tag{24}$$

3.3 Pseudo-codes of ICPSO

The detail steps of ICPSO are listed as follows:

Step1. Initializing each coordinate x_j^k and v_j^k sampling within $[x_{min}, x_{max}]$ and $[-v_{max}, v_{max}]$, respectively.

Step2. Computing the fitness value of each particle.

Step3. For $k'th$ dimensional value of $j'th$ particle, the personal historical best position p_j^k is updated as follows.

$$p_j^k = \begin{cases} x_j^k, & \text{if } f(\overrightarrow{x}_j) < f(\overrightarrow{p}_j) \ , \\ p_j^k & , \ \text{otherwise.} \end{cases} \tag{25}$$

Step 4. For $k'th$ dimensional value of $j'th$ particle, the global best position p_g^k is updated as follows.

$$p_g^k = \begin{cases} p_j^k, & \text{if } f(\overrightarrow{p}_j) < f(\overrightarrow{p}_g)\ , \\ p_g^k & , \text{ otherwise.} \end{cases} \tag{26}$$

Step5. Updating the accelerator vector with Eq.(8).

Step6. Updating the velocity and position vectors with Eq.(9) and Eq.(10).

Step7. If the criteria is satisfied, output the best solution; otherwise, goto step 2.

4 Integral Particle Swarm Optimization with Dispersed Accelerator Information

4.1 Predicted Accelerator Index

For integral controller, it controls the system according to the error feedback, therefore, integral-controlled particle swarm optimization (ICPSO) aims to increase the population diversity by adding two integral controllers. Therefore, to improve the performance of ICPSO, we need to give a roughly control for the trajectory of each particle.

For ICPSO, the most character is the addition of the accelerator item, and this new item conflicts the trajectory of each particle though it enhances the population diversity. Hence, one simple and intuitive method is to control the accelerator item so that we can obtain a roughly picture for the trajectory changes of each particle. Due to the complex problems, one restrict control may not meet the optimization tasks. Therefore, a predicted accelerator index is introduced by which we can control the particles' velocity trajectories smoothly.

Generally, in the first period, the accelerator item should be large enough to make a sufficient search to determine the region in which global optimum is fallen, then the accelerator item decreases to seek the exactly location of the global optimum. Following this idea, the predicted accelerator index is given in Figure 3.

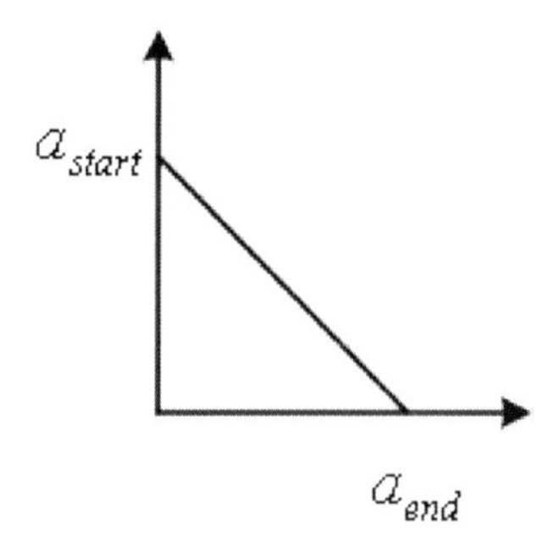

Fig. 3. Illustration for the Predefined Accelerator Index

The $x - axis$ coordinate denotes the generation, $y - axis$ represents the predicted accelerator value, and the predicted accelerator index is defined as follows:

$$a_{index}(t) = \begin{cases} 0 \quad , \quad \text{if } t > 0.8 \times T_{end}, \\ a_{start} - (a_{start} - a_{end}) \times \frac{t}{T_{end}}, \quad \text{otherwise} . \end{cases} \tag{27}$$

where T_{end} represents the largest iteration.

Because the accelerator is initialized within $[-a_{max}, a_{max}]$, while the constant a_{max} is always chosen as the upper bound of the domain $Upper$, the value a_{start} is selected as $Upper$.

For the choice of a_{end}, since the ICPSO algorithm is convergent, the limitation position of each particle is:

$$\lim_{t \to +\infty} x_{jk}(t+1) = \lim_{t \to +\infty} x_j^k + \lim_{t \to +\infty} v_j^k(t+1) \tag{28}$$

It implies $\lim_{t \to +\infty} v_j^k(t+1) = 0$, considering Eq.(8) may result

$$\lim_{t \to +\infty} v_{jk}(t+1) = \lim_{t \to +\infty} v_j^k + \lim_{t \to +\infty} a_j^k(t+1) \tag{29}$$

Hence,

$$\lim_{t \to +\infty} a_j^k(t+1) = 0 \tag{30}$$

Therefore, in this paper, $a_{end} = 0$.

However, even with this setting, there are still some particles can not stop in the final stage (the reason is partly due to the slow convergent speed of integral controller). This phenomenon implies that the predefined accelerator index should be 0 before the largest iteration. In this paper, this critical point is chosen as 80% of the largest iteration.

4.2 Phase Translation Principle of Cognitive Learning Factor

The predefined accelerator index only provides a weak control for each particle by translating among convergence phase and divergence phase. To choose the correct search pattern for each particle, one statistic value about current accelerator information is calculated, and is used to compare with the predefined accelerator index value. If this statistic value is larger than the predefined accelerator index value, the corresponding particle should make the convergence search to further decrease it's accelerator; otherwise, it may tend to make a divergence search to increase the accelerator value. This is the basic principle of this new algorithm – integral particle swarm optimization with dispersed accelerator information (IPSO-DAI).

The current proposed cognitive and social selection strategies have one common shortcoming: the parameter changing curve is predefined without any relationship between accelerator information and problem linkage [40][41][42][43]. This phenomenon makes the parameter automation is hardly to meet the requirements from the different problems.

In a IPSO-DAI algorithm, each particle analyzes the information gathered from its neighbors, then makes a movement guided by its decision. As a complex process, the individual decision plays an important role in animal collective behaviors. For example, for a group of birds (or fish), there exist many differences among individuals (particles). Firstly, in nature, there are many internal differences among birds, such as ages, catching skills, flying experiences, and muscles' stretching, etc. Furthermore, the current positions can also provide an important influence on individuals, e.g. individuals, lying in the side of the swarm, can make several choices differing from center others. Both of these differences mentioned above provide a marked contribution to the swarm complex behaviors.

For standard particle swarm optimization, each particle maintains the same flying (or swimming) rules according to (1), (2) and (3). At each iteration, the inertia weight w, cognitive learning factor c_1 and social learning factor c_2 are set the same values for all particles, thus the differences among particles are omitted. Since the complex swarm behaviors can emerge the adaptation, a more precise model, incorporated with the differences, can provide a deeper insight of swarm intelligence, and the corresponding algorithm may be more effective and efficient [39].

Since ICPSO aims to increase the population diversity, the computational efficiency is not very well. Therefore, in this paper, we will incorporate the dispersed method [39] into IPSO-DAI methodology to improve its performance.

For particle j, suppose $\overrightarrow{a}_j(t) = (a_j^1(t), a_j^2(t), ..., a_j^D(t))$ is the accelerator vector in iteration t, then, a statistical index – average accelerator of particle j at time t is computed by

$$a_{aver,j}(t) = \frac{\sum_{k=1}^{D} a_j^k(t)}{D} \tag{31}$$

This index is used to measure the average accelerator range for each particle. When compared with the predefined accelerator index, the following two cases occur:

Case1: the average accelerator of particle j is superior to the predicted accelerator index.

Because the predicted accelerator index is a predefined measure, this case implies the particle $j's$ accelerator is decreased slower than the expected degree. Therefore, a convergent setting for cognitive learning factor should be given to strength the convergent behavior.

Case2: the average velocity of particle j is worse than the predicted velocity index.

With the same reason, this case means the particle j should slow its velocity shorten ratio, in other words, a divergent setting for cognitive parameter should be provided to enhance its divergent behavior.

Similarly with ICPSO, the domain of cognitive coefficient is $[0.5, 2.5]$. Due to the dispersed control manner, the cognitive coefficient of each particle is selected according to the difference between the average accelerator information and predicted accelerator index. Since the general setting of cognitive parameter is 2.0 [42], in this paper, the cognitive parameter is chosen among $[0.5, 2.0]$ for the case1, while it is chosen from $[2.0, 2.5]$ for the case2.

Suppose the population size is m, and

$$a_{aver,worst}(t) = \max\{a_{aver,1}(t), a_{aver,2}(t), ..., a_{aver,m}(t)\} \tag{32}$$

$$a_{aver,best}(t) = \min\{a_{aver,1}(t), a_{aver,2}(t), ..., a_{aver,m}(t)\} \tag{33}$$

Let $c_{1,j}(t)$ denotes the cognitive learning factor of particle j at time t, then it is updated with the case1:

$$c_{1,j}(t) = \begin{cases} 0.5 & , \text{ if } v_{aver,worst} = v_{aver,best}, \\ 2.0 - 1.5 \times \frac{v_{aver,j} - v_{aver,best}}{v_{aver,worst} - v_{aver,best}}, & \text{otherwise} . \end{cases} \tag{34}$$

If case2 occurs, then the cognitive parameter is updated by

$$c_{1,j}(t) = \begin{cases} 2.0 & , \text{ if } v_{aver,worst} = v_{aver,best}, \\ 2.5 - 0.5 \times \frac{v_{aver,j} - v_{aver,best}}{v_{aver,worst} - v_{aver,best}}, & \text{otherwise} . \end{cases} \tag{35}$$

4.3 Time-Varying Social Learning Factor Adjustment Strategy

The current results [40][41][42][43] show the sum of these two accelerator coefficients (cognitive learning factor and social learning factor) should be a constant, e.g. 4.0, 5.976 and 3.0. Here, we chose the sum of $c_{1,j}$ and $c_{2,j}$ is 3.0. This setting only comes from the experimental tests, and the social coefficient adjustment strategy is defined as follows:

$$c_{2,j}(t) = 3.0 - c_{1,j}(t) \tag{36}$$

4.4 Mutation Strategy

To enhance the global capability, a mutation strategy [39], similarly with evolutionary computation, is introduced to enhance the ability escaping from the local optima.

At each time, particle j is uniformly random selected within the whole swarm, as well as the dimension k is also uniformly random selected, then, the v_j^k is changed as follows.

$$v_j^k = \begin{cases} 0.5 \times Upper \times r_1 & , \text{ if } r_2 < 0.5, \\ -0.5 \times Upper \times r_1 & \text{otherwise.} \end{cases} \tag{37}$$

where r_1 and r_2 are two random numbers generated with uniform distribution within $[0, 1]$.

4.5 Pseudo-codes of IPSO-DAI

The detail steps of IPSO-DAI are listed as follows:

Step1. Initializing each coordinate x_j^k and v_j^k sampling within $[x_{min}, x_{max}]$ and $[-v_{max}, v_{max}]$, respectively.

Step2. Computing the fitness value of each particle.

Step3. For $k'th$ dimensional value of $j'th$ particle, the personal historical best position p_j^k is updated as follows.

$$p_j^k = \begin{cases} x_j^k, & \text{if } f(\overrightarrow{x}_j) < f(\overrightarrow{p}_j) , \\ p_j^k & , \text{ otherwise.} \end{cases} \tag{38}$$

Step 4. For $k'th$ dimensional value of $j'th$ particle, the global best position p_g^k is updated as follows.

$$p_g^k = \begin{cases} p_j^k, & \text{if } f(\overrightarrow{p}_j) < f(\overrightarrow{p}_g) , \\ p_g^k & , \text{ otherwise.} \end{cases} \tag{39}$$

Step5. Computing the cognitive coefficient $c_{1,j}(t)$ of each particle according to Eq.(31)-(35).

Step6. Computing the social coefficient $c_{2,j}(t)$ of each particle with Eq.(36).

Step7. Updating the accelerator vector with Eq.(8) and Eq.(11);

Step8. Updating the velocity and position vectors with Eq.(9) and Eq.(10) in which the cognitive and social coefficients are changed with $c_{1,j}(t)$ and $c_{2,j}(t)$, respectively.

Step9. Making mutation operator described in section 3.4.

Step10. If the criteria is satisfied, output the best solution; otherwise, goto step 2.

4.6 Simulation Results

Test Functions

To fully testify the performance of IPSO-DAI and ICPSO, we employed a large set of standard benchmark functions [44] including five multi-modal functions with many local optima, and three multi-modal functions with a few local optima. The results of IPSO-DAI is compared with the standard PSO (SPSO), modified PSO with time-varying accelerator coefficients (MPSO-TVAC)[43] and the standard integral particle swarm optimization (IPSO)[38]. Furthermore, IPSO-DAI is applied to direct the orbits of discrete chaotic dynamical systems towards desired target region within a short time by adding only small bounded perturbations.

The chosen benchmarks are listed as follows, while more details can be found in reference [44]:

Sphere Model:

$$f_1(x) = \sum_{j=1}^{n} x_j^2$$

where $|x_j| \leq 100.0$, and

$$f_1(x^*) = f_1(0, 0, ..., 0) = 0.0$$

Schwefel Problem 2.22:

$$f_2(x) = \sum_{j=1}^{n} |x_j| + \prod_{j=1}^{n} |x_j|$$

where $|x_j| \leq 10.0$, and

$$f_2(x^*) = f_2(0, 0, ..., 0) = 0.0$$

Rosenbrock Function:

$$f_3(x) = \sum_{j=1}^{n-1} [100(x_{j+0} - x_j^2)^2 + (1 - x_j)^2]$$

where $|x_j| \leq 5.12$, and

$$f_3(x^*) = f_3(1, 1, ..., 1) = 0.0$$

Rastrigin Function:

$$f_4(x) = \sum_{j=1}^{n} [x_j^2 - 10cos(2\pi x_j) + 10]$$

where $|x_j| \leq 5.12$, and

$$f_4(x^*) = f_4(0, 0, ..., 0) = 0.0$$

Ackley Function:

$$f_5(x) = -20exp(-0.2\sqrt{\frac{1}{n}\sum_{j=1}^{n} x_j^2}) - exp(\frac{1}{n}\sum_{k=1}^{n} \cos 2\pi x_k) + 20 + e$$

where $|x_j| \leq 32.0$, and

$$f_5(x^*) = f_5(0, 0, ..., 0) = 0.0$$

Griewank Function:

$$f_6(x) = \frac{1}{4000}\sum_{j=1}^{n} x_j^2 - \prod_{j=1}^{n} \cos(\frac{x_j}{\sqrt{j}}) + 1$$

where $|x_j| \leq 600.0$, and

$$f_6(x^*) = f_6(0, 0, ..., 0) = 0.0$$

Penalized Function1:

$$f_7(x) = \frac{\pi}{30}\{10\sin^2(\pi y_1) + \sum_{i=1}^{n-1}(y_i - 1)^2[1 + 10\sin^2(\pi y_{i+1})]$$
$$+(y_n - 1)^2\} + \sum_{i=1}^{n} u(x_i, 10, 100, 4)$$

where $|x_j| \leq 50.0$, and

$$f_7(x^*) = f_7(1, 1, ..., 1) = 0.0$$

Penalized Function2:

$$f_8(x) = 0.1\{\sin^2(3\pi x_1) + \sum_{i=1}^{n-1}(x_i - 1)^2[1 + \sin^2(3\pi x_{i+1})]$$
$$+(x_n - 1)^2[1 + \sin^2(2\pi x_n)]\} + \sum_{i=1}^{n} u(x_i, 5, 100, 4)$$

where $|x_j| \leq 50.0$, and

$$f_8(x^*) = f_8(1, 1, ..., 1) = 0.0$$

Shekel's Foxholes Function:

$$f_9(x) = [\frac{1}{500} + \sum_{j=1}^{25} \frac{1}{j + \sum_{i=1}^{2}(x_i - a_{ij})^6}]^{-1}$$

where $|x_j| \leq 65.536$, and

$$f_9(x^*) = f_9(-32, -32) = 1.0$$

Hartman Family Function:

$$f_{10}(x) = -\sum_{i=1}^{4} c_i \exp[-\sum_{j=1}^{4} a_{ij}(x_j - p_{ij})^2]$$

where $x_j \in [0, 1]$, and

$$f_{10}(x^*) = f_{10}(0.114, 0.556, 0.852) = -3.86$$

Shekel's Family Function:

$$f_{11}(x) = -\sum_{i=1}^{10}[(x - a_i)(x - a_i)^T + c_i]^{-1}$$

where $x_j \in [0, 10]$, and

$$f_{11}(x^*) = f_{11}(a_1, a_2, a_{3,4}) = -10$$

Sphere model, Schwefel problem 2.22 and Rosenbrock are uni-model benchmarks, Rastrigin, Ackley, Griewank and two penalized functions are multi-modal benchmarks with many local optima, as well as Shekel's Foxholes Function, Hartman Family Function and Shekel's Family Function are multi-modal problems with a few local optima.

Experimental Setting

The coefficients of SPSO, MPSO-TVAC, ICPSO and IPSO-DAI are set as follows:

The inertia weight w is decreased linearly from 0.9 to 0.4 for all variants. Accelerator coefficients c_1 and c_2 are both set to 2.0 for SPSO, as well as in MPSO-TVAC and ICPSO, c_1 decreases from 2.5 to 0.5, while c_2 increases from 0.5 to 2.5. For IPSO-DAI, the cognitive learning factor and social learning factor are both adjusted as above mentioned manner. Total individuals are 100 for the first five benchmarks, as well as only 20 for the last three benchmarks, the velocity threshold v_{max} and the accelerator threshold a_{max} are both set to the upper bound of the domain. The dimensionality is 30, 100 and 300 for Rastrigin, Ackely, Griewank and two penalized functions, while 2 for Shekel's Foxholes Function, 3 for Hartman Family Function and 4 for Shekel's Family Function. For uni-modal benchmarks, the dimension is set to 30. In each experiment, the simulation run 30 times, while each time the largest iteration is $50 \times$ *dimension*.

Uni-Modal Functions

The comparison of the first three benchmars are listed in Tabl.1, in which *Fun.* represents the benchmark function, *Mean* denotes the average mean value, while *STD* denotes the standard variance.

Tab.1 shows that IPSO-DAI is not suit for uni-modal problems. For Sphere model, Schwefel problem 2.22 and Rosenbrock, ICPSO is always the best performance compared with SPSO, MPSO-TVAC and IPSO-DAI. This is partly because there is only one local optimum for uni-modal problems, in this case, there is no premature convergence phenomenon, and the local search can fast find the global optimum.

Table 1. Comparison Results for Final Three Benchmarks

Fun.	*Alg.*	Mean	STD
Sphere Model	SPSO	3.6606e-010	4.6577e-010
	MPSO-TVAC	4.1533e-028	2.2511e-027
	ICPSO	3.9903e-033	8.5593e-033
	IPSO-DAI	2.3519e-017	3.4807e-017
Schwefel Problem 2.22	SPSO	1.1364e-007	8.6198e-008
	MPSO-TVAC	8.9067e-010	3.8265e-009
	ICPSO	3.6417e-017	1.0102e-016
	IPSO-DAI	1.9136e-010	1.1546e-010
Rosenbrock Function	SPSO	5.6170e+001	4.3584e+001
	MPSO-TVAC	3.3589e+001	4.1940e+001
	ICPSO	2.2755e+001	1.9512e+001
	IPSO-DAI	3.4007e+001	2.7596e+001

Table 2. Comparison Results for Rastrigin Function

Dim.	*Alg.*	Mean	STD
30	SPSO	1.7961e+001	4.2276e+000
	MPSO-TVAC	1.5471e+001	4.2023e+000
	ICPSO	1.2138e+001	2.1390e+000
	IPSO-DAI	5.6899e+000	2.8762e+000
100	SPSO	9.3679e+001	9.9635e+000
	MPSO-TVAC	8.4478e+001	9.4568e+000
	ICPSO	7.8701e+001	7.3780e+000
	IPSO-DAI	4.4817e+001	1.2834e+001
300	SPSO	3.5449e+002	1.9825e+001
	MPSO-TVAC	2.7093e+002	3.7639e+001
	ICPSO	3.3172e+002	3.2496e+001
	IPSO-DAI	1.2302e+002	3.0750e+001

Table 3. Comparison Results for Ackley Function

Dim.	*Alg.*	Mean	STD
30	SPSO	5.8161e-006	4.6415e-006
	MPSO-TVAC	7.5381e-007	3.3711e-006
	ICPSO	9.4146e-015	2.9959e-015
	IPSO-DAI	1.3601e-012	9.0261e-013
100	SPSO	3.3139e-001	5.0105e-001
	MPSO-TVAC	4.6924e-001	1.9178e-001
	ICPSO	2.3308e+000	2.9163e-001
	IPSO-DAI	1.6494e-010	1.0251e-010
300	SPSO	2.8959e+000	3.1470e-001
	MPSO-TVAC	7.6695e-001	3.1660e-001
	ICPSO	6.7835e+000	5.1283e-001
	IPSO-DAI	2.7084e-003	5.9680e-003

Table 4. Comparison Results for Griewank Function

Dim.	*Alg.*	Mean	STD
30	SPSO	9.7240e-003	1.1187e-002
	MPSO-TVAC	9.4583e-003	1.6344e-002
	ICPSO	6.4048e-003	1.1566e-003
	IPSO-DAI	4.0226e-003	7.6099e-003
100	SPSO	2.1204e-003	4.3056e-003
	MPSO-TVAC	2.2598e-002	3.1112e-002
	ICPSO	8.5992e-003	1.6807e-002
	IPSO-DAI	1.4776e-002	2.8699e-002
300	SPSO	5.5838e-001	2.1158e-001
	MPSO-TVAC	3.9988e-001	2.4034e-001
	ICPSO	7.0528e-001	6.4000e-001
	IPSO-DAI	1.5201e-002	2.0172e-002

Table 5. Comparison Results for Penalized Function1

Dim.	*Alg.*	Mean	STD
30	SPSO	6.7461e-002	2.3159e-001
	MPSO-TVAC	1.8891e-017	6.9756e-017
	ICPSO	1.7294e-032	2.5762e-033
	IPSO-DAI	7.5097e-025	1.4271e-024
100	SPSO	2.4899e+000	1.2686e+000
	MPSO-TVAC	2.3591e-001	1.9998e-001
	ICPSO	8.4038e-002	1.0171e-001
	IPSO-DAI	1.2396e-019	1.8766e-019
300	SPSO	5.3088e+002	9.0264e+002
	MPSO-TVAC	4.2045e+000	3.0387e+000
	ICPSO	4.6341e-001	4.500e-001
	IPSO-DAI	4.0134e-006	9.4890e-006

Table 6. Comparison Results for Penalized Function2

Dim.	*Alg.*	Mean	STD
30	SPSO	5.4943e-004	2.4568e-003
	MPSO-TVAC	9.3610e-027	4.1753e-026
	ICPSO	1.7811e-032	1.1537e-032
	IPSO-DAI	4.5303e-024	9.8794e-024
100	SPSO	3.8087e+001	1.8223e+001
	MPSO-TVAC	3.7776e-001	6.1358e-001
	ICPSO	7.3122e-001	1.0264e+000
	IPSO-DAI	6.2095e-019	8.7984e-019
300	SPSO	3.2779e+004	4.4431e+004
	MPSO-TVAC	3.7343e+000	2.6830e+000
	ICPSO	7.4542e-001	1.3135e+000
	IPSO-DAI	3.2258e-005	2.5709e-005

Multi-modal Functions with Many Local Optima

The comparison results of Rastrigin, Ackley, Griewank and two penalized functions are listed in Tab.2-6, in which *Dim.* represents the dimension, *Mean* denotes the average mean value, while *STD* denotes the standard variance. To provide a more explanation, the dynamic comparison of these benchmarks for dimension 300 are plotted as Fig.4-8.

This set of benchmarks are regarded as the most difficult functions to optimize since the number of local minima increases exponentially as the function dimension increases. For Rastrigin, Ackley, Griewank, and two Penalized Functions, the performance of IPSO-DAI always superior to other three variants when solving high-dimensional numerical problems (e.g. dimensionality is 300). However, ICPSO is more suit for low dimension (=30) among SPSO, MPSO-TVAC, ICPSO and IPSO-DAI except for Griewank function.

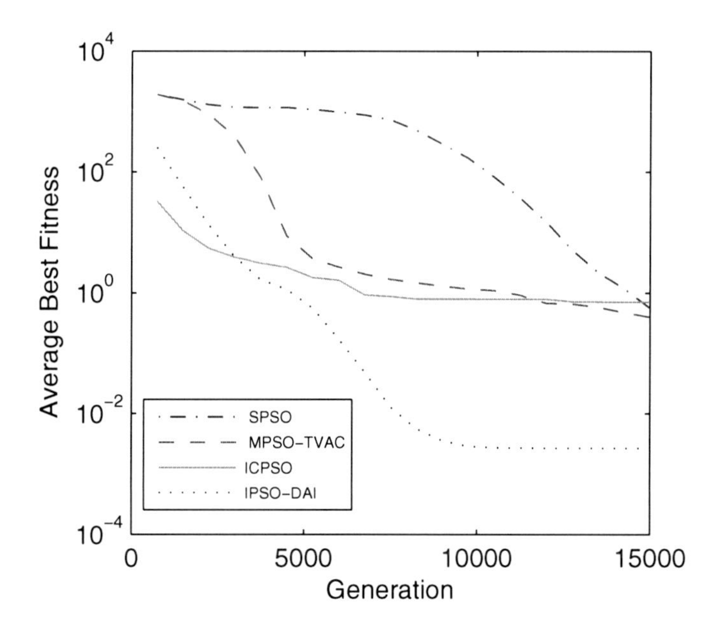

Fig. 4. Comparison Results on Rastrigin with Dimension 300

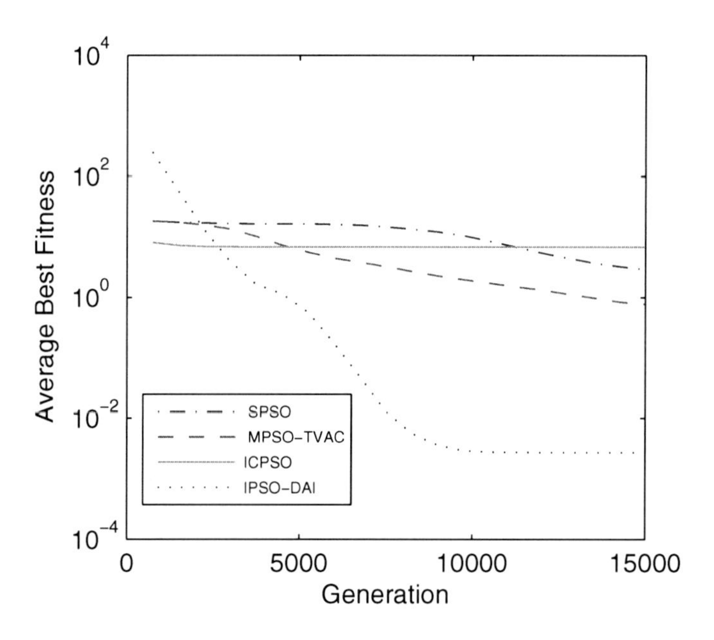

Fig. 5. Comparison Results on Ackley with Dimension 300

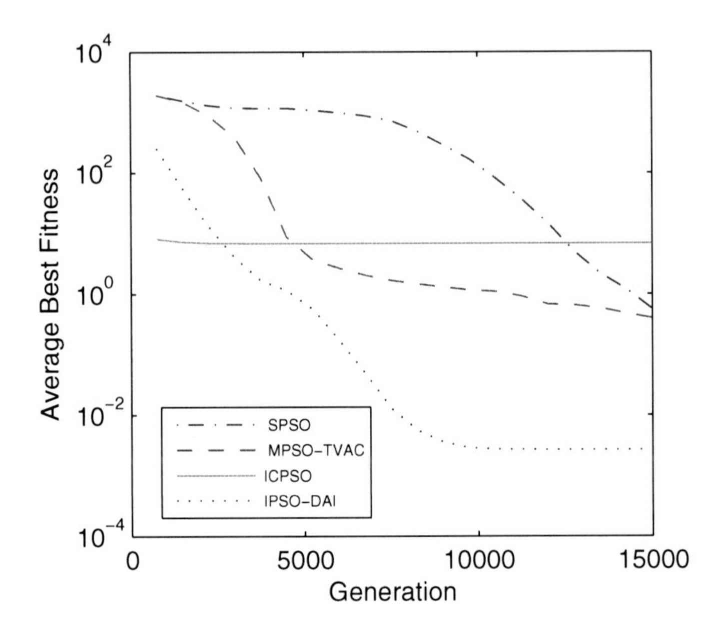

Fig. 6. Comparison Results on Griewank with Dimension 300

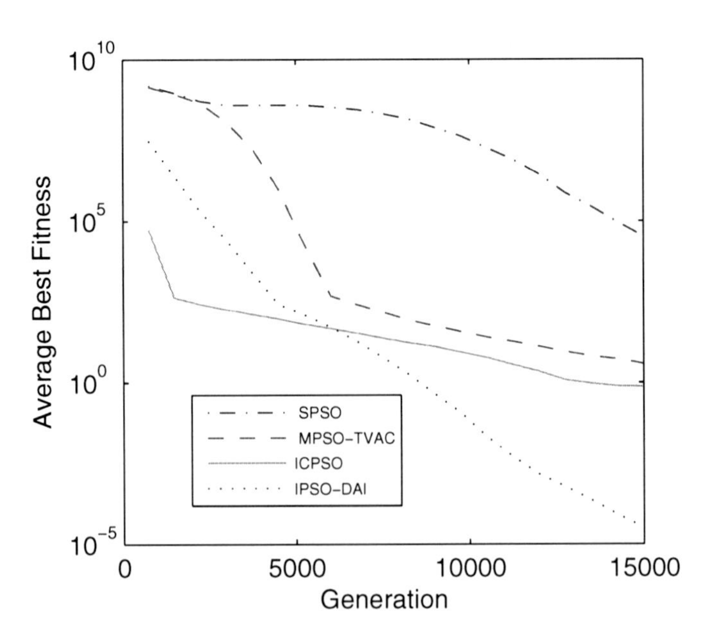

Fig. 7. Comparison Results on Penalized Function1 with Dimension 300

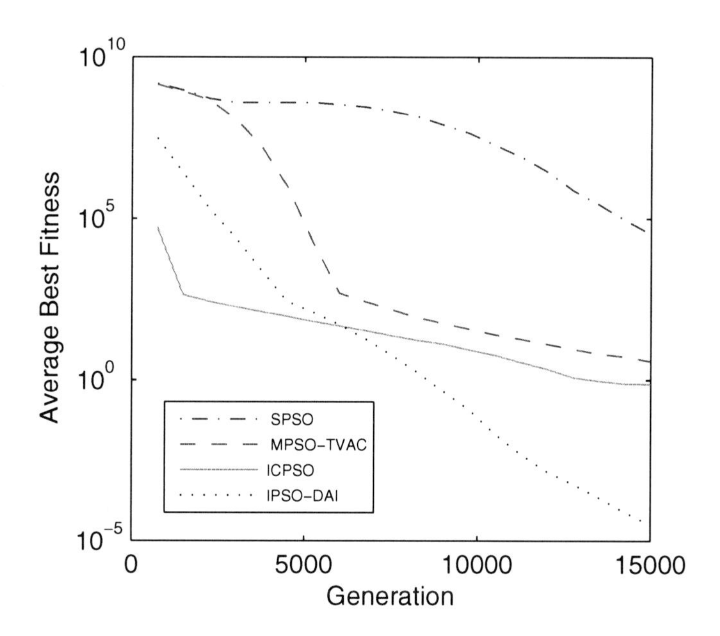

Fig. 8. Comparison Results on Penalized Function2 with Dimension 300

Multi-modal Functions with a Few Local Optima

The comparison results of last three benchmarks are listed in Tab.7, in which *Fun.* represents the benchmark function, *Mean* denotes the average mean value, while *STD* denotes the standard variance.

Because all dimensions of Shekel's Foxholes Function, Hartman Family Function and Shekel's Family Function are no larger than 4, Tab.6 shows that IPSO-DAI is also superior to other three variants when solving low dimensional

Table 7. Comparison Results for Final Three Benchmarks

Fun.	*Alg.*	Mean	STD
Shekel's Foxholes Function	SPSO	1.1967e+000	4.8098e-001
	MPSO-TVAC	1.9500e+000	2.1300e+000
	ICPSO	1.8565e+000	1.3899e+000
	IPSO-DAI	1.0973e+000	3.9953e-001
Hartman Family Function	SPSO	-3.8235e+000	4.3796e-002
	MPSO-TVAC	-3.8495e+000	2.1165e-002
	ICPSO	-3.8324e+000	4.1346e-002
	IPSO-DAI	-3.8593e+000	6.4523e-003
Shekel's Family Function	SPSO	-2.6803e+000	1.7567e+000
	MPSO-TVAC	-4.1664e+000	2.8045e+000
	ICPSO	-5.3892e+000	3.5067e+000
	IPSO-DAI	-6.0040e+000	3.6619e+000

multi-modal with a few local optima. To provide a more deep insight, an application for low dimensional multi-modal functions with a few local optima is proposed to further investigate the performance of IPSO-DAI.

Application of IPSO-DAI to Direct Orbits of Chaotic Systems

Directing orbits of chaotic systems is a multi-modal numerical optimization problem [45][46]. Consider the following discrete chaotic dynamical system:

$$\overrightarrow{x}(t+1) = \overrightarrow{f}(\overrightarrow{x}(t)), \quad t = 1, 2, ..., N \tag{40}$$

where state $\overrightarrow{x}(t) \in R^n$, $\overrightarrow{f} : R^n \to R^n$ is continuously differentiable.

Let $\overrightarrow{x}_0 \in R^n$ be an initial state of the system. If small perturbation $\overrightarrow{u}(t) \in R^n$ is added to the chaotic system, then

$$\overrightarrow{x}(t+1) = \overrightarrow{f}(\overrightarrow{x}(t)) + \overrightarrow{u}(t), \quad t = 0, 2, ..., N-1 \tag{41}$$

where $||\overrightarrow{u}(t)|| \leq \mu$, μ is a positive real constant.

The goal is to determine suitable $\overrightarrow{u}(t)$ so as to make $\overrightarrow{x}(N)$ in the ϵ−neighborhood of the target $\overrightarrow{x}(t)$, i.e., $||\overrightarrow{x}(N) - \overrightarrow{x}(t)|| < \epsilon$, where a local controller is effective for chaos control.

Generally, assume that $\overrightarrow{u}(t)$ acts only on the first component of $\overrightarrow{f}$, then the problem can be re-formulated as follows:

min $||\overrightarrow{x}(N) - \overrightarrow{x}(t)||$ by choosing suitable $\overrightarrow{u}(t)$, t=0,2,...,N-1

S.t.

$$\begin{cases} x_1(t+1) = f_1(\overrightarrow{x}(t)) + \overrightarrow{u}(t) \\ x_j(t+1) = \overrightarrow{f}_j(\overrightarrow{x}(t)) \qquad \text{j=2,3,...,n} \end{cases} \tag{42}$$

$$|u(t)| \leq \mu \tag{43}$$

$$\overrightarrow{x}(0) = \overrightarrow{x}_0 \tag{44}$$

As a typical discrete chaotic system, Hénon Map is employed as an example in this paper. Hénon May can be described as follows:

$$\begin{cases} x_1(t+1) = -px_1^2(t) + x_2(t) + 1 \\ x_2(t+1) = qx_1(t) \end{cases} \tag{45}$$

where $p = 1.4$, $q = 0.3$.

The target $\overrightarrow{x}(t)$ is set to be a fixed point of the system $(0.63135, 0.18941)^T$, $\overrightarrow{x}_0 = (0, 0)^T$, and $\overrightarrow{u}(t)$ is only added to $\overrightarrow{x}_1$ with the bound $\mu = 0.01$, 0.02 and 0.03. The population is 20, and the largest generation is 1000. Under the different values of N, Tab.8-10 list the mean objective value and the standard deviation value of 30 independent runs.

Table 8. Statistics Performance of IPSO-DAI with $\mu = 0.01$

N	*Alg.*	Mean	STD
7	SPSO	1.4621e-002	1.3349e-002
	MPSO-TVAC	1.4042e-002	5.1730e-004
	ICPSO	1.4173e-002	1.3432e-002
	IPSO-DAI	1.4058e-002	1.3064e-002
8	SPSO	3.5703e-003	1.1707e-003
	MPSO-TVAC	1.9464e-003	5.7341e-004
	ICPSO	2.2415e-003	1.1515e-003
	IPSO-DAI	2.1128e-003	1.1442e-003
9	SPSO	3.9608e-002	6.1158e-005
	MPSO-TVAC	1.4014e-002	7.2087e-002
	ICPSO	2.7385e-002	9.4852e-005
	IPSO-DAI	1.0387e-003	3.5334e-005

Table 9. Statistics Performance of IPSO-DAI with $\mu = 0.02$

N	*Alg.*	Mean	STD
7	SPSO	1.1919e-002	1.1304e-002
	MPSO-TVAC	1.1384e-002	2.9914e-004
	ICPSO	1.1751e-002	1.1206e-002
	IPSO-DAI	1.1525e-002	1.1149e-002
8	SPSO	8.3681e-004	3.4507e-005
	MPSO-TVAC	1.3811e-004	2.7768e-004
	ICPSO	3.3377e-004	1.7161e-006
	IPSO-DAI	8.9491e-005	3.5212e-008
9	SPSO	5.3174e-004	2.0244e-005
	MPSO-TVAC	8.5427e-005	8.7349e-005
	ICPSO	1.6105e-004	1.0441e-005
	IPSO-DAI	6.9653e-005	1.1674e-006

Table 10. Statistics Performance of IPSO-DAI with $\mu = 0.03$

N	*Alg.*	Mean	STD
7	SPSO	1.0416e-002	9.4072e-003
	MPSO-TVAC	9.7059e-003	9.0593e-003
	ICPSO	1.0171e-002	9.0770e-003
	IPSO-DAI	9.8517e-003	9.2056e-003
8	SPSO	6.4307e-004	8.3672e-006
	MPSO-TVAC	1.9522e-004	9.7019e-008
	ICPSO	1.1158e-004	1.7779e-006
	IPSO-DAI	5.4046e-005	7.5838e-009
9	SPSO	6.7555e-004	1.0643e-005
	MPSO-TVAC	1.2134e-004	2.5046e-007
	ICPSO	1.8667e-004	5.3299e-007
	IPSO-DAI	5.1206e-005	3.3442e-008

It is obviously IPSO-DAI is superior to other three variants significantly when solving multi-modal functions with a few local optima.

In one word, IPSO-DAI is suit for multi-modal problems especially with many local optima.

5 Conclusion

This chapter introduces a method to improve the population diversity by adding some controllers. As an example, integral-controlled particle swarm optimization is designed. However, the diversity is still not too small in the late stage, so, an enhanced ICPSO – ICPSO with dispersed accelerator information (IPSO-DAI) is presented. To testify the performance of IPSO-DAI, several benchmarks are used to compare in which three are uni-modal functions, five with multi-modal functions with many local optima, as well as three with multi-modal functions with a few local optima. Simulation results show IPSO-DAI may provide a best performance when compared with SPSO, MPSO-TVAC and ICPSO when solving high dimensional multi-modal problems especially with many local optima. However, IPSO-DAI is worse than ICPSO when solving uni-modal problems. Furthermore, IPSO-DAI is applied to direct the orbits of discrete chaotic dynamical systems, and achieves the best performance also. Further research topic includes the discrete IPSO-DAI and applications.

Acknowledgment

This paper are supported by Shanxi Science Foundation for Young Scientists under Grant 2009021017-2 and Natural Science Foundation of Shanxi under Grant 2008011027-2.

References

1. Kennedy, J., Eberhart, R.: Particle swarm optimization. In: Proceedings of ICNN 1995 - IEEE International Conference on Neural Networks, pp. 1942–1948. IEEE CS Press, Perth (1995)
2. Eberhart, R., Kennedy, J.: New optimizer using particle swarm theory. In: Proceedings of the Sixth International Symposium on Micro Machine and Human Science, pp. 39–43. IEEE CS Press, Nagoya (1995)
3. John, G., Lee, Y.: Multi-objective based on parallel vector evaluated particle swarm optimization for optimal steady-state performance of power systems. Expert Systems with Applications 36(8), 10802–10808 (2009)
4. Singh, N.A., Muraleedharan, K.A., Gomathy, K.: Damping of low frequency oscillations in power system network using swarm intelligence tuned fuzzy controller. International Journal of Bio-Inspired Computation 2(1), 1–8 (2010)
5. Lu, J.G., Zhang, L., Yang, H., Du, J.: Improved strategy of particle swarm optimisation algorithm for reactive power optimisation. International Journal of Bio-Inspired Computation 2(1), 27–33 (2010)

6. Begambre, O., Laier, J.E.: A hybrid Particle Swarm Optimization C Simplex algorithm (PSOS) for structural damage identification. Advances in Engineering Software 40(9), 883–891 (2009)
7. Chen, S., Hong, X., Luk, B.L., Harris, C.J.: Non-linear system identification using particle swarm optimisation tuned radial basis function models. International Journal of Bio-Inspired Computation 1(4), 246–258 (2009)
8. Lee, W.S., Chen, Y.T., Wu, T.H.: Optimization for ice-storage air-conditioning system using particle swarm algorithm. Applied Energy 86(9), 1589–1595 (2009)
9. Marinakis, Y., Marinaki, M., Doumpos, M., Zopounidis, C.: Ant colony and particle swarm optimization for financial classification problems. Expert Systems with Applications 36(7), 10604–10611 (2009)
10. Senthil, A.M., Ramana, M.G., Loo, C.K.: On the optimal control of the steel annealing processes as a two-stage hybrid systems via PSO algorithms. International Journal of Bio-Inspired Computation 1(3), 198–209 (2009)
11. Cura, T.: Particle swarm optimization approach to portfolio optimization. Nonlinear Analysis: Real World Applications 10(4), 2396–2406 (2009)
12. Parsopoulos, K.E., Kariotou, F., Dassios, G., Vrahatis, M.N.: Tackling magnetoencephalography with particle swarm optimization. International Journal of Bio-Inspired Computation 1(1/2), 32–49 (2009)
13. Xiao, R.B., Xu, Y.C., Amos, M.: Two hybrid compaction algorithms for the layout optimization problem. Biosystems 90(2), 560–567 (2007)
14. Wu, X.L., Cho, J.S., D'Auriol, B.J., et al.: Mobility-assisted relocation for self-deployment in wireless sensor networks. IEICE Transactions on Communications 90(8), 2056–2069 (2007)
15. Namasivayam, V., Gunther, R.: PSOAUTODOCK: A fast flexible molecular docking program based on swarm intelligence. Chemical Biology and Drug Design 70(6), 475–484 (2007)
16. Liu, H., Abraham, A., Choi, O., Moon, S.H.: Variable neighborhood particle swarm optimization for multi-objective flexible job-shop scheduling problems. In: Wang, T.-D., Li, X., Chen, S.-H., Wang, X., Abbass, H.A., Iba, H., Chen, G.-L., Yao, X. (eds.) SEAL 2006. LNCS, vol. 4247, pp. 197–204. Springer, Heidelberg (2006)
17. Yisu, J., Knowles, J., et al.: The landscape adaptive particle swarm optimizer. Applied Soft Computing 8(1), 295–304 (2008)
18. Arumugam, M.S., Rao, M.V.C.: On the improved performances of the particle swarm optimization algorithms with adaptive parameters, cross-over operators and root mean square (RMS) variants for computing optimal control of a class of hybrid systems. Applied Soft Computing 8(1), 324–336 (2008)
19. Ling, S.H., Iu, H.H.C., Chan, K.Y., Lam, H.K., Yeung, B.C.W., Leung, F.H.: Hybrid particle swarm optimization with wavelet mutation and its industrial applications. IEEE Transactions on Systems, Man, and Cybernetics, Part B: Cybernetics 38(3), 743–763 (2008)
20. Cui, Z.H., Zeng, J.C.: A guaranteed global convergence particle swarm optimizer. In: Tsumoto, S., Słowiński, R., Komorowski, J., Grzymała-Busse, J.W. (eds.) RSCTC 2004. LNCS (LNAI), vol. 3066, pp. 762–767. Springer, Heidelberg (2004)
21. Liu, H., Abraham, A.: Fuzzy turbulent particle swarm optimization. In: Proceedings of the Fifth International Conference on Hybrid Intelligent Systems, Brazil, pp. 445–450 (2005)
22. Liu, H., Abraham, A., Zhang, W.: A fuzzy adaptive turbulent particle swarm optimization. International Journal of Innovative Computing and Applications 1(1), 39–47 (2005)

23. Abraham, A., Liu, H.: Turbulent particle swarm optimization with fuzzy parameter tuning. In: Foundations of Computational Intelligence: Global Optimization, Studies in Computational Intelligence, vol. 3, pp. 291–312. Springer, Heidelberg (2009)
24. Vesterstrom, J., Thomsen, R.: A comparative study of differential evolution, particle swarm optimization, and evolutionary algorithms on numerical benchmark problems. In: Proceedings of the 2004 Congress on Evolutionary Computation (CEC 2004), vol. 2, pp. 1980–1987 (2004)
25. Li, X., Yao, X.: Tackling high dimensional nonseparable optimization problems by cooperatively coevolving particle swarms. In: Proceedings of Congress of 2009 Evolutionary Computation (CEC 2009), pp. 1546–1553 (2009)
26. Huang, T., Mohan, A.S.: Micro-particle swarm optimizer for solving high dimensional optimization problems (μPSO for high dimensional optimization problems). Applied Mathematics and Computation 181(2), 1148–1154 (2006)
27. Cui, Z.H., Cai, X.J., Zeng, J.C., Sun, G.J.: Particle swarm optimization with FUSS and RWS for high dimensional functions. Applied Mathematics and Computation 205(1), 98–108 (2008)
28. Cui, Z.H., Zeng, J.C., Sun, G.J.: A fast particle swarm optimization. International Journal of Innovative Computing, Information and Control 2(6), 1365–1380 (2006)
29. Zhan, Z.H., Zhang, J., Li, Y., Chung, H.S.H.: Adaptive particle swarm optimization. IEEE Transactions on System, Man & Cybernetics, Part B (2009), doi:10.1109/TSMCB.2009.2015956
30. Ghosh, S., Kundu, D., Suresh, K., Das, S., Abraham, A.: An Adaptive Particle Swarm Optimizer with Balanced Explorative and Exploitative Behaviors. In: Proceedings of the tenth International Symposium on Symbolic and Numeric Algorithms for Scientific Computing (2008)
31. Korenaga, T., Hatanaka, T., Uosaki, K.: Performance improvement of particle swarm optimization for high-dimensional function optimization. In: IEEE Congress on Evolutionary Computation, pp. 3288–3293 (2007)
32. Li, H., Li, L.: A novel hybrid particle swarm optimization algorithm combined with harmony search for high dimensional optimization problems. In: International Conference on Pervasive Computing, pp. 94–97 (2007)
33. Cai, X.J., Cui, Z.H., Zeng, J.C., Tan, Y.: Particle swarm optimization with self-adjusting cognitive selection strategy. International Journal of Innovative Computing, Information and Control 4(4), 943–952 (2008)
34. Cai, X.J., Cui, Z.H., Zeng, J.C., Tan, Y.: Performance-dependent adaptive particle swarm optimization. International Journal of Innovative Computing, Information and Control 3(6B), 1697–1706 (2007)
35. Cai, X.J., Cui, Y., Tan, Y.: Predicted modified PSO with time-varying accelerator coefficients. International Journal of Bio-inspired Computation 1(1/2), 50–60 (2009)
36. Upendar, J., Singh, G.K., Gupta, C.P.: A particle swarm optimisation based technique of harmonic elimination and voltage control in pulse-width modulated inverters. International Journal of Bio-Inspired Computation 2(1), 18–26 (2010)
37. Kumar, R., Sharma, D., Kumar, A.: A new hybrid multi-agen Cbased particle swarm optimisation technique. International Journal of Bio-Inspired Computation 1(4), 259–269 (2009)
38. Zeng, J.C., Cui, Z.H.: Particle Swarm Optimizer with Integral Controller. In: Proceedings of 2005 International Conference on Neural Networks and Brain, Beijing, pp. 1840–1842 (2005)

39. Cai, X.J., Cui, Z.H., Zeng, J.C., Tan, Y.: Dispersed particle swarm optimization. Information Processing Letters 105(6), 231–235 (2008)
40. Shi, Y., Eberhart, R.C.: A modified particle swarm optimizer. In: Proceedings of the IEEE International Conference on Evolutionary Computation, Anchorage, Alaska, USA, pp. 69–73
41. Shi, Y., Eberhart, R.C.: Parameter selection in particle swarm optimization. In: Proceedings of the 7th Annual Conference on Evolutionary Programming, pp. 591–600
42. Shi, Y., Eberhart, R.C.: Empirical study of particle swarm optimization. In: Proceedings of the Congress on Evolutionary Computation, pp. 1945–1950 (1999)
43. Ratnaweera, A., Halgamuge, S.K., Watson, H.C.: Self-organizing hierarchical particle swarm opitmizer with time-varying acceleration coefficients. IEEE Transactions on Evolutionary Computation 8(3), 240–255 (2004)
44. Yao, X., Liu, Y., Lin, G.M.: Evolutionary programming made faster. IEEE Transactions on Evolutionary Computation 3(2), 82–102 (1999)
45. Liu, B., Wang, L., Jin, Y.H., Tang, F., Huang, D.X.: Directing orbits of chaotic systems by particle swarm optimization. Chaos Solitons & Fractals 29, 454–461 (2006)
46. Wang, L., Li, L.L., Tang, F.: Directing orbits of chaotic dynamical systems using a hybrid optimization strategy. Physical Letters A 324, 22–25 (2004)

Particle Swarm Optimization for Markerless Full Body Motion Capture

Zheng Zhang[1], Hock Soon Seah[1], and Chee Kwang Quah[2]

[1] Nanyang Technological University, School of Computer Engineering, Singapore
zhang_zheng@pmail.ntu.edu.sg, ashsseah@ntu.edu.sg

[2] Republic Polytechnics, School of Sports, Health and Leisure, Singapore
quah_chee_kwang@rp.sg

Summary. The estimation of full body pose is a very difficult problem in Computer Vision. Due to high dimensionality of the pose space, it is challenging to search the true body configurations for any search strategy. In this chapter, we apply the stochastic Particle Swarm Optimization (PSO) algorithm to full body pose estimation problem. Our method fits an articulated body model to the volume data reconstructed from multiple camera images. Pose estimation is performed in a hierarchical manner with space constraints enforced into each sub-optimization step. To better address the dynamic pose tracking problem, the states of swarm particles are propagated according to a weak transition model. This maintains the diversity of particles and also utilizes the temporal continuity information. 3D distance transform is used for reducing the computing time of fitness evaluations. From the experiments, we demonstrate that by using PSO, our method can robustly fit the body model of a reference pose to the beginning frame, and can track fast human movements in multiple image sequences.

1 Introduction

Tracking of human body motion, also known as motion capture, is the problem of recovering the pose parameters of the human body from videos or image sequences. It has been used in a wide variety of applications. One important application is in films, animation or video games [31, 39], where motion capture is used for recording and mapping the movements of actors to animate characters. Another application is for sport analysis and biomechanics [10], where motion capture provides cost-effective solutions in applications such as rehabilitation and prevention of injuries, patient positioning, interactive medical training, or sport performance enhancement. Motion capture has also been applied to surveillance systems [9], for example, to analyse actions, activities, and behaviors both for crowds and individuals.

The main types of commercial motion capture technologies include magnetic, mechanical and optical systems, all of which require the subject to wear markers or devices on the body skin. Despite being popular, they have many disadvantages. Firstly, they require specific capture hardware and programs, with prohibitively high cost. Secondly, the requirment of wearing makers on the joints

B.K. Panigrahi, Y. Shi, and M.-H. Lim (Eds.): Handbook of Swarm Intelligence, ALO 8, pp. 201–220.
springerlink.com

is cumbersome and may introduce the skin artifact problem [5]. Thirdly, it is difficult to manipulate the motion data especially when problems occur. Due to emerging techniques and research in computer vision and computer graphics, markerless approach to motion capture has been developing fast these years. The markerless motion capture method, which estimates a subject's 3D pose from image sequences without the use of any marker or special device, is non compelling and low cost. However, despite many markerless motion capture approaches having been proposed, there are still a number of hard and often ill-posed problems, and it is still a highly active research domain in computer vision [13, 33, 38].

Recovering the 3D pose of human body from image data without using markers is a very challenging task. One of the major difficulties is the high number of degrees of freedom (DOFs) of the human body pose that needs to be recovered. Generally, at least 25 DOFs are needed for describing realistic human movements. Aside from the high DOFs, the sizes of the human body vary largely across individuals, with skins and clothings that may have large deformation in movements. Moreover, there are many ambiguities caused by self-occlusion or fast movements, which is even more ambiguous when a single camera is used.

Approaches of markerless human motion capture can be classified by the number of cameras they use, i.e., monocular or multi-view approach. Compared to monocular approaches, multi-view approaches have the advantage that they can deal better with occlusion and appearance ambiguity problems, leading to much more robust and accurate results. These approaches can be model-based or not. The model-based methods, which use a priori model of the subject to guide the pose tracking and estimation processes, can greatly simplify the problem and make the pose estimation more robust and accurate. Conventional model-based markerless motion capture methods typically use a predefined human body model which consists of a kinematical skeleton model and some form of shape representations. Simplified 2D or 3D shape primitives (2D stick-figure [28], blobs [43], superquadrics [14], superellipsoid [23], etc.) and sophisticated exact body shape models (deformable polygonal model or triangle mesh model [36, 4]) are usually used to represent the limbs or body shape. The typical framework of pose estimation is to fit the predefined models to the image features by using a deterministic or stochastic search strategy [4, 23, 11].

One important component of a markerless motion capture approach is the use of a search strategy. Bayesian filters and local optimization techniques, either stochastic or deterministic, are usually used. In this work we present the application of Particle Swarm Optimization (PSO) [24] to address this optimization problem. PSO is a stochastic search technique and has been proved to be an efficient method for many global optimization problems [35]. It has been shown to perform well on a great variety of nonlinear, non-differentiable and multimodal optimization problems [8]. With PSO algorithm, our approach fit an articulated body model to the 3D volume data reconstructed from multiple synchronized image sequences. To obtain robust results, we enforce space constraints into the matching-cost metric, and perform the pose search in a hierarchical way.

Experiments show that our approach is able to robustly capture complex and fast motion.

The rest of this chapter is organized as follows. In Sect. 2, we describe the problem and the related work. PSO and its application in dynamic optimization problem are then introduced in Sect. 3. Sect. 4 gives a detailed description of the use of PSO for pose initialization and tracking problem. Experiments and conclusion are separately given in Sect. 5 and Sect. 6.

2 Background

2.1 The Problem

The problem we address is to recover the full-body 3D human pose from multiple view sequences. Given a sequence of multiple camera images with recorded movement of a single person, our aim is to recover large scale body movements which include the movements of the head, arms, torso and legs. The problem is difficult due to high dimensionality of the pose space, deformable skin, loose clothing, and self-occlusion. The image observations typically have highly non-linear correlation with the body pose. To obtain robust and accurate results, we concentrate on model-based approaches, which exploit information of multiple images and can deal better with occlusion and ambiguities. A typical model-based approach uses an articulated human body model, which generally consists of a kinematic skeleton and a form of shape representation, to represent the observed subject. Pose estimation is usually performed in an analysis-by-synthesis framework in which the configuration of the model is continuously predicted and updated. For a model-based approach, the problem mainly covers the following two steps.

- **Pose Initialization**. The goal of pose initialization is to adjust the model to a configuration that is close to the true pose at the beginning frame. The model is defined beforehand, with its segment size parameters adjusted to the corresponding specific shape of the tracked subject. The adjustment can be performed manually or automatically.
- **Pose Estimation and Tracking**. The goal of this process is to recover 3D human poses over the image sequence. This process is usually integrated into an analysis-by-synthesis framework or performed on a per-frame basis.

Both of the two steps can be formulated as an optimization problem of which the model is fitted to the image data. The difference is that no initial pose or priori information is given for pose initialization, while an initial pose, good or erroneous, is provided from the previous frame for the pose estimation and tracking step. More formally, both of them can be described as the following weight minimization problem:

$$\min_{\boldsymbol{p}} \sum w_i E_i(\boldsymbol{Y}_i, \boldsymbol{X}_i(\boldsymbol{p})) \tag{1}$$

where w_i are weight factors used for balancing the impact of different objective errors E_i, $\boldsymbol{Y_i}$ denotes a set of priori information or observations, $\boldsymbol{X_i(p)}$ denotes a set of corresponding predictions using the kinematic body model, and $\boldsymbol{p}$ is the vector of pose parameters to be estimated.

2.2 Related Work

There are many methods that capture human motion from reconstructed 3D volume data [23, 32, 34]. The basic idea is to reconstruct the time-series volumes of the subject by using shape-from-silhouette or visual hull methods [26], and then acquire the internal motion from the volume sequences by fitting an articulated body model to the volume data. Our work is similar to this class of methods.

Large numbers of model-based approaches formulate the pose estimation as an optimization problem and use local optimization techniques to match an articulated model with image observations [4, 36, 23, 14, 6]. Gradient descent based optimization techniques such as Levenberg-Marquart [36, 6], Powell conjugate gradient algorithm [4] are usually used. Although these methods can reach accurate results if a good initial solution is provided, they cannot recover from errors and may easily meet tracking failures in cases of complex and fast motions.

Instead of using optimization techniques to search the true poses, many works address the problem in a probabilistic framework. These methods model the human motion as a stochastic process and use Bayesian filters to estimate the state or posterior distribution from a sequential observations perturbed by noises. Particle filter approaches are popular [11, ?, ?]. These methods can handle noise and are capable of recovering from errors. However, they need to find the system dynamic and observation models of human motion. These models, including the proposal transition and the likelihood function, are extremely difficult to build, hence they are usually specified in an ad-hoc fashion, which may lead to very poor tracking performances [12].

Evolutionary methods such as genetic algorithms also have been used for the problem [44, 41, 15, 18]. At present, few works use PSO to address the 3D human pose estimation problem. Robertson et al. [40] present a PSO-based approach to estimate the upper-body pose. In their work, a very simple upper-body skeleton model is fitted to 3D stereo data. Ivekovic and Trucco [19] employ a subdivision body model for estimating the pose of upper-body from multiple images. The PSO optimization is performed by minimizing the discrepancy between projected silhouettes of model and those extracted from multiple images. Ivekovic et al. [20] also present a similar work where they address upper-body pose estimation from multiple still images.

We apply PSO to full body pose estimation. Our method fits an articulated body model to the volume sequences reconstructed from multiple image data. Pose estimation is performed in a hierarchical manner with space constraints enforced into each sub-optimization step. To better address the dynamic pose tracking problem, the states of swarm particles are propagated according to a weak transition model. This maintains the diversity of particles and also utilizes the temporal continuity information. Based on PSO algorithm, our approach also addresses the pose initialization problem. With a little user interaction, the model can be fitted to the beginning volume frame automatically.

3 Particle Swarm Optimization

In this section we introduce the canonical PSO algorithm (Sect. 3.1). To deal better with motion tracking problem which in fact is a dynamic process, we describe the use of PSO for dynamic optimization problem in Sect. 3.2.

3.1 Canonical PSO

PSO is first published by Kennedy and Eberhart in 1995 [24]. It is a stochastic searching technique of population-based evolutionary computation, and has been successfully used for many hard numerical and combinatorial optimization problems [35, 8]. PSO is a relatively simple optimization algorithm that has roots in artificial life in general, and in bird flocking, fish schooling and swarming theory in particular. It is also related to genetic algorithms and evolutionary programming.

The basic idea behind PSO is the simulation of social behavior among particles "flying" through a multidimensional search space, based on the collective behavior observed in fish schooling or bird flocking. In detail, PSO maintains a swarm of particles each of which has its position and velocity in the multidimensional search space. The position is a candidate problem solution and the velocity directs the movement of the particle fly. Each particle also has a fitness value which can be computed through an objective function. At the beginning, each particle individual is initialized with a random position, a velocity and the fitness value are computed by the objective function with the particle's position used as input. At every time step, each particle is attracted towards the position that is affected by the best position $\boldsymbol{p}_i$ found by itself and the global best position $\boldsymbol{p}_g$ found by its neighborhood so far. In this work we use gbest topology (for "global best") where all particles are in a neighborhood. The velocity and position of each individual are iteractively updated according to the following two equations [8]. The swarm stops updating until a terminate criterion is met (usually a maximum number of iterations).

$$\boldsymbol{v}_{k+1}^i = \chi(\boldsymbol{v}_k^i + \boldsymbol{U}(0,\phi_1) \otimes (\boldsymbol{p}_i - \boldsymbol{x}_k^i) + \boldsymbol{U}(0,\phi_2) \otimes (\boldsymbol{p}_g - \boldsymbol{x}_k^i)) \tag{2}$$

$$\boldsymbol{x}_{k+1}^i = \boldsymbol{x}_k^i + \boldsymbol{v}_{k+1}^i \tag{3}$$

where $\boldsymbol{x}_k^i$ and $\boldsymbol{v}_k^i$ separately denote the position and velocity of the i-th particle at k-th iteration, χ is a constriction coefficient, $\boldsymbol{U}(0,\phi_i)$ represents a vector of random numbers uniformly distributed in $[0,\phi_i]$, $\otimes$ denotes a point-wise vector multiplication.

3.2 PSO for Dynamic Optimization

Most applications of PSO algorithms are only for static optimization problems. The problem of pose estimation and tracking from multiple sequences in fact are dynamic as the optimum pose changes over frames. PSO should be modified to better suit this problem.

We can use the normal PSO on a per-frame basis. However, this approach would be not so efficient because it does not utilize well the temporal continuity information of two consecutive frames, between which the change of the optimum pose vector is typically small. It would be better to continue on from the state of swarm at the previous frame when starting a new search for the coming frame, in contrast to completely reinitializing the swarm. However, there are two problems for continuously searching from the state of the old swarm: outdated memory and diversity loss [1, 2].

Outdated memory happens when the information of the swarm, including each particle's local best $\boldsymbol{p}_i$, the swarm's global best $\boldsymbol{p}_g$, and the corresponding fitnesses, may no longer be true due to the coming of a new frame. It may mislead the continuous search without any modification to the swarm information. This problem can be simply solved by first resetting each particle's best position to its current position and reevaluating their fitness, and then reset the swarm global best information, i.e., the global best position and its corresponding fitness.

Diversity loss is a more serious problem. As a stochastic searching technique, PSO maintains its searching ability by keeping the diversity of a swarm of particles which "fly" in the search space based on a metaphor of social interaction. Due to the convergence of previous optimization process, all particles may be close to the previous optimum position and the swarm has shrunk. If the new optimum position still lies within the collapsing swarm, the swarm as a whole may find the optimum efficiently as it still has sufficient diversity. However, if the current optimum position is outside of the swarm, the swarm may never find the true optimum due to the low velocities of the particles which will inhibit re-diversification and tracking.

Many works have proposed different strategies for solving the problem of diversity loss. Blackwell and Branke [1] summarize that the approaches can be grouped into three categories. The first category explores the re-diversification mechanisms. The example is the work of Hu and Eberhart [17]. They introduce an adaptive PSO and test many different re-diversifications that involve re-randomization of the entire, or part of the swarm. The second category attempts to maintain diversity throughout the run. This may be achieved by repulsion which can keep particles away from a premature convergence or orbit a nucleus of neutral particles. The third category is to maintain diversity with multi-populations. Blackwell and Branke [1, 2] propose an algorithm of multi-swarm PSO, with the aim of maintaining a multitude of swarms on different peaks. This approach work well in cases where the dynamic functions consist of several peaks.

In our method, we apply a sequential PSO framework [45] to maintain the diversity of the particles and make use of the temporal continuity information. Briefly, a Gaussian transition model is used for the particles' propagation. More detail is described in Sect. 4.5.

4 Our Approach

4.1 Overview

We propose methods for human body model pose initialization, pose estimation and tracking, based on a particle swarm optimization algorithm.

In our work, we use a parametric shape primitive called barrel model to represent body parts which are connected by joints in kinematic chains to describe the articulated body structure. The body model used is described in Sect. 4.2. We use 3D volume as the cue for pose initialization and tracking. In our work, volume data is obtained by using a volumetric visual hull method that is described in Sect. 4.3. We address the pose initialization problem by incorporating a set of user constraints into a volume based fitting process. Though losing automaticity as it needs the user to provide some information as constraints, this method is efficient and can obtain robust results. The pose initialization is described in Sect. 4.4. In Sect. 4.5, we describe the method for pose estimation and tracking. Pose estimation is performed in a hierarchical manner, with space constraints integrated into a PSO pose search algorithm. For pose tracking, particle diversity loss and pose propagation have to be addressed. Our method propagates the swarm particles according to a weak motion model, which enhances particle diversity for keeping the search ability of the swarm.

4.2 The Articulated Body Model

For a model-based markerless motion capture method, an articulated body model is needed. We use a body model consisting of 10 body segments, each of which is a barrel model for describing the shape and bone of the corresponding body part. Each body segment or the barrel model has a local coordinate system to define the locations of the points that are regularly sampled from surface points of the barrel shape. The body segments are connected by joints to form five open kinematic chains with a total of 29 DOFs. The root joint, the parent joint of every kinematic chain, is given three rotations and three translations to move freely in the 3D space. The head, shoulders and hips in the kinematic chains are modeled as ball joints that allow three rotations. Each elbow or knee is given two DOFs. The total DOF forms a 29-D pose vector that needs to be recovered.

The barrel model is built as a cylinder with two ends in different sizes. Its shape is controlled by three parameters: the length l, the top radius r_1, and the bottom radius r_2. For an arbitrary point $p(i,j)$ on the shape, its location $\boldsymbol{p}^l$ is given by

$$\boldsymbol{p}^l(i,j) = \begin{pmatrix} r\cos(i \times \delta_\theta) \\ -0.5l + j \times \delta_\eta \\ r\sin(i \times \delta_\theta) \end{pmatrix} \tag{4}$$

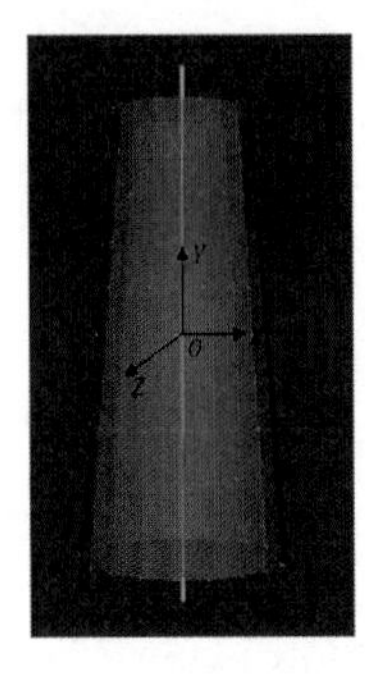

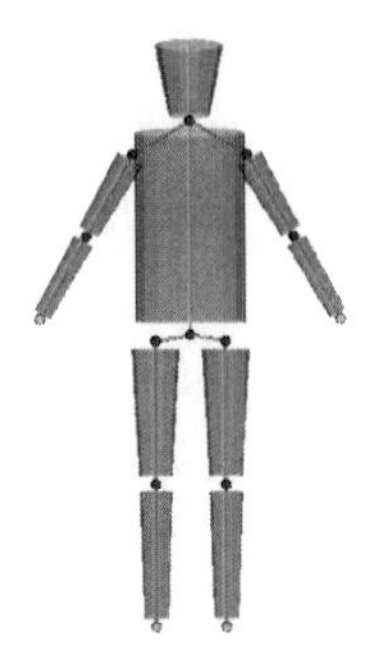

Fig. 1. (Left)The barrel model is the primitive element making up the articulated model. Each barrel model has a local coordinate system defining locations of sampling points (the white points) which are regularly sampled from the model surface. (Right) The articulated human body model is built from barrel models. It consists of 10 joints (red balls) with 29 DOFs and five open kinematic chains including torso-head, torso-left/right upperarm-left/right forearm, torso-left/right thigh-left/right calf.

where $r = (r_2 - r_1) \times (l - j \times \delta_\eta)/l + r_1$, δ_θ and δ_η separately denote the resolution of the latitude and the longitude, i.e., the height.

Given a specifc pose of the model, we need to explicitly compute the position of these model sampling points in the world coordinate system. Assuming point p on a body segment that is associated to a kinematic chain with a total of n configuration parameters $(\boldsymbol{\omega}, \boldsymbol{\theta})$ where $\boldsymbol{\omega}$ denots the six global parameters and $\boldsymbol{\theta}$ denotes other joint rotations, the position of the point $\boldsymbol{P}^w$ (homogeneous coordinate) associated with the world coordinate system can be computed by

$$\boldsymbol{P}^w(\boldsymbol{\theta}) = \boldsymbol{M}(\boldsymbol{\omega}) \prod_{i=1}^{n-6} \boldsymbol{M}(\theta_i) \boldsymbol{P}^w(\mathbf{0}) \tag{5}$$

where $\boldsymbol{P}^w(\mathbf{0})$ is the initial position of point p which is determined by the location $\boldsymbol{p}^l$ and the initial reference model configuration. For our model, as shown in Fig. 1, we set the reference configuration with arms at 30 degrees about the side of body. The homogeneous transformation matrices $\boldsymbol{M}(\boldsymbol{\omega})$ corresponds to the gobal rotation and translation. $\boldsymbol{M}(\theta_i)$ are other homogeneous transformation matrices specified by $\theta_i (i = 1, 2, \cdots, n-6)$.

4.3 Volume Reconstruction

Our method recovers body pose from volume data. We firstly reconstruct the 3D volume shape of the articulated body from image data and then fit the body model to the volume data in the 3D space to obtain the body pose. Working directly in 3D space brings more consistency to the tracking step and is not easily affected by self-occlusion. In addition, the volume data synthesizes all the information concerning camera parameters and background subtraction, thus it avoids repeated 3D-to-2D projections for each camera view and reduces algorithm complexity. To obtain the volume data, a multiple camera system is needed to be set up for capturing synchronized multiple image sequences. Silhouette data is then extracted from the raw images and used for volume reconstruction.

Silhouette Extraction

To extract the silhouettes, we use a background subtraction approach that identifies moving foreground objects from the background. The indoor capture environment has global or local illumination changes such as shadows and highlights. Shadows cast from foreground subject onto the environment are easily incorrectly classified as foreground. We adopt a RGB-color based background subtraction method [16] which can efficiently remove the shadow and highlight.

Shape-from-Silhouette and Visual Hulls

Shape-from-silhouette is a method of obtaining the visual hull which is the upperbound to the actual volume of the subject [27]. Several approaches have been proposed to compute visual hulls. They can be roughly separated into two categories: polyhedral and volumetric approach. Polyhedral approaches generate polyhedral visual hulls [30]. In this case, a 2D polygon approximation of each multi-view silhouette contour is firstly obtained, and then these silhouette polygons are back-projected into the space from their corresponding camera positions. The final object's visual hull is computed as the intersection of the silhouette cones. While in the case of volumetric approaches [42], the 3D space is divided into cubic volume elements (i.e., voxels) and a test is conducted for each voxel to decide whether the voxel lies inside or outside the object's visual hull. The test is performed by checking the overlap of each voxel's projections with all the multi-view silhouettes. The final reconstruction result is a volumetric representation of the object.

For our motion capture application, we use a volumetric approach to obtain the volume data. Volumeric approach is more robust since they do not rely on the silouette contours but on the occupied/non-occupied image regions. Additionally, its reconstruction can be easily limited to the desired accuracy. One problem is the high computational cost. Fortunately, the use of dedicated hardware and software acceleration techniques makes real-time volume reconstruction possible [25, 7].

We use a similar volume reconstruction method as that of Cheung et al. [7]. In contrast with their method, we only project the center rather than the eight corners of each voxel to the silhouette for the intersection test. The projection of the center point of each voxel is pre-computed and stored in a lookup table for computational speeding up. After the reconstruction, the internal voxels are removed and we can get the 3D surface voxel point cloud. Fig. 2 gives an illustration of this method. The method will output the reconstruction result that contains both inside and surface voxels of the whole volume of the subject. As the surface voxels are enough for the pose estimation application, a post-process is needed to remove the inside voxels. This can be done effectively by testing each voxel's 6-connected neighbors. If the type of any one of the neighbors is 0, the voxel is a surface voxel. This post-process can save much CPU time in rendering. Fig. 2 gives an illustration of this method. Fig. 4(b) shows an example of a volume reconstruction. The main steps of the method are described as follows:

1. Initialization: subdivide the 3D space of interest into $N \times N \times N$ equal sized cubes and compute the center point coordinate for each cube. Then build a lookup table for storing the pre-computed projection data.
2. For each camera view, perform a thorough scan:
   (1) For each cue:
   (2) If the type of the cube is "0", continue; Otherwise conduct the point-silhouette-intersection test.
   (3) If the projection of the center point is inside the silhouette, set the cube's type as "1"; otherwise, set as "0".
   (4) End for.
3. Remove the inside voxels.

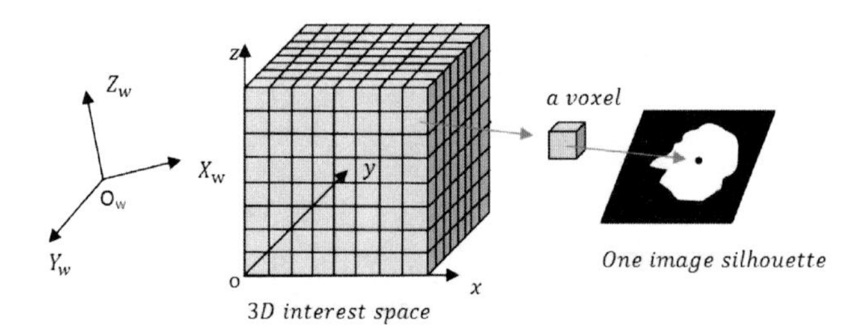

Fig. 2. The 3D interest space, i.e., the big cube, is divided into voxels of resolution N. The center point of each voxel that is indexed by the x-y-z values is projected to a silhouette for the point-silhouette-test. If the projected pixel is within all silhouettes of one multiple frame, the voxel belongs to the volume of the articulated body.

4.4 PSO for Pose Initialization

The goal of pose initialization is to recover 3D pose of a tracked subject from multiple still image data of the first frame. This problem is difficult as no temporal information can be used. Most approaches reduce the problem either by manually adjusting the model to the start frame, or by assuming that the subject's initial pose is known as a special start pose (e.g., [4, 22]). For a complete motion capture system, pose initialization problem needs to be solved.

Our approach applies PSO to fit the model to the reconstructed volume data of the starting frame. The fitting is performed by minimizing the sum of distance of the model points to the corresponding closest voxel. To make it feasible, the fit is implemented in a hierarchical search. We find that this strategy, however, may reach an erroneous pose if the PSO search is performed without any initial information. It is because in some cases the torso may be fitted to the thighs' voxel data that are difficult to be distinguished from the torso's voxels. Though better results may be reached if we run another search process, in our implementation we assume that the global translation of the body model is known. This can be done by interactively adjusting the body model to the voxel data in our system.

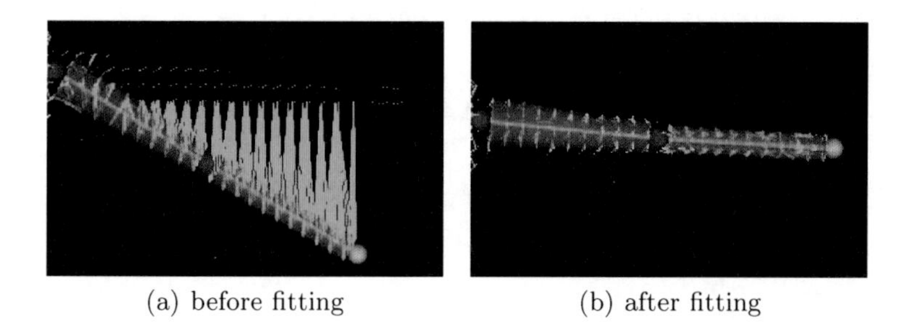

(a) before fitting (b) after fitting

Fig. 3. The illustration of fitting the right arm. The green lines denote the correspondences of the model points to their closest voxel points (white point set).

The hierarchical pose search is performed as a sequence of sub-optimizations on the body parts. We first compute the three global rotations of the torso. The three rotations of the head joint are then estimated using the same optimization method. After that, we turn to independently fit the four limbs each of which has five DOFs. Each sub-optimization process is performed by fitting the corresponding body model part using the volume cue. For example, if the sub-optimization is to compute the pose parameters of the root joint, the fitting should be performed on the torso of the body model. Or if the sub-optimization is for the joints of a leg, then the thigh and calf of the leg will be taken into account. The fitness function is given as follows:

$$E(\boldsymbol{x}) = \frac{1}{N} \sum_{\forall \boldsymbol{p} \in \boldsymbol{M}} \lambda \left\| \boldsymbol{p} - \mathcal{C}(\boldsymbol{p}, \mathcal{V}) \right\|^2 \quad (6)$$

where $\mathcal{M}$ is a set of N randomly sampled model points parameterized by pose parameters $\boldsymbol{x}$, and $\boldsymbol{p}$ is an arbitrary point from the set. $\mathcal{C}(\boldsymbol{p}, \mathcal{V})$ is the closest correspondence function which gives the closest voxel in the reconstructed voxel point set $\mathcal{V}$ to the model point $\boldsymbol{p}$. λ is a penalty factor for embedding space constraints which will be described in Sect. 4.5. Fig. 3 gives an illustration of the fit of one arm by using the objective function of Equ. 6.

4.5 PSO for Pose Estimation and Tracking

Objective Function

Similar to the pose initialization, we fit the articulated model to the volume in each frame during the pose estimation and tracking stage, by using the same hierarchical search method. The difference is that all pose parameters including the global translation are needed to be recovered. Volume is a strong cue which is sufficient to match the body model well against the images [22, 32]. For more robust results, here we incorporate the temporal information into the objective function

$$E(\boldsymbol{x}) = w_V E_V(\boldsymbol{x}) + w_T E_T(\boldsymbol{x}) \tag{7}$$

where w_V and w_T are two weight factors used to balance the influence of the two objective errors. The first term E_V is the same as that of Equ. 6 which pushes the body model to match the volume data. The second term E_T improves the temporal smoothness of the motion

$$E_T(\boldsymbol{x}) = \|\boldsymbol{x} - \acute{\boldsymbol{x}}\|^2 \tag{8}$$

where $\acute{\boldsymbol{x}}$ is the pose of the previous frame.

Constraints

To recover reliable poses, the fit is implemented by integrating space constraints into the object error function. Two space constraints are taken into account when computing the objective error $E_V(\boldsymbol{x})$. First, different body segments are not allowed to intersect in the space. Second, different model points should avoid taking the same closest feature point. The two constraints are integrated into the objective function by introducing the penalty factor λ. If the model point $\boldsymbol{p}$ lies in a voxel bin that has been occupied by other model points from a different body segment, the value of λ is set as λ_1; or else if the model point $\boldsymbol{p}$ has a closest feature voxel that has been taken by other points, λ is set as λ_2; otherwise it is set as 1. Both λ_1 and λ_2 are larger than 1; their actual values can be determined by experiments.

Fitness Evaluation Speedup

One main problem of the PSO based pose search is the time-consuming PSO execution. Usually the time to update the position and velocity of particles of PSO is much less than the time needed to compute the fitness value. Thus the computation time of the PSO is dominated by the fitness evluation.

In our pose estimation method, the most time consuming step of the fitness evluation is to find the closest correspondences between the model points and the voxels and to compute the distances of these point correspondences. We develop a novel and efficient method for this task. As each frame of the volume sequence can be considered as a 3D binary image, we apply a 3D distance transform (DT) to the binary volume, of which "1" represents the feature voxels (i.e., the surface voxels of human shape) and "0" represents empty voxels. The DT converts the 3D binary volume into a gray-level volume in which each voxel has a value equal to its distance to the closest feature voxel in the volume. The adopted 3D EDT computing method [21] can compute the Euclidean distance transform of a k-dimensional binary image in time linear in the total number of voxels.

The limitation of using DT to acquire the closest distance values is the spatial quantization of DT, which contains exact distance values only for points at the center of voxels. For a model point that lies in an arbitrary position of a voxel, we set the closest distance value approximately as the DT value of the voxel. Experiments show that the error caused by this approximation is acceptable, although more precision could be obtained via using a higher voxel resolution.

PSO for Dynamic Pose Tracking

As pose tracking from image sequences is a dynamic process, instead of using the normal PSO frame by frame, we maintain the same swarm for the whole tracking process. To overcome the loss of searching ability problem caused by the convergence of particles at previous frame, we randomly reinitialize each particle according to a Gaussian distribution

$$\boldsymbol{x}_{t+1}^{i} \sim \mathcal{N}(\boldsymbol{p}_{t}^{i}, \boldsymbol{\Sigma}) \tag{9}$$

where $\boldsymbol{p}_t^i$ is the best position of i-th particle at time step t. We assume $\boldsymbol{\Sigma}$ is a diagonal covariance matrix such that $\boldsymbol{\Sigma} = \boldsymbol{I}\delta_j^2$. The standard deviations δ_j are related to the frame rate of the captured image sequences as well as the body parts' motion. For simplicity, we set δ_j as a constant value δ and derive it by experiments.

In the view of Bayesian estimation, the Gaussian distribution can be taken as a transition or weak motion model according to which the PSO swarm propagates over the sequence. This method both utilizes the consecutive information between two frames and also keeps the particle diversity. Our use of PSO for the dynamic problem is similar to that of the work in [45] where they address object tracking problem.

5 Experiment Results

This section describes the experiments. We first give a short description of our multiple camera system. After that, we describe the results of the PSO-based method for pose initialization and pose tracking. We tune the PSO paramters by experiments to reach the best performance. All experiments are carried on the recorded multiple image sequences that cover two different subjects.

Capture System and Environemt

We set up our multiple camera system in an acquisition room of appropriate size, with static backround and stable illumination. This indoorS environment provides good condition for silhouette extraction. We use eight Point Gray cameras comprising three Flea 2 and five Grasshopper IEEE 1394b cameras. These CCD cameras are triggered to synchronously capture RGB color images at a resolution of 800×600 pixels at 15 frames per second (fps). The volume space of interest is set as $2.5 \times 2.5 \times 2.5m^3$. The cameras are calibrated using Zhang's method [46].

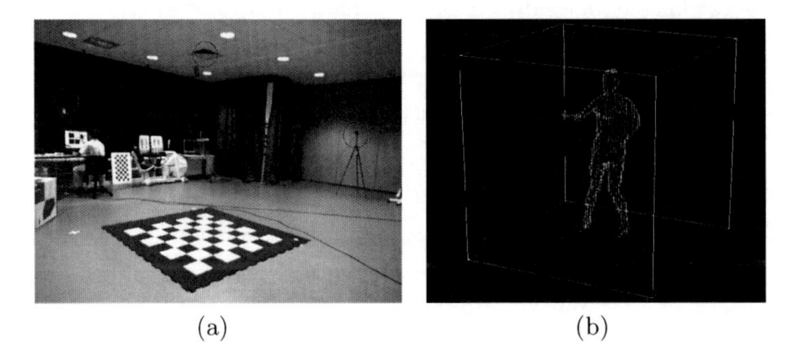

(a) (b)

Fig. 4. (a) The multiple camera system (only four cameras are shown) and capture environment. (b) Example of volume frame in which the cue denotes the interest space.

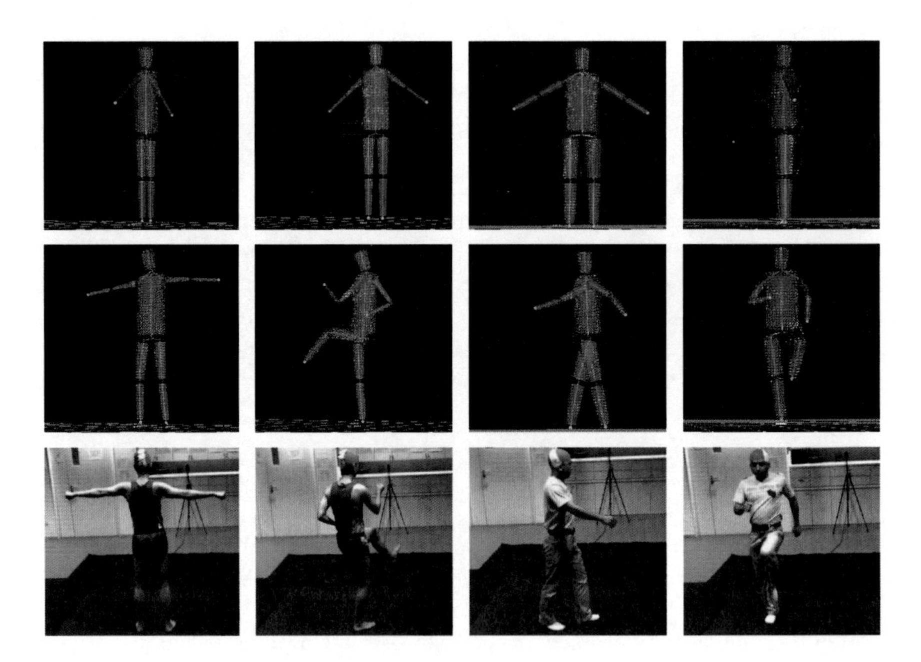

Fig. 5. Four example results of pose initialization. Above row: before pose initialization, the models are at the reference pose. For comparison, the voxels are shown in white color. Middle row: the corresponding pose initialization results. Bottom row: one view of the multiple raw images.

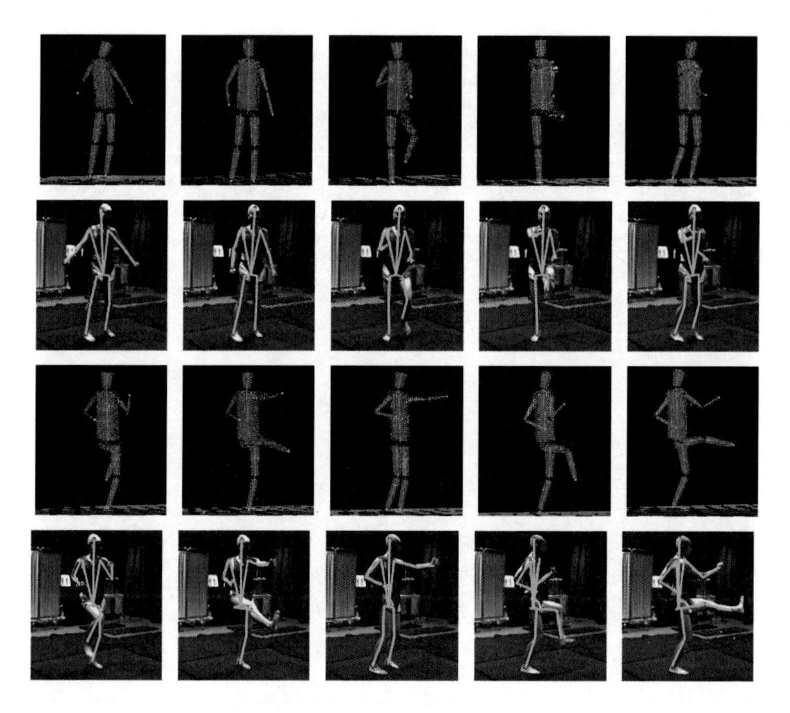

Fig. 6. Pose tracking example results of a sequence of one subject.

5.1 Pose Initialization Results

All experiments of pose initialization are implemented with χ set as 0.7298, ϕ_1 and ϕ_2 as 2.05 according to the Clerc's constriction PSO algorithm [8]. For the fitness evluation, the penalty values of λ_1 and λ_2 are separately set as 30 and 5. For all PSO execution, 20 particles are used. The maximum velocity V_{max} of each dimension can be set according to the corresponding joint angle ranges. For simplicity, we set all maximum velocity values as high as 90 degrees. This is acceptable for the constriction PSO which does not need V_{max} to damp the swarm's dynamics [37]. The stopping criterion is set as maximum iteration that is given as 40 for each 3-DOF sub-optimization (e.g., the sub-optmization for global rotations or the rotations of the head) and 60 for each 5-DOF sub-optimization (e.g., the sub-optimization for legs or arms).

Pose initialization is performed for each sequence captured on two subjects. In our current system, the body size parameters of the articulated body model for each subject are adjusted by hand. To show the performance of the PSO-based pose initialization, we fit the body model at the reference pose to an arbitrary frame which may not be the beginning frame. As described in Sect. 4.4, we put the model at the correct global position and let the PSO search all other pose parameters. Fig. 5 shows some example results. From the results, we can see that though the reference pose is quite far away from the correct body poses, PSO can robustly fit the model to the volume data to recover the initial poses.

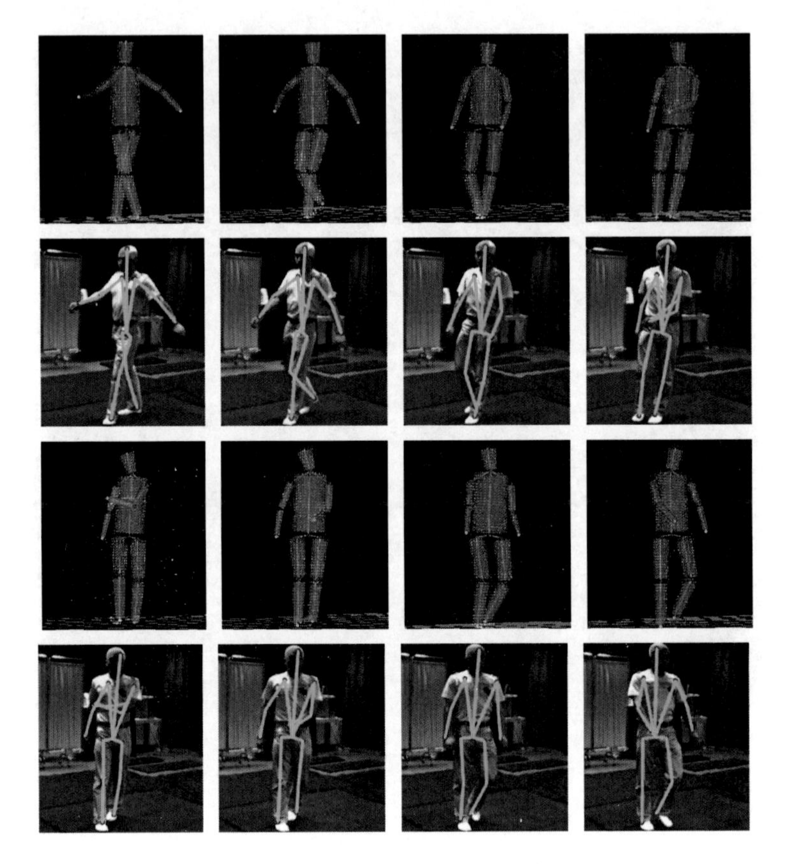

Fig. 7. Pose tracking example results of a squence of the second subject.

5.2 Pose Estimation and Tracking Results

For pose tracking, we set the same PSO parameters as that of the pose initialization experiments. The maximum iteration for each swarm is set to be the same as the pose initialization, except for the 6-DOFs sub-optmization of torso which is set as 80. The penalty factors λ_1 and λ_2 are separately set as 20 and 5 which are optimally determined by experiments. The objective function (Equ. 7) depends on two weight factors w_V and w_T. We experimented with random changes to the two parameters and use $w_V = 1$ and $w_T = 10$ for our sequences that are recorded at 15 fps. For sequences that have faster movements or low frame rate, w_T should be set to a smaller value. We set δ for swarm propagation (Equ. 9) as 15 and limit the maximum joint angle standard deviations to 40 degrees. All volume sequences for motion tracking are reconstructed at a voxel resolution of $80 \times 80 \times 80$. Fig. 6 and Fig. 7 show some pose tracking results of the two subjects.

The experiments demonstrate that by using PSO as the search method, we can successfully address the pose initialization and tracking problems. Based on PSO, our method robustly fits the body model to the beginning frame without a good initial pose, and tracks 3D human motion from the multiple image sequences.

We remark that the tracker would meet failures for the volume frames where the body limbs are close to the torso. This is a common problem for volume based method [6]. To obtain good tracking performances, the articulated model should be well suited the subject's body shape.

6 Conclusion

In this chapter, we apply PSO to full body motion capture without the use of markers. Due to the global search ability of PSO, pose initialization is done in a semi-automatic way, with a little user interaction. Though losing full automaticity as it needs the user to provide some information as constraints, this method is efficient and can obtain robust results. To track full body movements over several frames, our method fits an articulated body model to the volume data reconstructed from multiple images. Pose estimation is performed in a hierarchical way with space constraints enforced to each PSO-based sub-optimization step. To reduce the time of computing fitness values, 3D distance transform is used in a novel manner.

Compared with other optimizers like LM algorithm, PSO shows global search ability, which makes our tracker able to deal with fast movements. The stochastic particle-based search property also makes the tracker capable of recovering from tracking failures. As PSO is a derivative-free technique, our approach avoids computing the complex derivatives of trigonometric functions in the kinematic chains. In addition, as the PSO is readily parallelizable, a parallel PSO algorithm can be implemented for increasing tracking speed.

References

1. Blackwell, T., Branke, J.: Multi-swarm optimization in dynamic environments. In: Raidl, G.R., Cagnoni, S., Branke, J., Corne, D.W., Drechsler, R., Jin, Y., Johnson, C.G., Machado, P., Marchiori, E., Rothlauf, F., Smith, G.D., Squillero, G. (eds.) EvoWorkshops 2004. LNCS, vol. 3005, pp. 489–500. Springer, Heidelberg (2004)
2. Blackwell, T., Branke, J.: Multiswarms, exclusion, and anti-convergence in dynamic environments. In: IEEE Transactions on Evolutionary Computation, vol. 10, pp. 459–472 (2006)
3. Bray, M., Koller-Meier, E., Van Gool, L.: Smart particle filtering for high dimensional tracking. Computer Vision and Image Understanding 106(1), 116–129 (2007)
4. Carranza, J., Theobalt, C., Magnor, M.A., Seidel, H.-P.: Free-viewpoint video of human actors. In: ACM SIGGRAPH 2003 (2003)
5. Cerveri, P., Pedotti, A., Ferrigno, G.: Kinematical models to reduce the effect of skin artifacts on marker-based human motion estimation. Journal of Biomechanics 38(11), 2228–2236 (2005)
6. Cheung, G.K.M., Baker, S., Kanade, T.: Shape-from-silhouette of articulated objects and its use for human body kinematics estimation and motion capture. In: IEEE Conference on Computer Vision and Pattern Recognition (2003)

7. Cheung, G.K.M., Kanade, T., Bouguet, J.-Y., Holler, M.: A real time system for robust 3d voxel reconstruction of human motions. In: IEEE Conference on Computer Vision and Pattern Recognition (2000)
8. Clerc, M., Kennedy, J.: The particle swarm - explosion, stability, and convergence in a multidimensional complex space. IEEE Transaction on Evolutionary Computation 6, 58–73 (2002)
9. Collins, R.T., Lipton, A.J., Kanade, T.: A system for video surveillance and monitoring. Technical report, Robotics Institute, Carnegie Mellon University (1999)
10. Dariush, B.: Human motion analysis for biomechanics and biomedicine. Machine and Vision Application 14(4), 202–205 (2003)
11. Deutscher, J., Blake, A., Reid, I.: Articulated body motion capture by annealed particle filtering. In: IEEE Conference on Computer Vision and Pattern Recognition, vol. 2, pp. 126–133 (2000)
12. Gall, J., Rosenhahn, B., Brox, T., Seidel, H.-P.: Optimization and filtering for human motion capture. International Journal of Computer Vision (2008)
13. Gavrila, D.M.: The visual analysis of human movement: a survey. Computer Vision and Image Understanding 73(1), 82–98 (1999)
14. Gavrila, D.M., Davis, L.S.: 3-d model-based tracking of humans in action: a multi-view approach. In: IEEE Conference on Computer Vision and Pattern Recognition (1996)
15. Ho, S.-Y., Huang, Z.-B., Ho, S.-J.: An evolutionary approach for pose determination and interpretation of occluded articulated objects. In: Proceedings of the 2002 congresss on Evolutionary Computation, vol. 2, pp. 1092–1097 (2002)
16. Horprasert, T., Harwood, D., Davis, L.S.: A statistical approach for real-time robust background subtraction and shadow detection. In: IEEE ICCV, vol. 99, pp. 1–19 (1999)
17. Hu, X.H., Eberhart, R.C.: Proceedings of the 2002 congress on evolutionary computation. In: Adaptive particle swarm optimization: detection and response to dynamic systems, vol. 2, pp. 1666–1670 (2002)
18. Huang, S.S., Fu, L.C., Hsiao, P.Y.: A framework for human pose estimation by integrating data-driven markov chain monte carlo with multi-objective evolutionary algorithm. In: IEEE International Proceedings on Robotics and Automation, pp. 3748–3753 (2006)
19. Ivekovic, S., Trucco, E.: Human body pose estimation with PSO. IEEE Congress on Evolutionary Computation (2006)
20. Ivekovic, S., Trucco, E., Petillot, Y.R.: Human body pose estimation with particle swarm optimisation. Evolutionary Computation 16(4), 509–528 (2008)
21. Maurer Jr., C.R., Qi, R., Raghavan, V.: A linear time algorithm for computing exact euclidean distance transforms of binary images in arbitrary dimensions. IEEE Transactions on Pattern Analysis and Machine Intelligence 25, 265–270 (2003)
22. Kehl, R., Bray, M., Gool, L.V.: Full body tracking from multiple views using stochastic sampling. In: IEEE Conference on Computer Vision and Pattern Recognition (2005)
23. Kehl, R., Gool, L.V.: Markerless tracking of complex human motions from multiple views. Computer Vision and Image Understanding 104(2), 190–209 (2006)
24. Kennedy, J., Eberhart, R.: Particle swarm optimization. In: IEEE International Conference on Neural Networks (1995)
25. Ladikos, A., Benhimane, S., Navab, N.: Efficient visual hull computation for real-time 3d reconstruction using cuda. Computer Vision and Pattern Recognition Workshop 0, 1–8 (2008)

26. Laurentini, A.: The visual hull concept for silhouette-based image understanding. IEEE Trans. Pattern Anal. Mach. Intell. 16(2), 150–162 (1994)
27. Laurentini, A.: The visual hull concept for silhouette-based image understanding. IEEE Trans. Pattern Anal. Mach. Intell. 16(2), 150–162 (1994)
28. Leung, M.K., Yang, Y.-H.: First sight: A human body outline labeling system. IEEE Transactions on Pattern Analysis and Machine Intelligence 17(4), 359–377 (1995)
29. MacCormick, J., Isard, M.: Partitioned sampling, articulated objects, and interface-quality hand tracking. In: Vernon, D. (ed.) ECCV 2000. LNCS, vol. 1843, pp. 3–19. Springer, Heidelberg (2000)
30. Matusik, W., Buehler, C., McMillan, L.: Polyhedral visual hulls for real-time rendering. In: Proceedings of the 12th Eurographics Workshop on Rendering Techniques, pp. 115–126 (2001)
31. Menache, A.: Understanding Motion Capture for Computer Animation and Video Games. Morgan Kaufmann Publishers Inc., San Francisco (1999)
32. Mikic, I., Trivedi, M., Hunter, E., Cosman, P.: Human body model acquisition and tracking using voxel data. International Journal of Computer Vision (IJCV) 53, 199–223 (2003)
33. Moeslund, T.B., Hilton, A., Kruger, V.: A survey of advances in vision-based human motion capture and analysis. Comput. Vis. Image Underst. 104(2), 90–126 (2006)
34. Mundermann, L., Corazza, S., Andriacchi, T.P.: Accurately measuring human movement using articulated icp with soft-joint constraints and a repository of articulated models. In: IEEE Conference on Computer Vision and Pattern Recognition (2007)
35. Parsopoulos, K.E., Vrahatis, M.N.: Recent approaches to global optimization problems through particle swarm optimization. Natural Computing 1, 235–306 (2002)
36. Plankers, R., Fua, P.: Articulated soft objects for multiview shape and motion capture. IEEE Transactions on Pattern Analysis and Machine Intelligence 25(9), 1182–1187 (2003)
37. Poli, R., Kennedy, J., Blackwell, T.: Particle swarm optmization: an overview. Swarm Intelligence 1(1), 33–57 (2007)
38. Poppe, R.: Vision-based human motion analysis: An overview. Comput. Vis. Image Underst. 108(1-2), 4–18 (2007)
39. Pullen, K., Bregler, C.: Motion capture assisted animation: texturing and synthesis. In: SIGGRAPH 2002: Proceedings of the 29th Annual Conference on Computer Graphics and Interactive Techniques, pp. 501–508. ACM, New York (2002)
40. Robertson, C., Trucco, E., Ivekovic, S.: Dynamic body posture tracking using evolutionary optimisation. Electronics Letters 41, 1370–1371 (2005)
41. Shoji, K., Mito, A., Toyama, F.: Pose estimation of a 2d articulated object from its silhouette using a ga. In: 15th International Conference on Pattern Recognition, vol. 3, pp. 713–717 (2000)
42. Szeliski, R.: Rapid octree construction from image sequences. CVGIP: Image Understanding 58(1), 23–32 (1993)
43. Wren, C., Azarbayejani, A., Darrell, T., Pentland, A.: Pfinder: Real-time tracking of the human body. IEEE Transactions on Pattern Analysis and Machine Intelligence 19, 780–785 (1997)

44. X.Z., Liu, Y.C.: Generative tracking of 3d human motion by hierarchical annealed genetic algorithm. Journal of pattern recognition 41, 2470–2483 (2008)
45. Zhang, X.Q., Hu, W., Maybank, S., Li, X., Zhu, M.: Sequential particle swarm optimization for visual tracking. In: IEEE Conference on Computer Vision and Pattern Recognition (2008)
46. Zhang, Z.Y.: A flexible new technique for camera calibration. IEEE Transactions on Pattern Analysis and Machine Intelligence 22, 1330–1334 (1998)

An Adaptive Multi-Objective Particle Swarm Optimization Algorithm with Constraint Handling

Praveen Kumar Tripathi[1], Sanghamitra Bandyopadhyay[2], and Sankar Kumar Pal[2]

[1] Jaypee University of Information and Technology, Waknaghat Solan India
praveen.tripathi@juit.ac.in
[2] Machine Intelligence Unit, Indian Statistical Institute Kolkata India
{sanghami,sankar}@isical.ac.in

Summary. In this article we describe a Particle Swarm Optimization (PSO) approach to handling constraints in Multi-objective Optimization (MOO). The method is called Constrained Adaptive Multi-objective Particle Swarm Optimization (CAMOPSO). CAMOPSO is based on the Adaptive Multi-objective Particle Swarm Optimization (AMOPSO) method proposed in [1]. As in AMOPSO, the inertia and the acceleration coefficients are determined adaptively in CAMOPSO, while a penalty based approach is used for handling constraints. In this article, we first review some existing MOO approaches based on PSO, and then describe the AMOPSO method in detail along with experimental results on six unconstrained MOO problems [1]. Thereafter, the way of handling constraints in CAMOPSO is discussed. Its performance has been compared with that of the NSGA-II algorithm, which has an inherent approach for handling constraints. The results demonstrate the effectiveness of CAMOPSO for the test problems considered.

1 Introduction

Particle Swarm Optimization (PSO) is a relatively recent optimization algorithm that was proposed in 1995 [2]. It has its inspiration in the flocking behavior of birds and the schooling behavior of fishes. The advantages with PSO are its simplicity and efficiency, which has prompted its application for single objective optimization [3, 4]. This success motivated the researchers to apply it to multi-objective optimization (MOO) problems [5, 6]. The Multi-objective Optimization (MOO) basically covers many real-life optimization problems. It involves simultaneous optimization of more than one objective simultaneously. Often this task becomes challenging due to the inherent conflicting nature of the objectives involved in the optimization. Basically two approaches to solve the MOO problem can be considered, the classical approach and the evolutionary approach. The classical approach includes the operation research based techniques. In these cases, multi-objective optimization is handled by first converting the problem into a composite single objective optimization problem using

B.K. Panigrahi, Y. Shi, and M.-H. Lim (Eds.): Handbook of Swarm Intelligence, ALO 8, pp. 221–239.
springerlink.com

some weights. Although it makes the optimization problem very simple but often finding the appropriate weights for the corresponding objective becomes a challenging task. On the other hand, the population based nature of evolutionary approach to MOO allows the problem to be solved in its natural form. These techniques result in obtaining the different compromising solutions to the MOO problems in the population simultaneously at each iteration, thus ruling out the need for any weights. The area of multi-objective evolutionary algorithms (MOEA), has grown considerably from David Schaffers VEGA in 1984 [7], to the recent techniques like NSGA-II [8], SPEA-2 [9] and PESA-II [10]. The only disadvantage associated with MOEA is the high computational time. A relatively recent population based heuristic called Particle Swarm Optimization (PSO) was proposed in 1995 [2]. PSO is inspired by the swarming behavior of birds and the schooling behavior of fishes. This behavior is very simple to simulate and offers faster convergence than evolutionary algorithms. PSO has been found to be quite efficient in handling the single objective optimization problems [3, 11]. The simplicity and efficiency of PSO motivated researchers to apply it to the MOO problems since 2002. Some of these techniques can be found in [5]–[12].

This article addresses the issue of constraints handling in multi-objective optimization domain with PSO algorithm. The present article extends the concept of a multi-objective PSO, called adaptive multi-objective particle swarm optimization (AMOPSO) [1]. In AMOPSO the vital parameters of PSO i.e., inertia and acceleration coefficients, are adapted with the iterations, making it capable of effectively handling optimization problems of different characteristics. To overcome premature convergence, the mutation operator from [13] has been incorporated . In order to improve the diversity in the Pareto-optimal solutions, a method exploiting the nearest neighbor concept is used. This method for measuring diversity has an advantage that it needs no parameter specification, unlike the one in [12]. In this article, we first describe the AMOPSO algorithm [14] in detail. Some results comparing the performance of AMOPSO with other multi-objective PSOs viz., σ–MOPSO [14], NSPSO (Non-dominated sorting PSO) [15] and MOPSO [12] and also other multi-objective evolutionary methods viz., NSGA-II [8] and PESA-II [10] has been provided for six test problems using qualitative measures and also visual displays of the Pareto front.

Constraint handling is an important issue in MOO. In [8], the concept of constrained domination is proposed, where infeasible solutions are always discarded even when compared to a feasible but extremely poor solution. CAMOPSO, the version of AMOPSO that can handle constraints, incorporates a penalty based approach, as suggested for single objective genetic algorithm [16] and for multi-objective genetic algorithm in [17]. The motivation behind the penalty based approach has been to allow the solutions (particles) that are marginally out of the feasible search to return back into the feasible space. This approach will be very beneficial for the search space that has disconnected feasible space. Experimental results on five constrained test problems demonstrate the effectiveness of incorporating the penalty based constraint handling approach within the AMOPSO framework.

2 Multi-Objective Optimization

Considering a general optimization problem (minimization here) involving M objectives, it can be mathematically stated as in [18] :

$$\left.\begin{array}{rl} Minimize: & f(\mathbf{x}) = [f_i(\mathbf{x}), i = 1, \ldots, M] \\ subject\ to\ the\ constraints: & g_j(\mathbf{x}) \geq 0 \quad j = 1, 2, \ldots, J, \\ & h_k(\mathbf{x}) = 0 \quad k = 1, 2, \ldots, K, \end{array}\right\} \tag{1}$$

Here $\mathbf{x} = [x_1, x_2, \ldots, x_n]$, represents a candidate solution to the problem, where n is the dimension of the decision variable space. $f_i(\mathbf{x})$ is the ith objective function, $g_j(\mathbf{x})$ is the jth inequality constraint, and $h_k(\mathbf{x})$ is the kth equality constraint. Thus, MOO problem then in the above framework reduces to finding an $\mathbf{x}$ such that $f(\mathbf{x})$ is optimized.

As the notion of an optimum solution in multi-objective space is different compared to the single objective optimization, we need a different method for evaluation of the solutions. The concept of Pareto optimum, formulated by Vilfredo Pareto, provides such a method called *Pareto dominance* for such evaluation. The Pareto-dominance is defined as in [19], i.e., a vector $\overline{u} = (u_1, u_2, \ldots, u_M)$ is said to dominate a vector $\overline{v} = (v_1, v_2, \ldots, v_M)$ (denoted by $\mathbf{u} \preceq \mathbf{v}$), for a multi-objective minimization problem, if and only if

$$\forall i \in \{i, \ldots, M\}, \quad u_i \leq v_i \wedge \exists i \in \{1, \ldots M\} : u_i < v_i, \tag{2}$$

where M is the dimension of the objective space.

In other words, a solution $\overline{u}$ is said to dominate a solution $\overline{v}$ if the following conditions apply

a) solution $\overline{u}$ is not worse than solution $\overline{v}$ in any objective
b) solution $\overline{u}$ is strictly better than solution $\overline{v}$ in at least one objective

A solution $\overline{u} \in U$, where U is the universe, is said to be *Pareto Optimal* if and only if there exists no other solution $\overline{v} \in U$, such that $\overline{u}$ is dominated by $\overline{v}$. Such solutions ($\overline{u}$) are called *non-dominated solutions.* The set of all such non-dominated solutions constitutes the *Pareto-Optimal Set* or *non-dominated set.*

3 Particle Swarm Optimization (PSO)

The concept of PSO is inspired by the flocking behavior of the birds and the schooling behavior of fishes. It was first proposed by Kennedy in 1995 [2]. Like evolutionary algorithms PSO is also a population based heuristic, where the population of the potential solutions is called a *swarm* and each individual solution within the *swarm*, is called a *particle.* Each particle within the swarm is allowed to fly towards the optimum solution to simulate the swarming behavior. The flight of a particle in the swarm is influenced by its own experience and by the experience of the swarm.

Considering an N dimensional search space, an ith particle is associated with the position attribute $X_i = (x_{i,1}, x_{i,2}, \ldots, x_{i,N})$, the velocity attribute $V_i = (v_{i,1}, v_{i,2}, \ldots, v_{i,N})$ and the individual experience attribute $P_i = (p_{i,1}, p_{i,2}, p_{i,N})$ [4]. The position attribute (X_i) signifies the position of the particle in the search space, whereas the velocity attribute (V_i) is responsible for imparting motion to it. The P_i parameter stores the position (coordinates) corresponding to the particle's best individual performance. Similarly the experience of whole of the swarm is captured in the index g, which corresponds to the particle with the best overall performance in the swarm. The movement of the particle towards the optimum solution is governed by updating its position and velocity attributes. The velocity and position update equations are given as:

$$v_{i,j} = wv_{i,j} + c_1 r_1 (p_{i,j} - x_{i,j}) + c_2 r_2 (p_{g,j} - x_{i,j}) \tag{3}$$

$$x_{i,j} = x_{i,j} + v_{i,j}\chi \tag{4}$$

where $j = 1, \ldots, N$ and w, c_1, c_2, $\chi \geq 0$. w is the inertia weight, c_1 and c_2 the acceleration coefficients, χ is the constriction parameter and r_1 and r_2 are the random numbers generated uniformly in the range $[0, 1]$, responsible for imparting randomness to the flight of the swarm. The term $c_1 r_1 (p_{i,j} - x_{i,j})$ in Equation 3 is called *cognition* term whereas the term $c_2 r_2 (p_{g,j} - x_{i,j})$ is called the *social* term. The *cognition* term takes in to account only the particle's individual experience, whereas the *social* term signifies the interaction between the particles. The c_1 and c_2 values allow the particle to tune the cognition and the social terms respectively in the velocity update Equation 3. A larger value of c_1 allows the particle to explore a larger search space, while a larger value of c_2 encourages refinement around the global best solution.

4 Related Study

The simplicity and faster convergence of the PSO in solving the single objective optimization problems [3, 11], inspired its extension to the multi-objective problem domain. There have been several recent attempts to use PSO for multi-objective optimization [5, 12, 14, 20] and [21]–[22]. Some of these concepts have been surveyed briefly in this section. The dynamic neighborhood PSO [5] has been given for two objective MOO problems only. This concept assumes a considerable degree of prior knowledge in terms of the test problem properties. Here instead of a single *gbest*, a local *lbest* is obtained for each swarm member, that is selected from the closest two swarm members. The closeness is considered in terms of one of the objectives, while the selection of the optimal solution from the closest two is based on the other objective. The selection of the objectives for obtaining the closest neighbors and local optima is usually based on the knowledge of the problem being considered for optimization. Usually the simpler objective is considered for closest members computation. A single *pbest* solution is maintained for each member that gets replaced by the present solution only if the present solution dominates the *pbest* solution. In [22] two methods have been proposed to solve MOO using PSO. The first method uses *weighted aggregate approach*, whereas the second one is inspired by Schaffer's VEGA [7]. In the first approach, the algorithm needs to be run K times to get the K non-dominated solutions. Though this approach is of low computational cost, it needs multiple runs. The Vector Evaluated Particle Swarm Optimizer (VEPSO), uses one swarm corresponding to each objective. The best particle obtained for the second swarm is used to guide the velocities of the particles of the first swarm and vice-verse. The multi-objective particle swarm optimization algorithm (MOPSO) in [21] had its inspiration from the latest developments in MOEA. MOPSO maintains two archives, one for storing the globally non-dominated solutions and the other for storing the individual best solutions attained by each particle. MOPSO uses method inspired by [23] for maintaining diversity. The fitness assigned to each individual in the archive is computed on the basis of its density. The individual occupying less dense region gets higher fitness. This fitness is used in roulette wheel selection, to pick the *gbest* solution in velocity and position update Equations 3 and 4. In the local archive a solution gets replaced by the present solution, only if the former is dominated by the latter.

In [12], authors improved the aforementioned MOPSO by incorporating a mutation operator. The mutation operator boosts the exploring capability of the MOPSO. The article also addressed the constraint handling problem with MOO. The authors compared the proposed MOPSO with existing MOEAs, viz., NSGA-II [8] and PAES [23]. In [20] a concept has been given, that argues against the approach adopted in [21]. The authors suggested that the selection of the *gbest* from the archive, randomly on the basis of diversity as in [21] should be improved. The selection of the nearest dominating solution from the archive, for a particular particle in the swarm, has been used as its *gbest* solution. Since the process of finding the nearest dominating solution from the global archive is computationally complex, the concept of *domination tree* has been introduced. The concept of *turbulence* was also given that acts as a mutation operator on the velocity parameter of PSO. In [14] a methodology for MOPSO has been introduced. Authors have introduced a concept of σ, for selecting the best local guides [14]. σ values are assigned to the members of the archive as well as the swarm. For a particle in the swarm, a particle from the archive having the closest value of σ is chosen as the local guide. The authors have used the concept of *turbulence* in the decision space. σ–MOPSO has been compared with the MOPSO proposed in [20] and an MOEA, viz., SPEA-2 [9]. σ–MOPSO was found to be superior in terms of the convergence as well as diversity. The size of the archive has been kept constant. If the size of the archive exceeds its maximum limit, clustering is applied to truncate it. Since the aforementioned algorithm emphasizes closeness in σ values, it makes the selection pressure even higher, that may result in premature convergence. In [15] author has used non-dominated sorting in MOPSO. This concept has been inspired by NSGA-II [8] algorithm. Author has suggested the combined evaluation of the particles' personal bests and the offspring, instead of the single comparison between a particle's personal best and its offspring. If the size of the swarm be N, then a combined population of $2N$, comprising the swarm and its personal bests is first obtained. Then the non-dominated sorting of this population determines the particles of the next generation swarm. To ensure proper diversity amongst the solutions of the non-dominated solutions, two approaches namely *niche count* and *crowded distance* [8] methods are used. An archive containing the non-dominated solutions is also maintained, the best solutions from this archive in terms of diversity are selected as the global best for a particle. The details of this algorithm may be found in [15]. In [1] and [24] authors have addressed the issue of the adaptive parameters for the MOO problems in PSO domain.

In [24] the vital parameters of the PSO algorithms viz., inertia and acceleration coefficients have been made time variant, hence named as Time Variant Multi-objective Particle Swarm Optimization (TV-MOPSO) algorithm. The larger value of inertia and social acceleration coefficient encourages exploration where as the smaller value of inertia and the larger value of the cognition coefficient encourages exploitation. This issue is emphasized by allowing the inertia and the cognition coefficients to decrease with iterations, and the social coefficient to increase with the iterations. The performance of the proposed algorithm TV-MOPSO was compared with the existing MOEA and multi-objective PSOs. TV-MOPSO was found to perform better than most of the algorithms on different test problems, using different performance measures.

In [1] these vital parameters were treated as control parameters and were themselves optimized along with the usual parameters of the optimization problem. The present article extends this concept with the incorporation of a penalty based constraint handling mechanism.

The details of the aforementioned algorithms can be obtained from the respective references.

5 Adaptive Multi-Objective Particle Swarm Optimization: AMOPSO

This section describes the AMOPSO algorithm proposed in [1]. It is included here for the convenience of the readers.

In the initialization phase of AMOPSO, the particles of the swarm are assigned random values for the different coordinates, from the respective domains, for each dimension. Similarly the velocity for each particle is initialized to zero in each dimension. The Step 1 takes care of the initialization of AMOPSO. This algorithm maintains an archive for storing the best non-dominated solutions found in the flight of the particles. The size of the archive l_t at each iteration is allowed to attain a maximum value of N_a. Archive is initialized to contain the non-dominated solutions from the swarm. The Step 2 of AMOPSO deals with the flight of the particles within the swarm through the search space. The flight, given by Equations 3 and 4, is influenced by many vital parameters, which are explained below: In multi-objective PSO, the *pbest* stores the best

Algorithm AMOPSO [1]: O_f= AMOPSO(N_s,N_a,C,d)
/* N_s: size of the swarm, N_a: size of the archive, C: maximum number of iterations, d: the dimensions of the search space, O_f: the final output */

1. $t = 0$, randomly initialize S_0,
   /*S_t: swarm at iteration t */
   - initialize $x_{i,j}$,$\forall i$, $i \in \{1,\ldots,N_s\}$ and $\forall j$, $j \in \{1,\ldots,d\}$
     /* $x_{i,j}$: the jth coordinate of the ith particle */
   - initialize $v_{i,j}$, $\forall i$, $i \in \{1,\ldots,N_s\}$ and $\forall j$, $j \in \{1,\ldots,d\}$
     /* $v_{i,j}$: velocity of ith particle in j^{th} dimension */
   - $Pb_{i,j} \leftarrow x_{i,j}$,$\forall i$, $i \in \{1,\ldots,N_s\}$ and $\forall j$, $j \in \{1,\ldots,d\}$
     /* $Pb_{i,j}$: the jth coordinate of the personal best of the ith particle */
   - $A_0 \leftarrow$ *non_dominated*(S_0), $l_0 = |A_0|$
     /* returns the non-dominated solutions from the swarm*/
     /* A_t: archive at iteration t */
2. for $t = 1$ to $t = C$,
   - for $i = 1$ to $i = N_s$ /* update the swarm S_t */
     - /* updating the velocity of each particle */
       - $Gb \leftarrow$ *get_gbest*()
         /* returns the global best */
       - $Pb_i \leftarrow$ *get_pbest*()
         /* returns the personal best */
       - *adjust_parameters*(w_i, c_1^i, c_2^i)
         /* adjusts the parameters, w_i: the inertia coefficient, c_1^i: the local acceleration coefficient, and c_2^i: the global acceleration coefficient */
         $v_{i,j} = w_i v_{i,j} + c_1^i r_1(Pb_{i,j} - x_{i,j}) + c_2^i r_2(Gb_j - x_{i,j})$
         $\forall j$, $j \in \{1,\ldots,d\}$
     - /* updating coordinates */
       $x_{i,j} = x_{i,j} + v_{i,j}$
       $\forall j$, $j \in \{1,\ldots,d\}$
   - /* updating the archive */
     - $A_t \leftarrow$ *non_dominated*$(S_t \cup A_t)$
     - if $(l_t > N_a)$ *truncate_archive*()
       /* l_t: size of the archive */
   - mutate (S_t) /* mutating the swarm */
3. $O_f \leftarrow A_t$ and stop. /* returns the Pareto optimal front */

non-dominated solution attained by the individual particle. In AMOPSO the present solution is compared with the *pbest* solution, and it replaces the *pbest* solution only if it dominates that solution. In multi-objective PSO, often the multiple objectives involved in MOO problems are conflicting in nature thus making the choice of a single optimal solution difficult. To resolve this problem the concept of non-dominance is used [1]. Therefore instead of having just one individual solution as the global best a set of all the non-dominated solutions is maintained in the form of an *archive* [12]. Selecting a single particle from the archive as the *gbest* is a vital issue. There has been a number of attempts to address this issue, some of which may be found in [25]. In the concepts mentioned in [25], authors have considered non-dominance in the selection of the *gbest* solution. As the attainment of proper diversity is the second objective of MOO, it has been used in AMOPSO to select the most sparsely populated solution from the archive as the *gbest*. In AMOPSO the diversity measurement has been done using a novel concept. This concept is similar to the *crowding-distance* measure in [8]. The parameter (d_i) is computed as the distance of each solution to its immediate next neighbor summed over each of the M objectives. Density for all the solutions in the archive is obtained. Based on the density values as fitness, roulette wheel selection is done to select a solution as the *gbest*. The performance of PSO to a large extent depends on its inertia weight (w) and the acceleration coefficients (c_1 and c_2). In this article these are called *control parameters*. These parameters in AMOPSO have been adjusted using the function $adjust\_parameters(w_i, c_1^i, c_2^i)$.

Some values have been suggested for these parameters in the literature [4]. In most of the cases the values of these parameters were found to be problem specific, signifying the use of adaptive parameters [3, 4, 26]. In [1] the control parameters have been subjected to optimization through swarming, in parallel with that of the normal optimization variables. The intuition behind this concept is to evolve the control parameters also so that the appropriate values of these parameters may be obtained for a specific problem. Here the control parameters have been initially assigned some random values in the range suggested in [4]. They are then updated using the following equations:

$$v_{i,j}^c = w_i v_{i,j}^c + c_1^i r_1 (p_{i,j}^c - x_{i,j}^c) + c_2^i r_2 (p_{g,j}^c - x_{i,j}^c) \tag{5}$$

$$x_{i,j}^c = x_{i,j}^c + v_{i,j}^c \tag{6}$$

Here, $x_{i,j}^c$ is the value of the jth control parameter with the ith particle, whereas $v_{i,j}^c$, $p_{i,j}^c$ are the velocity and personal best for the jth control variable with i^{th} particle. $p_{g,j}^c$ is the global best for the jth control parameters. The previous iteration values of the control parameters have been used for the corresponding values of w_i, c_1^i and c_2^i. It is to be noted that in the above equations the values of j equal to, 1, 2 and 3, corresponds to the control parameters inertia weight, cognition acceleration coefficient and the global acceleration coefficient respectively.

The mutation operator of [13] has been used in AMOPSO to allow better exploration of the search space. Given a particle, a randomly chosen variable (one of the coordinate of the particle), say g_k, is mutated as given below:

$$g_k' : \begin{cases} g_k + \Delta(t, UB - g_k) \ if \ flip = 0, \\ g_k - \Delta(t, g_k - LB) \ if \ flip = 1. \end{cases} \quad (7)$$

where $flip$ denotes the random event of returning 0 or 1. UB denotes the upper limit of the variable g_k, while LB the lower limit. The function Δ is defined as:

$$\Delta(t, x) = x * (1 - r^{(1 - \frac{t}{max\_t})^b}) \quad (8)$$

where r is a random number generated in the range $[0, 1]$ and max_t is the maximum number of iterations, t is the iteration number. The parameter b determines the degree of dependence of mutation on the iteration number. Mutation operation on the swarm is done by the function $mutate(S_t)$ in AMOPSO at Step 2.

The selection of the *gbest* solution for the velocity update is done from this archive only. In AMOPSO the maximum size of the archive has been fixed to $N_a = 100$. The archive gets updated by the non dominated solutions of the swarm. All the dominated members from the archive are removed. Since the maximum size of the archive has been fixed, the density has been considered as in [12] to truncate the archive to the desired size. After running AMOPSO for a fixed number of iterations, the archive is returned as the resultant non dominated set.

6 CAMOPSO: Constraint Handling in AMOPSO

CAMOPSO incorporates constraint handling in MOO problems using the penalty based approach. In the constrained MOO problem formulated in Equation 1, J inequality and K equality constraints need to be satisfied. Constraints can be classified as hard or soft [18]. The hard constraints must always be satisfied, where as the soft constraints may be relaxed for the feasible solutions. Such hard equality constraints can be converted to soft inequality constraints [27]. For this reason only the constraints with inequalities have been considered.

In [28] constraints have been handled using a modified definition of the domination. The authors have introduced the notion of *constrained domination*. A solution i is said to constrained-dominate a solution j, if any of the following conditions is true:

a) solution i is feasible and solution j is not.
b) solutions i and j are both infeasible, but solution i has a smaller overall constraint violation.
c) solutions i and j are feasible and solution i dominates solution j.

In this approach an infeasible solution always gets eliminated when compared to a feasible one, irrespective of the goodness of the objective value. Therefore a poor feasible solution will be preferred as compared to a very good solution which is infeasible, although it may be very close to the feasible region. It may also be the case that the path to the best solutions lie through the infeasible space, whence this approach may be at a disadvantage.

In the present article a penalty based approach to constraint handling is incorporated that overcomes this problem to a large extent. This approach is motivated by the suggestions in [16] for single objective GAs. The Constrained MOO problem with only inequality constraints is now reformulated as follows:

$$\begin{array}{l} Minimize: f(\mathbf{x}) = [f_i(\mathbf{x}), i = 1, \ldots, M] \\ Subject\ to: g_j(\mathbf{x}) \geq 0 \quad j = 1, 2, \ldots, J. \end{array} \tag{9}$$

δ_j be the constraint violation corresponding to j^{th} inequality constraint then the penalty function $p(\mathbf{x})$, is defined as

$$p(\mathbf{x}) = n_c \times \max_j \{\delta_j\}, \forall\ [j = 1, 2, \ldots, J], \tag{10}$$

where n_c is given by

$$n_c = |\{j : g_j(\mathbf{x}) < 0\}|. \tag{11}$$

In other words if solution $\mathbf{x}$ is an infeasible solution that violates n_c inequality constraints, then the penalty function is n_c times the maximum amount of constraint violation, amongst these n_c constraints by $\mathbf{x}$. Thus the fitness of $\mathbf{x}$ corresponding to each objective now gets updated as

$$f(\mathbf{x}) = [f_i(\mathbf{x}) + p(\mathbf{x}), i = 1, \ldots, M] \tag{12}$$

The motivation behind this approach has been to give more priority to the feasible solutions, while degrading the infeasible solutions but not totally disqualifying them. Note that in most complex constrained search spaces, particularly with disconnected feasible regions, the infeasible solutions may also play a vital role.

7 Experimental Results

The effectiveness of AMOPSO has been already demonstrated on various standard test problems, that have known sets of Pareto optimal solutions and are characterized to test the algorithms on different aspects of performance in [1]. For the sake of completeness we reproduce those here in this article. AMOPSO has been compared with some MOEAs and MOPSOs. The MOEAs include NSGA-II and PESA-II, whereas the MOPSOs are MOPSO, σ-MOPSO and NSPSO. The codes for NSGA-II and MOPSO have been obtained from *http : //www.iitk.ac.in/kangal/codes.shtml* and *http : //www.lania.mx/ ccoello /EMOO/EMOOsoftware.html* respectively. The program for PESA-II has been obtained from the authors, whereas the other algorithms are implemented. The parameters used are: population/swarm size 100 for NSGA-II and NSPSO,

10 for PESA-II (as suggested in [10]), 50 for MOPSO, σ-MOPSO and AMOPSO, archive size 100 for PESA-II, σ-MOPSO, MOPSO and AMOPSO, number of iterations 250 for NSGA-II and NSPSO, 2500 for PESA-II, and 500 for MOPSO, σ-MOPSO and AMOPSO (to keep the number of function evaluations to 25000 for all the algorithms), cross-over probability 0.9 (as suggested in [8]) for NSGA-II and PESA-II, mutation probability inversely proportional to the chromosome length (as suggested in [8]), coding strategy binary for PESA-II (only binary version available) while real encoding is used for NSGA-II, NSPSO, σ-MOPSO, MOPSO and AMOPSO (PSO naturally operates on real numbers). The values of c_1 and c_2 have been used as 1 and 2 for σ-MOPSO and NSPSO, respectively (as suggested in [14] and [15]). The value of w has been used as 0.4 for σ-MOPSO whereas it has been allowed to decrease from 1.0 to 0.4 for NSPSO (as suggested in [14] and [15]). For AMOPSO the initial range of the values for c_1^i and c_2^i is $[0.5, 2.5]$ and that for w_i is $[0.0, 1.0]$ (as suggested in [26]). The values for the parameters of a particular algorithm have been used keeping in mind the suggestions in the respective literature. Usually a smaller size of the swarm is preferred. The relative performances of NSGA-II, PESA-II, σ-MOPSO, NSPSO, MOPSO and AMOPSO are evaluated on several test problems (i.e., two and three objectives), using some performance measures.

The constrained AMOPSO i.e., CAMOPSO has been compared with NSGA-II algorithm that has inbuilt mechanism for handling constrained problems. In this article we have included results on five test problems which are *constr*, *tnk*, *srn*, *ctp*1 and *ctp*2 [28].

7.1 Test Problems and Performance Measures

In this article the results on six standard unconstrained test problems as given in [1] have been reproduced for the sake of completion. Four of these test problems $ZDT1$, $ZDT2$, $ZDT3$, $ZDT4$ [29] are of two objectives, while the other two, i.e., $DLTZ2$ and $DLTZ7$ [30], are of three objcctives. The performance of the algorithms is evaluated with respect to the convergence measure Υ (viz., distance metric) and the diversity measure Δ [8] (viz., diversity metric). The Υ measure has been used for evaluating the extent of the convergence to the Pareto front, whereas Δ measure has been used for evaluating the diversity of the solutions on the non-dominated set. It should be noted that smaller the value of these parameters, the better will be the performance.

In this article we have included results on five constrained test problems which are *constr*, *tnk*, *srn*, *ctp*1 and *ctp*2 [28].

7.2 Results

The results include the initial results from [1] for the unconstrained test cases. The results are reported in terms of the mean and variance of the performance measures over 20 simulations in Tables 1 and 2 [1]. It can be seen that AMOPSO has resulted in better convergence on all the test problems in terms of Υ measure. PESA-II has also given the best convergence for the $ZDT2$ test

Table 1. Mean (M) and Variance (Var) of the Υ measures for the test problems

Υ Measure						
Algorithm	ZDT1	ZDT2	ZDT3	ZDT4	DLTZ2	DLTZ7
NSGA-II(M)	0.03348	0.07239	0.11450	0.51305	0.72775	0.04976
(VAR)	0.00475	0.03168	0.00794	0.11846	0.20400	0.00000
PESA-II(M)	0.00105	**0.00074**	0.00789	9.98254	0.03219	0.05139
(VAR)	0.00000	0.00000	0.00011	20.13400	0.00000	0.00000
σ-MOPSO(M)	0.01638	0.00584	0.10205	3.83344	0.03167	0.24494
(VAR)	0.00048	0.00000	0.00238	1.87129	0.00000	0.01126
NSPSO(M)	0.00642	0.00951	0.00491	4.95775	0.04938	0.05618
(VAR)	0.00000	0.00000	0.00000	7.43601	0.00000	0.00019
MOPSO(M)	0.00133	0.00089	0.00418	7.37429	0.82799	0.04986
(VAR)	0.00000	0.00000	0.00000	5.48286	0.01133	0.00000
AMOPSO(M)	**0.00099**	**0.00074**	**0.00391**	**0.40311**	**0.02024**	**0.02306**
(VAR)	0.00000	0.00000	0.00000	0.01259	0.00000	0.00000

Table 2. Mean (M) and Variance (Var) of the Δ measures for the test problems

Δ Measure						
Algorithm	ZDT1	ZDT2	ZDT3	ZDT4	DLTZ2	DLTZ7
NSGA-II(M)	0.39031	0.43077	0.73854	0.70261	0.97599	0.89149
(VAR)	0.00187	0.00472	0.01971	0.06465	0.00738	0.00052
PESA-II(M)	0.84816	0.89292	1.22731	1.01136	0.74808	0.74791
(VAR)	0.00287	0.00574	0.02925	0.00072	0.00093	0.00106
σ-MOPSO(M)	0.39856	0.38927	0.76016	0.82842	**0.60405**	0.94113
(VAR)	0.00731	0.00458	0.00349	0.00054	0.00194	0.00640
NSPSO(M)	0.90695	0.92156	0.62072	0.96462	0.72988	0.73812
(VAR)	0.00000	0.00012	0.00069	0.00156	0.00091	0.01002
MOPSO(M)	0.68132	0.63922	0.83195	0.96194	0.74808	0.87375
(VAR)	0.01335	0.00114	0.00892	0.00114	0.00093	0.08186
AMOPSO(M)	**0.31826**	**0.31996**	**0.53154**	**0.65060**	0.67273	**0.73201**
(VAR)	0.00060	0.00068	0.00036	0.00376	0.00087	0.00134

problem. Similarly, the values of the Δ measure in Table 2 show that AMOPSO is able to attain the best distribution of the solutions on the non-dominated front for all the test problems except on $DLTZ2$, where it is second to σ-MOPSO respectively.

To demonstrate the distribution of the solutions on the final non-dominated front, $ZDT3$ and $DLTZ2$ test problems have been considered as typical illustrations. Figures 1-2 show the resultant non dominated fronts corresponding to these test problems.

Figure 1 represents the final fronts obtained by the PSO algorithms for $ZDT3$ function. It can be seen from Figure 1(d) that σ-MOPSO has failed to converge

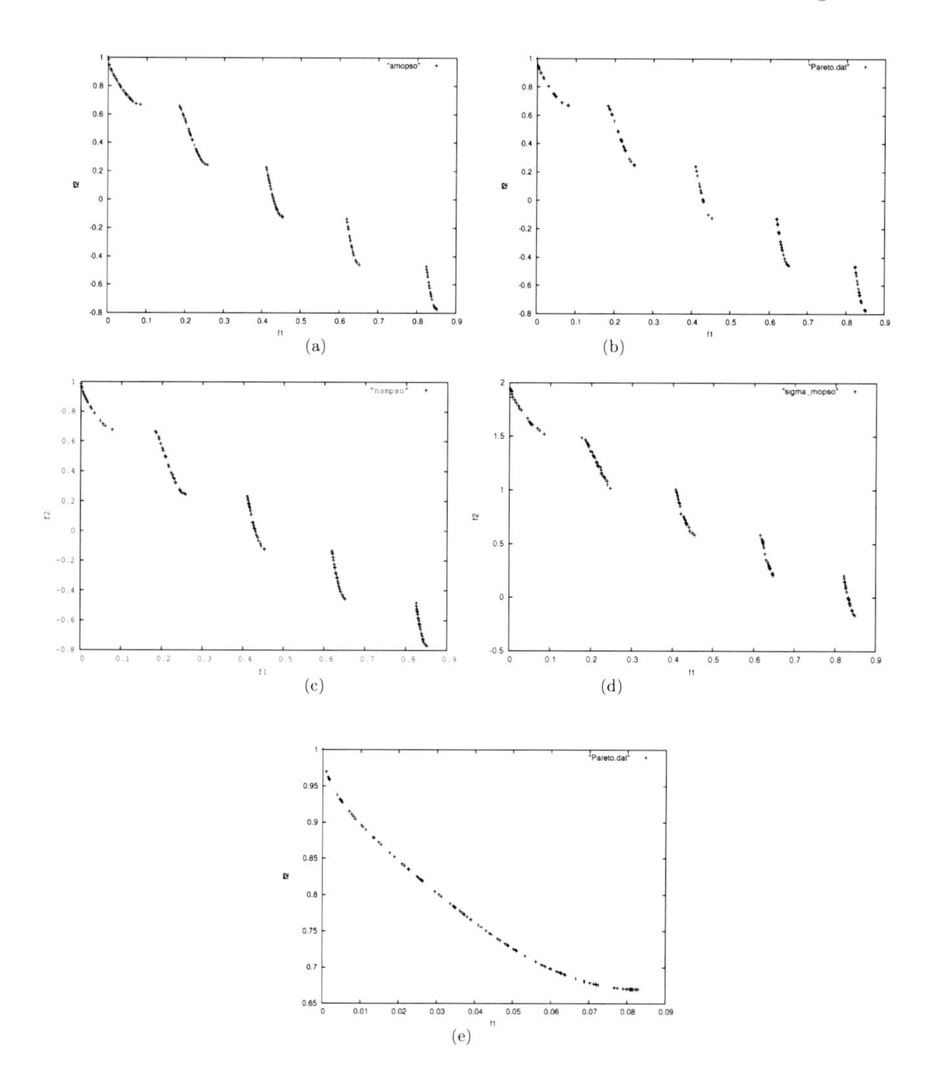

Fig. 1. Final Fronts of (a): AMOPSO, (b): MOPSO, (c): NSPSO and (d): σ-MOPSO (e): MOPSO (local convergence) on ZDT3

to the true Pareto-front properly. For this test problem, MOPSO is often found to converge to a local optimal front for this test function. Such an instance is shown in Figure 1(e), where MOPSO has been able to obtain only one front (not all the five) because of local optima problem. NSPSO has resulted in very good convergence, as evident from Figure 1(c), but its diversity is not as good as that of AMOPSO. Compared to all these algorithms, AMOPSO in Figure 1(a) has given better convergence and spread of the solutions on this test function. Figure 2 represents the final non-dominated fronts obtained by the algorithms on $DLTZ2$ test problem. MOPSO (Figure 2(b)) has failed to attain the full

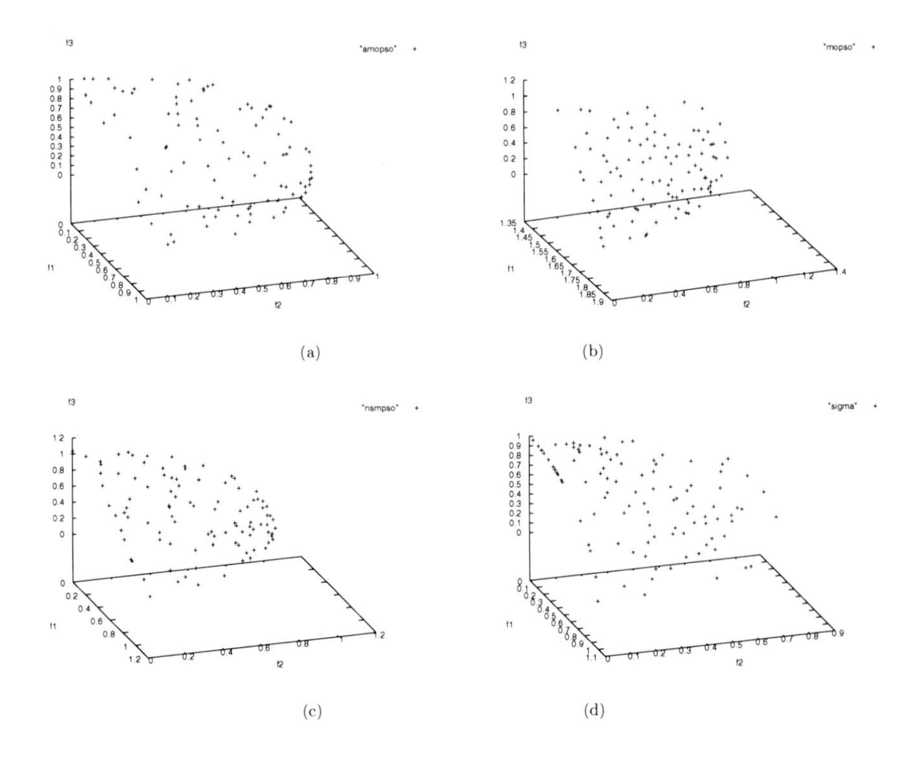

Fig. 2. Final Fronts of (a): AMOPSO, (b): MOPSO, (c): NSPSO and (d): σ-MOPSO on DLTZ2

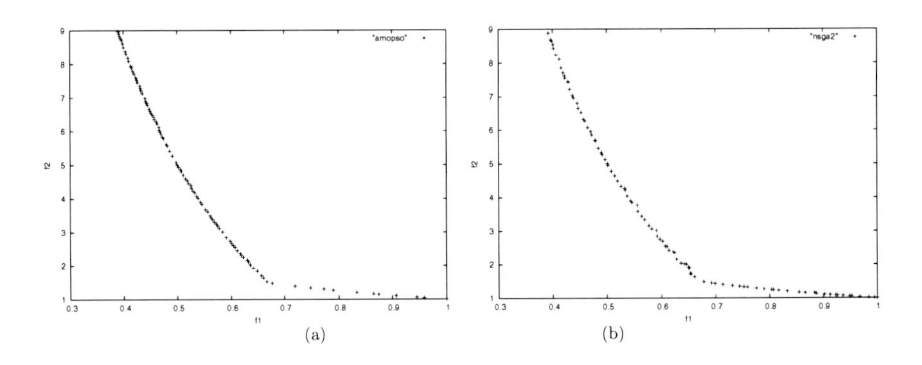

Fig. 3. Final Pareto-fronts of (a): AMOPSO and (b): NSGA-II on constr

non-dominated set. Similarly σ-MOPSO (Figure 2(d)) could not attain the non-dominated set properly. Although NSPSO has resulted in better shape of the Pareto front (Figure 2(c)), its convergence is not as good as that of AMOPSO as shown in Figure 2(a). These results in the form of tables and figures are available in an earlier reference [1].

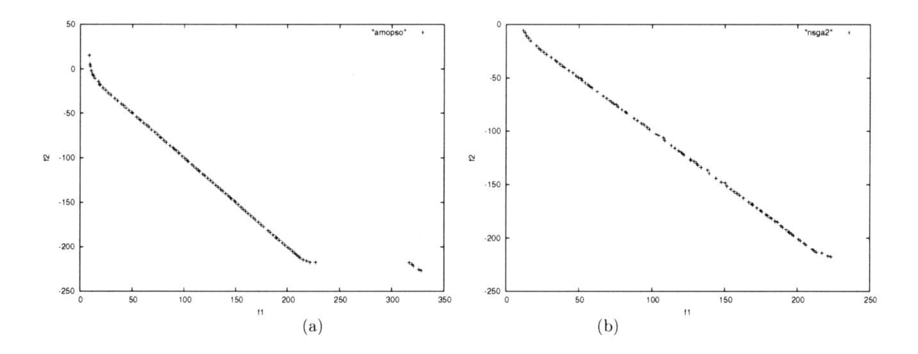

Fig. 4. Final Pareto-fronts of (a): AMOPSO and (b): NSGA-II on srn

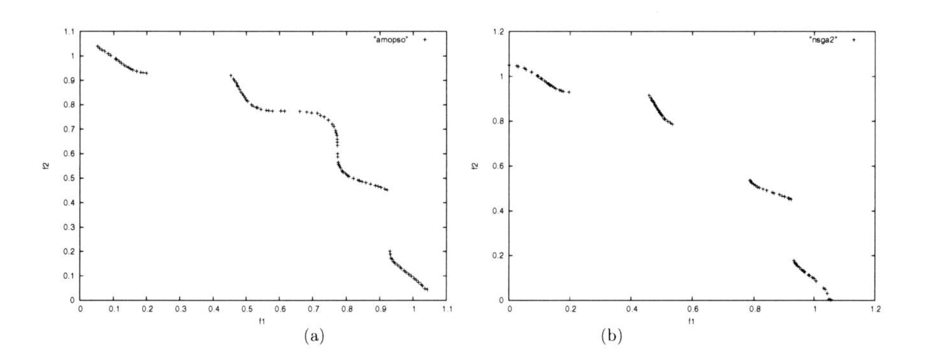

Fig. 5. Final Pareto-fronts of (a): AMOPSO and (b): NSGA-II on tnk

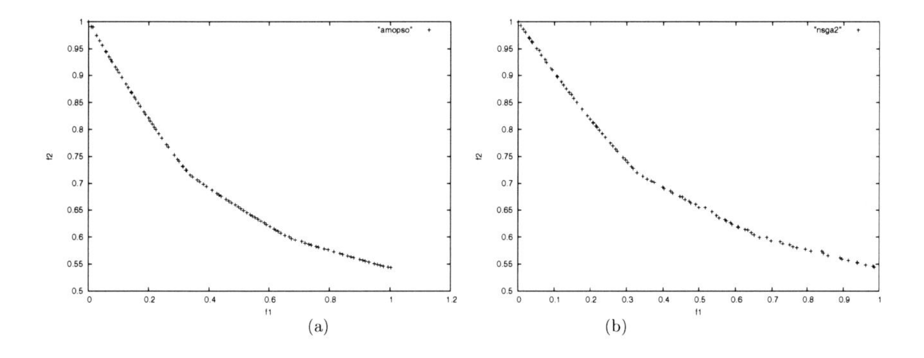

Fig. 6. Final Pareto-fronts of (a): AMOPSO and (b): NSGA-II on ctp1

Table 3 provides the comparative results of CAMOPSO and NSGA-II for the five constrained test problem considered here. The graphical representation of the Pareto fronts obtained by CAMOPSO and NSGA-II are provided in Figures 3-7. The higher value of purity measure whereas lower value of Sm

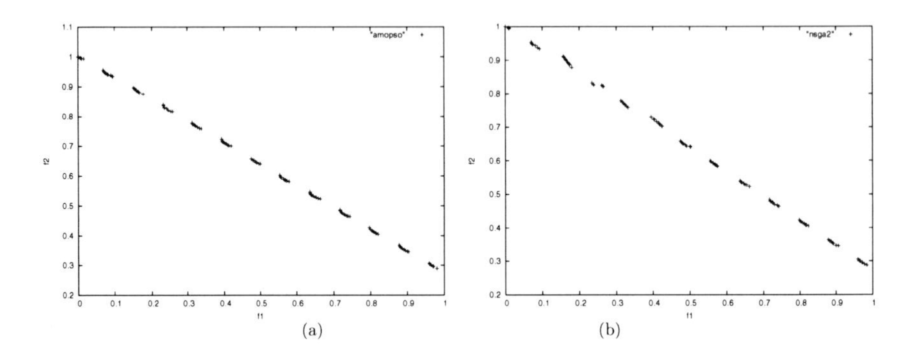

Fig. 7. Final Pareto-fronts of (a): AMOPSO and (b): NSGA-II on ctp2

Table 3. Results on Constrained test problems

Purity Measure					
Algorithm	constr	tnk	srn	ctp1	ctp2
NSGA-II	0.30	0.42	0.67	0.35	0.33
CAMOPSO	0.80	0.89	0.95	0.69	0.65
Sm Measure					
NSGA-II	0.082	0.040	0.012	0.010	0.021
CAMOPSO	0.057	0.037	0.010	0.009	0.019

signifies better performance. From the results reported in the Table 3 it is evident that the CAMOPSO has been successful in arriving at better results in all the five test functions considered in this article. CAMOPSO has been successful in obtaining better convergence as well as better diversity compared to NSGA-II. The performance on the test problem *tnk* can be seen from Figure 5, that represents the final non-dominated fronts obtained by CAMOPSO and NSGA-II for this problem. It can be clearly seen that the performance of CAMOPSO is better than NSGA-II. NSGA-II is found to fail to capture the true non-dominated front fully. Out of the five disconnected front segments NSGA-II obtained four segments, while CAMOPSO is successful in obtaining all the five segments.

8 Conclusions and Discussion

In the present article, a constrained multi-objective PSO algorithm, called CAMOPSO, has been presented. It incorporates constraint handling in AMOPSO, an earlier work for unconstrained MOO problems where the crucial control parameters were determined adaptively [1]. The motivation for the article has been to use the adaptive nature of the AMOPSO along with a penalty based constraints handling technique to address the constrained test problems of MOO nature. For the convenience of the readers we presented the description and the results from our previous work on AMOPSO [1].

The performance of CAMOPSO has been compared with that of NSGA-II for five constrained test problems. NSGA-II has been used for the comparison for the reason that it has an inherent mechanism for constraint handling through constrained domination. The results demonstrate the effectiveness of the CAMOPSO algorithm that integrates the penalty based constraint handling approach with the adaptively determined parameters of AMOPSO.The results demonstrate the effectiveness of the CAMOPSO algorithm that integrates the penalty based constraint handling approach with the adaptively determined parameters of AMOPSO. In future work we wish to incorporate more results for the constrained test problems, and investigate concepts for such problems.

Acknowledgements

This paper was done when one of the authors, S. K. Pal, was a J.C. Bose Fellow of the Government of India.

References

1. Tripathi, P.K., Bandyopadhyay, S., Pal, S.K.: Adaptive Multi-objective Particle Swarm Optimization Algorithm. In: IEEE Congress on Evolutionary Computation (CEC 2007), pp. 2281–2288 (2007)
2. Kennedy, J., Eberhart, R.: Particle Swarm Optimization. In: IEEE International Conference Neural Networks, pp. 1942–1948 (1995)
3. Engelbrecht, A.P.: Fundamentals of Computational Swarm Intelligence. John Wiley and Sons, USA (2006)
4. Bergh, F.V.D.: An Analysis of Particle Swarm Optimizers. PhD thesis, Faculty of Natural and Agricultural Science, University of Pretoria, Pretoria (2001)
5. Hu, X., Eberhart, R.: Multiobjective Optimization Using Dynamic Neighbourhood Particle Swarm Optimization. In: Proceedings of the 2002 Congress on Evolutionary Computation, part of the 2002 IEEE World Congress on Computational Intelligence, Hawaii, pp. 12–17. IEEE Press, Los Alamitos (2002)
6. Reyes-Sierra, M., Coello, C.A.C.: Multi-Objective Particle Swarm Optimizers: A Survey of The State-of-the-Art. International Journal of Computational Intelligence Research 2, 287–308 (2006)
7. Schaffer, J.D.: Some Experiments in Machine Learning using Vector Evaluated Genetic Algorithm. PhD thesis, Vanderbilt University, Nashville,TN (1984)
8. Deb, K., Pratap, A., Agarwal, S., Meyarivan, T.: A Fast and Elitist Multi-objective Genetic Algorithm: NSGA-II. IEEE Transactions On Evolutionary Computation 6, 182–197 (2002)
9. Zitzler, E., Laumanns, M., Thiele, L.: SPEA2: Improving the Strength Pareto Evolutionary Algorithm. Technical Report TIK-103, Computer Engineering and Network Laboratory (TIK), Swiss Fedral Institute of Technology (ETH), Gloriastrasse 35, CH-8092 Zurich, Swidzerland (2001)
10. Corne, D.W., Jerram, N.R., Knowles, J.D., Oates, M.J.: PESA-II: Region-based Selection in Evolutionary Multiobjective Optimization. In: Proceedings of the Genetic and Evolutionary Computing Conference (GECCO 2001), pp. 283–290. Morgan Kaufmann, San Francisco (2001)

11. van den Bergh, F., Engelbrecht, A.P.: A Study of Particle Swarm Optimization Particle Trajectories. Information Sciences 176, 937–971 (2006)
12. Coello, C.A.C., Pulido, G.T., Lechuga, M.S.: Handling Multiple Objectives With Particle Swarm Optimization. IEEE Transactions on Evolutionary Computation 8, 256–279 (2004)
13. Michalewicz, Z.: Genetic Algorithms + Data Structure = Evolution Programs. Springer, Heidelberg (1992)
14. Mostaghim, S., Teich, J.: Strategies for Finding Good Local Guides in Multi-objective Particle Swarm Optimization (MOPSO). In: Swarm Intelligence Symposium 2003, SIS 2003, Inidanapolis, Indiana, USA, pp. 26–33. IEEE Service Center, Los Alamitos (2003)
15. Li, X.: A Non-dominated Sorting Particle Swarm Optimizer for Multi-objective Optimization. In: Cantú-Paz, E., Foster, J.A., Deb, K., Davis, L., Roy, R., O'Reilly, U.-M., Beyer, H.-G., Kendall, G., Wilson, S.W., Harman, M., Wegener, J., Dasgupta, D., Potter, M.A., Schultz, A., Dowsland, K.A., Jonoska, N., Miller, J., Standish, R.K. (eds.) GECCO 2003. LNCS, vol. 2723, pp. 37–48. Springer, Heidelberg (2003)
16. Richardson, J.T., Palmer, M.R., Liepins, G., Hilliard, M.: Some Guidelines for Genetic Algorithms with Penalty Functions. In: Schaffer, J. (ed.) Proc. of the First Int'l Conf. on Genetic Algorithms, pp. 191–197 (1989)
17. Tripathi, P.K., Bandyopadhyay, S., Pal, S.K.: A Multi-objective Genetic Algorithm with Relatice Distance: Method, Performance Measure and Constrained Handling. In: International Conference on Computing: Theory and Applications (ICCTA 2007), Kolkata, India, pp. 315–319. IEEE Computer Society, Los Alamitos (2007)
18. Deb, K.: Multi-Objective Optimization using Evolutionary Algorithms. John Wiley and Sons, USA (2001)
19. Coello, C.A.C., Veldhuizen, D.A.V., Lamount, G.B.: Evolutionary Algorithms for Solving Multi-Objective Problems. Kluwer Academic Publishers, Dordrecht (2001)
20. Fieldsend, J., Singh, S.: A Multi-Objective Algorithm based upon Particle Swarm Optimization, an Efficient Data Structure and Turbulence. In: Proceedings of UK Workshop on Computational Intelligence (UKCI 2002), Bermingham, UK, vol. 2-4, pp. 37–44 (2002)
21. Coello, C., Lechuga, M.: MOPSO: A Proposal for Multiple Objective Particle Swarm Optimization. In: Proceedings of the 2002 Congress on Evolutionary Computation, part of the 2002 IEEE World Congress on Computational Intelligence, Hawaii, pp. 1051–1056. IEEE Press, Los Alamitos (2002)
22. Parsopoulos, K., Vrahatis, M.: Particle Swarm Optimization Method in Multiobjective Problems. In: Nyberg, K., Heys, H.M. (eds.) SAC 2002. LNCS, vol. 2595, pp. 603–607. Springer, Heidelberg (2003)
23. Knowles, J., Corne, D.: Approximating the Nondominated Front Using the Pareto Archived Evolution Strategy. Evolutionary Computation 8, 149–172 (2000)
24. Tripathi, P.K., Bandyopadhyay, S., Pal, S.K.: Multi-objective Particle Swarm Optimization with Time Variant Inertia and Acceleration Coefficients. Information Sciences 177, 5033–5049 (2007)
25. Alvarez-Benitez, J.E., Everson, R.M., Fieldsend, J.E.: A MOPSO Algorithm Based Exclusively on Pareto Dominance Concepts. In: EMO, pp. 459–473 (2005)
26. Ratnaweera, A., Halgamuge, S.K., Watson, H.C.: Self-Organizing Hierarchical Particle Swarm Optimizer with Time-Varying Acceleration Coefficients. IEEE Transactions On Evolutionary Computation 8, 240–255 (2004)
27. Deb, K.: Optimization For Engineering Design Algorithms and Examples. Prentice Hall, New Delhi (1995)

28. Deb, K., Pratap, A., Agarwal, S., Meyarivan, T.: A Fast and Elitist Multi-objective Genetic Algorithm: NSGA-II. Technical Report 200001, Kanpur Genetic Algorithms Laboratory (KanGAL),Indian Institute of Technology Kanpur, India (2000)
29. Zitzler, E., Deb, K., Thiele, L.: Comparison of Multiobjective Evolutionary Algorithms: Empirical Results. Evolutionary Computation Journal 8, 125–148 (2000)
30. Deb, K., Thiele, L., Laumanns, M., Zitzler, E.: Scalable Test Problems for Evolutionary Multi-Objective Optimization. Technical Report TIK-Technical Report No. 112, Institut fur Technische Informatik und Kommunikationsnetze,, ETH Zurich Gloriastrasse 35., ETH-Zentrum, CH-8092, Zurich, Switzerland (2001)

Multiobjective Particle Swarm Optimization for Optimal Power Flow Problem

M.A. Abido

Electrical Engineering Department
King Fahd University of Petroleum & Minerals
Dhahran 31261, Saudi Arabia

Abstract. A novel approach to multiobjective particle swarm optimization (MOPSO) technique for solving optimal power flow (OPF) problem is proposed in this chapter. The new MOPSO technique evolves a multiobjective version of PSO by proposing redefinition of global best and local best individuals in multiobjective optimization domain. A clustering algorithm to manage the size of the Pareto-optimal set is imposed. The proposed MOPSO technique has been implemented to solve the OPF problem with competing and non-commensurable cost and voltage stability enhancement objectives. The optimization runs of the proposed approach have been carried out on a standard test system. The results demonstrate the capabilities of the proposed MOPSO technique to generate a set of well-distributed Pareto-optimal solutions in one single run.

1 Introduction

In the past two decades, the problem of optimal power flow (OPF) has received much attention. It is of current interest of many utilities and it has been marked as one of the most operational needs. The OPF problem solution aims to optimize a selected objective function such as fuel cost via optimal adjustment of the power system control variables, while at the same time satisfying various equality and inequality constraints. The equality constraints are the power flow equations while the inequality constraints are the limits on control variables and the operating limits of power system dependent variables. The problem control variables include the real power generations, the generator bus voltages, the transformer tap settings, and the reactive power of switchable VAR sources, while the problem dependent variables include the load bus voltages, the generator reactive powers, and the line flows. Generally, the OPF problem is a large-scale highly constrained nonlinear nonconvex optimization problem.

A wide variety of optimization techniques have been applied to solving the OPF problems [1-18] such as nonlinear programming [1-5], quadratic programming [6-7], linear programming [8-10], Newton-based techniques [11-12], sequential unconstrained minimization technique [13], and interior point methods [14-15]. Generally,

B.K. Panigrahi, Y. Shi, and M.-H. Lim (Eds.): Handbook of Swarm Intelligence, ALO 8, pp. 241–268.
springerlink.com

nonlinear programming based procedures have many drawbacks such as insecure convergence properties and algorithmic complexity. Quadratic programming based techniques have some disadvantages associated with the piecewise quadratic cost approximation. Newton-based techniques have a drawback of the convergence characteristics that are sensitive to the initial conditions and they may even fail to converge due to the inappropriate initial conditions. Sequential unconstrained minimization techniques are known to exhibit numerical difficulties when the penalty factors become extremely large. Although linear programming methods are fast and reliable they have some disadvantages associated with the piecewise linear cost approximation. Interior point methods have been reported as computationally efficient; however, if the step size is not chosen properly, the sub-linear problem may have a solution that is infeasible in the original nonlinear domain [14]. In addition, interior point methods, in general, suffer from bad initial, termination, and optimality criteria and, in most cases, are unable to solve nonlinear and quadratic objective functions [15]. A comprehensive survey has been presented in [16] where more discussions on these techniques can be found.

Generally, most of these approaches apply sensitivity analysis and gradient-based optimization algorithms by linearizing the objective function and the system constraints around an operating point. Unfortunately, the problem of the OPF is a highly nonlinear and a *multimodal* optimization problem, *i.e.*, there exist more than one local optimum. Hence, local optimization techniques, which are well elaborated, are not suitable for such a problem. Moreover, there is no local criterion to decide whether a local solution is also the global solution. Therefore, conventional optimization methods that make use of derivatives and gradients are, in general, not able to locate or identify the global optimum. On the other hand, many mathematical assumptions such as convex, analytic, and differential objective functions have to be given to simplify the problem. However, the OPF problem is an optimization problem with, in general, non-convex, non-smooth, and non-differentiable objective functions. These properties become more evident and dominant if the effects of the valve-point loading of thermal generators and the nonlinear behavior of electronic-based devices such as FACTS are taking into consideration. Hence, it becomes essential to develop optimization techniques that are efficient to overcome these drawbacks and handle such difficulties.

Heuristic algorithms such as genetic algorithms (GA) [17] and evolutionary programming [18] have been recently proposed for solving the OPF problem. The results reported were promising and encouraging for further research in this direction. Unfortunately, recent research has identified some deficiencies in GA performance [19]. This degradation in efficiency is apparent in applications with highly *epistatic* objective functions, *i.e.*, where the parameters being optimized are highly correlated. In addition, the premature convergence of GA degrades its performance and reduces its search capability.

Generally, several objectives can be defined in OPF problem [20-24]. In [23-24], the OPF problem was converted to a single objective problem by linear combination of different objectives as a weighted sum. Unfortunately, this requires multiple runs as many times as the number of desired Pareto-optimal solutions. Furthermore, this method cannot be used to find Pareto-optimal solutions in problems having a non-convex Pareto-optimal front. On the contrary, the studies on evolutionary algorithms, over the past few years, have shown that these methods can be efficiently used to eliminate most of the difficulties of classical methods [25-35]. Since they use a population of solutions in their search, multiple Pareto-optimal solutions can be found in one single run. Generally, different multiobjective evolutionary algorithms have been widely applied and implemented to handle the multiobjective optimization problems in several disciplines. For example, nonlinear system identification [36], water pollution control [37], robust control and control system design [38-39], and automotive engine design [40]. The multiobjective evolutionary algorithms have been implemented recently to some power system optimization problems with impressive success such as environmental/economic power dispatch problem [41-46], VAR control problem [47-49], and optimal power flow [50].

Recently, the particle swarm optimization (PSO) has been proposed and introduced [51-55]. This technique combines social psychology principles in socio-cognition human agents and evolutionary computations. PSO has been motivated by the behavior of organisms such as fish schooling and bird flocking. Generally, PSO is characterized as simple in concept, easy to implement, and computationally efficient. Unlike the other heuristic techniques, PSO has a flexible and well-balanced mechanism to enhance and adapt the global and local exploration abilities. Researchers are paying more and more interest on PSO to solve multi-objective problems [56-70]. Changing a PSO to a multi-objective PSO (MOPSO) requires redefinition of global and local best individuals in order to obtain a front of optimal solutions in MOPSO. In multiobjective particle swarm optimization, there is no absolute global best, but rather a set of nondominated solutions. In addition, there may be no single local best individual for each particle of the swarm. Choosing the global best and local best to guide the swarm particles becomes nontrivial task in multiobjective domain.

Generally, most of the traditional approaches, used to solve the OPF problem, apply sensitivity analysis and gradient-based optimization algorithms by linearizing the objective function and the system constraints around an operating point. Unfortunately, the problem of the OPF is a highly nonlinear and a *multimodal* optimization problem, *i.e.*, there exist more than one local optimum. Hence, local optimization techniques, which are well elaborated, are not suitable for such a problem. Moreover, there is no local criterion to decide whether a local solution is also the global solution. Therefore, conventional optimization methods that make use of derivatives and gradients are, in general, not able to locate or identify the global optimum. On the other hand, many mathematical assumptions such as convex, analytic, and differential objective functions have to be given to simplify the problem. However, the OPF problem is an optimization problem with, in general, non-convex, non-smooth, and non-differentiable objective functions. These properties become more evident and dominant if the effects of the valve-point loading of thermal generators and the nonlinear behavior of electronic-based devices such as FACTS are taking into consideration. Hence, it becomes essential to develop optimization techniques that are efficient to overcome these drawbacks and handle such difficulties.

Recently, the particle swarm optimization (PSO) has been proposed and introduced. This technique combines social psychology principles in socio-cognition human agents and evolutionary computations. PSO has been motivated by the behavior of organisms such as fish schooling and bird flocking. Generally, PSO is characterized as simple in concept, easy to implement, and computationally efficient. Unlike the other heuristic techniques, PSO has a flexible and well-balanced mechanism to enhance and adapt the global and local exploration abilities. Investigators are paying more and more interest on PSO to solve multi-objective problems. Changing a PSO to a multi-objective PSO (MOPSO) requires redefinition of global and local best individuals in order to obtain a front of optimal solutions in MOPSO. In multiobjective particle swarm optimization, there is no absolute global best, but rather a set of nondominated solutions. In addition, there may be no single local best individual for each particle of the swarm. Choosing the global best and local best to guide the swarm particles becomes nontrivial task in multiobjective domain.

In this chapter, a new MOPSO technique will be developed and implemented to multiobjective OPF problem. Several objectives have been considered in this study. Namely, the considered objectives are minimization of fuel cost, minimization of real power transmission loss, enhancing the voltage stability, improving the load bus voltage profile, minimizing the emission impact, and minimizing the adjustments of problem control variables. Therefore, the problem under investigation will be handled as a multiobjective optimization problem with non-commensurable and competing objectives. Unlike the traditional methods, this work presents MOPSO technique that can deal with the problem as a true multiobjective optimization problem where more than one objective will be simultaneously handled as competing objectives. In addition, the proposed technique will provide the decision maker with several non-inferior solutions of the problem in a single run and preserve, at the same time, the diversity of Pareto optimal solutions. The most efficient optimal solution from the Pareto optimal set will be extracted and provided to the decision maker using fuzzy set theory. The effectiveness of the proposed techniques in obtaining the Pareto optimal set of solutions has been tested on different standard test systems. An optimal operation strategy will be recommended in view of the results obtained.

2 OPF Problem Formulation

The optimal power flow problem is formulated as a multiobjective optimization problem with the aim of optimizing some competing objective functions while satisfying several equality and inequality constraints. Generally the problem is formulated as follows.

2.1 Problem Objectives

Minimization of Fuel Cost: The generators cost curves are represented by quadratic functions with sine components. The superimposed sine components represent the rippling effects produced by the steam admission valve openings. The total *$/h* fuel cost $F(P_G)$ can be expressed as

$$J_1 = F(P_G) = \sum_{i=1}^{N} a_i + b_i P_{G_i} + c_i P_{G_i}^2 + \left| \, d_i \sin[e_i (P_{G_i}^{\min} - P_{G_i})] \, \right| \tag{1}$$

where N is the number of generators, a_i, b_i, c_i, d_i, and e_i are the cost coefficients of the ith generator, and P_{Gi} is the real power output of the ith generator. P_G is the vector of real power outputs of generators and defined as

$$P_G = [P_{G_1}, P_{G2}, ..., P_{G_N}]^T \tag{2}$$

***Enhancement of Voltage Stability*:** The power system ability to maintain constantly acceptable bus voltage at each bus under normal operating conditions, after load increase, following system configuration changes, or when the system is being subjected to a disturbance is a very important characteristic of the system. The non-optimized control variables may lead to progressive and uncontrollable drop in voltage resulting in an eventual wide spread voltage collapse.

In this study, voltage stability enhancement is achieved through minimizing the voltage stability indicator L-index [71-72] values at every bus of the system and consequently the global power system L-index.

For voltage stability evaluation, an indicator L-index is used. The indicator value varies in the range between 0 (the no load case) and 1 which corresponds to voltage collapse. The indicator uses bus voltage and network information provided by the load flow program.

For multi-node system

$$I_{bus} = Y_{bus} \times V_{bus} \tag{3}$$

By segregating the load buses (PQ) from generator buses (PV), equation (3) can be rewritten as

$$\begin{bmatrix} I_L \\ I_G \end{bmatrix} = \begin{bmatrix} Y_1 & Y_2 \\ Y_3 & Y_4 \end{bmatrix} \begin{bmatrix} V_L \\ V_G \end{bmatrix} \tag{4}$$

$$\begin{bmatrix} V_L \\ I_G \end{bmatrix} = \begin{bmatrix} H_1 & H_2 \\ H_3 & H_4 \end{bmatrix} \begin{bmatrix} I_L \\ V_G \end{bmatrix} \tag{5}$$

where:

V_L, I_L: Voltages and Currents for PQ buses

V_G, I_G: Voltages and Currents for PV buses

H_1, H_2, H_3, H_4: Submatrices generated from Y_{bus} partial inversion.

Let

$$\overline{V_{ok}} = \sum_{i=1}^{NG} H_{2ki} \cdot \overline{V_i} \tag{6}$$

where NG is the number of generators

$$H_2 = -Y_1 \times Y_2 \tag{7}$$

$$L_k = \left| 1 + \frac{V_{ok}}{V_k} \right| \tag{8}$$

L_k: L-index voltage stability indicator for bus k.

Stability requires that $L_k < 1$ and must not be violated on a continuous basis. Hence a global system indicator L describing the stability of the complete system is $L=L_{max}\{L_k\}$, where $\{L_k\}$ contains L indices of all load buses.

In practice L_{max} must be lower than a threshold value. The predetermined threshold value is specified at the planning stage depending on the system configuration and on the utility policy regarding the quality of service and the level of system decided allowable margin.

The objective is to minimize L_{max}, that is,

$$J_2 = L_{max} = \max\{L_k;\ k=1,2,\ldots,\text{number of buses}\} \tag{9}$$

2.2 Problem Constraints

Generation capacity constraint: Generator voltages V_G and real and reactive power outputs, P_G and Q_G, are restricted by their lower and upper limits as follows:

$$V_{G_i}^{\min} \le V_{G_i} \le V_{G_i}^{\max}, \quad i = 1,\ldots, NG \tag{10}$$

$$P_{G_i}^{\min} \le P_{G_i} \le P_{G_i}^{\max}, \quad i = 1,\ldots, NG \tag{11}$$

$$Q_{G_i}^{\min} \le Q_{G_i} \le Q_{G_i}^{\max}, \quad i = 1,\ldots, NG \tag{12}$$

where NG is number of generators.

Power balance constraint: These constraints represent typical load flow equations as follows.

$$P_{G_i} - P_{D_i} - V_i \sum_{j=1}^{NB} V_j [G_{ij} \cos(\delta_i - \delta_j) + B_{ij} \sin(\delta_i - \delta_j)] = 0, \tag{13}$$

$$Q_{G_i} - Q_{D_i} - V_i \sum_{j=1}^{NB} V_j [G_{ij} \sin(\delta_i - \delta_j) - B_{ij} \cos(\delta_i - \delta_j)] = 0, \tag{14}$$

where $i = 1,2,\ldots,NB$; NB is the number of buses; Q_{Gi} is the reactive power generated at i^{th} bus; P_{Di} and Q_{Di} are the ith bus load real and reactive power respectively; G_{ij} and B_{ij} are the transfer conductance and susceptance between bus i and bus j respectively; V_i and V_j are the voltage magnitudes at bus i and bus j respectively; δ_i and δ_j are the voltage angles at bus i and bus j respectively. This equality constraints are nonlinear equations that can be solved using Newton-Raphson method to generate a solution of the load flow problem.

Transformer constraints: Transformer tap T settings are bounded as follows:

$$T_i^{\min} \leq T_i \leq T_i^{\max}, \quad i = 1,\ldots, NT \tag{15}$$

where NT is the number of transformers.

Switchable VAR sources constraints: Switchable VAR compensations Q_C are restricted by their limits as follows:

$$Q_{ci}^{\min} \leq Q_{ci} \leq Q_{ci}^{\max}, \quad i = 1,\ldots, NC \tag{16}$$

where NC is the number of switchable VAR sources.

Load bus voltage constraints: These include the constraints of voltages at load buses V_L as follows:

$$V_{L_i}^{\min} \leq V_{L_i} \leq V_{L_i}^{\max}, \quad i = 1,\ldots, NL \tag{17}$$

where NL is the number of load buses.

Security constraints: for secure operation, the apparent power flow through the transmission line S_l is restricted by its upper limit as follows:

$$S_{l_k} \leq S_{l_k}^{\max}, \quad k = 1,\ldots, nl\,. \tag{18}$$

where nl is the number of transmission lines.

2.3 Problem Formulation

Aggregating the objectives and constraints, the problem can be mathematically formulated as a nonlinear constrained multiobjective optimization problem as follows.

$$\textit{Minimize}\ [J_1(\mathbf{x},\mathbf{u}), J_2(\mathbf{x},\mathbf{u})] \tag{19}$$

Subject to:

$$g(\mathbf{x},\mathbf{u}) = 0 \tag{20}$$

$$h(\mathbf{x},\mathbf{u}) \leq 0 \tag{21}$$

where:

x: is the vector of dependent variables consisting of slack bus power P_{G_1}, load bus voltages V_L, generator reactive power outputs Q_G. Hence, **x** can be expressed as

$$\mathrm{x}^T = [P_{G_1}, V_{L_1} \ldots V_{L_{NL}}, Q_{G_1} \ldots Q_{G_{NG}}] \tag{22}$$

u: is the vector of independent variables consisting of generator voltages V_G, generator real power outputs P_G except at the slack bus P_{G_1}, transformer tap settings T, and shunt VAR compensations Q_c. Hence, **u** can be expressed as

$$u^T = [V_{G_1} \dots V_{G_{NG}}, P_{G_2} \dots P_{G_{NG}}, T_1 \dots T_{NT}, Q_{c1} \dots Q_{c_{NC}}] \quad (23)$$

g: is the equality constraints.
h: is the inequality constraints.

3 Multiobjective Optimization

3.1 Principles and Definitions

Multiobjective optimization is a very important research topic for engineers, not only because of the multiobjective nature of most real-world problems, but also because there are still many open questions in this area. In fact, there is not even a universally accepted definition of "*optimum*" as in single-objective optimization [73-81].

The principles of multiobjective optimization are different from that in a single objective optimization. The main goal in a single objective optimization is to find the global optimal solution, resulting in the optimal value for the single objective function. However, in a multiobjective optimization problem, there is more than one objective function, each of which may have a different individual optimal solution. If there is sufficient difference in the optimal solutions corresponding to different objectives, the objective functions are often known as conflicting to each other.

Multiobjective optimization with such conflicting objective functions gives rise to a set of optimal solutions, instead of one optimal solution. The reason for the optimality of many solutions is that no one can be considered to be better than any other with respect to all objective functions. These optimal solutions have a special name of Pareto optimal solutions.

A general multiobjective optimization problem consists of a number of objectives to be optimized simultaneously with a number of inequality and equality constraints. It can be formulated as follows:

$$\textit{Minimize} \quad f_i(x) \quad i = 1, \dots, N_{obj} \quad (24)$$

$$\textit{Subject to} \quad \begin{cases} g_j(x) = 0 & j = 1, \dots, M \\ h_k(x) \le 0 & k = 1, \dots, K \end{cases} \quad \text{Constraints} \quad (25)$$

The f_i is the ith objective function, x is a vector that represents a solution, and N_{obj} is the number of objective functions.

3.2 Dominance and Pareto Optimal Solutions

For a multiobjective optimization problem, any two solutions x_1 and x_2 can have one of two possibilities- one dominates the other or none dominates the other. In a minimization problem, without loss of generality, a solution x_1 dominates x_2 *iff* the following two conditions are satisfied:

$$1.\ \forall i \in \{1, 2, \dots, N_{obj}\} : f_i(x_1) \le f_i(x_2) \quad (26)$$

$$2.\ \exists j \in \{1,\ 2,\ ...,N_{obj}\}: f_j(x_1) < f_j(x_2) \quad (27)$$

If any of the above conditions is violated, the solution x_1 does not dominate the solution x_2. If x_1 dominates the solution x_2, x_1 is called the nondominated solution within the set $\{x_1, x_2\}$. The solutions that are nondominated within the entire search space are denoted as *Pareto-optimal* and constitute the *Pareto-optimal set* or *Pareto-optimal front*.

Generally, the primary goals that a multiobjective optimization algorithm must achieve:

- Guide the search towards the Pareto-optimal region.
- Maintain population diversity in the Pareto-optimal front.

The first task is a natural goal of any optimization algorithm. The second task is unique to multiobjective optimization. Since no one solution in the Pareto-optimal set can be said to be better than the other, what an algorithm can do best is to find as many different Pareto-optimal solutions as possible.

4 Multiobjective Particle Swarm Optimization (Mopso)

4.1 Overview

A new evolutionary computation technique, called particle swarm optimization (PSO), has been proposed and introduced recently [51-55]. This technique combines social psychology principles in socio-cognition human agents and evolutionary computations. PSO has been motivated by the behavior of organisms such as fish schooling and bird flocking. Generally, PSO is characterized as simple in concept, easy to implement, and computationally efficient. Unlike the other heuristic techniques, PSO has a flexible and well-balanced mechanism to enhance the global and local exploration abilities.

Like evolutionary algorithms, PSO technique conducts search using a population of particles, corresponding to individuals. Each particle represents a candidate solution to the problem at hand. In a PSO system, particles change their positions by flying around in a multi-dimensional search space until a relatively unchanging positions has been encountered, or until computational limitations are exceeded. In social science context, a PSO system combines a social-only model and a cognition-only model [51]. The social-only component suggests that individuals ignore their own experience and adjust their behavior according to the successful beliefs of individuals in the neighborhood. On the other hand, the cognition-only component treats individuals as isolated beings. A particle changes its position using these models.

The advantages of PSO over other traditional optimization techniques can be summarized as follows: -

(a) PSO is a population-based search algorithm *i.e.*, PSO has implicit parallelism. This property ensures PSO to be less susceptible to getting trapped on local minima.

(b) PSO uses payoff (performance index or objective function) information to guide the search in the problem space. Therefore, PSO can easily deal with non-differentiable objective functions. Additionally, this property relieves PSO of assumptions and approximations, which are often required by traditional optimization methods.

(c) PSO uses probabilistic transition rules, not deterministic rules. Hence, PSO is a kind of stochastic optimization algorithm that can search a complicated and uncertain area. This makes PSO more flexible and robust than conventional methods.

(d) Unlike GA and other heuristic algorithms, PSO has the flexibility to control the balance between the global and local exploration of the search space. This unique feature of PSO overcomes the premature convergence problem and enhances the search capability.

(e) Unlike the traditional methods, the solution quality of the proposed approach does not rely on the initial population. Starting anywhere in the search space, the algorithm ensures the convergence to the optimal solution [55].

In multiobjective particle swarm optimization, a set of nondominated solutions must replace the single global best individual in the standard single objective PSO case. In addition, there may be no single local best individual for each particle of the swarm. Choosing the global best and local best to guide the swarm particles becomes nontrivial task in multiobjective domain. This work presents two-level of nondominated solutions approach to address these problems. In the proposed approach, elitism is also considered by copying any nondominated solution obtained to an external set in order to keep the new nondominated solutions obtained during generations. The external set is updated regularly to hold only the nondominated solutions. The basic definitions and the major steps of the proposed approach can be explained as follows.

4.2 MOPSO Algorithm

The basic elements of the proposed MOPSO technique are briefly stated and defined as follows [82-85]: -

- ***Particle,*** $X(t)$,: It is a candidate solution represented by an m-dimensional vector, where m is the number of optimized parameters. At time t, the jth particle $X_j(t)$ can be described as $X_j(t)=[x_{j,1}(t), \ldots, x_{j,m}(t)]$, where xs are the optimized parameters and $x_{j,k}(t)$ is the position of the jth particle with respect to the kth dimension, *i.e.*, the value of the kth optimized parameter in the jth candidate solution.

- ***Population,*** $pop(t)$,: It is a set of n particles at time t, *i.e.*, $pop(t)=[X_1(t), \ldots, X_n(t)]^T$.

- ***Particle velocity,*** $V(t)$,: It is the velocity of the moving particles represented by an m-dimensional vector. At time t, the jth particle velocity $V_j(t)$ can be described as $V_j(t)=[v_{j,1}(t), \ldots, v_{j,m}(t)]$, where $v_{j,k}(t)$ is the velocity component of the jth particle with respect to the kth dimension. The particle velocity in the kth dimension is limited by some maximum value, v_k^{max}. This limit enhances the local exploration of the problem space and it realistically simulates the incremental changes of human learning [51]. The maximum velocity in the kth dimension is characterized by the range of the kth optimized parameter and given by

$$v_k^{\max} = (x_k^{\max} - x_k^{\min}) / N \tag{28}$$

where N is a chosen number of intervals in the kth dimension.

- ***Inertia weight,*** $w(t)$,: It is a control parameter that is used to control the impact of the previous velocities on the current velocity. Hence, it influences the trade-off between the global and local exploration abilities of the particles. For initial stages of the search process, large inertia weight to enhance the global exploration is recommended while, for last stages, the inertia weight is reduced for better local exploration. An annealing procedure has been incorporated in order to make uniform search in the initial stages and very locally search in the later stages. A decrement problem for decreasing the inertia weight given as $w(t)=\alpha w(t-1)$, α is a decrement constant smaller than but close to 1, is proposed in this study.

- ***Nondominated local set,*** $S_j^*(t)$,: It is a set that stores the nondominated solutions obtained by the jth particle up to the current time. As the jth particle moves through the search space, its new position is added to this set and the set is updated to keep only the nondominated solutions.

- ***Nondominated global set,*** $S^{**}(t)$,: It is a set that stores the nondominated solutions obtained by all particle up to the current time. First, the union of all nondominated local sets is formed. Then, the nondominated solutions out of this union are members in the nondominated global set.

- ***External set***: It is an archive that stores a historical record of the nondominated solutions obtained along the search process. This set is updated continuously by applying the dominance conditions to the union of this set and the nondominated global set. Then, the nondominated solutions of this union are members in the updated external set.

- ***Local best,*** $X_j^*(t)$, and ***Global best,*** $X_j^{**}(t)$,: The individual distances between members in nondominated local set of the jth particle, $S_j^*(t)$, and members in nondominated global set, $S^{**}(t)$, are measured in the objective space. If $X_j^*(t)$ and $X_j^{**}(t)$ are the members of $S_j^*(t)$ and $S^{**}(t)$ respectively that give the minimum distance, they are selected as the local best and the global best of the jth particle respectively.

The size of the nondominated local set, the nondominated global set, and the external set could be extremely high due to accumulation of all nondominated solutions throughout the search. To keep the sizes of these sets manageable, clustering algorithm should be implemented.

5 MOPSO Implementation

5.1 Reducing Pareto Set by Clustering

The Pareto-optimal set can be extremely large or even contain an infinite number of solutions. In this case, reducing the set of nondominated solutions without destroying the characteristics of the trade-off front is desirable from the decision maker's point of view. An average linkage based hierarchical clustering algorithm [86] is employed to reduce the Pareto set to manageable size. It works iteratively by joining the adjacent clusters until the required number of groups is obtained. It can be described as:

> *Given a set P which its size exceeds the maximum allowable size N, it is required to form a subset P^* with the size N*

The algorithm is illustrated in the following steps.

Step 1: Initialize cluster set *C*; each individual $i \in P$ constitutes a distinct cluster.

Step 2: If number of clusters $\leq N$, then go to Step 5, else go to Step 3.

Step 3: Calculate the distance of all possible pairs of clusters. The distance d_c of two clusters c_1 and $c_2 \in C$ is given as the average distance between pairs of individuals across the two clusters

$$d_c = \frac{1}{n_1 . n_2} \sum_{i_1 \in c_1, i_2 \in c_2} d(i_1, i_2) \tag{29}$$

where n_1 and n_2 are the number of individuals in the clusters c_1 and c_2 respectively. The function d reflects the distance in the objective space between individuals i_1 and i_2.

Step 4: Determine two clusters with minimal distance d_c. Combine these clusters into a larger one. Go to Step 2.

Step 5: Find the centroid of each cluster. Select the nearest individual in this cluster to the centroid as a representative individual and remove all other individuals from the cluster.

Step 6: Compute the reduced nondominated set P^* by uniting the representatives of the clusters.

5.2 Best Compromise Solution

Fuzzy set theory has been implemented to derive efficiently a candidate Pareto-optimal solution for the decision makers. Upon having the Pareto-optimal set, the proposed approach presents a fuzzy-based mechanism to extract a Pareto-optimal solution as the best compromise solution. Due to imprecise nature of the decision maker's judgment, the *i*-th objective function of a solution in the Pareto-optimal set, F_i, is represented by a membership function μ_i defined as [46]

$$\mu_i = \begin{cases} 1, & F_i \le F_i^{\min}, \\ \dfrac{F_i^{\max} - F_i}{F_i^{\max} - F_i^{\min}}, & F_i^{\min} < F_i < F_i^{\max}, \\ 0, & F_i \ge F_i^{\max}. \end{cases} \tag{30}$$

where $F_i^{\max}$ and $F_i^{\min}$ are the maximum and minimum values of the i-th objective function respectively.

For each nondominated solution k, the normalized membership function μ^k is calculated as

$$\mu^k = \frac{\sum_{i=1}^{N_{obj}} \mu_i^k}{\sum_{j=1}^{M} \sum_{i=1}^{N_{obj}} \mu_i^j} \tag{31}$$

where M is the number of nondominated solutions. The best compromise solution is the one having the maximum of μ^k. As a matter of fact, arranging all solutions in Pareto-optimal set in descending order according to their membership function will provide the decision maker with a priority list of nondominated solutions. This will guide the decision maker in view of the current operating conditions.

5.3 MOPSO Computational Flow

In the proposed MOPSO algorithm, the population has n particles and each particle is an m-dimensional vector, where m is the number of optimized parameters. The computational flow of the proposed MOPSO technique can be described in the following steps.

***Step 1 (Initialization)*:** Set the time counter t=0 and generate randomly n particles, $\{X_j(0), j=1, \ldots, n\}$, where $X_j(0)=[x_{j,1}(0), \ldots, x_{j,m}(0)]$. $x_{j,k}(0)$ is generated by randomly selecting a value with uniform probability over the kth optimized parameter search space $[x_k^{min}, x_k^{max}]$. Similarly, generate randomly initial velocities of all particles, $\{V_j(0), j=1, \ldots, n\}$, where $V_j(0)=[v_{j,1}(0), \ldots, v_{j,m}(0)]$. $v_{j,k}(0)$ is generated by randomly selecting a value with uniform probability over the kth dimension $[-v_k^{max}, v_k^{max}]$. Each particle in the initial population is evaluated using the objective problems. For each particle, set $S_j^*(0)=\{X_j(0)\}$ and the local best $X_j^*(0)=X_j(0)$, $j=1, \ldots, n$. Search for the nondominated solutions and form the nondominated global set $S^{**}(0)$. The nearest member in $S^{**}(0)$ to $X_j^*(0)$ is selected as the global best $X_j^{**}(0)$ of the jth particle. Set the external set equal to $S^{**}(0)$. Set the initial value of the inertia weight $w(0)$.

***Step 2 (Time updating)*:** Update the time counter t=t+1.

***Step 3 (Weight updating)*:** Update the inertia weight $w(t)=\alpha\, w(t-1)$.

***Step 4 (Velocity updating)*:** Using the local best $X_j^*(t)$ and the global best $X_j^{**}(t)$ of each particle, j=1, …, n, the jth particle velocity in the kth dimension is updated according to the following equation:

$$v_{j,k}(t) = w(t)\, v_{j,k}(t-1) + c_1 r_1 (x^*_{j,k}(t-1) - x_{j,k}(t-1)) \\ + c_2 r_2 (x^{**}_{j,k}(t-1) - x_{j,k}(t-1)) \tag{32}$$

where c_1 and c_2 are positive constants and r_1 and r_2 are uniformly distributed random numbers in [0,1]. If a particle violates the velocity limits, set its velocity equal to the proper limit.

***Step 5 (Position updating)*:** Based on the updated velocities, each particle changes its position according to the following equation

$$x_{j,k}(t) = v_{j,k}(t) + x_{j,k}(t-1) \tag{33}$$

If a particle violates its position limits in any dimension, set its position at the proper limit.

***Step 6 (Nondominated local set updating)*:** The updated position of the jth particle is added to $S_j^*(t)$. The dominated solutions in $S_j^*(t)$ will be truncated and the set will be updated accordingly. If the size of $S_j^*(t)$ exceeds a prespecified value, the hierarchical clustering algorithm will be invoked to reduce the size to its maximum limit.

***Step 7 (Nondominated global set updating)*:** The union of all nondominated local sets is formed and the nondominated solutions out of this union are members in the nondominated global set $S^{**}(t)$. The size of this set will be reduced by hierarchical clustering algorithm if it exceeds a prespecified value.

***Step 8 (External set updating)*:** The external Pareto-optimal set is updated as follows.
(a) Copy the members of $S^{**}(t)$ to the external Pareto set.
(b) Search the external Pareto set for the nondominated individuals and remove all dominated solutions from the set.
(c) If the number of the individuals externally stored in the Pareto set exceeds the maximum size, reduce the set by means of clustering.

***Step 9 (Local best and global best updating)*:** The individual distances between members in $S_j^*(t)$, and members in $S^{**}(t)$, are measured in the objective space. If $X_j^*(t)$ and $X_j^{**}(t)$ are the members of $S_j^*(t)$ and $S^{**}(t)$ respectively that give the minimum distance, they are selected as the local best and the global best of the jth particle respectively.

***Step 10 (Stopping criteria)*:** If the number of iterations exceeds the maximum then stop, else go to step 2.

Fig. 1 shows the computational flow chart of the proposed MOPSO algorithm.

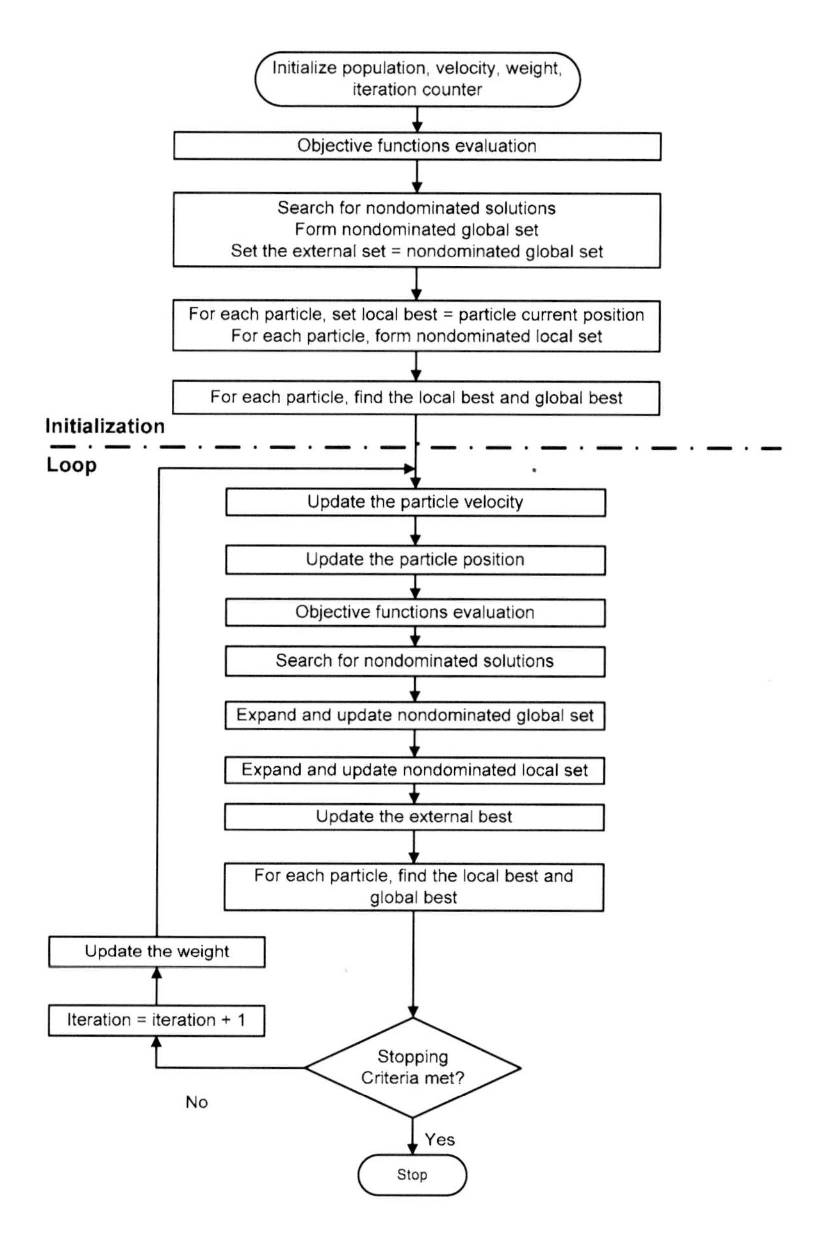

Fig. 1. Flow chart of the proposed MOPSO algorithm

5.4 Implementation

The proposed MOPSO based approach was implemented using FORTRAN language and the developed software program was executed on a 1.8-GHz Pentium 4 PC.

Initially, several runs have been done with different values of key parameters such as the initial inertia weight and the maximum allowable velocity. Other parameters are selected as:

Number of particles $n = 50$
Decrement constant $\alpha = 0.99$
$c_1 = c_2 = 2$
Maximum number of iterations = 1000
Nondominated local set size = 10
Nondominated global set size = 50

To satisfy the problem constraints, a procedure is imposed to check the feasibility throughout the search process. This ensures the feasibility of Pareto optimal solutions.

6 Results and Discussions

In this study, the standard IEEE 6-generator 30-bus test system shown in Fig. 2 is considered to assess the potential of MOPSO for solving the OPF problem. The power system considered has 30 buses with 41 transmission lines. This system contains 6 generators at buses 1, 2, 5, 8, 11, and 13 and 4 transformers at lines 6-9, 6-10, 4-12, and 27-28. Therefore, the total number of control variables in this case is 16. The lower limit of the generator voltages is 0.95 pu while the lower limit of transformer taps is 0.9 pu. The upper limit of both is set 1.1 pu. The values of fuel cost coefficients defined in Eq. (1) and generation limits are given in Table 1. The detailed line and load data of the test system is given in Appendix A.

Table 1. Generator fuel cost coefficients and power limits of IEEE 30-bus system

	G_1	G_2	G_5	G_8	G_{11}	G_{13}
a	0.0	0.0	0.0	0.0	0.0	0.0
b	200	175	100	325	300	300
c	37.5	175	625	83.4	250	250
d	0.0	0.0	0.0	0.0	0.0	0.0
e	0.0	0.0	0.0	0.0	0.0	0.0
P_G^{min}	0.50	0.20	0.15	0.10	0.10	0.12
P_G^{max}	2.00	0.80	0.50	0.35	0.30	0.40

To demonstrate the effectiveness of the proposed techniques, each of the objective functions will be optimized individually without considering the other objectives. This has been carried out in order to explore the extreme points of the trade-off surface and evaluate the diversity characteristics of the Pareto optimal solutions obtained by the proposed multiobjective evolutionary algorithms. The single objective PSO technique is implemented to solve the optimization problem associated with each objective.

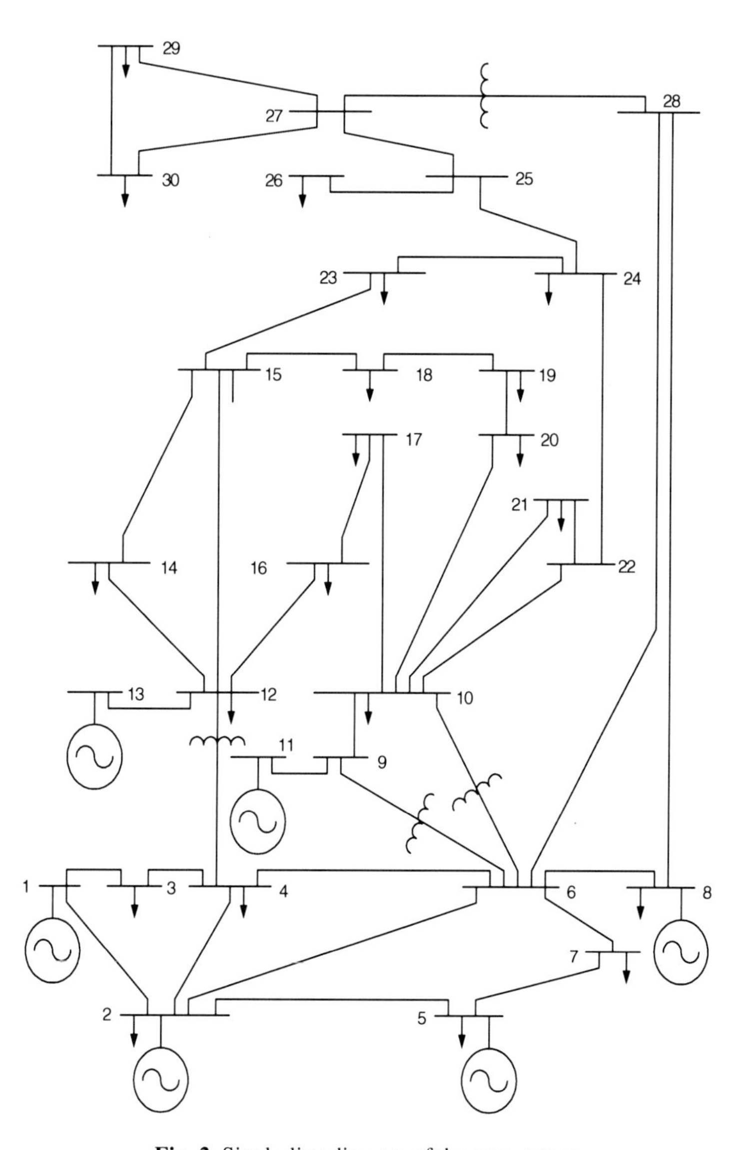

Fig. 2. Single-line diagram of the test system

In our implementation, the population size and the maximum number of iterations were selected as 50 and 500, respectively. The developed FORTRAN computer program used in this study was implemented on 3-GHz PC.

Convergences of different objective functions with test systems considered are shown in Figs. 3 and 4. The best solutions of different objectives when optimized individually are given in Table 2. The optimal values of the objective functions with and without optimization are given in Table 3.

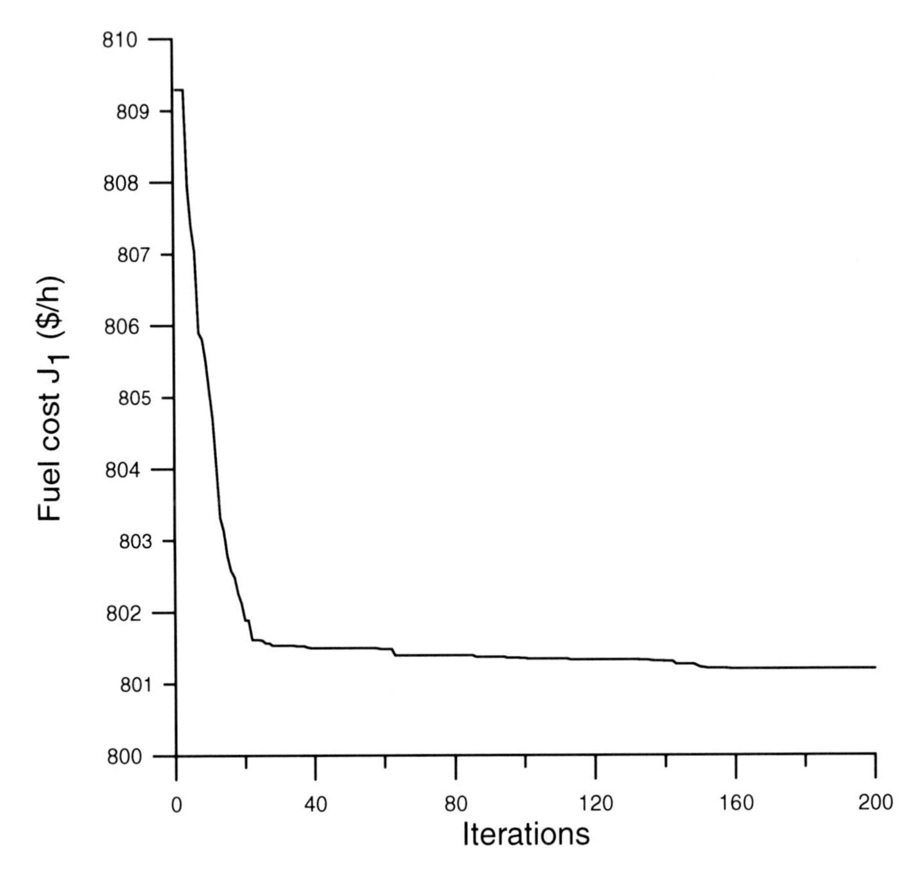

Fig. 3. Convergence of fuel cost minimization objective J_1

The developed MOPSO technique has been implemented to the test system in order to assess the effectiveness of the proposed approach to handle the multiobjective optimization problems. In this case J_1 and J_2 have been optimized simultaneously, i.e., minimization of fuel cost as well as minimization of the voltage stability enhancement objectives has been considered. The results of Pareto optimal front are shown in Fig. 5 for the test systems considered. It is quite clear that the problem is efficiently solved by the proposed MOPSO technique. The results also show that the obtained Pareto-optimal fronts have satisfactory diversity characteristics. The best compromise solution is extracted from the Pareto-optimal set obtained as described. The best compromise solution is given in Table 2.

The best objectives obtained by the proposed MOPSO are compared to those obtained by single objective PSO as given in Table 4. It is clear that the results are almost identical. It is also clear that the results of the proposed MOPSO is much better that that is recorded in [19]. This demonstrates that the search of the proposed approach span over the entire trade-off surface. In addition, the close agreement of the results shows clearly the capability of the proposed approach to handle multiobjective

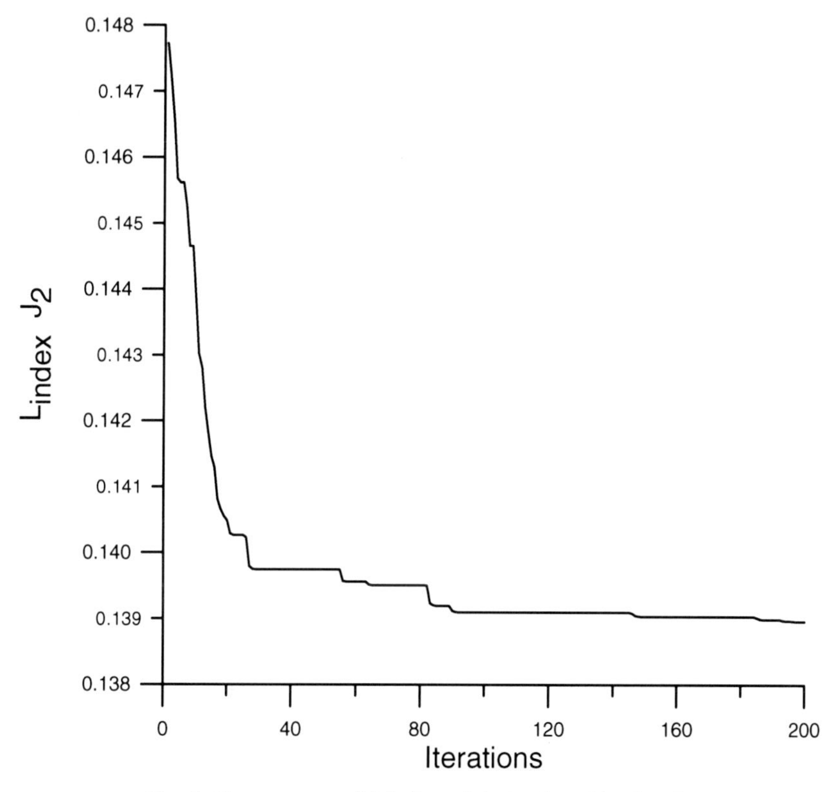

Fig. 4. Convergence of *L*-index minimization objective J_2

Table 2. The settings of control variables

	Initial	*Individual*		*Best Compromise*
		J_1	J_2	
P_{G1}	1.7656	1.7713	1.6495	1.7632
P_{G2}	0.4884	0.4880	0.3524	0.4884
P_{G5}	0.2151	0.2143	0.3418	0.2148
P_{G8}	0.2215	0.2136	0.3500	0.2157
P_{G11}	0.1214	0.1190	0.1003	0.1239
P_{G13}	0.1200	0.1200	0.1200	0.1200
V_{G1}	1.0500	1.0816	1.0832	1.0842
V_{G2}	1.0382	1.0628	1.0787	1.0674
V_{G5}	1.0114	1.0299	1.0348	1.0395
V_{G8}	1.0194	1.0356	1.0412	1.0424
V_{G11}	1.0912	1.1000	1.0300	1.0560
V_{G13}	1.0913	1.0794	1.0635	1.0749
$T_{6\text{-}9}$	0.9780	1.0002	0.9503	0.9773
$T_{6\text{-}10}$	0.9690	0.9591	0.9000	0.9000
$T_{4\text{-}12}$	0.9320	1.0144	0.9696	0.9916
$T_{27\text{-}28}$	0.9680	0.9640	0.9466	0.9490

Table 3. The best objectives when optimized individually

	Without optimization	*With optimization*
Fuel Cost J_1 **($/h)**	803.441	801.144
L_{index} J_2	0.1426	0.1379

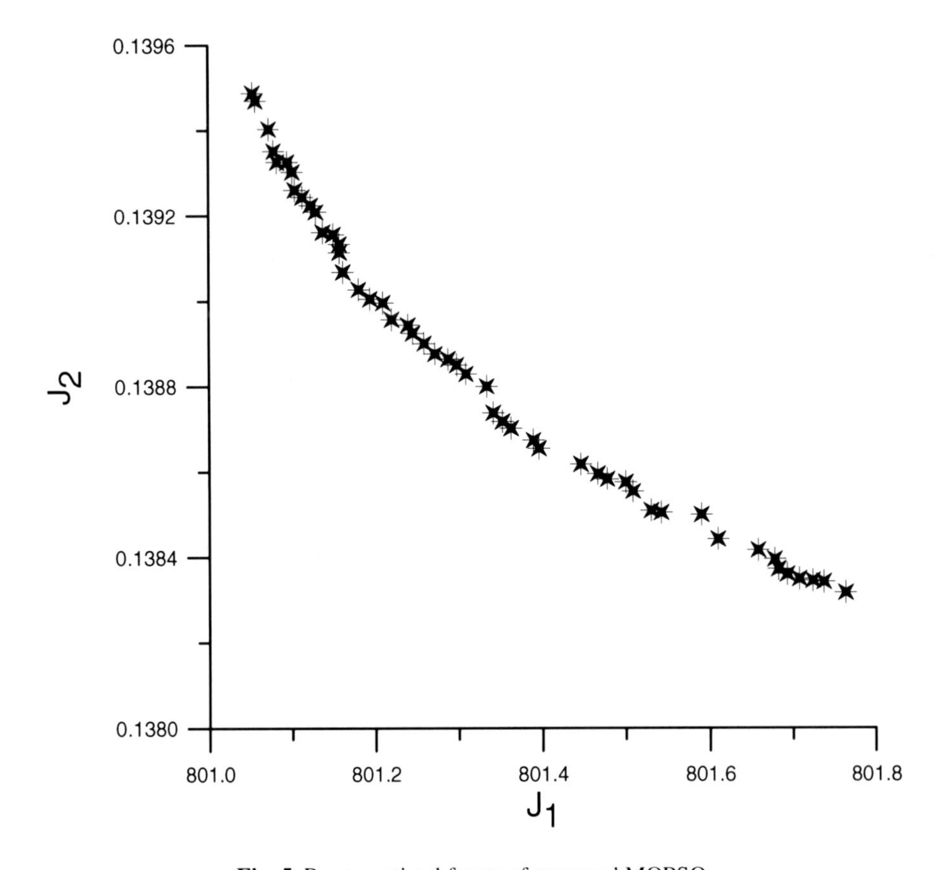

Fig. 5. Pareto optimal fronts of proposed MOPSO

Table 4. Best objectives of the proposed MOPSO vs. single objective PSO

J_1			J_2	
***Gradient Projection Method* [19]**	***PSO***	***Proposed MOPSO***	***PSO***	***Proposed MOPSO***
804.85	801.144	801.153	0.1379	0.1383

optimization problems as the best solution of each objective along with a manageable set of nondominated solutions can be obtained in one single run. The load bus voltage profile is given in Fig. 6 for the optimal settings of J_1, J_2, and best compromise solutions. It is clear that all the bus voltages are within the permissible limits. In addition,

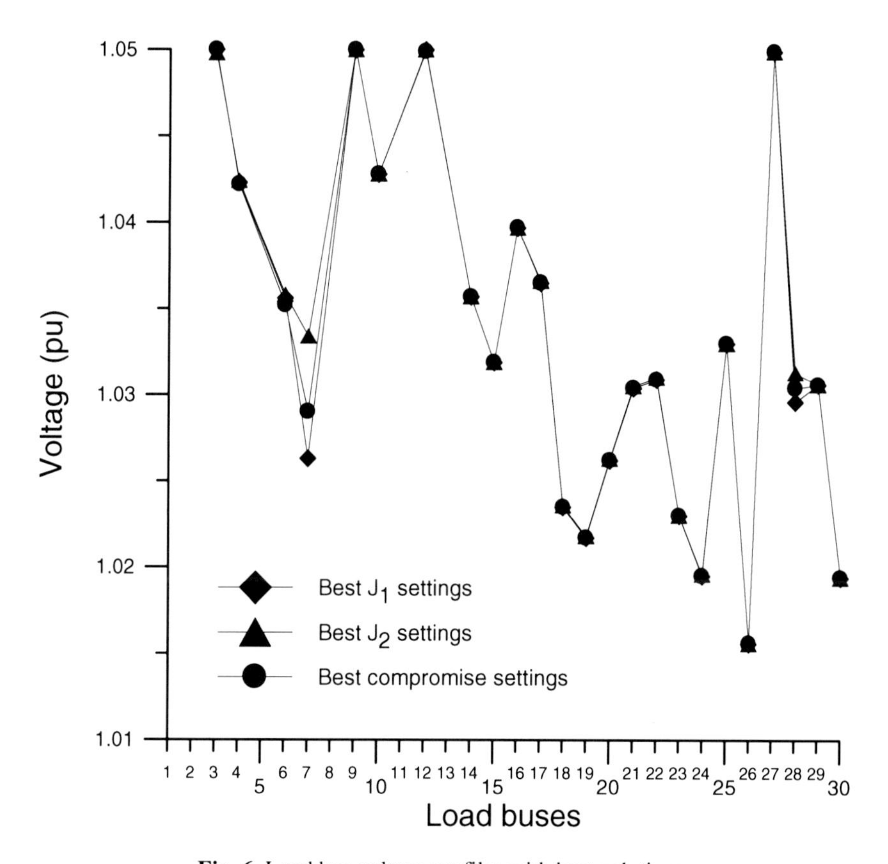

Fig. 6. Load bus voltage profiles with best solutions

no violations have been observed in all the dependent variables. It is worth mentioning that the run time of 0.631s per iteration is observed with the proposed MOPSO technique.

7 Conclusions

In this chapter, the single objective particle swarm optimization has been extended and a multiobjective version has been developed. In MOPSO the global best and local best has been replaced by nondominated global best and nondominated local best sets. Developed MOPSO has been implemented and successfully applied for solving the multiobjective OPF optimization problem. The OPF problem has been formulated as a multiobjective optimization problem with competing objectives. The proposed MOPSO technique has been compared with the single objective PSO to assess its potential and effectiveness. The simulation results demonstrate that the proposed MOPSO is an effective tool for solving multiobjective OPF problem where multiple Pareto optimal solutions, including the best solutions with respect to each objective, can be found in one simulation run. The results also show that the proposed MOPSO is capable of exploring efficiently the nondominated solutions of multiobjective optimization problems. The close agreement of the results shows clearly the capability of the proposed approach to handle multiobjective optimization problems as the best solution of each objective along with a manageable set of nondominated solutions can be obtained in one single run.

Acknowledgement

The author acknowledges the support of Electrical Power and Energy Systems Research Group, King Fahd University of Petroleum & Minerals, Saudi Arabia.

References

[1] Alsac, O., Stott, B.: Optimal Load Flow with Steady State Security. IEEE Trans. on Power Apparatus and Systems PAS-93, 745–751 (1974)
[2] Shoults, R., Sun, D.: Optimal Power Flow Based on P-Q Decomposition. IEEE Trans. on Power Apparatus and Systems PAS-101(2), 397–405 (1982)
[3] Happ, H.H.: Optimal Power Dispatch- A comprehensive Survey. IEEE Trans. on Power Apparatus and Systems PAS-96, 841–854 (1977)
[4] Mamandur, K.R.C., Chenoweth, R.D.: Optimal Control of Reactive Power Flow for Improvements in Voltage profiles and for Real Power Loss Minimization. IEEE Trans. on Power Apparatus and Systems PAS-100(7), 3185–3193 (1981)
[5] Habiabollahzadeh, H., Luo, G.X., Semlyen, A.: Hydrothermal Optimal Power Flow Based on a Combined Linear and Nonlinear Programming Methodology. IEEE Trans. on Power Systems PWRS-4(2), 530–537 (1989)
[6] Burchett, R.C., Happ, H.H., Vierath, D.R.: Quadratically Convergent Optimal Power Flow. IEEE Trans. on Power Apparatus and Systems PAS-103, 3267–3276 (1984)
[7] Aoki, K., Nishikori, A., Yokoyama, R.T.: Constrained Load Flow Using Recursive Quadratic Programming. IEEE Trans. on Power Systems 2(1), 8–16 (1987)
[8] Abou El-Ela, A.A., Abido, M.A.: Optimal Operation Strategy for Reactive Power Control. Modelling, Simulation & Control, Part A 41(3), 19–40 (1992)
[9] Stadlin, W., Fletcher, D.: Voltage Versus Reactive Current Model for Dispatch and Control. IEEE Trans. on Power Apparatus and Systems PAS-101(10), 3751–3758 (1982)
[10] Mota-Palomino, R., Quintana, V.H.: Sparse Reactive Power Scheduling by a Penalty-Function Linear Programming Technique. IEEE Trans. on Power Systems 1(3), 31–39 (1986)
[11] Sun, D.I., Ashley, B., Brewer, B., Hughes, A., Tinney, W.F.: Optimal Power Flow by Newton Approach. IEEE Trans. on Power Apparatus and Systems PAS-103(10), 2864–2875 (1984)
[12] Santos, A., da Costa, G.R.: "Optimal Power Flow Solution by Newton's Method Applied to an Augmented Lagrangian Function,". IEE Proc.-Gener. Transm. Distrib. 142(1), 33–36 (1995)
[13] Rahli, M., Pirotte, P.: Optimal Load Flow Using Sequential Unconstrained Minimization Technique (SUMT) Method under Power Transmission Losses Minimization. Electric Power Systems Research 52, 61–64 (1999)
[14] Yan, X., Quintana, V.H.: Improving an Interior Point based OPF by Dynamic Adjustments of Step Sizes and Tolerances. IEEE Trans. on Power Systems 14(2), 709–717 (1999)

[15] Momoh, J.A., Zhu, J.Z.: Improved Interior Point Method for OPF Problems. IEEE Trans. on Power Systems 14(3), 1114–1120 (1999)
[16] Momoh, J., El-Hawary, M., Adapa, R.: A Review of Selected Optimal Power Flow Literature to 1993 Parts I & II. IEEE Trans. on Power Systems 14(1), 96–111 (1999)
[17] Lai, L.L., Ma, J.T.: Improved Genetic Algorithms for Optimal Power Flow Under Both Normal and Contingent Operation States. Int. J. Electrical Power & Energy Systems 19(5), 287–292 (1997)
[18] Yuryevich, J., Wong, K.P.: Evolutionary Programming Based Optimal Power Flow Algorithm. IEEE Trans. on Power Systems 14(4), 1245–1250 (1999)
[19] Lee, K., Park, Y., Ortiz, J.: A United Approach to Optimal Real and Reactive Power Dispatch. IEEE Trans. on Power Apparatus and Systems 104(5), 1147–1153 (1985)
[20] Kessel, P., Glavitsch, H.: Estimating The Voltage Stability of a Power System. IEEE Trans. on Power Delivery 1(3), 346–354 (1986)
[21] Belhadj, C.A., Abido, M.A.: An optimized Fast Voltage Stability Indicator. In: IEEE Budapest Power Tech. 1999 Conference, Budapest, Hungary, BPT99-363-12 (August 29-September 2 1999)
[22] Tuan, T.Q., Fandino, J., Hadjsaid, N., Sabonnadiere, J.C., Vu, H.: Emergency Load Shedding to Avoid Risks of Voltage Instability Using Indicators. IEEE Trans. on Power Systems 9(1), 341–351 (1994)
[23] Abido, M.A.: Optimal power flow using tabu search algorithm. Electric Power Components & Systems 30(5), 469–483 (2002)
[24] Abido, M.A.: Optimal power flow using particle swarm optimization. International Journal of Electrical Power and Energy Systems 24(7), 563–571 (2002)
[25] Fogel, D.B.: Evolutionary Computation Toward a New Philosophy of Machine Intelligence. IEEE Press, Los Alamitos (1995)
[26] Fonseca, C.M., Fleming, P.J.: An Overview of Evolutionary Algorithms in Multiobjective Optimization. Evolutionary Computation 3(1), 1–16 (1995)
[27] Zitzler, E., Thiele, L.: An Evolutionary Algorithm for Multiobjective optimization: The Strength Pareto Approach, Swiss Federal Institute of Technology, TIK-Report, No. 43 (1998)
[28] Zitzler, E.: Evolutionary Algorithms for Multiobjective Optimization: Methods and Applications, Ph.D. Thesis, Swiss Federal Institute of Technology, Zurich (1999)
[29] Coello, C.A.C.: A Comprehensive Survey of Evolutionary-Based Multiobjective Optimization Techniques. Knowledge and Information Systems 1(3), 269–308 (1999)
[30] Zitzler, E., Laumanns, M., Thiele, L.: SPEA2: Improving the Strength Pareto Evolutionary Algorithm. In: Proceedings of EUROGEN 2001, Athens, Greece (September 2001)
[31] Zitzler, E., Thiele, L.: Multiobjective Evolutionary Algorithms: A Comparative Case Study and the Strength Pareto Approach. IEEE Trans. on Evolutionary Computation 3(4), 257–271 (1999)
[32] Zitzler, E., Thiele, L.: Multiobjective Optimization Using Evolutionary Algorithms – A Comparative Case Study. In: Parallel Problem Solving from Nature V, Amsterdam, September 1998, pp. 292–301. Springer, Heidelberg (1998)
[33] Deb, K., Pratap, A., Agarwal, S., Meyarivan, T.: A Fast and Elitist Multiobjective Genetic Algorithms: NSGA-II. EEE Trans. on Evolutionary Computation 6(2), 182–197 (2002)
[34] Schaffer, J.D.: Multi objective optimization with vector evaluated genetic algorithms, PhD Thesis, Vanderbilt University, Nashville, USA (1984)
[35] Zitzler, E., Deb, K., Thiele, L.: Comparison of Multiobjective Evolutionary Algorithms: Empirical Results. Evolutionary Computation 8(2), 173–195 (2000)
[36] Vázquez, K.R., Fleming, P.J.: Multiobjective Genetic Programming for Non-Linear System Identification. Electronics Letters 34(9), 930–931 (1998)

[37] Ritzel, J., Eheart, J.W., Ranjithan, S.: Using genetic algorithms to solve a multiple objective groundwater pollution containment problem. Water Resources Research 30(5), 1589–1603 (1994)
[38] Chipperfield, J., Dakev, N.V., Fleming, P.J., Whidborne, J.F.: Multiobjective robust control using evolutionary algorithms. In: IEEE International Conference on Industrial Technology, Shanghai, China (1996)
[39] Duarte, N.M., Ruano, A.E., Fonseca, C.M., Fleming, P.J.: Accelerating Multi-Objective Control System Design Using a Neuro-Genetic Approach. In: Congress on Evolutionary Computation, Piscataway, New Jersey, vol. 1, pp. 392–397. IEEE Service Center, Los Alamitos (July 2000)
[40] Chipperfield, J., Fleming, P.J.: Multiobjective Gas Turbine Engine Controller Design Using Genetic Algorithms. IEEE Transactions on Industrial Electronics 43(5) (October 1996)
[41] Abido, M.A.: Environmental/Economic Power Dispatch Using Multiobjective Evolutionary Algorithms. IEEE Trans. on Power Systems 18(4), 1529–1537 (2003)
[42] Abido, M.A.: A New Multiobjective Evolutionary Algorithm for Environmental/Economic Power Dispatch. In: IEEE Power Engineering Society Summer Meeting, Vancouver, Canada, July 15-19, pp. 1263–1268 (2001)
[43] Abido, M.A.: A Niched Pareto Genetic Algorithm for Multiobjective Environmental/Economic Dispatch. International Journal of Electrical Power and Energy Systems 25(2), 79–105 (2003)
[44] Abido, M.A.: A Novel Multiobjective Evolutionary Algorithm for Environmental/Economic Power Dispatch. Electric Power Systems Research 65(1), 71–81 (2003)
[45] Abido, M.A.: Multiobjective Environmental/Economic Power Dispatch Using Evolutionary Algorithms: A Comparative Study. In: IEEE Power Engineering Society Summer Meeting 2003, Toronto, Canada, July 13-18 (2003)
[46] Abido, M.A.: Multiobjective Evolutionary Algorithms for Electric Power Dispatch Problem. IEEE Trans. on Evolutionary Computations 10(3), 315–329 (2006)
[47] Abido, M.A., Bakhashwain, J.M.: Optimal VAR Dispatch Using a Multiobjective Evolutionary Algorithm. International Journal of Electrical Power & Energy Systems 27(1), 13–20 (2005)
[48] Abido, M.A., Bakhashwain, J.M.: A novel Multiobjective Evolutionary Algorithm for Optimal reactive power Dispatch problem. In: 10th IEEE International Conference on Electronics, Circuits, and Systems, ICECS 2003, Sharjah, United Arab Emirates, December 14-17, pp. 1054–1057 (2003)
[49] Abido, M.A.: Multiobjective Optimal VAR Dispatch Using Strength Pareto Evolutionary Algorithm. In: Proceedings of The IEEE World Congress on Computational Intelligence 2006, Vancouver, Canada, July 16-21, pp. 730–736 (2006)
[50] Abido, M.A.: Multiobjective Optimal power flow using strength Pareto Evolutionary Algorithm. In: Proceedings of the 39th International Universities Power Engineering Conference, UPEC 2004, Bristol, UK, September 6-8, vol. 8, pp. 457–461 (2004)
[51] Kennedy, J.: The Particle Swarm: Social Adaptation of Knowledge. In: Proceedings of the 1997 IEEE international Conference on Evolutionary Computation ICEC 1997, Indianapolis, Indiana, USA, pp. 303–308 (1997)
[52] Angeline, P.: Evolutionary Optimization versus Particle Swarm Optimization: Philosophy and Performance Differences. In: Proceedings of the 7th Annual Conference on Evolutionary Programming, pp. 601–610 (March 1998)
[53] Shi, Y., Eberhart, R.: Parameter Selection in Particle Swarm Optimization. In: Proceedings of the 7th Annual Conference on Evolutionary Programming, pp. 591–600 (March 1998)
[54] Ozcan, E., Mohan, C.: Analysis of a Simple Particle Swarm Optimization System. Intelligent Engineering Systems Through Artificial Neural Networks 8, 253–258 (1998)

[55] Abido, M.A.: Optimal Design of Power System Stabilizers Using Particle Swarm Optimization. IEEE Trans. on Energy Conversion 17(3), 406–413 (2002)
[56] Parsopoulos, K.E., Vrahatis, M.N.: Particle Swarm Optimization Method in Multiobjective Problems. In: Proceedings of the ACM 2002 Symposium on Applied Computing SAC 2002, pp. 603–607 (2002)
[57] Hu, X., Eberhart, R.: Multiobjective Optimization Using Dynamic Neighborhood Particle Swarm Optimization. In: Congress on Evolutionary Computation CEC 2002, vol. 2, pp. 1677–1681. IEEE Service Center, Piscataway (May 2002)
[58] Coello, C.A.C., Lechuga, M.S.: MOPSO: A Proposal for Multiple Objective Particle Swarm Optimization. In: Congress on Evolutionary Computation (CEC 2002), vol. 2, pp. 1051–1056. IEEE Service Center, Piscataway (May 2002)
[59] Fieldsend, J.E., Singh, S.: A Multi-Objective Algorithm based upon Particle Swarm Optimization, an Efficient Data Structure and Turbulence. In: Proceedings of the 2002 U.K. Workshop on Computational Intelligence, Birmingham, UK, September 2-4, pp. 37–44 (2002)
[60] Mostaghim, S., Teich, J.: Strategies for Finding Good Local Guides in Multiobjective Particle Swarm Optimization (MOPSO). In: Indianapolis, I.N. (ed.) Proceedings of 2003 IEEE Swarm Intelligence Symposium, Indianapolis, IN, USA, pp. 26–33 (April 2003)
[61] Hu, X., Eberhart, R., Shi, Y.: Particle Swarm with Extended Memory for Multiobjective Optimization. In: Proceedings of 2003 IEEE Swarm Intelligence Symposium, Indianapolis, IN, USA, April 2003, pp. 193–197 (2003)
[62] Li, X., et al.: A Nondominated Sorting Particle Swarm Optimizer for Multiobjective Optimization. In: Cantú-Paz, E., Foster, J.A., Deb, K., Davis, L., Roy, R., O'Reilly, U.-M., Beyer, H.-G., Kendall, G., Wilson, S.W., Harman, M., Wegener, J., Dasgupta, D., Potter, M.A., Schultz, A., Dowsland, K.A., Jonoska, N., Miller, J., Standish, R.K. (eds.) GECCO 2003. LNCS, vol. 2723, pp. 37–48. Springer, Heidelberg (2003)
[63] Mostaghim, S., Teich, J.: The Role of ε-Dominance in Multiobjective Particle Swarm Optimization Methods. In: Proceedings of IEEE Congress on Evolutionary Computation CEC 2003, Canberra, Australia, pp. 1764–1771 (2003)
[64] Lu, H.: Dynamic Population Strategy Assisted Particle Swarm Optimization in Multiobjective Evolutionary Algorithm design. IEEE Neural Network Society, IEEE NNS Student Research Grants 2002, Final reports (2003)
[65] Song, M.P., Gu, G.C.: Research on Particle Swarm Optimization: A Review. In: Proceedings of the 3rd International Conference on Machine Learning and Cybernetics, Shanghai, Chaina, August 26-29, pp. 2236–2241 (2004)
[66] Pulido, G.T., Coello Coello, C.A.: Using Clustering Techniques to Improve the Performance of a Multi-Objective Particle Swarm Optimizer. In: Deb, K., et al. (eds.) GECCO 2004. LNCS, vol. 3102, pp. 225–237. Springer, Heidelberg (2004)
[67] Parsopoulos, K.E., Tasoulis, D.K., Vrahatis, M.N.: Multiobjective Optimization Using Parallel Vector Evaluated Particle Swarm Optimization. In: Proc. IASTED Int. Conf. on Artificial Intelligence and Applications, as part of the 22nd IASTED Int. Multi-Conference on Applied Informatics, Innsbruck, Austria (2004)
[68] Mostaghim, S., Teich, J.: Covering Pareto-Optimal Fronts by Subswarms in Multiobjective Particle Swarm Optimization. In: Proceedings of IEEE Congress on Evolutionary Computation CEC 2004, Portland, Oregon, USA, June 19-23, pp. 1404–1411 (2004)
[69] Coello, C.A.C., Pulido, G.T., Lechuga, M.S.: Handling Multiple Objectives with Particle Swarm Optimization. IEEE Trans. on Evolutionary Computation 8(3), 256–279 (2004)
[70] Zhang, Y., Huang, S.: A Novel Multiobjective Particle Swarm Optimization for Buoys-Arrangement Design. In: Proceedings of IEEE/WIC/ACM International Conference on Intelligent Agent Technology (IAT 2004), Beijing, China, September 20-24, pp. 24–30 (2004)

[71] Belhadj, C.A., Abido, M.A.: An optimized Fast Voltage Stability Indicator. In: IEEE Budapest Power Tech. 1999 Conference, Budapest, Hungary, August 29-September 2, BPT99-363-12 (1999)
[72] Tuan, T.Q., Fandino, J., Hadjsaid, N., Sabonnadiere, J.C., Vu, H.: Emergency Load Shedding to Avoid Risks of Voltage Instability Using Indicators. IEEE Trans. on Power Systems 9(1), 341–351 (1994)
[73] Goldberg, D.E., Deb, K.: A Comparison of Selection Schemes Used in Genetic Algorithms. Foundations of Genetic Algorithms, 69–93 (1991)
[74] Goldberg, D.E., Richardson, J.: Genetic Algorithms with Sharing for Multimodal Function Optimization. In: Proceedings of the First International Conference on Genetic Algorithms and Their Applications, pp. 41–49 (1987)
[75] Fonseca, C.M., Fleming, P.J.: An overview of evolutionary algorithms in multiobjective optimization. Evolutionary Computation 3(1), 1–16 (1995)
[76] Fonseca, C.M., Fleming, P.J.: Multiobjective Optimization and Multiple Constraint Handling with Evolutionary Algorithms-Part I: A Unified Formulation. IEEE Transactions on Systems, Man, and Cybernetics, Part A: Systems and Humans 28(1), 26–37 (1998)
[77] Fonseca, C.M., Fleming, P.J.: Multiobjective Optimization and Multiple Constraint Handling with Evolutionary Algorithms-Part II: Application Example. IEEE Transactions on Systems, Man, and Cybernetics, Part A: Systems and Humans 28(1), 38–47 (1998)
[78] Srinivas, N., Deb, K.: Multiobjective Function Optimization Using Nondominated Sorting Genetic Algorithms. Evolutionary Computation 2(3), 221–248 (1994)
[79] Horn, J., Nafpliotis, N., Goldberg, D.E.: A Niched Pareto Genetic Algorithm for Multiobjective Optimization. In: Proceedings of the First IEEE Conference on Evolutionary Computation, IEEE World Congress on Computational Intelligence, vol. 1, pp. 67–72 (1994)
[80] Zitzler, E., Thiele, L.: An Evolutionary Algorithm for Multiobjective optimization: The Strength Pareto Approach. TIK-Report, No. 43 (1998)
[81] Coello, C.A.C., Christiansen, A.D.: MOSES: A Multiobjective Optimization Tool for Engineering Design. Engineering Optimization 31(3), 337–368 (1999)
[82] Abido, M.A.: Two-Level of Nondominated Solutions Approach To Multiobjective Particle Swarm Optimization. In: Proceedings of the 2007 Genetic and Evolutionary Computation Conference, GECCO 2007, London, UK, July 7-11, pp. 726–733 (2007)
[83] Abido, M.A.: Multiobjective Particle Swarm for Environmental/Economic Dispatch Problem. In: Proceedings of the 8th International Power Engineering Conference (IPEC 2007), Singapore, December 3-6, pp. 1894–1899 (2007)
[84] Abido, M.A.: Multiobjective Particle Swarm Optimization for Environmental/Economic Dispatch Problem, Accepted for Publication in Electric Power Systems Research (February 2009) (manuscript number: EPSR-D-08-00198)
[85] Abido, M.A.: Multiobjective Particle Swarm Optimization With Nondominated Local and Global Sets. Accepted for Publication in Journal of Natural Computation, A Special Issue on Swarm Intelligence (March 2009)
[86] Morse, J.N.: Reducing the size of the nondominated set: Pruning by Clustering. Computers and Operation Research 7(1-2), 55–66 (1980)

Appendix

The line and bus data of the IEEE 30-bus 6-generator system are given in Table A.1 and Table A.2 respectively.

Table A.1. IEEE 30-bus test system line data

Line	From	To	R	X	B	Max. MVA
1	1	2	0.01920	0.05750	0.02640	130.0
2	1	3	0.04520	0.18520	0.02040	130.0
3	2	4	0.05700	0.17370	0.01840	65.0
4	3	4	0.01320	0.03790	0.00420	130.0
5	2	5	0.04720	0.19830	0.02090	130.0
6	2	6	0.05810	0.17630	0.01870	65.0
7	4	6	0.01190	0.04140	0.00450	90.0
8	5	7	0.04600	0.11600	0.01020	70.0
9	6	7	0.02670	0.08200	0.00850	130.0
10	6	8	0.01200	0.04200	0.00450	32.0
11	6	9	0.00000	0.20800	0.00000	65.0
12	6	10	0.00000	0.55600	0.00000	32.0
13	9	11	0.00000	0.20800	0.00000	65.0
14	9	10	0.00000	0.11000	0.00000	65.0
15	4	12	0.00000	0.25600	0.00000	65.0
16	12	13	0.00000	0.14000	0.00000	65.0
17	12	14	0.12310	0.25590	0.00000	32.0
18	12	15	0.06620	0.13040	0.00000	32.0
19	12	16	0.09450	0.19870	0.00000	32.0
20	14	15	0.22100	0.19970	0.00000	16.0
21	16	17	0.08240	0.19230	0.00000	16.0
22	15	18	0.10700	0.21850	0.00000	16.0
23	18	19	0.06390	0.12920	0.00000	16.0
24	19	20	0.03400	0.06800	0.00000	32.0
25	10	17	0.03240	0.08450	0.00000	32.0
27	10	21	0.03480	0.07490	0.00000	32.0
28	10	22	0.07270	0.14990	0.00000	32.0
29	21	22	0.01160	0.02360	0.00000	32.0
30	15	23	0.10000	0.20200	0.00000	16.0
31	22	24	0.11500	0.17900	0.00000	16.0
32	23	24	0.13200	0.27000	0.00000	16.0
33	24	25	0.18850	0.32920	0.00000	16.0
34	25	26	0.25440	0.38000	0.00000	16.0
35	25	27	0.10930	0.20870	0.00000	16.0
36	28	27	0.00000	0.39600	0.00000	65.0
37	27	29	0.21980	0.41530	0.00000	16.0
38	27	30	0.32020	0.60270	0.00000	16.0
39	29	30	0.23990	0.45330	0.00000	16.0
40	8	28	0.06360	0.20000	0.02140	32.0
41	6	28	0.01690	0.05990	0.00650	32.0

Table A.2. IEEE 30-bus test system bus data

Bus #	IB	V	PG	PL	QL	P_G^{MIN}	P_G^{MAX}
1	1	1.0500	99.21	0.00	0.00	50.0	200.0
2	2	1.0450	80.00	21.70	12.70	20.0	80.0
3	3	1.0000	0.00	2.40	1.20	0.0	0.0
4	3	1.0000	0.00	7.60	1.60	0.0	0.0
5	2	1.0100	50.00	94.20	19.00	15.0	50.0
6	3	1.0000	0.00	0.00	0.00	0.0	0.0
7	3	1.0000	0.00	22.80	10.90	0.0	0.0
8	2	1.0100	20.00	30.00	30.00	10.0	36.0
9	3	1.0000	0.00	0.00	0.00	0.0	0.0
10	3	1.0000	0.00	5.80	2.00	0.0	0.0
11	2	1.0500	20.00	0.00	0.00	10.0	30.0
12	3	1.0000	0.00	11.20	7.50	0.0	0.0
13	2	1.0500	20.00	0.00	0.00	12.0	40.0
14	3	1.0000	0.00	6.20	1.60	0.0	0.0
15	3	1.0000	0.00	8.20	2.50	0.0	0.0
16	3	1.0000	0.00	3.50	1.80	0.0	0.0
17	3	1.0000	0.00	9.00	5.80	0.0	0.0
18	3	1.0000	0.00	3.20	0.90	0.0	0.0
19	3	1.0000	0.00	9.50	3.40	0.0	0.0
20	3	1.0000	0.00	2.20	0.70	0.0	0.0
21	3	1.0000	0.00	17.50	11.20	0.0	0.0
22	3	1.0000	0.00	0.00	0.00	0.0	0.0
23	3	1.0000	0.00	3.20	1.60	0.0	0.0
24	3	1.0000	0.00	8.70	6.70	0.0	0.0
25	3	1.0000	0.00	0.00	0.00	0.0	0.0
26	3	1.0000	0.00	3.50	2.30	0.0	0.0
27	3	1.0000	0.00	0.00	0.00	0.0	0.0
28	3	1.0000	0.00	0.00	0.00	0.0	0.0
29	3	1.0000	0.00	2.40	0.90	0.0	0.0
30	3	1.0000	0.00	10.60	1.90	0.0	0.0

A Multi-objective Resource Assignment Problem in Product Driven Supply Chain Using Quantum Inspired Particle Swarm Algorithm

Sri Krishna Kumar[1], S.G. Ponnambalam[2], and M.K. Tiwari[3,*]

[1] Wolfson School of Mechanical and Manufacturing Engineering, Loughborough University, UK
[2] School of Engineering, Monash University, Malaysia
[3] Department of Industrial Engineering and Management, India Institute of Technology, Kharagpur, India
mkt09@hotmail.com

Abstract. This chapter presents a novel approach that integrates the intangible factors with the tangible ones to model the resource assignment problem in a product driven supply chain. The problem has been mathematically modeled as a multi-objective optimization problem with the objectives of profit, quality, ahead time of delivery and volume flexibility. In this research, product characteristics have been associated with the design requirements of a supply chain. Different types of resources have been considered each differing in its characteristics, thereby providing various alternatives during the design process. The aim is to design integrated supply chains that maximizes the weighted sum of the objectives, the weights being decided by the desired product characteristics. The problem has been solved through a proposed Quantum inspired Particle Swarm Optimization (QPSO) metaheuristic. It amalgamates particle swarm optimization with quantum mechanics to enhance the search potential and make it suitable for integer valued optimization. The performance of the proposed solution methodology and its three variants has been authenticated over a set of test instances. The results of the above study and the insights derived through it validate the efficiency of the proposed model as well as the solution methodology on the problem at hand.

Keywords: Supply Chain, Resource assignment, Particle Swarm, Fuzzy Analytical Hierarchical Process, Multiobjective optimization.

1 Introduction

A supply chain can be defined as "a goal-oriented network of processes and stock-points used to deliver goods and services to customers" [1]. Amidst the global competition thriving among business entities, the urge to design and manage effective supply-chains for producing and delivering variety of products at low cost, high quality and short lead times is increasing day by day. The incessant pressure of global competition has forced product and service providers to transform and improve their

[*] Corresponding author.

B.K. Panigrahi, Y. Shi, and M.-H. Lim (Eds.): Handbook of Swarm Intelligence, ALO 8, pp. 269–292.
springerlink.com

operations and practices. More specifically, various strategic and operational tools are being adopted by the business entities to improve the performance of their supply chains.

A supply chain, in general, comprises of a number of business entities that include suppliers, manufacturers, distributors, vendors, etc. and their respective haulers and therefore represents a distributed system which require joint strategic planning to ensure effective performance [2]. The decisions that took place in a supply chain have been classified into four levels of hierarchy - strategic, tactical, operational, and real-time level. The most important among these levels is the strategic level, which deals with the resource assignment, flow of goods, cost-effective location of facilities etc., as the success of the decisions taken at other levels depends to a large extent upon the decisions that took place at this level. However, the decision making at this level is a difficult process as it involves the resolution of complex resource allocation problem which involves the selection of different business entities (out of the available ones) based on various tangible and intangible factors.

This chapter focuses on the strategic decision level of the supply chain and thereby considers the resource assignment problem in a product driven supply chain. The operational effectiveness of such a supply chain is mainly determined by the involvement of tangible (equipments, plants, fleets, hardware) and intangible factors (organizational processes, skill, know-how, quality, and reputation). Since tangible factors are measured in quantitative terms, these can easily be incorporated into the design consideration of the resource assignment problem. Intangible factors like quality, although cannot be measured directly, a good number of methodologies [3, 4, 5] are suggested in the literature to do away with this limitation. Most of these methodologies assign relative scores to each of the available alternatives and rank them on the basis of these scores. However, integration of intangible factors with the tangible ones is still a challenge faced by the research community. In this chapter, a multi-objective optimization model has been formulated in order to integrate these two factors into the resource assignment problem of a product driven supply chain. This chapter introduces the application of mathematical tools of fuzzy set theory [6] in decision making model because traditional AHP employs crisp values to score the criteria, which is not an exact and realistic method to specify the decision maker's feeling. The suggested approach facilitates the notion of expressing decision maker's opinion in linguistic variables such as high, medium or low which are thereafter translated into mathematical form by presenting them in a form of triangular trapezoidal fuzzy numbers.

The proposed fuzzy multi-criteria decision making model will evaluate the different criteria associated with the resources and assign the operational quality to each resource. Evaluation criteria as specified in the supply chain council (SCOR) model have been used in this chapter [7]. After assigning the quality to resources, a mathematical multi-objective optimization model related to cost, ahead time of delivery, quality and volume flexibility can be established for the resource assignment problem of a product driven supply chain.

In order to solve the aforementioned optimization problem, this chapter proposes a novel metaheuristic, Quantum inspired Particle Swarm Optimization (QPSO), for constrained optimization. The skeleton of the proposed QPSO metaheuristic encompasses a proposed encoding schema which amalgamates the quantum philosophy [8] with the particle swarm strategy [9]. The encoding schema offers advantages like thorough representation of search space, reduced dependence on population size, and enhanced search capability [10] while the characteristics of particle swarm aid in efficiently orienting the search towards potential regions of the search space. The metaheuristic and its three variants have been applied to solve the aforementioned supply chain resource allocation problem and the results were thoroughly analyzed to draw useful insights. The chapter has been prepared with the following objectives:

- To mathematically model the complex product driven resource allocation problem as a multiobjective optimization problem considering tangible (profit, advance time of delivery and volume flexibility) and intangible factors (quality).
- To validate the efficacy and applicability of the proposed particle swarm methodology on the problem at hand as well as to validate the model and performance of solution methodology.
- To perform sensitivity analysis and draw useful insights beneficial to both practitioners and researchers.

The underlying problem and the proposed methodology will be detailed in the sections to follow. The rest of the chapter is organized as follows. Section 2 deals with the literature review while Section 3 mathematically models the problem. Section 4 provided a brief overview of the Particle Swarm Optimization metaheuristic. The proposed solution methodology is detailed in Section 5, followed by results and discussion in Section 6. Finally, conclusions and future considerations are detailed in Section 7.

2 Related Work

A comprehensive study of the literature reveals that recently much work has been targeted to the area of supply chain design and modeling. [11] design a production, distribution, and vendor network for a global supply chain model. Their model addressed the issues related to reengineering of existing supply chain, however it did not consider the efficiency of supply chain network processes. [12] proposed a model of supply chain network which is composed of three stages: supplier network, producer network and distributor network, and defined the nature of relationship between each stage. An integrated multi-objective supply chain model for strategic and operational supply chain planning has been proposed by [13, 14], in that they presented a strategic production-distribution model for supply chain design with consideration of bills of materials for operational processes. For an extensive review of strategic production-distribution models in a global supply chain environment, readers are referred to [15].

According to [16], models of supply chain design and analysis can be categorized as deterministic analytical model, stochastic analytical model, game-theoretic model and simulation model. They suggested the importance of incorporating intangible factors into design consideration and conclude that nowadays recent research drift is to develop new methodologies for evaluating intangible factors in the mathematical model of a given supply chain. An Analytical Network Process (ANP) method has been utilized by [4] to explore the relationships among logistics, supply chain environment and logistics principles. [17] demonstrated the implementation of analytic hierarchy process (AHP) in supply chain development and covers the seven pre-defined analytical phases for evaluation process. [18] proposed a design framework of value chain network based on Data Envelopment Analysis (DEA) and multi-criteria decision models. However, their model do have few limitations related to design of entire supply chain network. [5] presented the product driven supply chain and utilized AHP to evaluate suppliers under different supply chain strategies, and preemptive goal programming (PGP) to select suppliers.

While scanning the literature, it can easily be revealed that there is no holistic method to measure the operational quality of resources in product driven supply chain. Although few attempts have been made in the literature to incorporate the intangible factors into design consideration, still researchers are investigating the new efficient approaches that can cope with these demerits. Other important observations that can be obtained from the existing body of literature are;

1. Intangible factors like the quality of a product can be an effective measure to determine the operational performance of a product driven supply chain. However, till now these factors are not integrated in the design consideration of resource assignment problems at strategic decision making level.
2. Since, a supply chain design problem is *NP-hard* [19], there is a need for the development of an efficient solution methodology that can identify the optimal or near-optimal solution in minimum computational time.

The objective of this research is to mathematically formulate the design criteria of product driven supply chain. The underlying problem is solved using a quantum inspired particle swarm methodology proposed in this research. It is based on particle swarm metaheuristic, initially proposed by [20], which is a population based search strategy. Recent researches on quantum inspired evolutionary algorithms [21] have shown that quantum based techniques produces good results for combinatorial optimization problems. The fundamentals of quantum mechanics have been integrated with particle swarm to make it more suitable for integer optimization problem and to enhance its properties. The proposed modification further reduces the computational complexity by reducing the swarm size requirements.

3 The Mathematical Model

The assignment of resource capabilities at strategic level is a leading aspect affecting the performance of supply chains. In this chapter, a product driven supply chain problem comprising of four different echelons viz. suppliers, haulers, manufacturers and vendors has been considered (Figure 1). All of these echelons are linked together with a forward flow of material and backward flow of information. At the strategic level, this chapter aims at reducing ahead time of delivery and cost while, improving the quality of product and volume flexibility. The typical network structure of supply chain conceived in this chapter is aimed at the integrated design of supply chains for the manufacturing and delivery of different grades of a product. The selection of an appropriate type of supply chain is assumed to be driven by the characteristics of product an organization is manufacturing. The supply chain model formulated in this study constitute of different types of suppliers, manufacturers, their respective

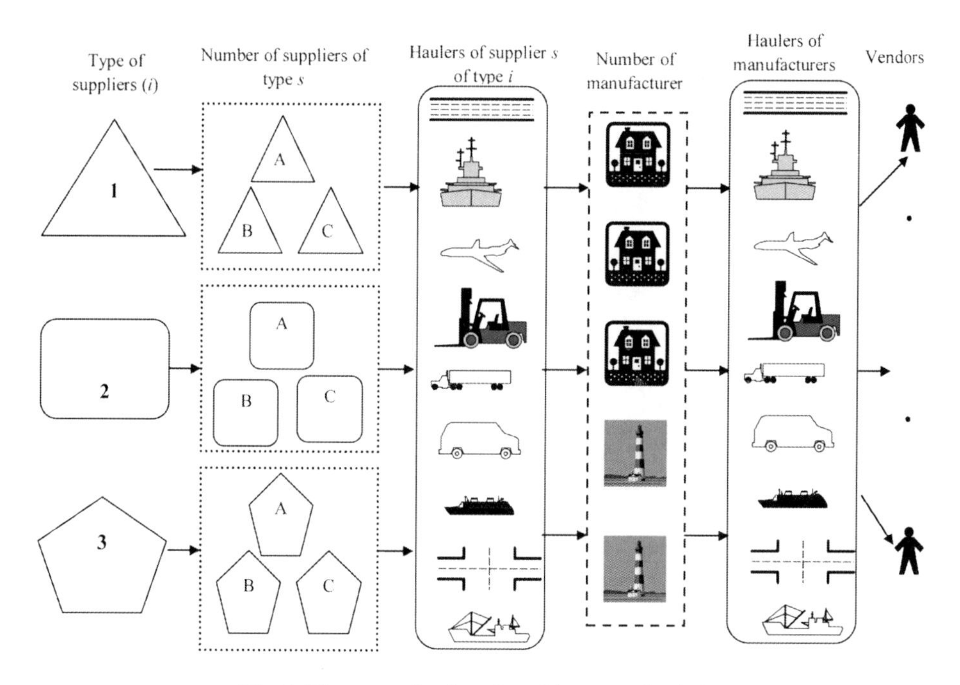

Fig. 1. The network of product driven supply chain

haulers, and different vendor locations at which the final product is to be delivered. The scenario considered here is based on the design of one to one business process chain where, each vendor desires a specific grade of product which is produced and delivered to the selected manufacturer, each manufacturer only order parts from one supplier, and each supplier and manufacturer can select a hauler available to them to transport their products.

The optimization of resource assignment is to identify the optimal upstream suppliers, manufacturers, downstream vendors and their corresponding haulers so that a supply chain ensuring certain profit, lead time of delivery, quality and volume flexibility can be designed. In particular, supply chains with varying weights of the objectives have to be designed with appropriate allocation of resources. Following discussion elaborates the quality assignment of resources by fuzzy based AHP approach, thereafter notations, decision variables and related constraints are detailed. Finally, an integrated mathematical model of profit, ahead time of delivery, operational quality and volume flexibility has been explained.

3.1 Assignment of Operational Quality to Resources

After the assignment of resources, their operational quality is the main intangible factor for measuring the effectiveness of product driven supply chain and has to be incorporated in the design model. As mentioned earlier, this chapter determines the quality of each resource by fuzzy based multi-criteria decision making model. Here, the objective is to design n-number of supply chains with varying level of quality. In each supply chain, vendor set the desired weight of objectives. All the suppliers, manufacturers and haulers are ranked on the basis of proposed fuzzy analytical hierarchy procedure. The performance metrics used in this chapter for selecting the suppliers are adopted from supply chain operations reference (SCOR) model [22].

Owing to the inefficient and unreliable weighted sum method as suggested in the literature, an approach that generalizes the aggregating schemes into some "ideal" criterion using mathematical tools of fuzzy set theory has been proposed. This approach is in accordance with the spirit of fuzzy set theory and undoubtedly it is fruitful in a many applications. A brief introduction of suggested approach is discussed in the following sub-section.

3.1.1 Steps of Fuzzy Based Multi-criteria Decision Making Model

There are seven steps in the proposed approach;

Step 1: Form a weighting team of decision makers that can evaluate the performance metrics.

Step 2: Determine the evaluation criteria of suppliers, manufacturers and haulers. These criteria are adopted from supply chain operation reference model.

Step 3: Set proper linguistic scales (e.g. very high, high, medium, low, and very low) and ask decision makers to give their opinion on the relative importance of the evaluation criteria by pair-wise comparison (Table 1).

Step 4: Convert the linguistic variables into triangle fuzzy numbers to get every member's fuzzy reciprocal matrix.

Step 5: Aggregate the decision maker's fuzzy reciprocal matrices by geometric means and form the final aggregated fuzzy reciprocal matrix.

Step 6: Take the geometric row means of every performance criteria and normalize it to get its local weight.

Table 1. Linguistic scale and its triangle fuzzy number conversion

Performance measure used by decision makers	Trapezoidal fuzzy scale
Very high (9:1)	(8,9,9)
Between very high and high (7:1)	(6,7,8)
High (5:1)	(4,5,6)
Between high and medium (3:1)	(2,3,4)
Medium (2:1)	(1/2,1,2)
Exactly equal (1:1)	(1,1,1)
Between medium and low (1:3)	(1/4,1/3,1/2)
Low (1:5)	(1/6,1/5,1/4)
Between low and very low (1:7)	(1/8,1/7,1/6)
Very low (1:9)	(1/9,1/9,1/8)

Step 7: Translate each final triangle fuzzy score of performance criterion into a crisp value for ranking purposes. There are several de-fuzzification methods in literature, this chapter, however employs [23] formula for its proven superior performance.

Based on the aforementioned steps, quality associated with each supplier, manufacturer and hauler after the assignment of resources can be determined. The aggregate quality of the supply chain can then be calculated by taking the average of the quality of all the members involved in it. The next section details the notations, decision variables and constraints modeled for the underlying problem.

3.2 List of Notations and Decision Variables

The following notations and decision variables have been widely used in the chapter.

Indices	
$s = 1,2,\ldots, S_i$	Suppliers
$i = 1,2,\ldots,S$	Type of suppliers
$h=1,2,\ldots,SHM_{si}$	Type of haulers of supplier s of type i
$m = 1,2,\ldots,M$	Manufacturers
$j=1,2,\ldots,MH_m$	Haulers of manufacturer m
$v = 1,2,\ldots,V$	Vendors

Notations

AT_v	Ahead time of delivery of products ordered by vendor v
C_v	Cost of products ordered by vendor v
DMV_{mv}	Distance between manufacturer m and vendor v
DSM_{sim}	Distance from supplier s of type i to manufacturer m
DT_v	Maximum delivery time of products ordered by vendor v
M	Total number of manufacturers
MCM_m	Manufacturing cost of unit-batch products produced by manufacturer m
MCS_{imv}	Manufacturing cost of parts supplied by suppliers of type i and ordered by manufacturer m for products ordered by vendor v
MCV_{mv}	Manufacturing cost of products produced by manufacturer m and ordered by vendor v
MF_v	Manufacturer flexibility
MHF_v	Hauler's flexibility with respect to manufacturer

MH_m	Total number of haulers of manufacturer m
MS_{si}	Manufacturing cost of unit batch products supplied by supplier s of type i.
MTM_m	Manufacturing time of unit-batch products produced by manufacturer m
MT_{mv}	Manufacturing time of products produced by manufacturer m and ordered by vendor v
MTS_{si}	Manufacturing time of unit batch products supplied by supplier s of type i.
MTV_{simv}	Manufacturing time of parts supplied by supplier s of type i and ordered by manufacturer m for products ordered by vendor v
$norm(.)$	Represents normalized objective (.)
Obj	Final integrated objective function
OCV_v	Operational cost of vendor v during the period of OT_v
OT_v	Observation period of vendor v
PCM_m	Maximum production capacity of manufacturer m during the observation period OT
PCS_{si}	Maximum production capacity of supplier s of type i during the observation period OT
P_v	Profit of vendor v
Q_v	Operational quality of supply chains of vendor v
S	Total number of type of suppliers
SF_v	Supplier Flexibility
SHF_v	Hauler's flexibility with respect to supplier
SHM_{si}	Total number of haulers of supplier s of type i
S_i	Total number of suppliers of type i
$TCHM_{jm}$	Transportation cost of unit-mileage and unit batch transportation transported by hauler j of manufacturer m
$TCHM_{mj}$	Maximum transportation capability of hauler j of manufacturer m during the observation period OT
TCH_{sih}	Maximum transportation capacity of hauler h of supplier s of type i during the observation period OT
$TCHS_{sih}$	Transportation cost of unit-mileage and unit batch transportation transported by hauler h of supplier s of type i.

TCS_{imv}	Transportation cost of parts supplied by suppliers of type i and ordered by manufacturer m for products ordered by vendor v
TCV_{mv}	Transportation cost of products produced by m manufacturer and ordered by vendor v
$TTHM_{jm}$	Transportation time of unit-mileage transported by hauler j of manufacturer m
$TTHS_{sih}$	Transportation time for unit-mileage transported by hauler h of supplier s of type i.
TT_{mjv}	Transportation time of products produced by manufacturer m, transported by hauler j and ordered by vendor v
TTV_{sihmv}	Transportation time of parts supplied by supplier s of type i, transported by hauler h and ordered by manufacturer m for products ordered by vendor v
UP_v	Unit price of product of vendor v
V	Total number of vendors
VF_v	Volume Flexibility of vendor v
Vol_v	Volume of products ordered by vendor v during the OT_v period
W_{1v}	Weight assigned by vendor v to profit objective
W_{2v}	Weight assigned by vendor v to ahead time of delivery objective
W_{3v}	Weight assigned by vendor v to quality objective
W_{4v}	Weight assigned by vendor v to volume flexibility objective

Decision Variables

$$a_{siv} = \begin{cases} 1, & \text{When supplier } s \text{ of type } i \text{ is selected in a supply chain of vendor } v \\ 0, & \text{Otherwise} \end{cases}$$

$$y_{sihmv} = \begin{cases} 1, & \text{When hauler } h \text{ of supplier } s \text{ of type } i \text{ is connected with manufacturer } m \text{ of vendor } v \\ 0, & \text{Otherwise} \end{cases}$$

$$l_{mjv} = \begin{cases} 1, & \text{When hauler } j \text{ of manufacturer } m \text{ is connected with vendor } v \\ 0, & \text{Otherwise} \end{cases}$$

$$b_{mv} = \begin{cases} 1, & \text{When manufacturer } m \text{ is selected for vendor } v \\ 0, & \text{Otherwise} \end{cases}$$

Based on the network graph shown in figure 1, the multi-objective optimization model for resource assignment problem in product driven supply chain can be written as:

$$Max\ Obj = \sum_{v=1}^{V} \left(w_{1v} \cdot norm(P_v) + w_{2v} \cdot norm(AT_v) + w_{3v} \cdot norm(Q_v) + w_{4v} \cdot norm(VF_v) \right) \quad (1)$$

Where, $w_{1v} + w_{2v} + w_{3v} + w_{4v} = 1 \qquad \forall \ v$

3.3 Proposed Multi-objective Optimization Model

In this section, individual objectives i.e. profit, ahead time of delivery, quality and volume flexibility are mathematically modeled. Operational quality assignment to resources is done by the approach suggested in section 3.1.1. Details of remaining objective formulation are given below:

3.3.1 Profit

The profit of product driven supply chains conceived here is given by the difference of price of the product and total cost incurred in its manufacturing and delivery (equation 2). Total cost in the four echelon supply chain is calculated at each stage and can be mathematically represented by equations 3, 4, 5, 6 and 7.

$$P_v = UP_v \cdot Vol_v - C_v \quad (2)$$

$$C_v = \sum_{m=1}^{M} \left(\sum_{i=1}^{S_i} (MCS_{imv} + TCS_{imv}) + MCV_{mv} + TCV_{mv} \right) + OCV_v \qquad \forall \ v \quad (3)$$

$$MCS_{imv} = \sum_{s=1}^{S} Vol_v \cdot MS_{si} \cdot a_{siv} \qquad \forall \ v, i \quad (4)$$

$$TCS_{imv} = \sum_{s=1}^{S} \sum_{h=1}^{SHM_{si}} Vol_v \cdot TCHS_{sih} \cdot DSM_{sim} \cdot y_{sihmv} \qquad \forall \ v, i, m \quad (5)$$

$$MCV_{mv} = Vol_v \cdot MCM_m \cdot b_{mv} \qquad \forall \ v, m \quad (6)$$

$$TCV_{mv} = \sum_{j=1}^{MH_m} Vol_v \cdot TCHM_{jm} \cdot DMV_{mv} \cdot l_{mjv} \qquad \forall \ v, m \quad (7)$$

3.3.2 Ahead Time of Delivery

Ahead time of delivery refers to the time left ahead of the scheduled delivery of the product. It can be mathematically modeled as the difference between the observation period of all vendors and the maximum delivery time available. Equation (8) denotes the ahead-time of delivery calculation whereas, the maximum delivery time can be determined from equation 9, 10, 11, 12 and 13.

$$AT_v = OT_v - DT_v \qquad \forall \; v \tag{8}$$

$$DT_v = \max_i \left(MTV_{simv} + TTV_{sihmv} \right) + MT_{mv} + TT_{mjv} \qquad \forall \; s,i,h,m,v \tag{9}$$

$$MTV_{simv} = \sum_{s=1}^{S} \left(Vol_v \cdot MTS_{si} \cdot a_{siv} \right) \qquad \forall \; i,m,v \tag{10}$$

$$TTV_{sihmv} = \sum_{s=1}^{S} \sum_{h=1}^{SHM_{si}} \left(TTHS_{sih} \cdot DSM_{sim} \cdot y_{sihmv} \right) \qquad \forall \; i,m,v \tag{11}$$

$$MT_{mv} = \sum_{m=1}^{M} \left(Vol_v \cdot MTM_m \cdot b_{mv} \right) \qquad \forall \; v \tag{12}$$

$$TT_{mjv} = \sum_{j=1}^{MH_m} \left(TTHM_{jm} \cdot DMV_{mv} \cdot l_{mjv} \right) \qquad \forall \; m,j,v \tag{13}$$

3.3.3 Volume Flexibility

It can be defined as the minimum of the difference between suppliers, haulers, and manufacturer's capacity with total volume of products ordered by v vendor (equation 18). First, individual flexibility of supplier, hauler and manufacturer will be calculated from equation 14, 15, 16 and 17 and thereafter, the minimum of the obtained flexibility would be considered as the final volume flexibility of the product driven supply chain considered in this chapter.

$$SF_v = \sum_{i=1}^{S_i} \sum_{s=1}^{S} \left(a_{si} \cdot PCS_{si} - Vol_v \right) \qquad \forall \; v, i \tag{14}$$

$$SHF_v = \sum_{i=1}^{S_i} \sum_{s=1}^{S} \sum_{h=1}^{SHM_{si}} \left(x_{sih} \cdot TCH_{sih} - Vol_v \right) \qquad \forall \; v \tag{15}$$

$$MF_v = \sum_{m=1}^{M} \left(b_{mv} \cdot PCM_m - Vol_v \right) \qquad \forall \; v \tag{16}$$

$$MHF_v = \sum_{m=1}^{M} \sum_{j=1}^{MH_m} \left(l_{mjv} \cdot TCHM_{mj} - Vol_v \right) \qquad \forall \; v \tag{17}$$

Therefore, volume flexibility can be written as

$$VF_v = \min\left(SF_v, SHF_v, MF_v, MHF_v \right) \tag{18}$$

3.3.4 Constraints

Since, the proposed mathematical model formulates a one to one product driven supply chain, following constraints (equation 19, 20, 21 and 22) are modeled in order to check that only one supplier, manufacturer, and their respective single hauler will be selected from the available resources. Equation 23 and 24 ensures that the continuity would remain in the supply chain at different echelons. If a particular type of supplier is not selected in the process chain, it implies that in the next echelon that supplier would not be taken into account.

$$\sum_{s=1}^{S} a_{siv} = 1 \qquad \forall \; v, i \qquad (19)$$

$$\sum_{s=1}^{S} y_{sihmv} = 1 \qquad \forall \; v, i, m \qquad (20)$$

$$\sum_{m=1}^{M} b_{mv} = 1 \qquad \forall \; v \qquad (21)$$

$$\sum_{j=1}^{MH_m} l_{mjv} = 1 \qquad \forall \; v, m \qquad (22)$$

$$y_{sihmv} = 0, \text{ if } a_{\text{siv}} = 0 \qquad \forall \; s, i, v, h \qquad (23)$$

$$l_{mjv} = 0, \text{ if } b_{\text{mv}} = 0 \qquad \forall \; m, j, v \qquad (24)$$

$$C_v \le Vol_v \cdot UP_v \qquad \forall \; v \qquad (25)$$

$$DT_v \le OT_v \qquad \forall \; v \qquad (26)$$

$$\sum_{v=1}^{V} Vol_v \cdot a_{siv} \le PCS_{si} \qquad \forall \; s, i \qquad (27)$$

$$\sum_{v=1}^{V} Vol_v \cdot y_{sihmv} \le TCH_{sih} \qquad \forall \; s, i, h, m \qquad (28)$$

$$\sum_{v=1}^{V} Vol_v \cdot b_{mv} \le PCM_m \qquad \forall \; m \qquad (29)$$

$$\sum_{v=1}^{V} Vol_v \cdot l_{mjv} \le TCHM_{mj} \qquad \forall \; m, j \qquad (30)$$

Equation 25 depicts that the cost of the products ordered by v vendor should be less than or equal to the total cost of the products manufactured by the resultant supply chain of v vendor. Equation 26 restricts the maximum delivery time of the products ordered by v vendor to be less than their observation period. It is also imperative to ensure that the maximum capacity of suppliers, their respective haulers, and manufacturers should be more than the volume of the products ordered by vendor v. This constraint at each echelon is modeled in equations 27, 28, 29 and 30 respectively. In the next section, proposed solution methodology has been detailed along with the explanation of implementation procedure to the underlying problem.

4 Overview of Particle Swarm Methodology

Particle Swarm [20] is a recently proposed metaheuristic which mimics the social-psychological model of social influence and social learning of the particles. In Particle swarm, a population of candidate problem solutions, randomly initialized in a high dimensional search space, discovers optimal regions of the space through individual's emulation of successful particles of the swarm. In PSO, each individual, named as *particle*, of the population, named as *swarm*, adjusts its motion towards its previous best position and towards the best position attained by any member of its topological neighborhood [24]. The position vector and velocity vector of the ith particle in n dimensional search space are represented as $X_i = (x_{i1}, x_{i2}, ..., x_{in})$ and $Vel_i = (v_{i1}, v_{i2}, ..., v_{in})$, respectively. During initialization, each particle is randomly placed in the search space as candidate solution and evaluated by a user defined fitness function. Let us say that the best position attained by any particle i at time t be $M_i = (m_{i1}, m_{i2}, ..., m_{in})$, and the global best particle found so far till time t be $P_g = (p_{g1}, p_{g2}, ..., p_{gn})$. Then, the new position and velocity of the particle is calculated using the equations:

$$vel_{in} = W \otimes vel_{in} \oplus c_1 \otimes rand1(.) \otimes (m_{in} - x_{in}) \oplus c_2 \otimes rand2(.) \otimes (p_{gn} - x_{in}) \quad (31)$$

$$x_{in} = x_{in} \oplus vel_{in} \quad (32)$$

Where, W is the inertial weight, c_1 and c_2 are acceleration coefficients and $rand1(.)$ and $rand2(.)$ are random numbers in the range [0, 1]. The first part of the equation represents the inertial velocity component which is a multiple of previous velocity of the particle and an inertial weight factor. The second part denotes the cognitive search component by the virtue of which the particle is motivated to move towards the regions of its best found position so far. The third part is known as the social search component, which directs the particle to move towards the global optimum. Finally, new position of the particle is calculated by adding the velocity component to its position (equation 32).

Initially, a population of particles is generated with random positions and velocities. The fitness of each particle is evaluated according to the chosen objective function. During each generation, the positions and velocities of the particles are updated using equation 31 and 32, and the fitness values are calculated. Individual best and/or Global best are updated in case a particle finds better position than the previously stored position in its memory or the global best position respectively. The combined effect of the cognitive and social search component imparts high exploratory potential and faster convergence to the search.

However, the high speed of convergence, as obtained by PSO, severely marks its ability to converge towards global optimal and induces premature convergence [25], and as such hinders the objective of using it as a solution methodology. Further, the design of PSO makes it particularly suitable for real valued optimization and do not fit the requirements of the underlying problem which, in essence, is an integer programming problem. In order to circumvent all such demerits and incorporate some enhanced features this approach proposes a Quantum inspired Particle Swarm Optimization (QPSO) metaheuristic for optimizing the problem at hand.

5 The Proposed QPSO Metaheuristic

The proposed metaheuristic derives its governing traits from two ideologies: conduct of particles in the swarm and quantum mechanics. While the socio-cognitive behavior of particle acts as the state transition mechanism and enables the search to move towards optimum, quantum inspired characteristics provide stochastic characteristics and ensures the convergence towards global optimum in long run. The proposed QPSO is characterized by rapid convergence and global search capability, simultaneously. The agglomeration of principles from quantum physics in a traditional particle swarm approach offers distinguished advantages like:

1. One individual of QPSO can represent many states at the same time. Such property reduces the dependence of the search on the population size as small population can also represent the entire search space effectively while maintaining the required diversity and transition pressure. This on the other hand reduces the computational burden.
2. QPSO can be utilized for optimizing complex combinatorial optimization and integer programming problems.
3. It ensures positive variation kernel and in association with state transition mechanism of PSO (cognitive and social search components) leads the search towards global optimality.

The proposed approach differs from the traditional particle swarm algorithm in its representation of particle's position which instead of being deterministic is taken as stochastic in this study. A fitness based probability update mechanism is also devised and incorporated to enhance the probability of selection of those members whose selection have shown better performance.

A. Encoding Schema: The proposed encoding schema, to certain extent, is based on the concept of qubits and superposition of states of quantum mechanics. In quantum system, elementary information unit is the qubit. In the proposed approach, real numbers representing probabilities of selection of an entity (a modified form of qubits) are used for encoding. The technique is illustrated through an example shown in figure 2. Consider a simple problem in which we have to choose 2 suppliers out of the 6 available ones for two different supply chains. The selection of two supplier is carried out in two phase. The stochastic encoding scheme proposed for this is shown in figure 2(a). The six bit string represents six different suppliers with the values in those bit strings representing the probability scores of selection. For example, the value 0.41 in the first bit of the string shown in figure 2(a) denotes the probability score of selection of first supplier. In the first phase, these values are normalized to sum to one, as shown in figure 2(b). The values obtained represent the normalized probability of selection of each supplier and the first supplier could be selected from it using the roulette wheel selection strategy. Let us suppose that supplier 2 is selected as our first supplier. Now, we are left with 5 suppliers (figure 2(c)) from which we have to choose the second supplier. This constitute the second phase in which the individual probability of selection of the remaining suppliers (figure 2(d)) is again calculated and thereafter the roulette wheel selection is applied. Thus, the number of phases depends

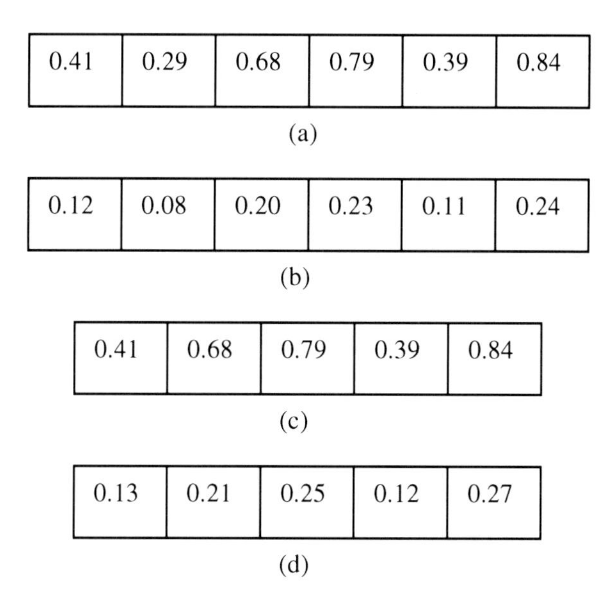

Fig. 2. String representation and decoding schema

upon the total number of supply chain we have to design. Evaluation of objective function can be done once the process of string decoding is over. This methodology apart from providing other benefits ensures that constraints associated with the problem are not violated. The proposed methodology is generic and with suitable modifications can be applied over a wide range of problems.

B. Fitness based Update rule: This mechanism has been devised to provide proportional benefits to those selections or allocation which produce better results. For this, after evaluation, the fitness value of the decoded string is compared with the fitness value stored in the particle's memory. In case, current fitness value is found to be better than the value stored in the memory, the probability values in the bits corresponding to the selected ones is proportionately increased using the equations,

$$Incr_{ij} = \frac{Fit_i}{PN \times C} \tag{33}$$

$$x'_{ij} = x_{ij} + Incr_{ij} \tag{34}$$

where, x_{ij} represents those bits which have been selected, x'_{ij} represents its modified value, $Incr_{ij}$ denotes the amount of increment in those bits which are selected, Fit_i is the fitness value of the particle i after decoding, PN is the phase number, and C is a constant which is used here as a scaling factor. In this research, the aforementioned approach has been implemented on the underlying resource allocation problem with its three other proposed variants namely, QPSO with random inertial weights

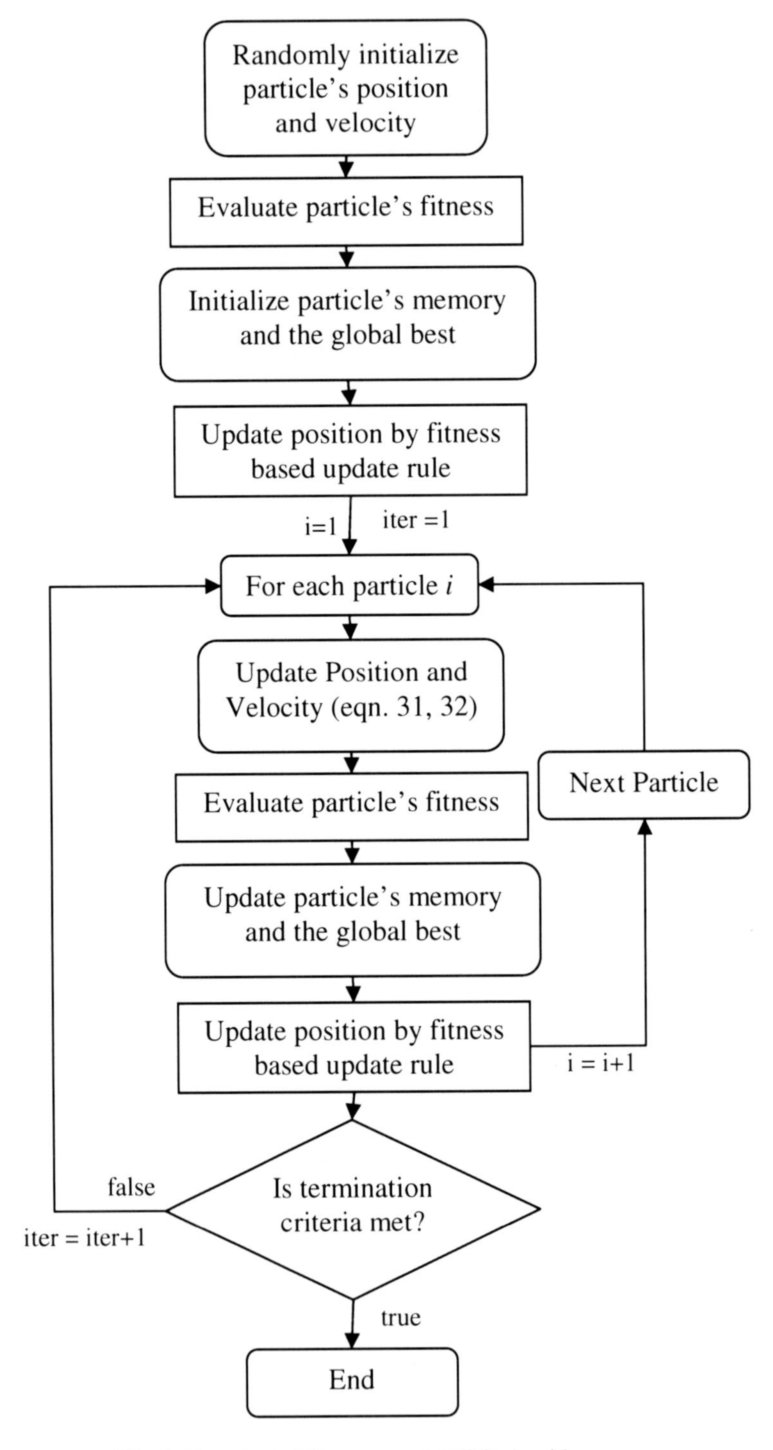

Fig. 3. Flowchart of the proposed QPSO algorithm

(QPSO-RandIW), QPSO with time varying inertial weights(QPSO-TVIW) and QPSO with time varying acceleration coefficients(QPSO-TVAC). The concept of variants is derived from [26] (see [26] for detail discussion) as the modifications done by them over PSO have earlier shown better performance. The flowchart of the proposed QPSO approach is given in Figure 3.

6 Results and Discussions

In this section, results of the proposed methodology and its variants on the resource allocation problem have been detailed. Eight instances have been simulated which correspond to the problem faced by a leading steel manufacturing group. The problem encompasses the design of supply chains within the available resources for the manufacture of several similar kinds of product. The problem design has been detailed in table 2. All the problem instances conceived consist of three class of resources (available suppliers, manufacturers and their respective haulers) which are classified as *A*, *B* and *C* respectively. Resources of Class *A* have high level of quality, and therefore corresponding manufacturing cost, and lower manufacturing time whereas, Class *B* resources are having these values comparatively inferior and finally the least values are assigned to the resources of Class *C*. The product driven supply chains to be designed are divided into three categories: quality driven (maximization of quality), cost driven (minimization of cost or maximization of profit) and homogenous one. The weights given to the objectives of quality, cost, flexibility and ahead time of delivery are taken as (0.4, 0.2, 0.2, 0.2), (0.1, 0.4, 0.25, 0.25) and (0.25, 0.25, 0.25, 0.25) respectively for the aforementioned three kinds of supply chain taken in this study.

Table 2. Details of problem instances undertaken in the case study

Sr. No	Type of suppliers (S_i)	Number of suppliers of each type (S)	Number of haulers for each suppliers (SHM_{si})	Number of manufacturers (M)	Number of haulers for each manufacturers (MH_m)	Number of supply chains to design	Type of supply chain
1	2	3	3	5	5	1	Q
2	3	5	5	5	5	1	C
3	3	5	5	5	5	2	Q, C
4	2	6	6	10	10	3	Q,C,H
5	3	6	6	10	10	3	Q,C,H
6	4	8	6	10	10	3	Q,C,H
7	3	10	10	10	10	3	Q,C,H
8	4	10	10	15	15	3	Q,C,H

* Q = Quality Driven, C = Cost Driven, H = Homogenous.

The proposed quantum inspired particle swarm optimization algorithm and its three variants namely, QPSO-RandIW, QPSO-TVIW, and QPSO-TVAC have been implemented over all the problem instances. The parameter setting used for all the algorithms have been listed in table 3 and the obtained results are provided in table 4. As is evident from table 3, all the algorithms utilized similar kind of parameter settings to ensure that the comparison is being carried out on similar grounds. The fitness function is taken as the weighted sun of normalized objective function values, the weight being assigned as per supply chains to be designed. The number of fitness function evaluations has been taken as the termination criteria. Hundred independent runs were performed for each of the problem instance with all the four algorithms concerned. Table 4 lists the mean and standard deviation values of the fitness of the best particle found.

Table 3. Parameter Settings for the compared approaches

Sr. No.	QPSO	QPSO-RandIW	QPSO-TVIW	QPSO-TVAC
Population Size	25	25	25	25
Inertial Weight	0.75	Randomly generated between 0.5 and 1.	Varied between 0.4 to 0.9.	0.75
Acceleration Coefficient	2.0, 2.0	2.0, 2.0	2.0,2.0	varied in the range (2.5-0.5), (0.5-2.5)
No. of fitness evaluations	25,000	25,000	25,000	25,000

Table 4. Results for all algorithms on benchmark problems (mean of 100 runs and standard deviations (Std. dev)).

Problem No.	QPSO		QPSO-RandIW		QPSO-TVIW		QPSO-TVAC	
	Mean	Std. dev.	Mean	Std. dev.	Mean	Std. dev.	Mean	Std. dev.
1	0.9696	0.001285	0.9690	0.001293	0.9712	0.001280	**0.9716**	**0.001279**
2	0.9461	**0.001352**	0.9456	0.001368	**0.9479**	0.001358	0.9473	0.001361
3	1.5161	0.001794	1.5142	0.001802	1.5191	**0.001765**	**1.5224**	0.001782
4	2.0875	0.002123	2.0901	0.002067	2.0965	0.002064	**2.0992**	**0.002041**
5	2.0962	0.002186	2.0957	0.002193	2.1006	0.002164	**2.1073**	**0.002169**
6	2.1105	0.002265	2.1098	0.002295	**2.1148**	0.002234	2.1119	**0.002221**
7	2.2405	0.002652	2.2450	0.002673	2.2468	0.002651	**2.2482**	**0.002619**
8	2.2460	0.002681	2.2461	0.002679	2.2449	**0.002666**	**2.2482**	0.002696

It is evident from Table 4 that results obtained by all the four algorithms are comparable; however, QPSO-TVAC shows the best performance with highest values of average fitness and minimum standard deviation than the rest of the algorithms for majority of the problem instances considered. The low values of standard deviation as obtained by the QPSO-TVAC demonstrate its high repeatability and good convergence characteristics. The second best results were shown by QPSO-TVIW, followed

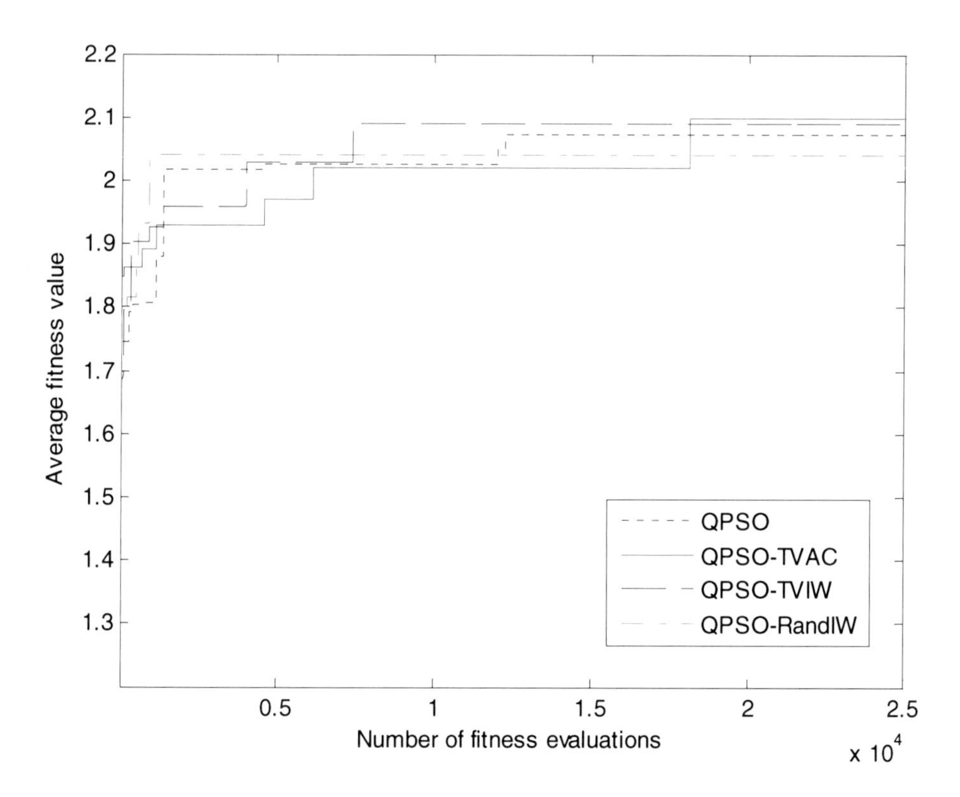

Fig. 4. Convergence trend of the proposed algorithms (Problem instance 5)

by QPSO and the worst performance registered was that of QPSO-RandIW. The convergence trend as obtained with the four algorithms (on problem instance 1) is shown in figure 4.

The percentage contributions of different Class (A, B and C) of resources in the optimized design of different grades of supply chain for the test instances considered are reported in Table 5. It is evident from the table that in the design of quality driven supply chain and cost driven supply chain, the major contributions are from the resources of Class A and C respectively. The results indicate that the quality objective in a supply chain can only be fulfilled when most of the partners are quality conscious. However, restrictions enforced by cost criteria limit their proportion in the optimum design. The results also reveal that in the homogenous design of supply chains, resources of Class B occupy major proportions as contrary to the general assumption of equal proportions of all three classes of resources in its optimum design.

In order to gain deeper insights into the results obtained, sensitivity analysis is performed. The problem instance 5 (number is chosen due to its intermediate size) with the objective of designing a single supply chain. Three scenarios were designed for performing the analysis. Apart from testing the performance of the proposed methodology, these scenarios are taken to test the effectiveness of the model formulation by evaluating profit, quality, flexibility, ahead time of delivery and the tradeoffs among them.

Table 5. Percentage contribution of different types of members in the optimum supply chain configuration of different problem instances

	Type of design	% contribution of Class A	% contribution of Class B	% contribution of Class C
Problem instance 1	Q	0.79	0.16	0.05
Problem instance 2	C	0.184	0.298	0.518
Problem instance 3	Q	0.82	0.153	0.027
	C	0.267	0.245	0.488
Problem instance 4	Q	0.74	0.13	0.13
	C	0.123	0.317	0.56
	H	0.258	0.456	0.286
Problem instance 5	Q	0.71	0.21	0.08
	C	0.292	0.333	0.375
	H	0.25	0.42	0.33
Problem instance 6	Q	0.78	0.15	0.07
	C	0.238	0.471	0.291
	H	0.187	0.281	0.532
Problem instance 7	Q	0.8	0.13	0.07
	C	0.198	0.419	0.383
	H	0.158	0.246	0.596
Problem instance 8	Q	0.73	0.17	0.1
	C	0.236	0.435	0.329
	H	0.134	0.265	0.601

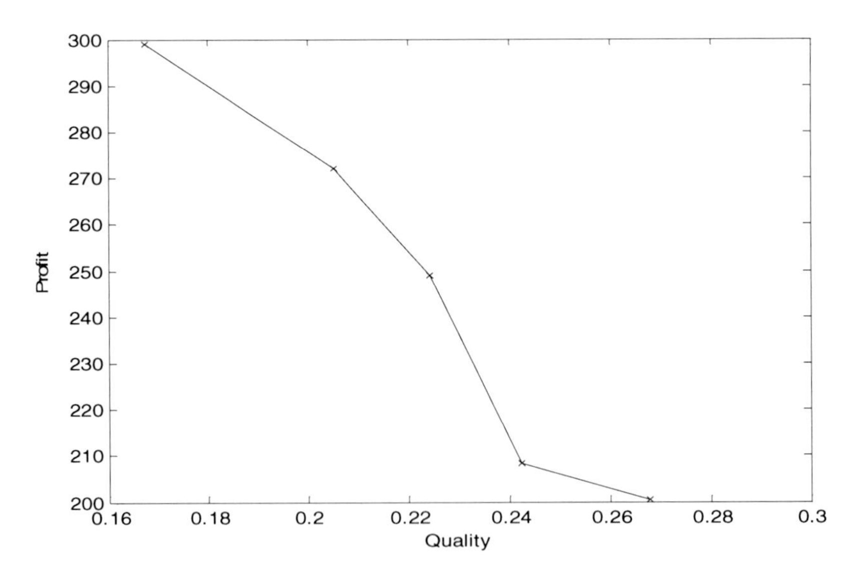

Fig. 5. Trade-off between profit vs. quality

In the first scenario, the weights for the objectives of profit and quality were varied while keeping those for flexibility and ahead time of delivery constant. The price of the product at different levels of the quality is kept constant and thus the results also represent the tradeoff between cost and quality (figure 5). The result shows that with an increase in quality, the profit decreases, but the decrease in profit is not linear indicating that quality increments affect the cost more dramatically and the price of the product should be aptly decided to overcome the additional cost incurred due to quality improvements. The aforementioned fact is responsible for the need to optimize the design of supply chain for ensuring the maintenance of minimum quality standards required by the customers at an affordable price.

In the second and third scenario, the effect of ahead time variations and volume flexibility on the profit is examined and the results of the relationship are given in figures 6 and 7 respectively. It appears from the figures that the profit decreases (as cost increases) with the increase in ahead time of delivery and for a fixed quantity of demand the addition of more capacity in terms of volume flexibility leads to decreased profits (due to increased costs). However, before deriving any conclusions about the distribution of profit with volume flexibility and ahead time of delivery, consideration of time-varying demand, different ahead times of distributed customer locations, and backorder cost needs to taken in account. These observations only indicate that the increase in value of the aforementioned two objectives lead to an increase in cost and their effects, combined with practical scenarios like distributed customer zones and varying ahead times are responsible for the introduction of concepts like consolidation hubs (to reduce the transportation cost, ahead time of delivery and flexibility requirements) and distribution centers for efficient operation in global supply chains. Moreover, the results of the above study also validate the efficiency of the proposed model as well as the solution methodology on the problem at hand.

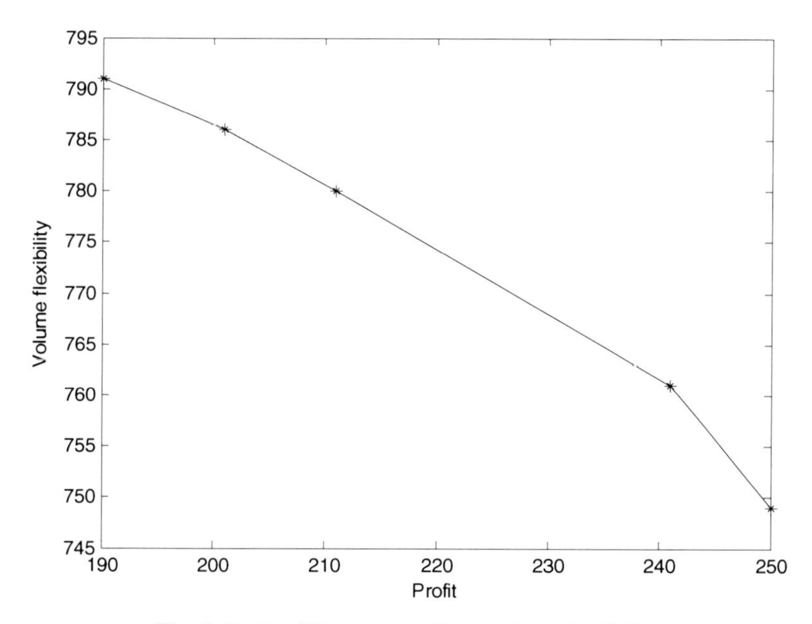

Fig. 6. Trade-off between profit vs. volume flexibility

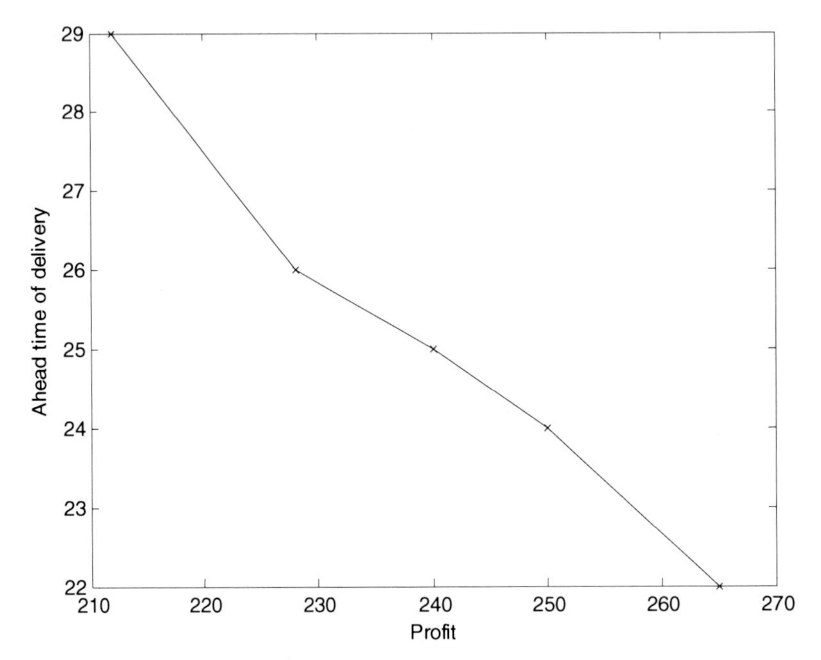

Fig. 7. Trade-off between profit and ahead time of delivery

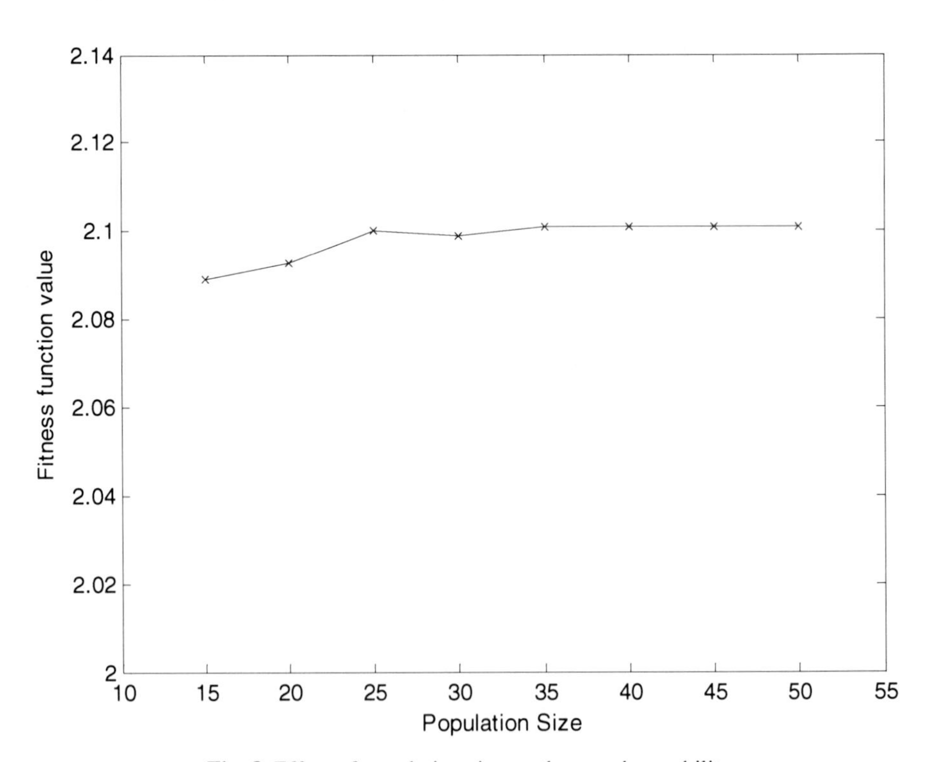

Fig. 8. Effect of population size on the search capability

Finally, the effect of population size variations on the performance of the proposed methodology is studied and the obtained results are shown in figure 8. The probabilistic representation, being able to represent the entire search space with comparatively much lesser number of particles, is expected to relieve the burden of adjusting the population size with the search space dimensions. The same is validated by our analysis which shows that the results obtained with smaller population size are quite comparable with the results obtained with larger population size.

7 Conclusions and Next Steps

In this chapter, a model of resource allocation for a product driven supply chain has been conceived. The supply chain is modeled as a network, nodes of which represent a resource in the network. The objective of the study was to design a supply chain with specific requirements and to maximize the integrated objective comprising of quality, profit, volume flexibility and ahead time of delivery. In order to solve the problem, a novel strategy based on particle swarm optimization and quantum mechanics is devised. The performance of the proposed methodology and its three variants were tested over eight benchmark instances where the obtained results have showed their efficiency in solving the underlying problem. A few general observations based on the sensitivity analysis were also presented which further proved the efficacy of the proposed model and the solution methodology.

This research raises several questions, of which, few most important that requires further work are listed. First, the model has been formulated for a one-to-one supply chain and need to be enhanced to include the possibility of multiple resource selection at each stage. For instance, a vendor may desire to procure a raw material from two different suppliers in order to have more flexibility and minimize disruptions. Second, inclusion of other intangible objectives as customer service satisfaction level, skills etc. is worth to be explored.

References

[1] Hopp, W.J.: Supply Chain Science. McGraw Hill, New York (2006)

[2] Poirier, C., Quinn, F.: How are we doing? A survey of supply chain progress, Supply Chain Management Review, 24–31 (2004)

[3] Saaty, T.L.: Fundamentals of Decision Making and Priority Theory with Analytic Hierarchy Process. RWS Publications, Pittsburgh (1994)

[4] Meade, L., Sakris, J.: Strategic analysis of logistics and supply chain management systems using the analytical network process. Transportation Research 1998; Part E 34(3), 201–215 (1998)

[5] Wang, G., Huang, S.H., Dismukes, J.P.: Product-driven supply chain selection using integrated multi-criteria decision making methodology. International Journal of Production Economics 91(1), 1–15 (2004)

[6] Zadeh, L.: Fuzzy logic and its application to approximate reasoning. Information Processing 74, 591–594 (1974)

[7] Supply Chain Council. Supply chain operations reference model—Overview of SCOR Version 6.0. Supply Chain council Pittsburgh, PA (2003)

[8] Rieffel, E., Polak, W.: An introduction to quantum computing for non-physicists (January 2000) (arxive.org; quantph/9809016 v2)
[9] Kennedy, J., Eberhart, R.C.: Swarm Intelligence. Morgan Kaufmann Publishers, San Francisco (2001)
[10] Grover, L.: A Fast Quantum Mechanical Algorithm for Database Search. In: Proceedings of the 28th Annual ACM Symposium on the Theory of Computing, pp. 212–219 (1996)
[11] Arntzen, B.C., Brown, G.G., Harrision, T.P., Trafton, L.L.: Global supply chain management at digital equipment corporation. Interfaces 25(1), 69–93 (1995)
[12] Erenguc, S.S., Simpson, N.C., Vakharia, A.J.: Integrated production/distribution planning in supply chains: An invited review. European Journal of Operational Research 115, 219–236 (1999)
[13] Sabri, E.H., Beamon, B.M.: A multi-objective approach to simultaneous strategic and operational planning in supply chain design. OMEGA The International Journal of Management Science 28, 581–598 (2000)
[14] Yan, H., Yu, Z., Cheng, T.C.E.: A strategic model for supply chain design with logical constraints. Computers and Operations Research 30(14), 2135–2155 (2003)
[15] Govil, M., Proth, J.M.: Supply Chain Design and Management: Strategic and Tactical Perspectives. Academic Press, London (2002)
[16] Beamon, B.M.: Supply chain design and analysis: models and methods. International Journal of Production Economics 71(1-3), 145–155 (1998)
[17] Korpela, J., Kylaheiko, K., Lehmusvaara, A., Tuominen, M.: An analytic approach to production capacity allocation and supply chain design. International Journal of Production Economics 78(2), 187–195 (2002)
[18] Talluri, S., Baker, R.C., Sakris, J.: A framework for design efficient value chain networks. International Journal of Production Economics 62(1), 133–144 (1999)
[19] Amiri, A.: Designing a distribution network in a supply chain system: formulation and efficient solution procedure. European Journal of Operation research 171(2), 567–576 (2006)
[20] Kennedy, J., Eberhart, R.C.: Particle Swarm Optimization. In: Proceedings of IEEE International Conference on Neural Networks, pp. 1942–1948 (1995)
[21] Han, K., Kim, J.: Quantum-inspired evolutionary algorithm for a class of combinatorial optimization. IEEE Transactions on Evolutionary Computation 6(6), 580–593 (2002)
[22] Huang, S., Sheoran, S., Keskar, H.: Computer assisted supply chain configuration based on supply chain operations reference (SCOR) model. Computer and Industrial Engineering 48, 377–394 (2005)
[23] Opricovic, S., Tzeng, G.H.: Defuzzification for a fuzzy multi-criteria decision model. International Journal of Uncertainty, Fuzziness and Knowledge-based Systems 11(5), 635–652 (2003)
[24] Kennedy, J., Mendes, R.: Neighborhood topologies in fully informed and best of neighborhood Particle Swarms. IEEE Transactions on Systems, Man, and Cybernetics, Part – C: Applications and Reviews 36(4), 515–519 (2006)
[25] Parsopoulos, K.E., Vrahatis, M.: On the computation of all global minimizers through particle swarm optimization. IEEE Transactions on Evolutionary Computation 8(3), 211–224 (2004)
[26] Ratnaweera, A., Halgamuge, S.K., Watson, H.C.: Self-organizing hierarchical particle swarm optimizer with time varying acceleration coefficients. IEEE Transactions on Evolutionary Computation 8(3), 240–255 (2004)

Part B
Bee Colony Optimization

Honeybee Optimisation – An Overview and a New Bee Inspired Optimisation Scheme

Konrad Diwold[1], Madeleine Beekman[2], and Martin Middendorf[1]

[1] Department of Computer Science,
Universität Leipzig, Germany
{kdiwold,middendorf}@informatik.uni-leipzig.de

[2] Behaviour and Genetics of Social Insects Lab and
Center for Mathematical Biology
School of Biological Sciences, The University of Sydney,
Sydney, NSW 2006, Australia
mbeekman@bio.usyd.edu.au

Abstract. In this chapter we discuss honey bee optimisation algorithms, which constitute a new trend in the field of swarm intelligence. As the name suggests this class of algorithms is based on the behaviour of honeybees. Current algorithms are based on either of two principles: foraging or mating. Algorithms based on mating utilize the behavioral principles of polyandry found in honey bees and algorithms based on foraging apply the principles of collective resource exploration/exploitation of bee colonies in the context of optimisation.

After reviewing the biological foundations, the existing bee optimisation algorithms will be outlined. We also discuss the potential of bee nest-site selection as a source for new bee-inspired optimization algorithms. A detailed model based on the honeybee nest-site selection process found in nature is described and empirically tested regarding its optimisation behaviour. Building on this model a new algorithmic scheme for bee-inspired optimization algorithms – Bee Nest-Site Selection Scheme (BNSSS) – is proposed.

Keywords: bee algorithms, nature-inspired algorithms, combinatorial optimisation, function optimization.

1 Introduction

Bee-inspired algorithms are a new type of algorithm that has emerged in the field of swarm intelligence in recent years. These algorithms attempt to utilize the principles underlying the collective behaviour of honeybees, and have already been applied to various domains such as robotics [35, 80], network routing [66, 30], multi-agent systems [52] and optimisation [10].

Honeybees are eusocial insects that live in colonies composed of a single queen and up to several thousand workers. The collective behaviour of the colony as

B.K. Panigrahi, Y. Shi, and M.-H. Lim (Eds.): Handbook of Swarm Intelligence, ALO 8, pp. 295–327.
springerlink.com

a whole [33] enables it to solve several complex tasks such as maintaining a constant temperature in the hive [40], keeping track of changing foraging conditions [104, 14, 15] and selecting the best possible nest site [93]. It is remarkable that all of these tasks do not depend on a central control but are achieved using principles of self-organisation.

This chapter focuses on the growing number of optimisation algorithms that are inspired by the behaviour of honeybee colonies. Such bee-inspired optimisation methods have been proposed for various problem domains such as the optimisation of continuous functions [42, 76, 75, 109]), data mining [31, 61], genetic algorithms [41, 48], vehicle routing [100, 36, 60], image analysis [67], and protein structure prediction [6].

Although these algorithms draw their inspiration from honeybee behaviour, they are based on different concepts. In general, one can distinguish between two main classes of honeybee optimisation algorithms: algorithms that utilize genetic and behavioural mechanisms underlying the bee's mating behaviour, and algorithms that take their inspiration from the bee's foraging behaviour.

The first class of optimisation algorithms makes use of the fact that a honeybee colony comprises a large number of individuals that are genetically heterogeneous due to the queen mating with multiple males (see next section). Many of the mating-inspired algorithms extend existing optimisation algorithms from the field of evolutionary computation [29] by introducing bee-inspired operators for mutation or crossover. Other algorithms in this class evolve populations of solutions by imitating a bee's maiden flight.

For the second class of optimisation algorithms, foraging in honeybees is interesting as the underlying decentralized decision-making processes enable a colony to balance exploitation of known food sources with exploration for new–and potentially better–food sources in a dynamic environment [12]. Algorithms based on foraging usually use artificial bees to search for solutions and thus associate solutions with food sources. Depending on the number of food sources (solutions) found and on their quality, a subset of the artificial bee population will explore the environment (search space) by finding new food sources (creating new solutions), while the remaining bees exploit the environment around the found food sources in order to try to find even better food sources (i.e., they perform local search operations in order to improve a found solution).

In this chapter we introduce a third possible class of optimisation algorithms based on nest-site selection, which is another characteristic behaviour of honeybee colonies. In contrast to foraging, where bees can typically forage at different locations simultaneously, nest-site selection always involves the selection of a single new site. Finding a good nest site is vital for the continued survival of the colony and the corresponding decision-making process should be flexible enough to allow the discovery of superior new nest sites during later stages of the selection process. This makes nest-site selection of particular interest for dynamic optimisation problems, in which the problem instance is likely to change during the optimisation process.

In this chapter we review current honeybee optimisation algorithms and describe the underlying biological principles of honeybee behaviour. Additionally we discuss whether the honeybees' nest-site selection process is a useful inspiration for the design of new optimisation algorithms.

The chapter is structured as follows. Section 2 provides an overview of the biological principles underlying the mating, foraging, and nest-site selection behaviour of honeybees. An overview of current optimisation algorithms based on mating and foraging behaviour is given in Section 3. Section 4 discusses the potential of the nest-site selection process in the design of optimisation algorithms. In section 5 the Bee Nest-Site Selection Scheme (BNSSS) is introduced. A summary and conclusions are given in Section 6.

2 The Nature of the Honeybee

This section briefly outlines the biology of honeybees, focusing on those aspects relevant to optimisation algorithms. We will particularly describe the process by which honeybees select a new nest site as we will use nest-site selection as the basis for new optimisation algorithms.

2.1 Mating

Honeybee colonies contain a single queen mated to a large number of males and thousands of workers. The queen is normally the only individual that reproduces in the colony whereas the workers clean the nest, forage for food and feed the brood. Virgin queens and males (drones) are produced during the reproductive season when new colonies are formed. Males gather at conspicuous landmarks in the environment ("drone congregation areas") to which virgin queens are attracted. Queens mate with an average of 20 or so males in flight [70] and store a lifetime's worth of sperm in a sperm storage device called a spermatheca.

Because of multiple mating by the queen, workers within a colony differ genetically. When workers that do not share the same father differ in their task threshold, genetically based task specialization results. Imagine that workers sired by father A will be the first to start foraging for pollen when the pollen stores are running low. Because they will collect pollen before the "pollen foraging" threshold of other workers is reached (e.g., those not sired by father A), the majority of pollen foragers will be workers that share the same father. This genetic diversity is thought to enable a colony to respond resiliently to changes in the environment [68]. Empirical work has shown that honeybee colonies comprising a genetically diverse work force indeed perform better [40, 63, 62].

2.2 Foraging Behavior

Workers not only forage for pollen, they also collect nectar which is stored in the colony and becomes honey. While pollen is used rather rapidly as it is fed to the developing brood, nectar is stored to allow the colony to survive periods when forage is not available. Because only colonies that contain sufficient honey stores are able to survive the winter [87], honeybees have evolved a unique mechanism that allows them to recruit nestmates to food sources found: the dance language [106]. The use of the dance language enables a colony to rapidly exploit and monopolise profitable food sources while almost ignoring those that are of mediocre quality [13].

The honeybees' dance encodes information about the direction and distance of the food source found, and up to 7 dance followers [98], potential recruits, are able to extract this information upon which they will leave the colony and try to locate the advertised food source. During a typical dance the dancer strides forward vigorously shaking her body from side to side [99]. This is known as the "waggle phase" of the dance. After the waggle phase the bee makes an abrupt turn to the left or right, circling back to start the waggle phase again. This is known as the "return phase". At the end of the second waggle, the bee turns in the opposite direction so that with every second circuit of the dance she will have traced the famous figure-of-eight pattern of the waggle dance [106].

The most information-rich phase of the dance is the waggle phase. During the waggle phase the bee aligns her body so that the angle of deflection from vertical is similar to the angle of the goal from the sun's current azimuth. Distance information is encoded in the duration of the waggle phase. Dances for nearby targets have short waggle phases, whereas dances for distant targets have protracted waggle phases.

Honeybees modulate their waggle dance depending on the profitability of the food source found. The more profitable the food source, the "livelier" the dance and the longer they dance for [84]. As a result, bees dancing for highly profitable sites attract more dance followers than those that dance for mediocre sites. The dance language enables a honeybee colony to track the constantly changing foraging conditions [14, 15, 104].

2.3 Nest-Site Selection

Honeybees also use their dance language to communicate the location of potential nest sites. When a colony reproduces, the old queen leaves the colony with about a third of the colony's workers while one of her daughters will inherit the old nest. The homeless bees, called a swarm, now need to locate potential nest sites, normally hollows in trees or crevices in buildings, and choose the best possible site out of all options available.

A swarm of honeybees deciding on a new home is one of the most impressive examples of decentralised decision-making in animals as only about 5% of the bees in the swarm take part in the decision-making process [86]. Several hundred scout bees fly from the swarm cluster to search out tree cavities and other potential dwelling places. The dozen or so scouts that find suitable cavities assess the quality of the site found for characteristics such as volume, height, aspect of the entrance, and entrance size [81, 85]. After returning to the swarm the scout performs a waggle dance if she has rated the site of sufficient quality to be considered. Dance followers use the information encoded in the dance to locate the advertised site, which they then independently evaluate for quality.

The number of dance circuits in the first dance performed by a returning scout is positively correlated with the scout's perception of the site's quality. After completing her dance, the scout leaves the swarm to re-evaluate the nest site before returning again and dancing another time for the same site. Each time an individual scout dances for the same nest site after having re-evaluated that site, she reduces the number of dance circuits by a fixed (approx 17 dance circuits [94]) number of waggle runs, regardless of the site's quality [91]. This means that high quality sites are advertised for longer than poor quality sites because the initial number of circuits is higher. Thus, over time more individuals are recruited to high quality sites compared to sites of lower quality.

While inspecting a potential nest site, a scout estimates the number of other scouts that are also evaluating the site. If this number exceeds a threshold ("quorum") the scout returns to the swarm and signals that the quorum has been reached by "piping", an auditory signal produced by wing vibration [88]. This "piping" signal informs other swarm members to prepare for flight as a decision on the new site has been made [105]. Finally, when the swarm is prepared to travel to its new nest site, scouts from the chosen nest site run excitedly through the swarm, breaking up its structure and inducing other bees to take off. Although the process of swarm guidance is not completely understood, it is thought that the scouts guide the swarm by flying rapidly through the swarm in the direction of the nest site [11, 38, 51, 82]. A new colony is established.

3 Bee Inspired Algorithms

As outlined above, current bee-inspired optimisation approaches are based on one of two behaviours found in bees: foraging or mating. This section outlines some of the existing algorithms based on these behaviours. For previous reviews of bee optimisation methods and related techniques, the interested reader should refer to [10] and [44].

3.1 Mating-Based Optimisation Algorithms

Mating-based optimisation algorithms draw their inspiration from the mating behaviour of honeybee queens. As outlined in the last section, young queens mate with multiple drones during their maiden fight, which leads to mixed paternity within a bee colony. Since mixed paternity has been shown to improve a colony's performance, researchers have been inspired to develop optimisation algorithms that exploit the principles underlying the queen's mating behaviour. These algorithms are also related to the area of evolutionary computing as they rely on the principles of selection, crossover, and mutation.

Honeybee Mating Optimisation Algorithm

One example of a genetically based optimisation algorithm is the honeybee mating optimisation algorithm (HBMO) developed by Abbass [1, 2] for discrete optimisation problems. The HBMO algorithm contains four main (artificial) entities called queens, drones, brood, and workers. The algorithm operates in two stages: maiden flight and brood development. Both stages are executed alternately until a stopping criterion is satisfied.

The HBMO operates with several queens. At the beginning of the maiden flight each queen is equipped with a single randomly generated reference solution. In addition to this reference solution a queen has a flight speed s, an energy e, and a limit for the amount of sperm (i.e., sample solutions from drones) she can store in her spermatheca (i.e., a pool of new sample solutions). A queen stops her maiden flight either when her energy is depleted or when a maximum number of drone solutions has been collected. During each step of her maiden flight a queen Q encounters a drone D. She will absorb the drone's solution with a probability of

$$p(Q, D) = e^{-d/s} \tag{1}$$

where d represents the fitness difference between the queen's reference solution and the drone's solution and s represents the flight speed of the queen. As can be seen from $p(Q, D)$ a queen is very likely to accept a drone's solution if the solution is better than the queen's reference solution or if her speed is high. After each step the queen's flight speed and energy are decreased which results in the queen becoming more selective with respect to absorbing potential drone solutions over the course of her flight.

The maiden flight is completed when a maximum number of drone solutions has been collected, or when the queen's energy is depleted. At this point, a queen will mate with a drone solution randomly selected from those in her spermatheca. Mating involves the application of a crossover operation to the selected drone solution and the queen's reference solutions, and results in a single offspring solution. In addition a mutation operation might be applied to the offspring solution. The survival of the offspring depends on the quality of this offspring reference solution.

The offspring solutions of the queens are nursed by the workers during the brood development stage. A worker represents a local search heuristic and nursing corresponds to the application of this heuristic trying to improve the offspring solution. Then, before a new maiden flight stage is started the least fit queens are replaced with the fittest offspring until no offspring is fitter than the least fit queen. Again, this process is repeated until a stopping criterion is satisfied.

Initially the HBMO algorithm was proposed for solving the Boolean satisfiability problem (SAT) [1, 2] and has since been adapted for several other problems such as water reservoir management [4, 37], data clustering [31, 61] and vehicle routing [60].

Honey Bee Inspired Evolutionary Computation

Other approaches that are based on honeybee mating utilise bee-inspired operators within existing evolutionary computation algorithms, see for example Sato and Hagiwara's bee system [79], Jung's queen-bee evolution [41] or Karci's bee-inspired genetic crossover operator [48]. As these methods extend well-known optimisation methods we will not go into further detail here.

3.2 Foraging Behaviour Based Approaches

Foraging behaviour based approaches take inspiration from the mechanisms underlying the foraging process in honeybees. Besides the experimental studies outlined in the last section, several theoretical models support and outline the effectiveness of the honeybee's decentralized decision-making process when foraging [83, 19, 95, 97, 28, 13].

Sherman and Visscher [95] investigated when waggle-dance recruitment is beneficial. Their results suggest that this recruitment increases the amount of food a colony can collect when resources are scarce. A recent study by Dornhaus et al. [28] suggests that the recruitment dance is especially beneficial if resources are few in number and of variable quality. Beekman and Lew [13] found that recruitment is most beneficial if the average success of locating new food patches falls below the average success of recruits. Additionally they showed that communication facilitates the rapid exploitation of highly profitable food sources when several food sources of different quality are present. Thus, the bees' dance communication regulates the trade-off between exploitation and exploration.

These studies underline the usefulness of honeybee foraging behaviour in terms of optimisation in a dynamic environment in which resources are sparse and differ in quality, as is the case in many problem domains of optimisation. Moreover, the above mentioned studies outline the importance of direct communication between the bees. Inspired by these findings, direct information transfer plays an important role in algorithms based on honeybee foraging which are outlined below. This is in contrast to ant colony optimisation algorithms that rely on indirect communication via artificial pheromones [].

The Artificial Bee Colony Algorithm (ABC))

The Artificial Bee Colony algorithm (ABC) was introduced by Karaboga [42, 9] for function optimisation. Each solution (i.e., a position in the search space) represents a potential food patch and the solution quality corresponds to the food patch's quality. Agents (artificial bees) search and exploit the food sources in search space.

The ABC uses three types of agents: employed bees, onlooker bees, and scouts. Employed bees (EB) are associated with the current solutions of the algorithm. In every step of the algorithm an EB tries to improve the solution it represents using a local search step, after which it will try to recruit onlooker bees (OBs) for its current position. OBs select among the promoted positions according to their quality, meaning that better solutions will attract more OBs. Once an OB has selected an EB and thus a solution it tries to optimise the EB's position by means of a local search step. An EB updates its position if an OB it recruited was able to spot a better position, otherwise it remains on its current position. In addition, an EB will abandon its position if it was not able to improve its position for a certain number of steps. When an EB abandons its position it becomes a scout, meaning that it selects a random position in the search space and becomes employed at that position.

The algorithm can be described in more detail as follows: given a dim dimensional function F and a population of n agents, $n_e = n/2$ EBs and $n_o = n/2$ OBs. The algorithm is initialized by placing EB i ($i \in n_e$) on a random location θ_i in the search space. $F(\theta_i)$ is then the quality of the position of EB i.

In every iteration, each EB tries to improve its location using a local search step. First, EB i calculates a new a candidate solution

$$\theta_i^* = \theta_i + rand(-1,1)(\theta_i - \theta_k) \tag{2}$$

where θ_k corresponds to the position of another randomly chosen EB with index k ($i \neq k$) and $rand(-1,1)$ constitutes a random number between -1 and 1 drawn from a uniform distribution. Note that formula 2 is typically not applied for all dimensions of θ_i. While the number of dimensions that are taken into account in the case of a constraint optimisation problem depends on a parameter called the perturbation rate (see Karaboga and Basturk [45] for more details), only one dimension is taken into account for unconstrained optimisation problems. The dimension(s) to be altered are randomly chosen. After a new candidate solution is calculated a greedy selection mechanism is used in order to decide if θ_i should be discarded

$$\theta_i = \begin{cases} \theta_i & \text{if } F(\theta_i) > F(\theta_i^*) \\ \theta_i^* & \text{else} \end{cases} \tag{3}$$

After each EB has updated its position, each OB chooses one of the current solutions. A standard roulette wheel selection [29]

$$P_i = \frac{F(\theta_i)}{\sum_{k=i}^{n_e} F(\theta_k)} \tag{4}$$

is used, and better solutions attract more OBs. After choosing a solution an OB tries to improve the solution using the same mechanism as outlined in Eq. 2. The EB that corresponds to this solution updates its position if a better position is found by the OB.

The algorithm keeps track of how many steps an EB has been at the same solution. If the number of steps spent on the same position reaches a certain value $limit$ the EB abandons its position and scouts for a new position, which corresponds to choosing a random position in search space. The parameter thus controls the exploitation/exploration rate of the system. In [47] the impact of this parameter $limit$ was investigated, and found to depend on the problem's dimensionality and the number of employed bees in the system, with an optimal value given as $limit = n_e \cdot dim$. In a very recent study [5] this suggestion was reexamined. It was concluded that small colonies should use a value $limit > n_e \cdot dim$, as they need more time to search in the vicinity of the EBs' solutions than large colonies. Aderhold et al. [3] studied the influence of population size and the OB/EB ratio on the optimisation behaviour of ABC. Their study suggests

Algorithm 1. Artificial Bee Colony (ABC)

```
1: place each employed bee on a random position in the search space
2: while stopping criterion not met do
3:    for all EBs do
4:       if steps on same position == limit then
5:          choose random position in search space
6:       else
7:          try improve position (according to Eq. 2)
8:          if better position found then
9:             change position
10:            reset steps on same position
11:         end if
12:      end if
13:   end for
14:   for all OBs do
15:      choose position of employed bee (according to Eq. 4)
16:      try improve position (according to Eq. 2)
17:   end for
18: end while
```

that both population size and OB/EB ratio are dependent on the optimisation goals. For a better understanding the ABC algorithm is outlined in Algorithm 1.

The ABC has been used in several problem domains such as unconstrained numerical optimisation [46, 47], constrained numerical optimisation [45], the training of neuronal networks [43] and protein structure prediction [6].

Bees Algorithm (BA)

The Bees Algorithm (BA) was introduced by Pham et al. in 2005 [76] as an optimisation method for continuous and combinatorial function optimisation.

As in the ABC the population of bees is divided into two groups: scouts and recruits. While scouts are responsible for the exploration of the search space the recruits try to exploit (i.e., improve) found solutions via local search. The algorithm depends on a set of parameters which will be outlined briefly below.

The optimisation process starts by assigning each of n scout bees to a random position in the search space. A scout's fitness corresponds to the quality of the position (solution) it currently occupies. The best $m \leq n$ scouts are selected and the rest are discarded (selected scouts are referred to as selected bees). The selected bees are further partitioned according to their fitness into e elite selected bees and the $m - e$ non-elite selected bees.

Each selected bee is assigned a number of recruits, and how many depends on the solution quality of the bee. Each elite bee receives neq recruits, each non-elite bee nsp recruits.

Each recruit performs a local search step at its assigned position according to

$$x_j^* = (x_j - ngh) + (rand(0,1) * ngh * 2) \tag{5}$$

with ngh denoting the search patch size. The best improvement of a selected bee's solution will replace this solution. If none of the solutions found by the recruits yields an improvement over the selected bee's solution, the solution is maintained. The scout population is filled up with these m solutions and $n - m$ random solutions and the algorithm repeats until a stop criterion is satisfied. It should be noted that the BA algorithm was recently improved [78] by introducing more local search methods such as mutation, creep, crossover, interpolate, and extrapolate, that can be used by recruits to improve given selected solutions. The algorithm underlying a standard BA is outlined in Algorithm 3.2 according to [76].

Algorithm 2. Bees Algorithm (BA)

1: place each bee on a random position in the search space
2: evaluate the fitness of the population
3: **while** stop criterion not satisfied **do**
4: select solutions for a local search (exploitation)
5: assign bees to commit local search on selected solutions and evaluate fitness
6: for each solution select the best improvement
7: replace remaining solutions with random solutions (scout)
8: **end while**

The BA has been applied to various engineering problems, such as the training of neural networks [76, 77, 73, 72], controller formation [74], image analysis [67] and job multi-objective optimisation [75].

Bee Colony Optimisation Algorithm (BCO)

The Bee Colony optimisation Algorithm (BCO) [100] constitutes a generalized and improved version of the Bee System algorithm [56]. Both algorithms were designed to tackle combinatorial optimisation problems. As the two algorithms are basically identical we will treat them as one in the following.

BCO divides the optimisation process into $I \geq 1$ iterations, where I is a parameter set by the user. During each iteration B virtual bees try to construct a solution for the given problem. Due to the combinatorial nature of the problems BCO tackles, solutions are constructed as a consecutive extension of initial partial solutions. To do so each iteration is divided into a finite sequence of $m \geq 1$ stages $ST = st_1, st_2, \ldots, st_m$.

During a stage s_j a bee will extend its current partial solution by adding an available partial solution. In the BCO terminology extending a current solution with a partial solution is called forward pass. How a forward pass is implemented depends on the underlying problem. In [56] the BCO (then called Bee System) was used to solve the traveling salesman problem and the Logit model [24] was used to decide how to extend partial solutions.

After each bee has performed a forward pass, a backward pass is performed. This means that all bees compare their current partial solutions. On the basis of this comparison bees decide whether or not to keep their current partial solution, promote it to other bees, or abandon it. Bees that give up their current partial solution will choose one of the solutions promoted by other bees. The backward pass ends the stage. The sequence of stages leads to an iterative solution buildup where bad partial solutions will be abandoned and the search will focus on promising partial solutions.

At the end of each iteration, it is determined whether the best solution found in that iteration should become the new global best solution. The underlying algorithm is outlined in Algorithm 3.

Algorithm 3. Bee Colony optimisation (BCO)

```
1: initialization
2: for all I iterations do
3:   for all m stages do
4:     for all B bees do
5:       forward pass: choose partial solution
6:     end for
7:     for all B bees do
8:       backward pass: exchange information about partial solutions with bees in
         nest
9:     end for
10:  end for
11:  if best solution obtained in iteration is global best, update best-known solution
12: end for
```

The BCO has been used to solve problems in traffic and transportation [56, 57, 58, 100, 103, 102, 101].

The Bee Colony-Inspired Algorithm (BCiA)

The bee colony-inspired algorithm (BCiA) was recently introduced by Häckel and Dippold [36] for the vehicle routing problem with time windows (VRPWTW). Given a number of customers that have to be supplied with goods within a certain time window, optimising the VRPWTW requires finding a route schedule that minimizes the associated costs (number of vehicles needed and total tour length). In order to avoid conflicts between the two optimisation objectives, BCiA operates in two stages – in the first it tries to reduce the number of vehicles needed for a valid solution and in the second it tries to minimize the total tour length. Instead of a single population of virtual bees, BCiA uses 2 populations P_1 and P_2 of bees, operating in stages 1 and 2, respectively. The principles used in BCiA are similar to those used in the ABC [9], but ABC was designed for numerical optimisation problems whereas BCiA tackles discrete optimisation problems.

BCiA uses a virtual bee population that consists of three bee types: employed bees (EBs), follower bees (FBs), and scouts. The separation of roles within a population is similar to ABC: EBs are associated to current solutions, FBs try to improve those solutions (similar to OBs in ABC), and scouts provide the population with new solutions (this is done by EBs in ABC where they are also called scouts when providing a new solution).

The two populations P_1,P_2 of BCiA each consist of n_{eb} EBs, n_{fb} FBs and n_{scout} scouts. Initially the EBs of both populations are initialized with random solutions. Each iteration of the algorithm is divided into two stages. In the first stage the first population P_1 tries to improve its solutions. This is done similarly to the ABC, i.e., each FB chooses an EB based on its fitness with respect to the first optimisation goal F_1.

FBs then try to improve an EB's solution by constructing a new solution. During the construction process a new solution is constructed taking the EB's solution into account. The details of the process depend on the specific optimisation problem and will not be discussed here (the interested reader is referred to the publication itself [36]). If the solution found by an FB has better quality than the EB's current solution, the latter is replaced by the FB's solution.

After the FBs try to improve the current solutions in the populations, the scouts create new solutions. Scouts do not use a reference solution when generating a new solution, but apart from this the generation process is identical to the one used by the FBs. The best scout solutions will replace the worst EBs if their quality is better. After the improvement step a solution exchange between the two populations P_1 and P_2 is initiated. Each EB in P_1 that is not yet present in P_2 and has a better quality regarding the optimisation goal of P_2 (i.e., F_2) than the worst EB in population P_2 is added to P_2, while the worst EB in P_2 is deleted.

The second stage of the iteration is then executed. It follows the same sequence of actions as in the first stage but uses population P_2 and the optimisation goal F_2. At the end of each iteration the age (i.e., number of iterations the solution was not improved) of all solutions (i.e., EBs) is checked. Solutions that exceed a certain age are exchanged (similar to ABC). Any old solutions of the population P_1 are substituted by new scout-generated solutions, and if any solutions in P_2 are abandoned, they are substituted by the best solution with respect to F_2 from population P_1. The substitution in P_2 only happens if the solution in P_1 contains the best (i.e., smallest) number of vehicles known so far, otherwise no substitution takes place.

BCiA terminates when a stop criterion is satisfied. The algorithm is outlined in algorithm 3.2.

Bee Colony Optimisation Algorithm (BCOA)

Introduced by Chong et al. in 2006 [22, 23], the Bee Colony Optimisation Algorithm (BCOA) was originally proposed for the job shop scheduling problem.

BCOA consists of a population of n foragers. During each iteration each forager f_i constructs a solution for the given optimisation problem. The foragers then promote their solutions to each other. Based on the quality of its own solution, a forager can decide to keep its previous solution or abandon it and adopt that of another forager. After each forager has decided, it will create a

Algorithm 4. The bee colony-inspired algorithm (BCiA)

```
1: initialize populations
2: while stop criteria not met do
3:    for all i ∈ {1,2} do
4:       for all FBs ∈ P_i do
5:          choose EB ∈ P_i
6:          construct new solution regarding F_i using EB
7:       end for
8:       update EBs ∈ P_i according to the solutions found by the FBs
9:       for all scouts ∈ P_i do
10:         construct a new solution with respect to F_i
11:         exchange worst EB if better solution is found by scout
12:      end for
13:      if i equals 1 then
14:         update EBS in P_2 according to P_1
15:      else
16:         update EBS in P_1 according to P_2
17:      end if
18:   end for
19:   check age of solutions and replace them if age exceeds limit
20: end while
```

new solution based on its current solution. The general principle of BCOA has similarities to the BCO which was outlined above.

Each iteration in the BCOA can be divided into two phases: the dancing phase and the foraging phase. During an iteration each forager constructs a solution for the given problem (how will be explained later). Then each forager f_i ($i \in [1, n]$) returns to the hive and performs a waggle dance with a certain probability p. Let $Pf_i = 1/C^i_{max}$ denote the profitability rating of the solution a dancing forager f_i is trying to promote, where C^i_{max} represents the fitness of f_i's current solution. The average profitability rating of all dancing foragers is given by $Pf_{colony} = 1/n_d \sum_{i \in F_d} Pf_i$, with n_d corresponding to the number of dancing bees, and F_d the set of dancing bees.

The waggle dance of forager f_i will last for $D = d_i \cdot A$ steps with $d_i = Pf_i / Pf_{colony}$ depending on the profitability rating of the obtained solution (e.g., make span, tour length) and $< A$ denoting a waggle dance scaling factor. Each forager also attempts to follow a randomly selected dance of another forager with probability r_i, with r_i depending on the profitability rating of the solution found (see Table 1) (i.e., foragers that found a solution with high profitability rating are unlikely to follow another forager's dance).

The BCOA has been extended for the travelling salesman problem [55, 107, 108] and a recent modification of the algorithm for feature selection problems has also been proposed [96].

Table 1. Look-up Table for adjusting r_i according to the profitability rating

Profitability Rating	r_i
$Pf_i < 0.9 \times Pf_{colony}$	0.60
$0.9 \times Pf_{colony} \leq Pf_i \leq 0.95 \times Pf_{colony}$	0.20
$0.95 \times Pf_{colony} \leq Pf_i \leq 1.15 \times Pf_{colony}$	0.02
$1.15 \times Pf_{colony} \leq Pf_i$	0.00

The Virtual Bee Algorithm (VBA)

The Virtual Bee Algorithm (VBA) algorithm scheme for numerical optimisation was introduced by Yang in 2005 [109]. It proposes function optimisation via a set of virtual bees that are initialized on random positions in a given search space. Each position of the search space is assigned virtual food, such that the food quality corresponds to the value of the function to be optimised at a given position. Virtual bees will explore the search space and communicate found food patches to other bees. Bees that receive information about other food patches will incorporate this information into their search behaviour. Please note that Yang gives only a very schematic description of the algorithm. The exact details of how communication, search, and incorporation of solutions obtained from other individuals is handled is not explained in [109]. The VBA was tested on two 2 dimensional test functions and the author claims that it outperformed a standard genetic algorithm. As the article lacks detailed information on the proposed algorithm it is hard to validate these findings.

4 The Optimisation Potential of Nest-Site Selection

As described in the last section, current bee-inspired optimisation algorithms are based on two behaviors found in honeybees – mating and foraging. We already alluded to the fact that from an optimisation point of view the honeybee's nest-site selection process exhibits many useful features with respect to dynamic and noisy optimisation problems.

As explained in Section 2, a colony will produce a homeless swarm when a young queen is being reared and the old queen leaves the established nest with about one third of the workers. This swarm now needs to find a new nest site. At first glance one might think that the process underlying nest-site selection is similar to foraging, but there are fundamental differences. During foraging, the main problem for the colony is to find the distribution of scouts and foragers over the available food patches that optimises the balance between exploration and exploitation. In contrast, nest-site selection constitutes a decision process, as the swarm has to decide on one nest site by solving the best-of-n-problem [93].

Foraging bees exploit finite resources and the distribution of foragers over the known food sources has to be traded off against an exploration for new food sources. A driving force in this process is the depletability of resources as it steers the exploration to exploitation ratio of a colony. This makes the foraging paradigm a useful concept in the context of capacity management of limited resources (e.g., the foraging-based server allocation algorithm by Nakrani and Tovey [66]).

When dealing with a classical optimisation problem a given solution will not change its quality, optimisation algorithms based on foraging thus always need to associate a given solution with some artificial depletion mechanism to steer the exploration/exploitation behavior of the algorithm. During nest-site selection bees face the same problem, i.e., the quality of potential nest sites remains the same no matter how often they are revisited by scouts. By the means of dance attrition (i.e., a decrease in number of dances for a potential nest site over consecutive visits of the site) bees can overcome this problem.

When trying to find a new nest site, bees face a speed-accuracy trade-off. A decision needs to be quickly as a swarm is vulnerable to predation and inclement weather, but not so fast that the swarm settles for a sub-optimal nest site. Hence, the decision-making process has to account for temporal delays in nest site discoveries and needs to exhibit sufficient flexibility in order to incorporate late discovered nest sites into the decision-making process. This is a situation also faced in dynamic or noisy optimisation problems.

Several studies, both experimental [92, 93, 91, 89, 90] and theoretical [20, 71, 65, 39, 54], have investigated nest-site selection in honeybees. A model has also recently been used to explore the applicability of nest-site selection to an optimisation context [27].

In this section we describe a model for the honeybee nest-site selection process and experiments that show its potential for solving optimisation problems as first introduced in [27]. Clearly, a biological model of real bee behaviour can not directly be used as an optimisation algorithm for function optimisation. Therefore, we propose a Bee Nest-Site Selection Scheme (BNSSS) that can be used as a framework to design algorithms for function optimisation.

In Subsection 4.1 the model for honeybee nest site is described. Experimental results are described in Subsection 4.2. The BNSSS framework is described in Section 5.

4.1 A Spatial Nest-Site Selection Model

The nest-site selection model of [27] is an extension of the individual-based nest-site selection model for honey bees developed by Janson et al. [39]. The extension incorporates spatial features, which allow for the study of the effects of the locations of different potential nest sites and varying scouting behavior on the nest-site selection process. In the following the original model and its extensions are outlined.

The simulation of the extended model for the nest-site selection process in honeybees operates in discrete time-steps where each time step represents 1 second of real time. Note that in the original model a step size of 5 seconds was used. As we are interested in simulating scouting behavior of bees, this temporal resolution is too coarse. Real bees are able to travel with a maximum speed of 5 meters per second [11], meaning that they can travel a maximum distance of

25 meters in a time-step of 5 seconds. In a spatial simulation this would make it very likely for a bee to miss a potential nest site by simply flying over it. Thus a smaller timescale is used in the extended model.

The virtual bees in the simulation are invoked in random order. Some behaviors such as "scouting", "assessing" and "missing" are associated with a mean duration time T_E. The exact duration is determined by $T(E) = \lambda \cdot TE$, where $\lambda = \mu/10$ is a scalar factor, with μ being drawn from a chi-square distribution $\chi^2(10)$. Note that this leads to an expected value of 1 for λ.

A virtual bee has an internal state and at every simulation-step each bee acts according to its current internal state. The possible states are

- REST: The bee is on the swarm but currently not involved in nest-site selection
- SEARCH: The bee is on the swarm and tries to find a dance to follow
- SCOUT: The bee searches the surroundings for potential nest sites
- ASSESS: The bee is at a potential nest site and assesses its quality
- DANCE: The bee is on the swarm and dances for its preferred site
- FOLLOW: The bee is on the swarm and has found a dance and follows it
- RECRUITED: The bee flies to the nest site advertised in the dance it followed
- MISS: The bee misread the dance and searches the surrounding of the swarm unsuccessfully before returning to the swarm

A state diagram that shows how the state of a bee can be changed is depicted in figure 1). In the following the behaviour that corresponds to the different states is explained in more detail.

Resting

The probability that a resting bee will start to search for a dance is given by $P_{rest} = 0.002$ per second. This is equal to the probability that a bee will return to resting after having tried to find a dance. This probability is based on empirical studies [21, 12].

Searching

The probability that a searching bee will locate a dance for a nest site depends on the number of dances currently performed in the swarm D and is given by $P_{find} = 0.005 \cdot D$. If a bee is able to find a dance it is assigned to one of the available dances in a random uniform fashion. A bee will only follow a dance if it has less than 7 other followers. The probability that it will start to follow a dance is given by $P_{follow} = 0.2^{min\{2,f\}}$, where f denotes the number of bees already following the dance.

The longer a bee unsuccessfully tries to locate a dance, the higher the probability that it will start scouting. The probability that a searching bee will switch to scouting is given by

$$P_{scout}(t) = \frac{t^2}{t^2 + \theta^2}$$

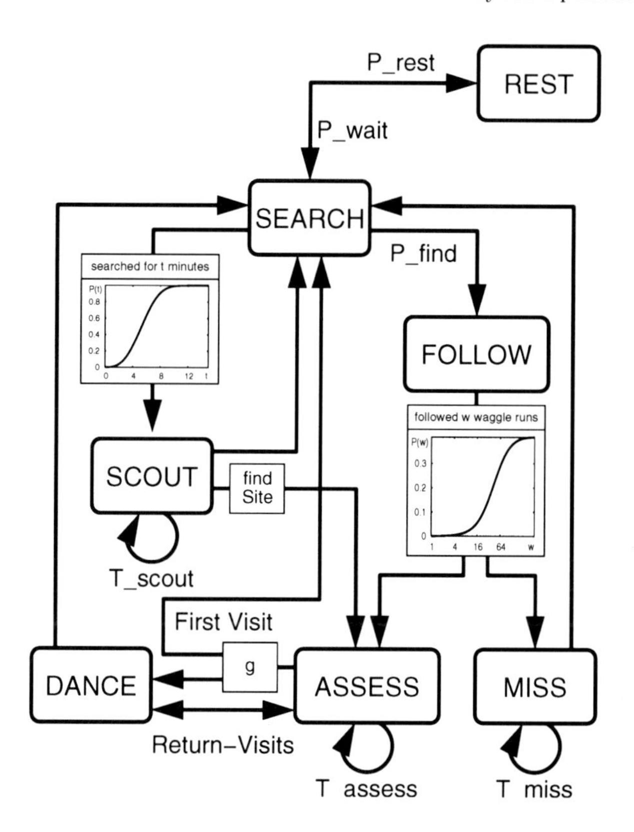

Fig. 1. State diagram of the decision making process underlying nest-site selection. Reprint from Janson et al. [39]

where t denotes the number of time steps of unsuccessful searching and $\theta = 4000$. This behaviour modulates the exploration/exploitation rate of the swarm. If very few potential nest sites have been found and thus the number of dancing bees is low, bees searching for a dance are likely to become scouts. When many sites have been found and therefore dances are abundant, a searching bee is likely to find a dance to follow and will become a recruit instead of a scout.

Scouting

As Lindauer [53] observed, bees usually scout the surroundings for about 20 minutes before returning to the swarm. Thus in the model a bee scouts for $T_{scout} = 1200$ time steps. Note that the scouting mechanism of the extended model differs fundamentally from the mechanism used in the original by Janson et al [39]. In the original model the success of a scouting trip was modeled using probability distributions because the model has no search space. This means that after T_{scout} scouting iterations a bee either finds a potential nest site or not. If it finds a nest site it starts assessing the site, otherwise it returns to the swarm.

The earliest a nest site can thus be discovered is $T(Scout)$ iterations after the first bee started to scout.

In the extended model the scouts move through a 2-dimensional environment in search of potential nest sites. A bee is able to spot a potential nest site if the target subtends a bee's visual angle greater or equal to α_{min} [50, 34]. A low visual angle makes it more likely to spot a distant target than a high visual angle. The visual angle α_{min} can range between five and fifteen degrees.

The diameter of a bee nesting box is around 40 cm. Given an assumed minimal angle of 8 degrees, a scout is able to spot a potential nest site (e.g., nesting box, tree, etc.) up to a distance of approximately 280 cm. If a scout finds a nest site it will immediately start assessing it. The scouting process can be divided into two phases:

1. scouting: a bee will scout as long as it is able to be back at the swarm after T_{scout} time steps.
2. returning: if the remaining scouting time is smaller or equal to the time needed to return to the swarm a scout return to the swarm.

Scouting Strategy

In our model scouts use an intermittent search strategy, as proposed by Janson et al. [39]. Indeed, instead of moving continuously, many animals show intermittent locomotion [49, 17, 16]. In our model an intermittent search strategy is realized as follows: scouts choose a random location within a search area that is defined by the range of locations that are reachable within one third of its available scouting time T_{scout}. After reaching the chosen location a scout will perform a correlated random walk (CRW) to search for a potential nest site [110, 7]. Various species such as ants, beetles and butterflies have been shown to perform CRWs [25]. As a result, CRW has been used to reproduce movement patterns from various experimental data (e.g., [18, 26]). In contrast to pure random walk (i.e., Brownian motion), CRW incorporates directional persistence in movement patterns. Given a position and a direction, directional persistence can be achieved by limiting the angular displacement of the direction between successive steps.

For the scouts' movement a CRW with a fixed movement length of 1 meter per step is used. Angular displacement is achieved by adding directional noise which is drawn from a wrapped Cauchy distribution [8]. Wrapped Cauchy distributions contain a shape parameter ρ $(0 \leq \rho \leq 1)$ which controls the directional persistence. If $\rho = 0$ the resulting walk is uncorrelated. In contrast $\rho = 1$ results in total correlation, which means that no noise is added to the direction. For the simulation runs a correlation parameter value $\rho = 0.5$ is used resulting in slightly correlated movement steps.

Flying towards a Destination

A scout flying towards a destination travels with a speed of 5 meters per second. If the distance to the destination is smaller than 5 meters, the bee is placed on the destination, otherwise it will travel 5 meters in the direction of the destination. In order to prevent bees from flying in straight lines, angular noise was added from a uniform random distribution η_{fly} $(-22.5 \leq \eta \leq 22.5)$. Because a bee aligns its flying direction each time step, it is guaranteed to arrive at the destination.

Site Assessment

If a scout successfully locates a potential nest site it will assess it. Nest-site assessment in real bees usually lasts for about 10 minutes [53] which corresponds to 600 time-steps in the simulation. Each nest site S has a certain quality Q_S, which in the simulation corresponds to a natural number in the range [0-100]. The quality of a nest site S is perceived by a bee during the assessment. Quality is always perceived with some noise, thus $Q(S) = Q_S + \delta$, with δ $N(0, \sigma^2)$ drawn from a normal distribution with a standard deviation of $\sigma = 10$. If the perceived quality $Q(S)$ exceeds a bee's quality threshold Φ, the bee dances for the nest site when it returns home. Otherwise it switches to searching after it returns home. As in the original model, the threshold Φ is set to 50 for all individuals in the simulation. After a bee has completed the assessment of a nest site it flies back to the swarm.

Dancing

If a bee discovers a potential nest site S (i.e., $Q(S) > \Phi$), it dances for it after returning to the swarm. The number of waggle runs performed during a dance depends on two factors, the perceived quality of the nest site Q(S) and the number of consecutive visits to the nest site. Based on empirical data [91], the simulated bees perform $Q(S)$ waggle runs after their first visit to the nest site and $Q(S) - 16(k-1)$ after returning for the kth time. If $Q(S) - 16(k-1) \leq 0$ it will stop dancing for this site and switch to searching.

The distance and the direction to the potential nest site have been incorporated into the dance by assuming that a waggle phase lasts 2.4 seconds per kilometer and 1.5 seconds are added for the return phase [32]. Thus a single dance for a potential nest site located 1000 meters away from the swarm takes 3.9 seconds.

Following

A searching bee that has found a dance and was able to follow it, follows the dance until the dancer ceases dancing. If the bee had previously visited the advertised site, it will find that site again. Otherwise the number of waggle runs w followed determines the probability of correctly locating the advertised nest site. The probability of finding a nest site is

$$P_{findSite}(w) = \frac{s(w)}{1.5 \cdot u(w) + s(w)}$$

where w denotes the number of waggle runs a bee followed, $u(w) = 1 - 1/\sqrt{(w+1)}$ represents the distribution of unsuccessful bees and $s(w) = w^2/(w^2 + \Theta)$ with $\theta = 60$ represents the distribution of successful bees. This probability is based on experimental data [64].

Successfully Recruited to Nest Site

If a bee has been successfully recruited for a potential nest site and correctly read the dance it followed, it flies towards the proposed nest site. After reaching the nest site S it starts to assess it. In case the assessment is successful (i.e., $Q(S) > \Phi$), the bee returns to the swarm and starts to dance for the nest site. Otherwise, it returns to the swarm and starts to search for new dances.

Missing the Advertised Nest Site

If a bee is not able to read a dance correctly it will not be able to find the nest site that was advertised in the dance. In such cases, the bee flies the same distance as the advertised site, but in the wrong direction. This is achieved by adding a maximum of 5 degree noise drawn from a uniform random distribution to the actual direction towards the advertised nest site. After reaching the wrong location a bee searches the surroundings for 400 seconds.

4.2 Experiments

Using the model outlined above, three experiments are described that outline the optimisation potential of the honeybee's nest-site selection process. The model was implemented in Java using the MASON multi-agent simulation library [59]. Unless stated otherwise the simulation runs used the parameter values mentioned in the last subsection. All results are averaged over 10 runs. The number of bees is $n = 500$ (this is similar to the number of real honey bees that take part in the decision-making process for the nest site).

Experiment 1: Nest-Site Selection in a Dynamic Environment

This experiment tests how the nest-site selection process performs in an environment where the quality of the sites changes over time. In an environment with two nest sites located equally far from the swarm but which differ in quality, the number of scouts would build up quickly at the higher quality nest site if the quality of the nest sites remains the same. In this experiment however the quality of the nest sites is swapped at regular intervals. While such a situation is unlikely to occur in nature, changing optima are ubiquitous in dynamic optimisation problems.

The environment contains two potential nest sites $n1$, $n2$ that are located in opposite directions 150 meters away from the swarm's position. Site $n1$ is initialized with a good quality $q_{good} = 75$ while $n2$ is initialized with a bad quality $q_{bad} = 45$. The simulation runs for 32 hours corresponding to 115200 simulation steps. At an interval of 28800 simulation steps (i.e., every 8 hours) the qualities of the nest sites are swapped. This leads to a total amount of 3 quality switches over the whole simulation run.

Given that there is a probability that a scout will not find a given nest site, it is possible that the swarm discovers one nest site only during the simulation, or perhaps even none at all. In addition, a swarm can "forget" a low quality nest site if no dance for that site is sustained prior to the switch in nest-site quality. When this happens the site needs to be rediscovered after the qualities have been swapped. To ensure that bees are aware of both nest sites each time their quality is switched, a random bee is chosen that starts dancing for the nest site that was of low quality but switched to high quality.

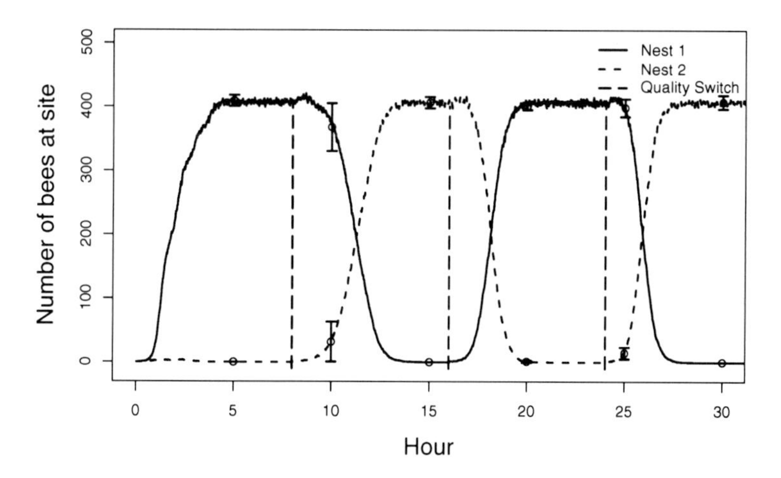

Fig. 2. Average number of bees assessing a nest site where the site qualities are swapped every 28800 simulation steps. Error bars show standard deviation.

Figure 2 depicts the average number of bees at each nest site over 10 simulation runs. The swarm quickly adapts to changes in nest-site quality. It is clear that the process is rather slow as it takes the swarm approximately 2 hours to adapt to the change in quality. However, this is not necessarily a disadvantage as it makes a swarm resilient towards noise. Even though in real bees the quality of a nest site is most likely to remain constant, the discovery of a new nest site also constitutes a change in the swarm's environment. Without the ability to react to changes in the environment, a swarm could get stuck in a suboptimal solution if it finds a nest site of mediocre quality early in the decision-making process. In terms of optimisation, adapting to a dynamic environment is an interesting aspect, as it can be applied to the detection of changing locations of the optima in problems with dynamic fitness functions.

Experiment 2: Nest-Site Selection in a Noisy Environment

This experiment tests whether the swarm is capable of selecting a stable mediocre quality nest site and disregard a site of sometimes high but very unstable quality.

The number of bees and the number and position of the potential nest sites is the same as in Experiment 1, but the quality of nest site $n2$ is kept constant at mediocre quality $q_{mediocre} = 55$ whereas the quality of nest site $n1$ changes at an interval of 1800 simulation steps (i.e., every 30 minutes) alternately between good $q_{good} = 75$ and very bad $q_{vbad} = 35$. Nest site $n1$ is initialized with a good quality q_{good}. A simulation run lasted for 32 hours corresponding to 115200 simulation steps. To ensure that the swarm is aware of each nest site, a random bee starts dancing for each nest site at the beginning of the simulation.

Figure 3 shows the average number of bees at the two nest sites over 10 simulation runs. As can be seen, the swarm is able to direct most scouts towards the stable mediocre nest site. At the start of a simulation the number of bees builds up quickly at both nest sites. This is caused by the fact that one bee starts dancing for each nest site when the simulation is started. However as bees begin to revisit the nest sites, more bees are recruited towards the mediocre stable site. This is due to the revisit behavior of honeybees. Even though many bees will initially promote nest site $n1$ more strongly than nest site $n2$ due to better quality, the ongoing revisiting will cause many bees to abandon the unstable site and choose the stable site. nest site $n1$ will never be completely abandoned due to the fact that some visiting bees will always experience it as a very good nest site and thus revisit it. In general this experiment demonstrates that the nest-site selection mechanism is to some extent resilient towards noise.

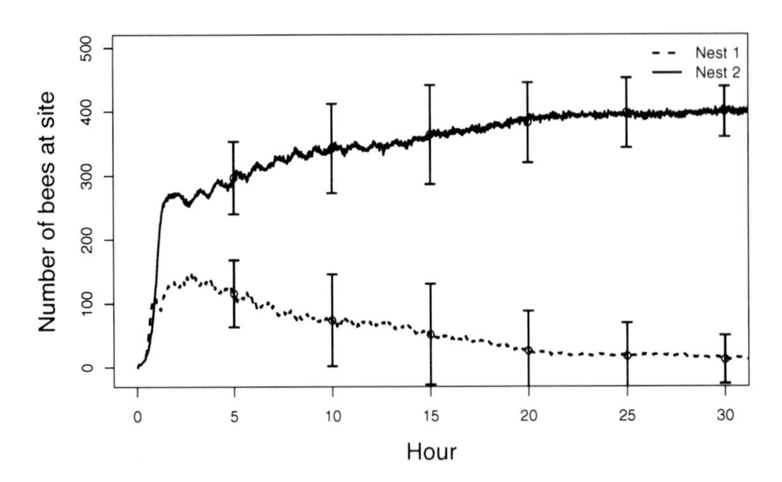

Fig. 3. Average number of bees assessing a nest site when the nest site of high quality changes in quality. The quality of nest site $n1$ changes each 1800 simulation steps between $q_{good} = 75$ and $q_{vbad} = 35$, whereas the quality of nest site $n2$ is constant at $q_{mediocre} = 55$. Error bars show standard deviation.

Experiment 3: Function Optimisation via Iterative Nest-Site Selection

When searching for a new nest site bees typically have to decide between several discovered nest sites. In case of the European honeybee *Apis mellifera*, the number of possible good nest sites is limited as they live in cavities. The swarm needs to ensure that it decides for the best site possible so that it becomes unlikely that the nest site turns out to be of insufficient quality, forcing the swarm to move again. However, for bee species that live in the open such as the Dwarf honey bee *Apis florea*, the quality of the nest site appears to be less important and the swarm has the chance to "upgrade" if its initial decision was suboptimal [69].

Thus it interesting to see if an iterative nest-site selection process as found in *Apis florea* can lead to an optimisation in an environment with many potential nest sites. In this experiment it is assumed that the swarm's environment corresponds to the search space of a continuous function. Each position in the environment constitutes a potential nest site, and its quality corresponds to a value of the function at that position. The test functions used in the experiment and their associated parameter values are given in Table 2. Initially the swarm is placed at position [-20,-20] for the Sphere function and [-10,-10] for the Booth function after which the nest-site selection process was started.

Table 2. Test functions and domain space range (R). The dimension of each function is 2.

		R
Sphere	$f_{sp}(\mathbf{x}) = \sum_{i=1}^{n} x_i^2$	$[-25; 25]^n$
Booth	$f_{bt}(\mathbf{x}) = (x_1 + 2x_2 - 7)^2 + (2x_1 + x_2 - 5)^2$	$[-10; 10]^n$

For this experiment the scouting behavior of the bees has been changed as the first version of the extended model is orientated to the behaviour of the European honeybee *Apis mellifera* where a scout assesses a nest site for a certain period of time before returning to the swarm. Because in this experiment each location corresponds to a potential nest site, scouts would immediately start to assess sites after a single scouting step. To overcome this, a scout remembers the best position it encountered during its scouting trip. If the quality of that position is better than the current location of the swarm it starts dancing for that site.

The quality of a newly discovered site depends on the quality difference regarding the current location of the swarm. If a scout discovers a nest site that is x% better than the swarm's current location this site is assigned quality x.

Bees that followed a dance for a potential nest site become recruits for this nest site and will fly towards it. If they encounter a better site on their way to this nest site, they abandon the recruitment process and become scouts. Recruits that do not find an advertised nest site also become scouts.

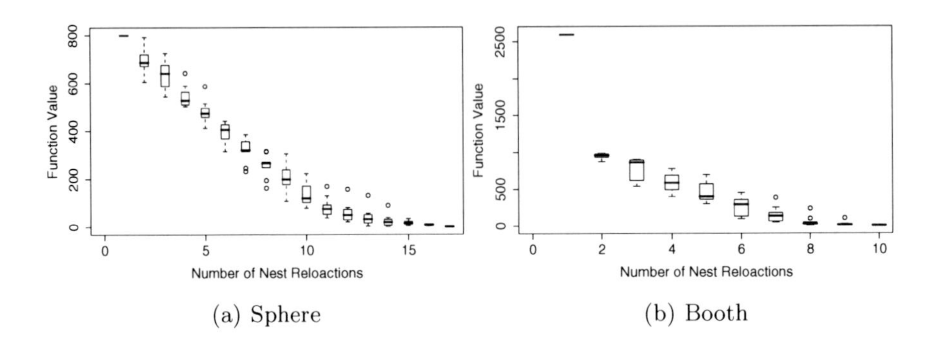

(a) Sphere (b) Booth

Fig. 4. Boxplots of the quality of the occupied nest site over several relocations for the two test functions.

Nest sites are assessed by recruits and returning bees for a certain amount of time. During that time a bee counts the number of other bees present at the site. If the number of bees at a site reaches a given quorum $q = 10$ the swarm is relocated to this new site and the nest-site selection process is restarted. The parameter values used in this experiment are: step size $step = 0.1$, scouting $T_{scout} = 100$, and assessment time $T_{assess} = 20$. A simulation run is stopped when no swarm relocation occurs within 3600 simulation steps.

The changes in the quality of the found nest sites for both test functions over several nest site relocations is depicted in Figure 4. As can be seen, the bees are able to iteratively optimise the position of the swarm within the search space. The optimisation process is limited by several factors. Since scout time T_{scout} and step size $step$ are fixed, scouts are only able to explore a certain range around the swarm's current location whereas a fixed step size prevents scouts from finding better solutions as they are likely to fly over them. This is especially the case when the swarm is close to the global optimum when scouts should actually search at a finer scale. Another limiting factor is the quality assessment. Remember that the quality of a newly found nest site is determined according to the potential improvement with respect to the current location of the swarm. To make an algorithm based on nest-site selection applicable to real optimisation problems, the swarm needs to become more sensitive to small quality differences to identify better potential nest sites when the swarm comes closer to the location of an optimum.

The decision-making process underlying the optimisation is slow. The higher the quorum q of bees needed at a potential nest before the swarm changes to this site, the slower is the optimisation process. The quorum mechanism could however also prove to be useful in terms of optimisation, because the existence of a quorum prevents a premature convergence onto a local minimum by slowing down the decision-process, thus giving better sites a higher chance to be discovered and thus entering the decision-making process. Another potential benefit of the quorum mechanism is that it requires bees to revisit and reassess a given site several times which is important for dynamic or noisy optimisation functions.

Algorithm 5. Bee Nest-Site Selection Scheme (BNSSS)

1: place bees on a random home position (swarm location) in the search space
2: initialize parameters Φ, Φ', Ψ, d, and d'
3: **while** stopping criterion not satisfied **do**
4:   **for all** scouts **do**
5:     $k = h$
6:     **repeat**
7:       the scout flies to a random position x with maximum distance d to its current home
8:       **if** $f(x) \geq \Phi$ **then**
9:         the scout performs k local search steps to find an improved location
10:         set $k = 0$
11:       **else**
12:         set $k = k - 1$
13:       **end if**
14:     **until** $k <= 0$
15:   **end for**
16:   **for all** followers **do**
17:     randomly assign the follower to one of the scout where the probability depends on the quality of the location of the scout
18:     **repeat**
19:       the follower flies to a random position x with maximum distance d' from the location of the scout it is assigned to
20:       **if** $f(x) \geq \Phi'$ **then**
21:         the follower performs k local search steps to find an improved location
22:         set $k = 0$ and stops
23:       **else**
24:         set $k = k - 1$
25:       **end if**
26:     **until** $k <= 0$
27:     **if** the follower could not find a location x with $f(x) \geq \Phi'$ **then**
28:       the follower abandons the scout
29:     **end if**
30:   **end for**
31:   **if** the swarm has found a location that is better than its home location **then**
32:     then its new home location is the best of these locations
33:   **else if** there exists a scout which has more than Ψ followers assigned to it **then**
34:     the swarm is assigned to the scout or one of its followers which has the best location
35:   **else**
36:     the swarm is assigned a new randomly chosen home location or it stays at its current location
37:   **end if**
38:   update d, d', Φ, Φ' and Ψ
39: **end while**

---

5 The BNSSS Scheme

Since some aspects of the bee nest site selection model are relevant for real bees but are not useful for a function optimisation algorithm we present here a scheme — called Bee Nest-Site Selection Scheme (BNSSS) — for the design of optimisation algorithms. The BNSSS, described in Algorithm 5, is provided as a framework into which details have to be added when a specific algorithm is designed. For example the values of d, d', Φ, Φ', and Ψ have to be defined. Φ, Φ', and Ψ should depend on the quality of the locations that have been found already. The values for d and d' might decrease during the run of the algorithm so that the swarm concentrates on a small area of the search space. In contrast, for dynamic optimization functions it might also be necessary to increase the values of d and d' at points in the decision-making process when it is found that the function to be optimized has changed. In addition, for noisy optimization functions it might be suitable to set $d' = 0$ so that the location of a scout is evaluated several times.

The BNSSS scheme is here given for a single swarm of bees but a multi-swarm version is also possible.This would require defining how the different swarms cooperate – for example, the swarms might be implemented to remain a certain distance from each other in order to cover different parts of the search space.

6 Conclusion

This chapter provided an overview of bee-inspired optimization algorithms. The existing algorithms are based on two behaviours found in bees: mating and foraging. Algorithms based on mating behaviour are closely related to evolutionary computing and exploit the fact that the polyandry common in honeybees increases a colony's fitness, as it leads to several different worker phenotypes and increases the task distribution in a colony. In particular, we described the Honeybee Mating Optimisation Algorithm (HBMO) [1, 2]. Foraging-based algorithms comprise population based optimization methods that take inspiration from the decentralized exploration/exploitation behaviour of a bee colony while it is gathering food resources. In particular we described the Artificial Bee Colony Algorithm (ABC) [42, 9], the Bees Algorithm (BA) [76], the Bee System algorithm [56] and Bee Colony optimisation Algorithm (BCO) [100], the Bee Colony-Inspired Algorithm (BCiA) [36], the Bee Colony Optimisation Algorithm (BCOA) [22, 23] and the The Virtual Bee Algorithm (VBA) [109].

We also presented nest site selection as a third type of honeybee behaviour that is useful in the context of optimisation. Nest-site selection involves the active discovery of potential sites by scout bees and a decision among these found sites. A model for the nest-site selection process of real bees [27] was presented. The results of three experiments show that the selection process is successful in a dynamic and noisy environment and that the process can detect/decide on the best stable solution even when better but noisier solutions are present. Additionally it was shown that an iterative application of the selection process can lead to function optimisation.

Finally we proposed an algorithmic scheme — called "Bee Nest-Site Selection Scheme" (BNSSS) — that is inspired by the nest-site selection model and can be used to design optimization algorithms.

Acknowledgments

This work was supported by the Human Frontier Science Program, Research Grant "Optimization in natural systems: ants, bees and slime moulds". We are grateful to Cliodhna Quigley for assistance with the manuscript.

References

1. Abbass, H.A.: Marriage in honeybees optimization (MBO): A haplometrosis polygynous swarming approach. In: Proceedings of the Congress on Evolutionary Computation, pp. 207–214 (2001)
2. Abbass, H.A.: A single queen single worker honey bees approach to 3-sat. In: Proceedings of the Genetic and Evolutionary Computation Conference GECCO, pp. 807–814 (2001)
3. Aderhold, A., Diwold, K., Scheidler, A., Middendorf, M.: Artificial bee colony optimization: A new selection scheme and its performance. Accepted for International Workshop on Nature Inspired Cooperative Strategies for Optimization (NICSO 2010) (2010)
4. Afshar, A., Haddad, O.B., Marino, M.A., Adams, B.J.: Honey-bee mating optimization (hbmo) algorithm for optimal reservoir operation. Journal of the Franklin Institute 344(5), 452–462 (2007)
5. Akay, B., Karaboga, D.: Parameter tuning for the artificial bee colony algorithm. In: Nguyen, N.T., Kowalczyk, R., Chen, S.-M. (eds.) ICCCI 2009. LNCS(LNAI), vol. 5796, pp. 608–619. Springer, Heidelberg (2009)
6. Bahamish, H.A.A., Abdullah, R., Salam, R.A.: Protein tertiary structure prediction using artificial bee colony algorithm. In: Proceedings of the Asian International Conference on Modelling & Simulation, vol. 0, pp. 258–263 (2009)
7. Bartumeus, F., Da Luz, M.G.E., Viswanathan, G.M., Catalan, J.: Animal search strategies: a quantitative random-walk analysis. Ecology 86(11), 3078–3087 (2005)
8. Baschelet, E.: Circular Statistics in biology. Academic Press, New York (1981)
9. Basturk, B., Karaboga, D.: An artificial bee colony (ABC) algorithm for numeric function optimization. In: IEEE Swarm Intelligence Symposium (2006)
10. Baykasoglu, A., Oezbakir, L., Tapkan, P.: Artificial Bee Colony Algorithm and Its Application to Generalized Assignment Problem. In: Swarm Intelligence: Focus on Ant and Particle Swarm Optimization, pp. 113–144. Itech Education and Publishing (2007)
11. Beekman, M., Fathke, R.L., Seeley, T.D.: How does an informed minority of scouts guide a honey bee swarm as it flies to its new home? Animal Behavior 71(1), 161–171 (2006)
12. Beekman, M., Gilchrist, A.L., Duncan, M., Sumpter, D.J.T.: What makes a honeybee scout? Behavioral Ecology and Sociobiology 61, 985–995 (2007)
13. Beekman, M., Lew, J.B.: Foraging in honeybeeswhen does it pay to dance? Behavioral Ecology 19, 255–262 (2008)

14. Beekman, M., Ratnieks, F.L.W.: Long range foraging by the honeybee apis mellifera l. Functional Ecology 14, 490–496 (2000)
15. Beekman, M., Sumpter, D.J.T., Seraphides, N., Ratnieks, F.L.W.: Comparing foraging behaviour of small and large honey bee colonies by decoding waggle dances made by foragers. Functional Ecology 18, 829–835 (2004)
16. Bell, W.J.: Searching behavior patterns in insects. Annual Reviews of Entomology 35, 447–467 (1990)
17. Benichou, O., Coppey, M., Moreau, M., Suet, P.-H., Voituriez, R.: Optimal search strategies for hidden targets. Physical Review Letters 94, 198101 (2005)
18. Bergman, C., Schaefer, J.A., Luttich, S.N.: Caribou movement as correlated random walk. Oecologia 123, 364–374 (2000)
19. Biesmeijer, J.C., de Vries, H.: Modelling collective foraging by means of individual behaviour rules in honey-bees. Behavioral Ecology and Sociobiology 44, 109–124 (1998)
20. Britton, N.F., Franks, N.R., Pratt, S.C., Seeley, T.D.: Deciding on a new home: how do honeybees agree? Proceedings of the Royal Society 269, 1383–1388 (2002)
21. Camazine, S., Visscher, P.K., Finley, J., Vetter, R.S.: House-hunting by honey bee swarms: collective decisions and individual behaviors. Insectes societies 46, 348–360 (1999)
22. Chong, C.S., Low, M.Y.H., Sivakumar, A.I., Gay, K.L.: A bee colony optimization algorithm to job shop scheduling. In: Proceedings of the 2006 Winter Simulation Conference (2006)
23. Chong, C.S., Low, Y.H.M., Sivakumar, A.I., Gay, K.L.: Using a bee colony algorithm for neighborhood search in job shop scheduling problems. In: Proceedings of the 21st European Conference on Modeling and Simulation (ECMS) (2007)
24. Cramer, J.S.: The origins and development of the logit mode (2003)
25. Crist, T.O., Guertin, D.S., Wiens, J.A., Milne, B.T.: Animal movement in heterogeneous landscapes: an experiment with elodes beetles in shortgrass prairie. Functional Ecology 6, 536–544 (1992)
26. Crone, E.E., Schultz, C.B.: Old models explain new observations of butterfly movement at patch edges. Ecology 89, 2061–2067 (2008)
27. Diwold, K., Beekman, M., Middendorf, M.: Bee nest site selection as an optimization process (2009) (manuscript)
28. Dornhaus, A., Kluegl, F., Oechslein, C., Puppe, F., Chittka, L.: Benefits of recruitment in honey bees: effects of ecology and colony size in an individual-based model. In: Behavioral Ecology (2006)
29. Eiben, A.E., Smith, J.E.: Introduction to Evolutionary Computing. Springer, Heidelberg (2003)
30. Farooq, M.: Bee-Inspired Protocol Engineering: From Nature to Networks. Natural Computation Series. Springer, Heidelberg (2008)
31. Fathian, M., Amiri, B., Maroosi, A.: Application of honey-bee mating optimization algorithm on clustering. Applied Mathematics and Computation 190(2), 1502–1513 (2007)
32. Gardner, K.E., Seeley, T.D., Calderone, N.W.: Do honeybees have two discrete dances to advertise food sources? Animal Behaviour 75, 1291–1300 (2008)
33. Giardina, I.: Collective behavior in animal groups: theoretical models and empirical studies. HFSP Journal 2, 205–219 (2008)
34. Giurfa, M., Vorobyev, M., Kevan, P., Menzel, R.: Detection of coloured stimuli by honeybees: minimum visual angles and receptor specific contrasts. Journal of Comparative Physiology A 178, 699–709 (1996)

35. Gordon, N., Wagner, I.A., Brucks, A.M.: Discrete bee dance algorithms for pattern formation on a grid. In: Proceedings of the 2003 IEEE/WIC International Conference on Intelligent Agent Technology (IAT 2003), p. 545 (2003)
36. Häckel, S., Dippold, P.: The bee colony-inspired algorithm (bcia): a two-stage approach for solving the vehicle routing problem with time windows. In: Proceedings of the 11th Annual Conference on Genetic and Evolutionary Computation, pp. 25–32 (2009)
37. Haddad, O.B., Afshar, A., Marino, M.A.: Design-operation of multi-hydropower reservoirs: Hbmo approach. Water Resources Management 22, 1709–1722 (2008)
38. Janson, S., Middendorf, M., Beekman, M.: Honey bee swarms: How do scouts guide a swarm of uninformed bees? Animal Behaviour 70, 349–358 (2005)
39. Janson, S., Middendorf, M., Beekman, M.: Searching for a new home – scouting behavior of honeybee swarms. Behavioral Ecology 18, 384–392 (2007)
40. Jones, J.C., Myerscough, M.R., Graham, S., Oldroyd, B.P.: Honey bee nest thermoregulation: diversity promotes stability. Science 305, 402–404 (2004)
41. Jung, S.H.: Queen-bee evolution for genetic algorithms. Electronics Letters 39(6), 575–576 (2003)
42. Karaboga, D.: An idea based on honey bee swarm for numerical optimization. Technical Report TR06, Erciyes University, Engineering Faculty, Computer Engineering Department (2005)
43. Karaboga, D., Akay, B., Ozturk, C.: Artificial Bee Colony (ABC) optimization algorithm for training feed-forward neural networks. In: Torra, V., Narukawa, Y., Yoshida, Y. (eds.) MDAI 2007. LNCS (LNAI), vol. 4617, pp. 318–329. Springer, Heidelberg (2007)
44. Karaboga, D., Akay, B.: A survey: algorithms simulating bee swarm intelligence. Artificial Intelligence Review (2009)
45. Karaboga, D., Basturk, B.: Artificial Bee Colony (ABC) optimization algorithm for solving constrained optimization problems. In: Foundations of Fuzzy Logic and Soft Computing, pp. 789–798. Springer, Heidelberg (2007)
46. Karaboga, D., Basturk, B.: A powerful and efficient algorithm for numerical function optimization: artificial bee colony (ABC) algorithm. Journal of Global Optimization 39, 459–471 (2007)
47. Karaboga, D., Basturk, B.: On the performance of artificial bee colony (ABC) algorithm. Applied Soft Computing 8, 687–697 (2008)
48. Karcı, A.: Imitation of Bee Reproduction as a Crossover Operator in Genetic Algorithms. In: Zhang, C., W. Guesgen, H., Yeap, W.-K. (eds.) PRICAI 2004. LNCS (LNAI), vol. 3157, pp. 1015–1016. Springer, Heidelberg (2004)
49. Kramer, D.L., McLaughlin, R.L.: The behavioral ecology of intermittent locomotion. American Zoology 41, 137–153 (2001)
50. Kugler, H.: Blütenoekologische Untersuchungen mit Hummeln. VI Planta, Arch wiss Bot 19, 781–789 (1933)
51. Latty, T., Duncan, M., Beekman, M.: High bee traffic disrupts transfer of directional information in flying honeybee swarms. Animal Behaviour 78, 117–121 (2009)
52. Lemmens, N., Tuyls, K.: Stigmergic landmark foraging. In: Proceedings of Autonomous Agents and Multi-Agent Systems (AAMAS), pp. 497–504 (2009)
53. Lindauer, M.: Schwarmbienen auf Wohnungssuche. Zeitschrift für vergleichende Physiologie 37, 263–324 (1955)

54. List, C., Elsholtz, C., Seeley, T.D.: Independence and interdependence in collective decision making: an agent-based model of nest-site choice by honeybee swarms. Philosophical Transactions of the Royal Society of London series B 364, 755–762 (2009)
55. Lu, X., Zhou, Y.: A Novel Global Convergence Algorithm: Bee Collecting Pollen Algorithm. In: Advanced Intelligent Computing Theories and Applications. With Aspects of Artificial Intelligence, pp. 518–525. Springer, Heidelberg (2008)
56. Lucic, P., Teodorovic, D.: Bee system: Modeling combinatorial optimization transportation engineering problems by swarm intelligence. In: Preprints of the TRISTAN IV Triennial Symposium on Transportation Analysis, pp. 441–445 (2001)
57. Lucic, P., Teodorovic, D.: Transportation modeling: an artificial life approach. In: Proceedings of the 14th IEEE International Conference on Tools with Artificial Intelligence, pp. 216–223 (2002)
58. Lucic, P., Teodorovic, D.: Computing with bees: Attacking complex transportation engineering problems. International Journal on Artificial Intelligence Tools 12(3), 375–394 (2003)
59. Luke, S., Balan, G.C., Panait, L., Cioffi-Revilla, C., Paus, S.: Mason: A Java multi-agent simulation library. In: Proceedings of the Agent 2003 Conference on Challenges in Social Simulation (2003)
60. Marinakis, Y., Marinaki, M., Dounias, G.: Honey bees mating optimization algorithm for large scale vehicle routing problems. Natural Computing (2009)
61. Marinakis, Y., Marinaki, M., Matsatsinis, N.: A Hybrid Clustering Algorithm Based on Honey Bees Mating Optimization and Greedy Randomized Adaptive Search Procedure. In: Learning and Intelligent Optimization, pp. 138–152. Springer, Heidelberg (2008)
62. Mattila, H.R., Seeley, T.D.: Genetic diversity in honey bee colonies enhances productivity and fitness. Science 317, 362–364 (2007)
63. Mattila, H.R., Burke, K.M., Seeley, T.D.: Genetic diversity within honeybee colonies increases signal production by waggle-dancing foragers. Proceedings of the Royal Society of London series B 275, 809–816 (2008)
64. Mautz, D.: Der Kommunikationseffekt der Schwänzeltänze bei Apis mellifera carnica. Zeitschrift für vergleichende Physiologie 72, 192–220 (1971)
65. Myerscough, M.R.: Dancing for a decision: a matrix model for nest-site choice by honeybees. Proceedings of the Royal Society of London series B 270, 577–582 (2003)
66. Nakrani, S., Tovey, C.: On honey bees and dynamic server allocation in internet hosting centers. Adaptive Behavior 12, 223–240 (2004)
67. Olague, G., Puente, C.: The honeybee search algorithm for three-dimensional reconstruction. In: Rothlauf, F., Branke, J., Cagnoni, S., Costa, E., Cotta, C., Drechsler, R., Lutton, E., Machado, P., Moore, J.H., Romero, J., Smith, G.D., Squillero, G., Takagi, H. (eds.) EvoWorkshops 2006. LNCS, vol. 3907, pp. 427–437. Springer, Heidelberg (2006)
68. Oldroyd, B.P., Fewell, J.H.: Genetic diversity promotes homeostasis in insect colonies. Trends in Ecology and Evolution 22, 408–413 (2007)
69. Oldroyd, B.P., Gloag, R.S., Even, N., Wattanachaiyingcharoen, W., Beekman, M.: Nest-site selection in the open-nesting honey bee apis florea. Behavioral Ecology and Sociobiology 62, 1643–1653 (2008)
70. Palmer, K.A., Oldroyd, B.P.: Evolution of multiple mating in the genus apis. Apidologie 31, 235–248 (2000)

71. Passino, K.M., Seeley, T.D.: Modeling and analysis of nest-site selection by honeybee swarms: the speed and accuracy trade-off. Behavioral Ecology and Sociobiology 59, 427–442 (2006)
72. Pham, D.T., Ghanbarzadeh, A., Koc, E., Otri, S.: Application of the bees algorithm to the training of radial basis function networks for control chart pattern recognition. In: Proceedings of the 5th CIRP International Seminar on Intelligent Computation in Manufacturing Engineering (CIRP ICME 2006), pp. 711–716 (2006)
73. Pham, D.T., Soroka, A.J., Koc, E., Ghanbarzadeh, A., Otri, S., Packianather, M.: Optimising neural networks for identification of wood defects using the bees algorithm. In: Proceedings of the 4th International IEEE Conference on Industrial Informatics. INDIN, pp. 1346–1351 (2006)
74. Pham, D.T., Darwish, A.H., Eldukhri, E.E.: Optimisation of a fuzzy logic controller using the bees algorithm. International Journal of Computer Aided Engineering and Technology 1, 250–264 (2009)
75. Pham, D.T., Ghanbarzadeh, A.: Multi-objective optimisation using the bees algorithm. In: Proceedings of IPROMS 2007 Conference (2007)
76. Pham, D.T., Ghanbarzadeh, A., Koc, E., Otri, S., Rahim, S., Zaidi, M.: The bees algorithm - a novel tool for complex optimisation problems. In: Proceedings of IPROMS 2006 Conference, pp. 454–461 (2006)
77. Pham, D.T., Koc, E., Ghanbarzadeh, A., Otri, S.: Optimisation of the weights of multi-layered perceptrons using the bees algorithm. In: IMS 2006 Intelligent Manufacturing Systems Conference (2006)
78. Pham, D.T., Pham, Q.T., Ghanbarzadeh, A., Castellani, M.: Dynamic optimisation of chemical engineering processes using the bees algorithm. In: 17th IFAC World Congress COEX, pp. 6100–6105 (2008)
79. Sato, T., Hagiwara, M.: Bee system: finding solution by a concentrated search. In: Proceedings of the IEEE International Conference on Systems, Man, and Cybernetics, 1997. Computational Cybernetics and Simulation, pp. 3954–3959 (1997)
80. Schmickl, T., Crailsheim, K.: Trophallaxis within a robotic swarm: bio-inspired communication among robots in a swarm. Autonomous Robots 25, 171–188 (2008)
81. Schmidt, J.O.: Hierarchy of attractants for honey bee swarms. Journal of Insect Behavior 14, 469–477 (2001)
82. Schultz, K.M., Passino, K.M., Seeley, T.D.: The mechanism of flight guidance in honeybee swarms: subtle guides or streaker bees? Journal of Experimental Biology 211, 3287–3295 (2008)
83. Seeley, T.D., Camazine, S., Sneyd, J.: Collective decision-making in honey bees: how colonies choose among nectar sources. Behavioral Ecology and Sociobiology 28, 277–290 (1991)
84. Seeley, T.D., Mikheyev, A.S., Pagano, G.J.: Dancing bees tune both duration and rate of waggle-run production in relation to nectar-source profitability. Journal of Comparative Physiology A 186, 813–819 (2000)
85. Seeley, T.D., Morse, R.A.: Nest site selection by the honey bee, apis-mellifera. Insectes Sociaux 25, 323–337 (1978)
86. Seeley, T.D., Morse, R.A., Visscher, P.K.: The natural history of the flight of honey bee swarms. Psyche 86, 103–113 (1979)
87. Seeley, T.D., Visscher, P.K.: Survival of honeybees in cold climates: the critical timing of colony growth and reproduction. Ecological Entomology 10, 81–88 (1985)

88. Seeley, T.D., Visscher, P.K.: Choosing a home: how the scouts in a honey bee swarm perceive the completion of their group decision making. Behavioral Ecology and Sociobiology 54, 511–520 (2003)
89. Seeley, T.D., Visscher, P.K.: Group decision making in nest-site selection by honey bees. Apidologie 35, 101–116 (2004)
90. Seeley, T.D., Visscher, P.K.: Quorum sensing during nest-site selection by honeybee swarms. Behavioral Ecology and Sociobiology 56, 594–601 (2004)
91. Seeley, T.D.: Consensus building during nest–site selection in honey bee swarms: the expiration of dissent. Behavioral Ecology and Sociobiology 53, 417–424 (2003)
92. Seeley, T.D., Buhrman, S.C.: Group decision making in swarms of honeybees. Behavioral Ecology and Sociobiology 45, 19–31 (2001)
93. Seeley, T.D., Buhrman, S.C.: Nest–site selection in honey bees: how well do swarms implement the "best-of-n" decision rule. Behavioral Ecology and Sociobiology 49, 416–427 (2001)
94. Seeley, T.D., Visscher, P.K.: Sensory coding of nest-site value in honeybee swarms. Journal of Experimental Biology 211, 3691–3697 (2008)
95. Sherman, G., Visscher, P.K.: Honeybee colonies achieve fitness through dancing. Nature 419, 920–922 (2002)
96. Subbotin, S.A., Oleinik, A.A.: Multiagent optimization based on the bee-colony method. Cybernetics and Systems Analysis 45, 177–186 (2009)
97. Sumpter, D.J.T., Pratt, S.C.: A modelling framework for understanding social insect foraging. Behavioral Ecology and Sociobiology 53, 131–144 (2003)
98. Tautz, J., Rohrseitz, K.: What attracts honeybees to a waggle dancer? Journal of Comparative Physiology A 183, 661–667 (1998)
99. Tautz, J., Rohrseitz, K., Sandeman, D.C.: One-strided waggle dance in bees. Nature 382, 32 (1996)
100. Teodorovic, D., Dell'Orco, M.: Bee colony optimization - a cooperative learning approach to complex transportation problems. In: Advanced OR and AI Methods in Transportation. Proceedings of the 10th Meeting of the EURO Working Group on Transportation, pp. 51–60 (2005)
101. Teodorovic, D., Dell'Orco, M.: Mitigating traffic congestion: solving the ride-matching problem by bee colony optimization. Transportation Planning and Technology 31, 135–152 (2008)
102. Teodorovic, D., Lucic, P.: Schedule synchronization in public transit by fuzzy ant system. Transportation Planning and Technology 28, 47–77 (2007)
103. Teodorovic, D., Lucic, P., Markovic, G., Dell'Orco, M.: Bee colony optimization: Principles and applications. In: Proceedings of the 8th Seminar on Neural Network Applications in Electrical Engineering (NEUREL), pp. 151–156 (2006)
104. Visscher, P.K., Seeley, T.D.: Foraging strategy of honeybee colonies in a temperate deciduous forest. Ecology 63, 1790–1801 (1982)
105. Visscher, P.K., Seeley, T.D.: Coordinating a group departure: who produces the piping signals on honeybee swarms? Behavioral Ecology and Sociobiology 61, 1615–1621 (2007)
106. von Frisch, K.: The dance language and orientation of bees. Harvard University Press, Cambridge (1967)
107. Wong, L., Low, M.Y.H., Chong, C.S.: A bee colony optimization algorithm for traveling salesman problem. In: Proceedings of the Asia International Conference on Modelling & Simulation, vol. 0, pp. 818–823 (2008)

108. Wong, L.P., Low, M.Y.H., Chong, C.S.: An efficient bee colony optimization algorithm for traveling salesman problem using frequency-based pruning. In: Proceedings of the 7th IEEE International Conference on Industrial Informatics (INDIN 2009), pp. 775–782 (2009)
109. Yang, X.-S.: Engineering optimizations via nature-inspired virtual bee algorithms. In: Mira, J., Álvarez, J.R. (eds.) IWINAC 2005. LNCS, vol. 3562, pp. 317–323. Springer, Heidelberg (2005)
110. Zollner, P.A., Lima, S.L.: Search strategies for landscape–level interpatch movements. Ecology 80(3), 1019–1030 (1999)

Parallel Approaches for the Artificial Bee Colony Algorithm

Rafael Stubs Parpinelli[1,2], César Manuel Vargas Benitez[2], and Heitor Silvério Lopes[2]

[1] Santa Catarina State University (UDESC) - Joinville - SC - Brazil
parpinelli@joinville.udesc.br
[2] Federal University of Technology Paraná (UTFPR) - Curitiba - PR - Brazil
cbenitez@cpgei.ct.utfpr.edu.br, hslopes@utfpr.edu.br

Abstract. This work investigates the parallelization of the Artificial Bee Colony Algorithm. Besides a sequential version enhanced with local search, we compare three parallel models: master-slave, multi-hive with migrations, and hybrid hierarchical. Extensive experiments were done using three numerical benchmark functions with a high number of variables. Statistical results indicate that intensive local search improves the quality of solutions found and, thanks to the coevolution effect, the multi-population approaches obtain better quality with less computational effort. A final comparison between models was done analyzing the trade-offs between quality of solution and processing time.

Keywords: Swarm Intelligence, Parallelism, Bee Foraging, ABC Algorithm.

1 Introduction

In the beginning of the research in the area of Swarm Intelligence, the mainstream paradigms were Ant Colony Optimization[1] [9] and Particle Swarm Optimization[2] [16]. ACO is inspired by the foraging behavior of ants and the PSO is motivated by the simulation of fish schools and bird flocks social behavior. Over the years, both methods have been applied successfully in a vast range of problems [8]. Notwithstanding, in recent years several other swarm intelligence algorithms have appeared, and, amongst them, those inspired by specific behaviors of honey bees, such as bees foraging [12, 19] and bees mating [11]. Currently, there are a variety of algorithms inspired by the bee foraging behavior found in literature, such as: Bee System [21], Honey Bee Algorithm [17], BeeHive [29], Virtual Bee Algorithm [30], Bee Colony Optimization [27], Bees Swarm Optimization [10], Bees Algorithm [19], Honey Bee Foraging [2] and the Artificial Bee Colony Algorithm [12].

This work focuses on the Artificial Bee Colony Algorithm (ABC), which was first proposed by Karaboga [12] for solving multi-dimensional and multi-modal

[1] ACO repository: http://iridia.ulb.ac.be/~mdorigo/ACO/
[2] PSO Repository: http://www.particleswarm.info

B.K. Panigrahi, Y. Shi, and M.-H. Lim (Eds.): Handbook of Swarm Intelligence, ALO 8, pp. 329–345.
springerlink.com

optimization problems. The ABC algorithm is a population-based algorithm inspired by the foraging behavior of bees. In this metaphor, bees are the possible solutions to the problem, and they fly within the environment (the search space) to find the best food source location (best solution).

Two remarkable features of such swarm-based systems are self-organization and decentralized control that leads to an emergent behavior. Emergent behavior is a property that emerges through interactions among system components (bees) and it is not possible to be achieved by any of the components of the system acting alone [4]. Bees interactions occur through a waggle dance performed inside the hive. It is used to recruit other bees to exploit a food source. The quality of a food source is proportional to the intensity of the waggle dance performed by a bee. Hence, best food sources lead in a more intense waggle dance, and this can reinforce the exploitation of the best food locations.

To date, the Artificial Bee Colony Algorithm (ABC) is one of the most widely used bee algorithm for problem solving. Some successful applications found in the literature using it include: generalized assignment problem [3], energy distribution network configuration [24], neural networks training [15], multi-objective problems [18], template matching in digital images [7]. A recent work [14] compared the ABC algorithm performance against other evolutionary computation algorithms (Genetic Algorithm, Particle Swarm Optimization, Differential Evolution and Evolution Strategies) upon several benchmark functions. Results showed that the performance of the ABC was better than or similar to those of the other algorithms. Another relevant work concerning the ABC algorithm analyzed the tuning of control parameters [1]. More about the ABC algorithm can be found in the repository[3].

It is known that population-based metaheuristics such as Genetic Algorithms [6][26], PSO [28] and ACO [5][25] can explore efficiently the use of parallel concepts to speed up the search process. In [28] the authors used a master-slave model to compare both synchronous and asynchronous PSO versions in an engineering problem. A PSO master-slave model was also used by [22] to solve a biomechanical problem that involves finding the kinematic structure of an ankle joint model. In [25] both master-slave and the island models of an ACO were compared using several instances of the well-known Traveling Salesman Problem.

To the best of our knowledge, to date, no application that explores parallelization methods for the ABC algorithm was found in the literature. Hence, this work aims at applying to the ABC algorithm concepts of parallel processing widely used in other population-based metaheuristics. This work compares the performance of the standard Karaboga model [14] with an Enhanced Sequential version (ES-ABC) and other three parallel models, namely: Master-Slave ABC (MS-ABC); Multi-Hive ABC with migrations (MH-ABC); and Hybrid Hierarchical ABC (HH-ABC).

Experiments were done using three well-known mathematical functions. An important feature is that the test functions are purposely set with a high number of variables in order to turn the problem solving process extremely difficult to all

[3] ABC Repository: `http://mf.erciyes.edu.tr/abc/`

models. Therefore, one can evaluate the benefits of the parallelized approaches for the ABC algorithm in comparison with the regular sequential version.

2 Bee Foraging Behavior

Most social insects, such as ants and bees, spend most of their life in foraging for food. Honey bee colonies have a decentralized system to collect the food and can adjust the searching pattern precisely so that to enhance the collection of nectar [23].

Bees can estimate the distance from the hive to food sources by measuring the amount of energy consumed when they fly, besides the direction and the quality of the food source. This information is shared with their nest-mates by trophallaxis (direct contact) and by performing a waggle dance. The exchange of information among bees is the most important occurrence in the formation of the collective knowledge. The dance floor is the place in the hive where the coming back forager bees perform the waggle dance to recruit more foragers. Bees that decide foraging without any guidance from other bees are called scouts. Bees that attend to the waggle dance at the dance floor can decide which food source to go based on its quality. The quality of a food source is proportional to the quantity of nectar found there, and this information is transmitted by changing the intensity of the waggle dance and through antennae contacts. The better the food source, the more intense is the dance and the contacts [20]. Therefore, each forager bee can behave in three different ways after unloading the food: it can perform the waggle dance to recruit more foragers to the same food source; it can abandon the food source due to loss of available resources; or it can directly return to foraging.

The basic idea concerning the algorithms based on the bee foraging behavior is that foraging bees have a potential solution to an optimization problem in their memory (i.e., a configuration for the problem decision variables). This potential solution corresponds to the location of a food source and has an aggregated quality measure (i.e., value of the objective function). The food source quality information is exchanged through the waggle dance that probabilistically biases other bees to exploit food sources with higher quality. Based on this behavior, the ABC algorithm was proposed, and it is described in the next section.

3 Artificial Bee Colony Algorithm

The ABC algorithm [12] works with a swarm of n solutions x (food sources) of dimension d that are modified by the artificial bees. The bees aim at discovering places of food sources v (locations in the search space) with high amount of nectar (good fitness).

In the ABC algorithm there are three types of bees: the scout bees that randomly fly in the search space without guidance; the employed bees that exploit the neighborhood of their food sources selecting a random solution to be perturbed; and the onlooker bees that are placed on the food sources by using a probability based selection process. As the nectar amount of a food source increases, the probability value P_i with which the food source is preferred by onlookers increases too.

If the nectar amount of a new source is higher than that of the previous one in their memory, they update the new position and forget the previous one. If a solution is not improved by a predetermined number of trials controlled by the parameter *limit*, then the food source is abandoned by the corresponding employed bee and it becomes a scout bee.

Each cycle of the search consists of moving the employed and onlooker bees onto the food sources and calculating their nectar amounts; and determining the scout bees and directing them onto possible food sources. The ABC pseudo-code is shown in Algorithm 1.

The ABC algorithm attempts to balance exploration and exploitation by combining local search methods, carried out by employed and onlooker bees, with global search methods, managed by scout bees.

Algorithm 1. Pseudo-code of Artificial Bee Colony Algorithm (ABC).

1: Parameters: n, *limit*
2: Objective function $f(x)$, $x = [x_1, x_2, ..., x_d]^T$
3: Initialize the food positions randomly x_i $i = 1, 2, ..., n$
4: Evaluate fitness $f(x_i)$ of the individuals
5: **while** stop condition not met **do**
6: Employed phase:
7: Produce new solutions with k, j and ϕ at random
8: $v_{ij} = x_{ij} + \phi_{ij} \cdot (x_{ij} - x_{kj})$ $k \in \{1, 2, ..., n\}, j \in \{1, 2, ..., d\}, \phi \in [0,\ 1]$
9: Evaluate solutions
10: Apply greedy selection process for the employed bees
11: Onlooker phase:
12: Calculate probability values for the solutions x_i
13: $P_i = \frac{f_i}{\sum_{j=i}^{n} f_j}$
14: Produce new solutions from x_i selected using P_i
15: Evaluate solutions
16: Apply greedy selection for the onlookers
17: Scout phase:
18: Find abandoned solution: If limit exceeds, replace it with a new random solution
19: Memorize the best solution achieved so far
20: **end while**
21: Postprocess results and visualization

4 Implementations of the ABC

This section describes the models derived from the standard ABC algorithm: an enhanced sequential version of the standard ABC algorithm, and three parallel models. These models are detailed in the following sections.

4.1 Enhanced Sequential Model

In the "employed phase" of the standard ABC algorithm (see Algorithm 1), a single element of the solution vector is perturbed in order to generate a new solution. An enhanced version of the ABC was devised, named ES-ABC (Enhanced Sequential ABC), in which that phase is extended for a predefined number of times. In fact, such procedure improves the local search capability of the algorithm, allowing a better exploration of the neighborhood of a given solution in the multidimensional search space. In the ES-ABC, that phase becomes a greedy search because once a new solution generated by the perturbation is evaluated, it replaces the original solution if its fitness is better. The number of times the procedure is repeated is a user-defined parameter. Obviously, the number of fitness evaluation increases significantly, leading to a higher computational cost (when compared with the standard ABC algorithm). On the other hand, an improvement in the quality of solutions is expected, thanks to an enhanced local search. However, depending on the fitness landscape of the function under optimization, the intensive local search provided by this approach can lead to local optima.

4.2 Parallel Master-Slave Model

The MS-ABC model is a global single-population system where a master process divides the computational effort into several slave processes, each one running in a different processor. The MS-ABC model uses one core for each process, master or slave. The slaves do their assigned job and send back the results to the master process. Figure 1 illustrates the master process (ellipse with solid line) with three slave processes (ellipses with dashed lines). The purpose of dividing the computational effort by the master process is to distribute to each slave a predefined number of function evaluations.

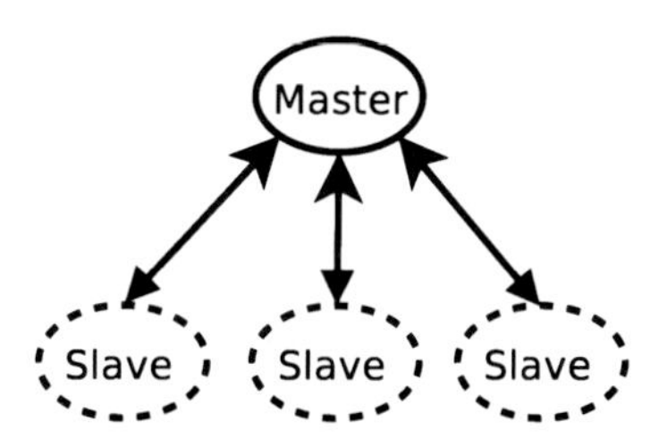

Fig. 1. Master-slave model.

The master coordinates the initialization of the swarm, the iterations of the algorithm phases (employed, onlooker and scout), and the distribution of bees (solutions) to slaves. In Algorithm 1 the bold and italic instructions are where the parallelization occurs that is, basically, in the function evaluations. Slaves, in turn, receive a number of bees, compute the objective function, and return the value for each bee. Recall that, for each bee (solution) received by a slave processor, a local search is performed (in the same way as in the ES-ABC model described above).

4.3 Parallel Multi-Hive Model

The multi-hive model (MH-ABC) is a multiple-population coarse-grained system that uses two or more hives that are initialized at the same time with different random seeds. Each population is seen as an island. The hives work independently from each other, and each one runs an ABC algorithm, but a migration process occurs periodically. The MH-ABC approach uses one processing core for each hive. Two versions were implemented: one uses the basic ABC (MH-ABC) and the other uses the ES-ABC (MH-ES-ABC). Migrations take place between hives, and are defined by a migration policy that includes some parameters defined by the user. The two most important parameters are: migration gap, the number of generations between successive migrations; and the migration rate, number of individuals that will migrate at each migration event. Besides, it is also defined some criteria for selection of immigrants and substitution of emigrants. Although the topology and connectivity between islands has some influence in the convergence of the algorithm and its quality [26], we preferred to use the simplest approach: a unidirectional ring. That is, each island sends immigrants to only one other island, and each island receives emigrants only from one other island. Figure 2 illustrates the multi-hive model with four hives.

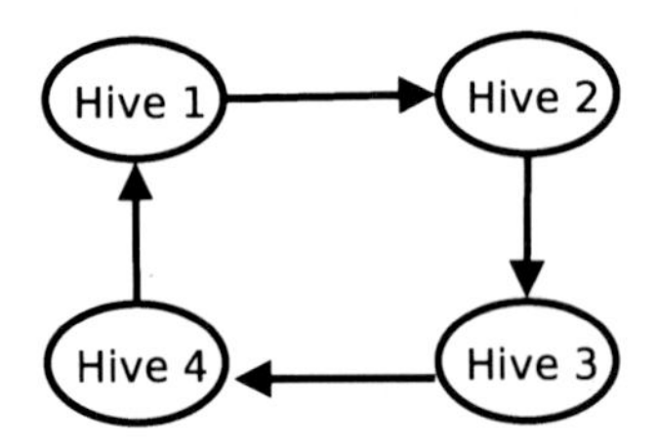

Fig. 2. Multi-hive model.

4.4 Parallel Hybrid Hierarchical Model

The HH-ABC has two levels: at the upper level it has multiple-population coarse-grained islands (such as the MH-ABC), and at the lower level it has single-population master-slaves (such as the MS-ABC). These two levels can be seen as an hierarchy of hives. The objective of such hybrid approach is to join the advantages of the two models previously mentioned in a single approach. In the HH-ABC, there are more than two hives working independently with a migration policy; and for each hive there is a master process that segregates the computational effort to various slave processes. Figure 3 illustrates the HH-ABC model with four hives at the upper level (ellipses with solid lines) and three slaves in each hive at the lower level (ellipses with dashed lines.

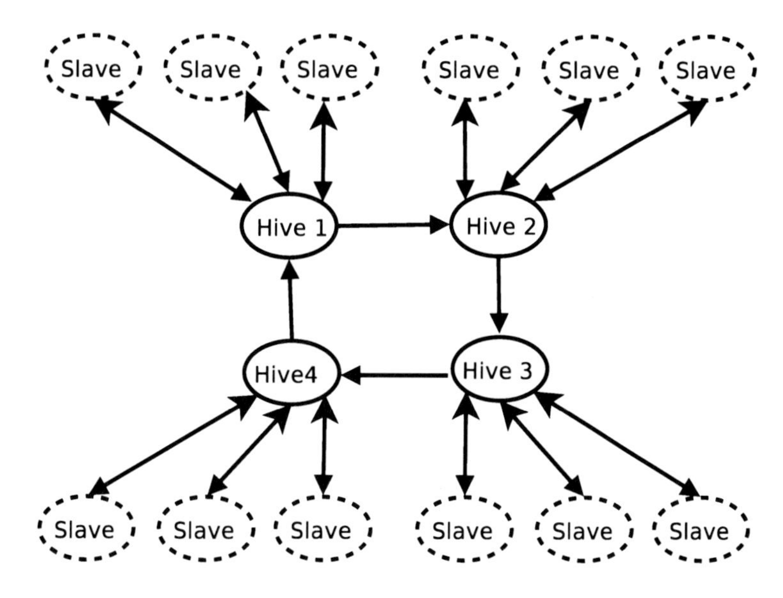

Fig. 3. Hierarchical model.

5 Experiments

All experiments were done using a cluster of networked computers interconnected by a gigabit Ethernet switch. Each node of the cluster was composed of a PC with core-2 Quad processor at 2.8 MHz, 2 Gbytes of RAM and running Suse Linux operating system. All models were implemented using MPICH2 library[4].

The ABC algorithm was applied to three well-known test functions, extensively used in the literature as benchmark [14][13] (see Table 1).

Table 1. Numerical benchmark functions

Function	Ranges	Minimum value
$f_1(\mathbf{x}) = \sum_{i=1}^{n}(x_i^2 - 10\cos(2\pi x_i) + 10)$	$-5.12 \leq x_i \leq 5.12$	$f_1(\mathbf{0}) = 0$
$f_2(\mathbf{x}) = \frac{1}{4000}\left(\sum_{i=1}^{n} x_i^2\right) - \left(\prod_{i=1}^{n}\cos\left(\frac{x_i}{\sqrt{i}}\right)\right) + 1$	$-600 \leq x_i \leq 600$	$f_2(\mathbf{0}) = 0$
$f_3(\mathbf{x}) = \sum_{i=1}^{n-1}\left(0.5 + \frac{\sin^2\left(\sqrt{x_{i+1}^2 + x_i^2}\right) - 0.5}{\left(0.001\left(x_{i+1}^2 + x_i^2\right) + 1\right)^2}\right)$	$-100 \leq x_i \leq 100$	$f_3(\mathbf{0}) = 0$

The first function ($f_1(\mathbf{x})$) is the Rastrigin function that is a multi-modal function and is based on the Sphere function with the addition of cosine modulation to produce many local minima. The locations of the minima are regularly distributed. The main difficulty in finding optimal solutions to this function is that an optimization algorithm can be easily trapped in a local optimum on its way towards the global optimum. $\mathbf{x}$ is defined in the range of $[-5.12, 5.12]$ and the

[4] MPICH2: `www.mcs.anl.gov/research/projects/mpich2/`.

global minimum value for $f_1(\mathbf{x})$ is 0 and the corresponding global optimum solution is $\mathbf{x}_{opt} = (x_1, x_2, \ldots, x_n) = (0, 0, \ldots, 0)$. Surface plot and contour lines of $f_1(\mathbf{x})$ are shown in Fig. 4.

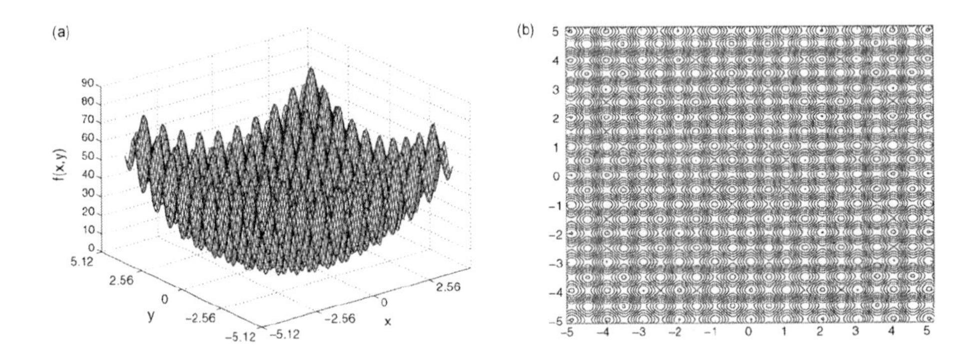

Fig. 4. Rastrigin function: (a) surface plot and (b) contour lines.

The second function ($f_2(\mathbf{x})$) is the Griewank function that is strongly multi-modal, because its number of local optima increases with the dimensionality. $\mathbf{x}$ is defined in the range of $[-600, 600]$ and the global minimum value for $f_2(\mathbf{x})$ is 0 and the corresponding global optimum solution is $\mathbf{x}_{opt} = (x_1, x_2, \ldots, x_n) = (0, 0, \ldots, 0)$. Surface plot and contour lines of $f_2(\mathbf{x})$ are shown in Fig. 5.

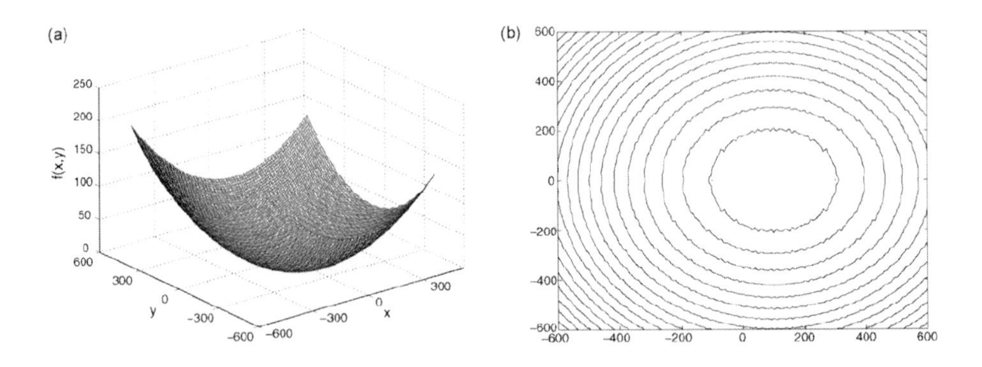

Fig. 5. Griewank function: (a) surface plot and (b) contour lines.

The third function ($f_3(\mathbf{x})$) is the generalized Schaffer function that is strongly multi-modal. $\mathbf{x}$ is defined in the range of $[-100, 100]$ and the global minimum value for $f_3(\mathbf{x})$ is 0 and the corresponding global optimum solution is $\mathbf{x}_{opt} = (x_1, x_2, \ldots, x_n) = (0, 0, \ldots, 0)$. Surface plot and contour lines of $f_3(\mathbf{x})$ are shown in Fig. 6.

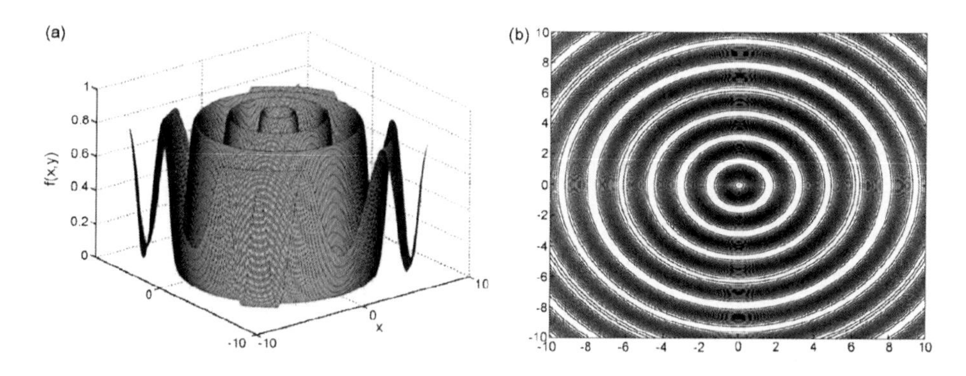

Fig. 6. Generalized Schaffer function: (a) surface plot and (b) contour lines.

In [13], experiments were done using these functions with different dimensionalities ranging from 5 to 100. The number of variables (dimensions) for all functions in our experiments was arbitrarily set to 200. This high number of variables turns the problem solving process extremely difficult to all models (and, possibly, to any heuristic optimization method). This is because we are interested in evaluating the behavior of the parallel models, regarding processing time and quality of solutions. In all experiments, the maximum number of iterations and swarm size were set to 10,000 and 1,000 bees, respectively. For all models, the stop condition is the maximum number of iterations.

Since the ABC is a stochastic algorithm, every experiment was repeated 100 times with different random seeds. The performance of each approach takes into account the best found solution and the processing time. The average function values and the standard deviations of the best solutions were recorded. The average time and the standard deviations of the elapsed time were also recorded.

In the ES-ABC model the extension of the employed phase was arbitrarily set to 50 times for each employed bee. This fact, by itself, makes the computational load high enough to demand parallelization of the algorithm.

The MS-ABC model was run with 10 slaves and, therefore, each slave has to process 100 bees.

Both MH-ABC and MH-ES-ABC models use 4 islands with an unidirectional ring topology. This number of islands was set due two facts: availability of hardware (so as to enforce one processor to each process); and the expectance of observing the coevolution effects as soon as possible. From the literature of parallel evolutionary algorithms, it is known that a small number of islands in a ring topology accelerate the coevolution process, thus requiring a reduced number of generations/iterations.

The migration gap was set to 10% of the number of iterations, that is, a migration event takes place at every 1,000 iterations. The reason to set this value for the migration gap is the same as for the number of islands. The migration rate was set to 2, such that the best solution and a random chosen solution of each island are migrated to the next island in the topology. Two randomly chosen solutions of an island are substituted by the incoming immigrants. Two additional tests were made changing the amount of bees per island. One test uses 250 bees per island (MH-ABC$_1$ and MH-ES-ABC$_1$) and another uses 1,000 bees per island (MH-ABC$_2$ and MH-ES-ABC$_2$). The reason for these two approaches is that the ABC and ES-ABC models uses 1,000 bees and, if we divide this population into four hives, we have 250 bees per hive. The subscribed index 1 and 2 only indicates different population sizes.

At the upper level, the HH-ABC model uses the same configurations of the MH-ABC model, that is, four islands with a unidirectional ring topology. Conversely, at the lower level, it uses the MS-ABC configurations, that is, one master and 10 slaves. In the same way as before, two tests were done, using 250 and 1,000 bees per hive (corresponding to HH-ABC$_1$ and HH-ABC$_2$, respectively). Hence, each slave has to process 25 bees and 100 bees each one for HH-ABC$_1$ and HH-ABC$_2$, respectively.

Table 2 shows the approximated computational effort, measured as the number of function evaluations for one iteration. This approximation takes into account the function evaluations performed in one iteration carried out by employed and onlooker bees. It should be kept in mind that the number of evaluations corresponding to the scout bees is modulated by the parameter *limit*.

Table 2. Approximated function evaluations number of each model for one iteration.

Model	Evaluations
ABC	1,000
MH-ABC$_1$	1,000
MH-ABC$_2$	4,000
ES-ABC	25,500
MS-ABC	25,500
MH-ES-ABC$_1$	25,500
MH-ES-ABC$_2$	102,000
HH-ABC$_1$	25,500
HH-ABC$_2$	102,000

6 Results and Analysis

The statistical results of 100 runs obtained by ABC, MH-ABC$_{1,2}$, ES-ABC, MS-ABC, MH-ES-ABC$_{1,2}$ and HH-ABC$_{1,2}$ are shown in Table 3. Each column shows the average of the best values obtained in each run and the mean of the elapsed time in minutes by each model for each function, followed by the respective standard deviations.

Both multi-hive versions without local search (MH-ABC$_1$ and MH-ABC$_2$) achieved a better performance that the standard ABC, for functions f_1, f_2, and f_3. Possibly, this is due to the coevolution effect in an interconnected multi-population approach. Immigrants inject diversity in the population, thus allowing a more effective search. The effect of coevolution is illustrated in Fig. 7 for function f_1. The figure shows the fitness value of the best individual at each

Table 3. Statistical results obtained by all models using three test functions.

Model	f_1 Quality	f_1 Time (min)	f_2 Quality	f_2 Time (min)	f_3 Quality	f_3 Time (min)
ABC	13.7 ± 2.5	4.0 ± 0.1	$8.4 \times 10^{-7} \pm 0.0$	4.5 ± 0.2	231.7 ± 1.8	26.4 ± 1.9
MH-ABC$_1$	12.7 ± 2.4	1.0 ± 0.0	$5.1 \times 10^{-7} \pm 0.0$	1.1 ± 0.0	230.1 ± 2.7	6.7 ± 0.0
MH-ABC$_2$	10.5 ± 1.6	4.1 ± 0.2	$6.8 \times 10^{-8} \pm 0.0$	4.4 ± 0.2	230.4 ± 2.8	27.0 ± 0.2
ES-ABC	2.0 ± 1.1	129.7 ± 1.2	$4.6 \times 10^{-15} \pm 0.0$	144.8 ± 1.2	206.5 ± 0.8	654.3 ± 0.3
MS-ABC	3.9 ± 2.3	59.6 ± 6.5	$4.7 \times 10^{-15} \pm 0.0$	62.6 ± 7.5	208.9 ± 1.4	296.1 ± 45.1
MH-ES-ABC$_1$	8.1 ± 2.5	33.6 ± 1.6	$4.3 \times 10^{-4} \pm 0.0$	34.8 ± 0.6	209.3 ± 1.1	94.2 ± 5.5
MH-ES-ABC$_2$	2.4 ± 0.5	130.3 ± 1.4	$4.2 \times 10^{-15} \pm 0.0$	145.5 ± 1.0	207.4 ± 1.0	668.4 ± 0.7
HH-ABC$_1$	9.8 ± 1.9	28.0 ± 0.5	$5.3 \times 10^{-4} \pm 0.0$	28.1 ± 0.6	208.3 ± 1.0	87.3 ± 0.0
HH-ABC$_2$	2.0 ± 1.6	86.3 ± 0.9	$3.3 \times 10^{-16} \pm 0.0$	87.5 ± 0.9	206.9 ± 1.9	316.8 ± 0.6

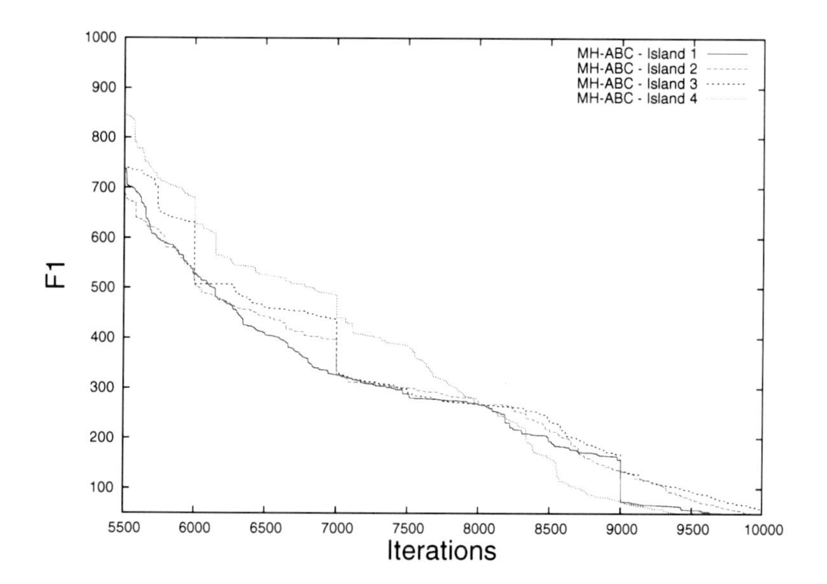

Fig. 7. Coevolution for function f_1.

population, between iterations 5,000 to 10,000. The migration procedure that occurs at each 1,000 iterations can be viewed at iterations 6,000; 7,000; 8,000; and 9,000. When an individual of good quality arrives to a population, it not only improves the local best solution, but also, through recombination, induces a further improvement of the quality. The difference between MH-ABC$_1$ and MH-ABC$_2$ is the size of their population: the former has 250 bees per hive, and the latter, 1,000 bees per hive. Results shown in Table 3 indicates that the higher the population in each hive, the better the results. However the computational cost increases linearly with the size of the population. In particular, the results obtained by MH-ABC$_2$, that has the same computational effort needed by ABC, have shown to be better than both ABC and MH-ABC$_1$, concerning quality, with time consumption similar to the ABC.

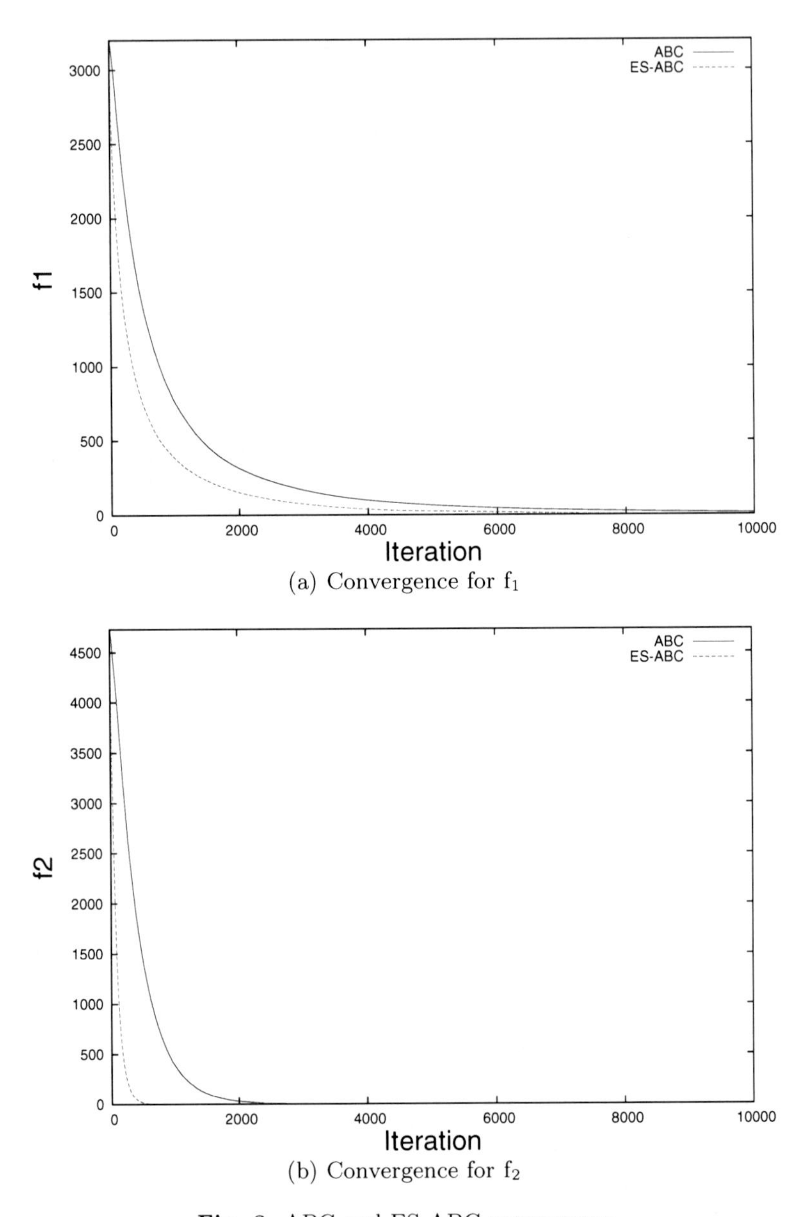

(a) Convergence for f_1

(b) Convergence for f_2

Fig. 8. ABC and ES-ABC convergence.

It is possible to observe in Table 3 that the quality obtained by the ES-ABC is better than the quality obtained by the standard ABC – recall that we are dealing with a minimization problem. The same holds for the multi-hive versions of the standard ABC. On the other hand, the increased number of function evaluations in the ES-ABC leads to a much higher computational cost, that is, the processing time, thus justifying the need for parallel computing. Figure 8 shows the convergence of ABC and ES-ABC for both functions, f_1 and f_2. We observed that the convergence of f_3 is similar to function f_1. Each point in the figure represents the averaged fitness value for each iteration over 100 runs. In these plots we observe that ES-ABC converges faster than ABC, thanks to its enhanced local search. This fact suggests that the local search is beneficial to the algorithm for optimizing the test functions.

Comparing the MS-ABC and the ES-ABC, it is noticed that the quality of solutions was almost the same (differences are not statistically significant) in both approaches for the three functions. However, the master-slave model achieved a speed up of 2.17 for function f_1, 2.20 for function f_2, and 2.2 for function f_3. Recall that a speed up higher than one suggests that the parallelization of the algorithm decreases the overall computational cost. It is important to notice that the speed up can be significantly different when optimizing other functions or using other running parameters. The computed efficiency for the master-slave model is 0.21 for function f_1, 0.22 for function f_2, and 0.22 for function f_3. These values suggest that the processors are not being fully used all the time. In fact, speed up and efficiency are a direct consequence of the balance between the processing load of the slaves and the communication load between master and slaves.

Concerning the elapsed time between MH-ES-ABC$_1$ and ES-ABC, the former run 4 times faster than the latter. It should be noticed that the quality was not as good as the quality obtained by the comparison between MH-ABC$_1$ and ABC. This can be explained by the intensive local search with less bees, possible leading to a premature convergence to local minima. When using the same computational effort, as in MH-ES-ABC$_2$, the quality obtained was better than that obtained by the ES-ABC and both with similar processing times in both functions.

Comparing the hierarchical approaches each other (HH-ABC$_1$ and HH-ABC$_2$), again, using a larger population yields significantly improvement in the quality of solutions, but at the expense of a proportional increment of processing time. On the other hand, comparing the hierarchical approaches with the multi-hive ones, the former ones always need less computational effort than the latter ones, for similar (or slightly better) quality of solutions. This fact suggests that the hierarchical approaches can lead to a good trade-off between quality of solution and processing time.

To synthesize results of all experiments in a single plot, so as to be possible to observe all trade-offs between computational cost and quality of solutions, Figure 9 shows a graphical distribution for f_1 concerning all tested models (for f_2 and f_3 the distribution are similar). The analysis of this Pareto-like plot indicates that the best approach, considering at the same time, the minimization of quality (this is a minimization problem) and minimization of processing time, is the point closest to the origin, that is, MH-ES-ABC$_1$, closely followed by HH-ABC$_1$.

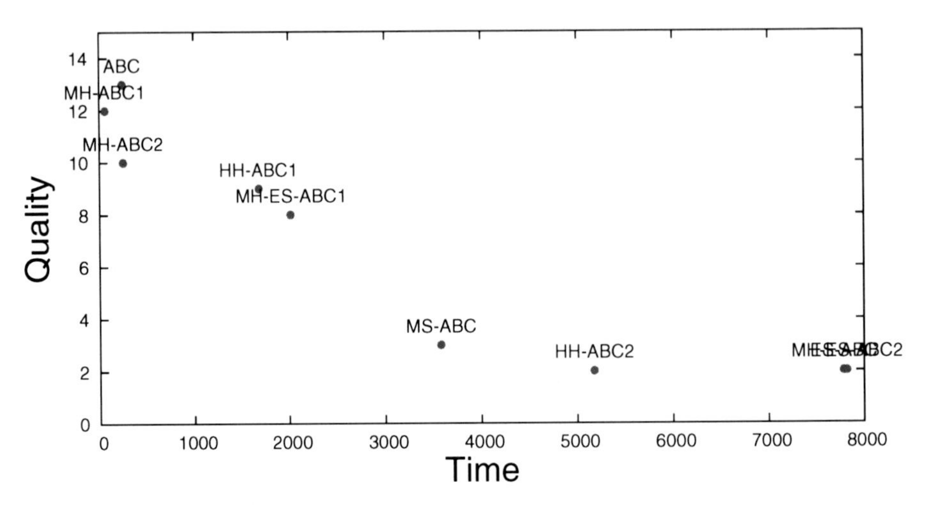

Fig. 9. Distribution concerning Quality x Time for f_1.

7 Conclusions and Future Work

This work analyzed the application of parallel computation to the ABC algorithm [12]. The case study was done using three mathematical functions. They were purposely set with a high number of variables in order to turn the problem solving process challenging to all tested versions of the algorithm. Therefore, one can evaluate the benefits of the parallelized approaches with respect to the regular sequential version. Besides a sequential model enhanced with local search (ES-ABC), three parallel models were proposed: a master-slave approach that divides the processing load into several processors; a multi-hive approach that promotes periodic migrations between independent subpopulations; and a hierarchical approach that hybridizes the two former models.

Experiments showed that intensive local search improves the quality of solutions. However the increment of computational cost may require the use of parallel processing. Future work will focus on more efficient local search strategies, so as to decrease the number of fitness evaluations. Also, more research will be done towards devising strategies to avoid stagnation in local minima.

The number of bees in each hive have a significant influence in the results. According to our experiments, a larger number of bees leads to better results than when using a smaller populations. Again, the increase of computational cost for large populations suggests the use of parallel models.

The effect of coevolution in the population could be clearly observed in the multi-hive models. As a consequence, better results with less computational effort could be achieved. Future work will investigate other topologies for connecting hives, as well a deep study of the migration policy between hives.

Since the experiments were performed using a cluster computing environment (the parallelization cannot be simulated in a single-core computer), the communication overhead (due to the communication between processes) takes significant influence upon the computational time, as discussed in Section 6. To overcome such drawback, a future trend of research is the use of another hardware platform, such as those based on General-Purpose Graphics Processing Units (GPGPU)[5] that are devices with high-performance multi-core processors capable of very high computation and data throughput.

Future work will also investigate parallel versions of other Swarm Intelligence methods, such as ACO and PSO, so as to compare their results with the parallel ABC, for the same optimization problems dealt here.

Overall, the conclusions obtained in the experiments can be generalized with care for other problems and control parameters. However, they firmly indicate the benefits of parallelization and the proposed models. Finally, we believe that the directions pointed out in this study can be useful not only for the ABC algorithm and its variants, but also for other swarm intelligence paradigms.

Acknowledgements. The authors would like to thank the reviewers for the fruitful comments. This work is supported by UDESC (Santa Catarina State University) and FUMDES program, as well as by the Brazilian National Research Council (CNPq) under research grant no. 309262/2007-0.

References

1. Akay, B., Karaboga, D.: Parameter tuning for the artificial bee colony algorithm. In: 1st International Conference on Computational Collective Intelligence - Semantic Web, Social Networks & Multiagent Systems (October 2009)
2. Baig, A.R., Rashid, M.: Honey bee foraging algorithm for multimodal & dynamic optimization problems. In: GECCO 2007: Proceedings of the 9th Annual Conference on Genetic and Evolutionary Computation, p. 169 (2007)
3. Baykasoğlu, A., Ozbakir, L., Tapkan, P.: Artificial bee colony algorithm and its application to generalized assignment problem. In: Chan, F.T.S., Tiwari, M.K. (eds.) Swarm Intelligence: Focus on Ant and Particle Swarm Optimization, December 2007, pp. 532–564. Itech Education and Publishing (2007)
4. Bonabeau, E., Dorigo, M., Theraulaz, G.: Swarm intelligence: from natural to artificial systems. Oxford University Press, Oxford (1999)
5. Bullnheimer, B., Kotsis, G., Strauss, C.: Parallelization Strategies for the Ant System. In: High Performance Algorithms and Software in Nonlinear Optimization, pp. 87–100. Kluwer, Dordrecht (1998)
6. Cantú-Paz, E.: A survey of parallel genetic algorithms. Calculateurs Paralleles, Reseaux Et Systems Repartis 10 (1998)
7. Chidambaram, C., Lopes, H.S.: A new approach for template matching in digital images using an artificial bee colony algorithm. In: World Congress on Nature and Biologically Inspired Computing (NaBIC 2009) (2009)
8. Clerc, M.: Particle Swarm Optimization. ISTE Press (2006)
9. Dorigo, M., Stützle, T.: Ant Colony Optimization. MIT Press, Cambridge (2004)

[5] GPGPU: `http://gpgpu.org/`

10. Drias, H., Sadeg, S., Yahi, S.: Cooperative bees swarm for solving the maximum weighted satisfiability problem. In: IWAAN International Work Conference on Artificial and Natural Neural Networks, pp. 318–325 (2005)
11. Haddad, O.B., Afshar, A.: Mbo algorithm, a new heuristic approach in hydrosystems design and operation. In: 1st International Conference on Managing Rivers in the 21st Century, pp. 499–504 (2004)
12. Karaboga, D.: An idea based on honey bee swarm for numerical optimization. Technical report, Erciyes University, Engineering Faculty, Computer Engineering Department (2005)
13. Karaboga, D., Akay, B.: Artificial bee colony (abc), harmony search and bees algorithms on numerical optimization. In: IPROMS 2009 Innovative Production Machines and Systems Virtual Conference (2009)
14. Karaboga, D., Akay, B.: A comparative study of artificial bee colony algorithm. Applied Mathematics and Computation 214, 108–132 (2009)
15. Karaboga, D., Ozturk, C.: Neural networks training by artificial bee colony algorithm on pattern classification. Neural Network World 19(3), 279–292 (2009)
16. Kennedy, J., Eberhart, R.C.: Swarm Intelligence. Morgan Kaufmann, San Francisco (2001)
17. Nakrani, S., Tovey, C.: On honey bees and dynamic allocation in an internet server colony. In: Proceedings of 2nd International Workshop on the Mathematics and Algorithms of Social Insects (2003)
18. Pawar, P.J., Rao, R.V., Shankar, R.: Multi-objective optimization of electrochemical machining process parameters using artificial bee colony (abc) algorithm. In: Advances in Mechanical Engineering (AME 2008) (December 2008)
19. Pham, D.T., Ghanbarzadeh, A., Koc, E., Otri, S., Rahim, S., Zaidi, M.: The bees algorithm - a novel tool for complex optimisation problems. In: Proceedings of IPROMS, pp. 454–461 (2006)
20. Reinhard, J., Srinivasan, S.: The Role of Scents in Honey Bee Foraging and Recruitment. In: Food Exploitation by Social Insects: Ecological, Behavioral, and Theoretical Approaches, vol. 1, pp. 165–182. CRC Press, Boca Raton (2009)
21. Sato, T., Hagiwara, M.: Bee system: Finding solution by a concentrated search. In: Proceedings of the IEEE International Conference on Systems, Man, and Cybernetics, vol. 4(C), pp. 3954–3959 (1997)
22. Schutte, J.F., Reinbolt, J.A., Fregly, B.J., Haftka, R.T., George, A.D.: Parallel global optimization with the particle swarm algorithm. Journal of Numerical Methods in Engineering 61, 2296–2315 (2003)
23. Seeley, T.: The Wisdom of the Hive. Harvard University Press (1995)
24. Srinivasa, R.R., Narasimham, S.V.L., Ramalingaraju, M.: Optimization of distribution network configuration for loss reduction using artificial bee colony algorithm. International Journal of Electrical Power and Energy Systems Engineering (IJEPESE) 1(2) (2008)
25. Stützle, T.: Parallelization strategies for ant colony optimization. In: Proceedings of PPSN-V, Fifth International Conference on Parallel Problem Solving from Nature, pp. 722–731. Springer, Heidelberg (1998)
26. Tavares, L.G., Lopes, H.S., Erig Lima, C.R.: A study of topology in insular parallel genetic algorithms. In: World Congress on Nature and Biologically Inspired Computing (2009)
27. Teodorovic, D., Dell'Orco, M.: Bee colony optimization - a cooperative learning approach to complex transportation problems. In: Advanced OR and AI Methods in Transportation, pp. 51–60 (2005)

28. Venter, G., Sobieszczanski-Sobieski, J.: A parallel particle swarm optimization algorithm accelerated by asynchronous evaluations. In: 6th World Congresses of Structural and Multidisciplinary Optimization (June 2005)
29. Wedde, H.F., Farooq, M., Zhang, Y.: Beehive: An efficient fault-tolerant routing algorithm inspired by honey bee behavior. In: Dorigo, M. (ed.) Ant Colony Optimization and Swarm Intelligence, pp. 83–94. Springer, Berlin (2004)
30. Yang, X.-S.: Engineering optimizations via nature-inspired virtual bee algorithms. In: Mira, J., Álvarez, J.R. (eds.) IWINAC 2005. LNCS, vol. 3562, pp. 317–323. Springer, Heidelberg (2005)

Bumble Bees Mating Optimization Algorithm for the Vehicle Routing Problem

Yannis Marinakis[1] and Magdalene Marinaki[2]

[1] Technical University of Crete, Department of Production Engineering and Management, Decision Support Systems Laboratory, 73100 Chania, Greece
marinakis@ergasya.tuc.gr

[2] Technical University of Crete, Department of Production Engineering and Management, Industrial Systems Control Laboratory, 73100 Chania, Greece
magda@dssl.tuc.gr

Abstract. Recently, a number of swarm intelligence algorithms based on the behaviour of the bees have been presented. These algorithms are divided, mainly, in two categories according to the bees' behaviour in the nature, the foraging behaviour and the mating behaviour. The most important approaches that simulate the foraging behaviour of the bees are the Artificial Bee Colony algorithm, the Virtual Bee algorithm, the Bee Colony Optimization algorithm, the BeeHive algorithm, the Bee Swarm Optimization algorithm and the Bees algorithm. Contrary to the fact that there are many algorithms that are based on the foraging behaviour of the bees, the main algorithm proposed based on the mating behaviour is the Honey Bees Mating Optimization algorithm. This chapter introduces a new algorithmic nature inspired approach based on Bumble Bees Mating Optimization for successfully solving the Vehicle Routing Problem. Bumble Bees Mating Optimization algorithm is a new population-based swarm intelligence algorithm that simulates the mating behaviour that a swarm of bumble bees perform. Two sets of benchmark instances are used in order to test the proposed algorithm with very satisfactory results.

Keywords: Bumble Bees Mating Optimization, Vehicle Routing Problem, Swarm Intelligence.

1 Introduction

In the last years, several biological and natural processes have been influencing the methodologies in science and technology in an increasing manner. Nature inspired intelligence becomes increasingly popular and a significant number of methods driven by concepts from nature/biology have been developed. When the task is optimization within complex domains of data or information, the methods that represent successful animal and micro-organism team behaviour are the most popular nature inspired approaches. Such nature inspired approaches

B.K. Panigrahi, Y. Shi, and M.-H. Lim (Eds.): Handbook of Swarm Intelligence, ALO 8, pp. 347–369.
springerlink.com

are the Particle Swarm Optimization ([43]) (inspired by birds flocks or fish schools), the artificial immune systems (that mimic the biological one ([18])), or the ant colonies (ants foraging behaviors gave rise to Ant Colony Optimization ([20])), etc.

Recently, a number of swarm intelligence algorithms, based on the behaviour of the bees have been presented ([7, 42]). These algorithms are divided, mainly, in two categories according to their behaviour in the nature, the foraging behaviour and the mating behaviour. The most important approaches that is inspired of the foraging behaviour of the bees are the Artificial Bee Colony (ABC) Algorithm proposed by ([41]), the Virtual Bee Algorithm proposed by ([90]), the Bee Colony Optimization Algorithm proposed by ([83]), the BeeHive algorithm proposed by ([88]), the Bee Swarm Optimization Algorithm proposed by ([21]) and the Bees Algorithm proposed by ([68]). The Artificial Bee Colony algorithm ([40, 41]) is, mainly, applied in continuous optimization problems and simulates the waggled dance behaviour that a swarm of bees perform during the foraging process of the bees. In this algorithm, there are three groups of bees, the employed bees (bees that determine the food source (possible solutions) from a prespecified set of food sources and share this information (waggle dance) with the other bees in the hive), the onlookers bees (bees that based on the information that they take from the employed bees they search for a better food source in the neighborhood of the memorized food sources) and the scout bees (employed bees that their food source has been abandoned and they search for a new food source randomly). The Virtual Bee Algorithm ([90]) is, also, applied in continuous optimization problems. In this algorithm, the population of bees is associated with a memory, a food source, and then all the memories communicate between them with a waggle dance procedure. The whole procedure is similar with a genetic algorithm and it has been applied on two function optimization problems with two parameters. In the BeeHive ([88]) algorithm, a protocol inspired from dance language and foraging behaviour of honey bees is used. In the Bees Swarm Optimization ([21]), initially a bee finds an initial solution (food source) and from this solution the other solutions are produced with certain strategies. Then, every bee is assigned in a solution and when they accomplish their search, the bees communicate between them with a waggle dance strategy and the best solution will become the new reference solution. To avoid cycling the authors use a tabu list. In the Bees Algorithm ([68]), a population of initial solutions (food sources) are randomly generated. Then, the bees are assigned to the solutions based on their fitness function. The bees return to the hive and based on their food sources, a number of bees are assigned to the same food source in order to find a better neighborhood solution. In the Bee Colony Optimization ([83]) algorithm a step by step solution is produced by each forager bee and when the foragers return to the hive a waggle dance is performed by each forager. The other bees follow the foragers based on the fitness function of the best forager bee. This algorithm looks like the Ant Colony Optimization ([20]) algorithm but it does not use at all the concept of pheromone trails.

The main algorithm proposed based on the marriage behaviour is the Honey Bees Mating Optimization Algorithm (HBMO), that was presented ([1, 2]). Since then, it has been used on a number of different applications ([3, 23, 37, 50, 55, 57, 58, 51, 56, 82]). The Honey Bees Mating Optimization algorithm simulates the mating process of the queen of the hive. The mating process of the queen begins when the queen flights away from the nest performing the mating flight during which the drones follow the queen and mate with her in the air ([1, 3]). The algorithm is a swarm intelligence algorithm since it uses a swarm of bees where there are three kinds of bees, the queen, the drones and the workers. There is a number of procedures that can be applied inside the swarm. In the honey bees mating optimization algorithm, the procedure of mating of the queen with the drones is described. First, the queen is flying randomly in the air and, based on her speed and her energy, if she meets a drone then there is a possibility to mate with him. Even if the queen mates with the drone, she does not create directly a brood but stores the genotype (with the term "genotype" we mean some of the basic characteristics of the drones, i.e. part of the solution) of the drone in her spermatheca and the brood is created only when the mating flight has been completed. A crossover operator is used in order to create the broods. In a hive the role of the workers is simply the brood care (i.e. to feed them with the "royal jelly") and, thus, they are only a local search phase in the honey bees mating optimization algorithm. And thus, this algorithm combines both the mating process of the queen and one part of the foraging behavior of the honey bees inside the hive. If a broods is better (fittest) than the queen, then this brood replaces the queen.

In this chapter, a new algorithm that simulates the mating behavior of the Bumble bees, the **Bumble Bees Mating Optimization (BBMO)**, is presented. This algorithm is a population-based swarm intelligence algorithm that simulates the mating behavior that a swarm of bumble bees perform. The proposed algorithm was applied for the solution of the Vehicle Routing Problem (VRP) with remarkable results since in most of the instances used, the algorithm found the best known solution. A hybridized initial version of the algorithm was presented in ([59]) for clustering. The rest of the chapter is organized as follows. In the next section a description of the vehicle routing problem is presented. In section 3 the proposed algorithm, the Bumble Bees Mating Optimization for the Vehicle Routing Problem (BBMOVRP), is presented and analyzed in detail. Computational results are presented and analyzed in section 4 while in the last section conclusions and future research are given.

2 The Vehicle Routing Problem

The **Vehicle Routing Problem (VRP)** or the **capacitated vehicle routing problem (CVRP)** is often described as the problem in which vehicles based on a central depot are required to visit geographically dispersed customers in order to fulfill known customer demands. Let $G = (V, E)$ be a graph where $V = \{i_1, i_2, \cdots i_n\}$ is the vertex set (i_1 refers to the depot and the customers are indexed $i_2, \cdots, i_n$) and $E = \{(i_l, i_m) : i_l, i_m \in V\}$ is the edge set. Each customer must be assigned to exactly one of the k vehicles and the total size of deliveries for customers assigned to each vehicle must not exceed the vehicle capacity (Q_k). If the vehicles are homogeneous, the capacity for all vehicles is equal and denoted by Q. A demand q_l and a service time st_l are associated with each customer node i_l. The demand q_1 and the service time st_1 which are refered to the demand and service time of the depot are set equal to zero. The travel cost and the travel time between customers i_l and i_m is c_{lm} and tt^k_{lm}, respectively, and T_k is the maximum time allowed for a route of vehicle k. The problem is to construct a low cost, feasible set of routes - one for each vehicle. A route is a sequence of locations that a vehicle must visit along with the indication of the service it provides, where the variable x^k_{lm} is equal to 1 if the arc (i_l, i_m) is traversed by vehicle k and 0 otherwise. The vehicle must start and finish its tour at the depot. In the following we present the mathematical formulation of the VRP ([10]):

$$J = \min \sum_{l=1}^{n}\sum_{m=1}^{n}\sum_{k=1}^{K} c_{lm} x^k_{lm} \tag{1}$$

s.t.

$$\sum_{i_l=1}^{n}\sum_{k=1}^{K} x^k_{lm} = 1, \quad i_m = 2, \cdots, n \tag{2}$$

$$\sum_{i_m=1}^{n}\sum_{k=1}^{K} x^k_{lm} = 1, \quad i_l = 2, \cdots, n \tag{3}$$

$$\sum_{i_l=1}^{n} x^k_{lf} - \sum_{i_m=1}^{n} x^k_{fm} = 0, k = 1, \cdots, K \quad i_f = 1, \cdots, n \tag{4}$$

$$\sum_{i_l=1}^{n} q_l \sum_{i_m=1}^{n} x^k_{lm} \leq \quad Q_k, k = 1, \cdots, K \tag{5}$$

$$\sum_{i_l=1}^{n} st^k_l \sum_{i_m=1}^{n} x^k_{lm} + \quad \sum_{i_l=1}^{n}\sum_{i_m=1}^{n} tt^k_{lm} x^k_{lm} \leq T_k, \quad k = 1, \cdots, K \tag{6}$$

$$\sum_{i_m=2}^{n} x^k_{1m} \leq 1, \qquad k = 1, \cdots, K \tag{7}$$

$$\sum_{i_l=2}^{n} x^k_{l1} \leq 1, \qquad k = 1, \cdots, K \tag{8}$$

$$X \in S \tag{9}$$

$$x^k_{lm} = 0 \text{ or } 1, \qquad \text{for all } i_l, i_m, k \tag{10}$$

Objective function (1) states that the total distance is to be minimized. Equations (2) and (3) ensure that each demand node is served by exactly one vehicle. Route continuity is represented by (4), i.e. if a vehicle enters in a demand node, it must exit from that node. Equations (5) are the vehicle capacity constraints and (6) are the total elapsed route time constraints. Equations (7) and (8) guarantee that vehicle availability is not exceeded.

The vehicle routing problem was first introduced by Dantzig and Ramser (1959) ([17]). As it is an NP-hard problem, a large number of approximation techniques were proposed. These techniques are classified into two main categories: Classical heuristics that were developed mostly between 1960 and 1990 ([4, 9, 10, 12, 13, 19, 25, 26, 31, 48, 49, 65, 87]) and metaheuristics that were developed in the last fifteen years. Metaheuristic algorithms are classified in categories based on the used strategy. Tabu Search strategy is the most widely used technique for this problem and a number of researchers have proposed very efficient variants of the standard Tabu Search algorithm ([6, 15, 28, 66, 74, 75, 77, 85, 89]). Very interesting and efficient algorithms based on the concept of Adaptive Memory, according to which a set of high quality VRP solutions (elite solutions) is stored and, then, replaced from better solutions through the solution process, have been proposed ([76, 78, 79]). Simulated annealing ([66]), record to record travel ([34, 46]), greedy randomized adaptive search procedure ([71]) and threshold accepting algorithms ([80, 81]) are, also, applied efficiently in the VRP. In the last ten years a number of nature inspired metaheuristic algorithms have been applied for the solution of the Vehicle Routing Problem. The most commonly used nature inspired methods for the solution of this problem are genetic algorithms ([5, 8, 62, 70]), ant colony optimization ([11, 72, 73]), honey bees mating optimization ([55, 56]), bee colony optimization ([36]), particle swarm optimization algorithm ([52, 53]) and other evolutionary techniques ([16, 63, 64]). The reader can find more detailed descriptions of these algorithms in the survey papers ([9, 10, 24, 29, 30, 45, 44, 54, 78]) and in the books ([32, 33, 67, 84]).

3 Bumble Bees Mating Optimization Algorithm for the Vehicle Routing Problem

3.1 Bumble Bees Behavior

Bumble bees are social insects that form colonies consisting of the queen, many workers (females) and the drones (males). Queens are the only members of the nest to survive from one season to the next, as they spend the winter months hibernating in a protected underground overwintering chamber. Upon emerging from hibernation, a queen collects pollen and nectar from flowers and searches for a suitable nest site and when she finds such a place, she prepares wax pots to store food and wax cells into which eggs are laid ([35, 91, 92, 93, 94]).

The bumble bee queen can lay fertilized or unfertilized eggs. The fertilized eggs have chromosomes from the queen and a male or males she mated with the previous year and they develop into workers while the unfertilized eggs contain chromosomes from the queen alone and they develop into males. After the emergence of the first workers, the queen no longer forages as the workers take over the responsibilities of collecting food (foragers) and the queen remains in the nest laying eggs and tending to her young. Some workers, also, remain in the nest and help raise the brood (household workers). Males do not contribute in collecting food or helping rear young as the sole purpose of the males is to mate with the queens. Bumble bee workers are able to lay haploid eggs when the queen's ability to suppress the workers' reproduction diminishes. These eggs are developed into viable male bumble bees ([35, 91, 92, 93, 94]).

A few days after the males leave the nest, new queens will emerge. After new queens and males have gone, the colony begins to deteriorate. The founder queen stops laying eggs and grows weak from old age while the remaining workers continue to forage for food but only for themselves. Away from the colony, the new queens and males live off nectar and pollen and spend the night on flowers or in holes. The queens are eventually mated (often more than once), the sperm from the mating is stored in spermatheca and she searches for a suitable location for diapause. Three different mating behaviors exist in bumble bees. The first mating behavior is where a male perches on a tall structure and waits for queens to fly by and he will pursue them for mating once one queen is spotted. The second mating behavior is when males create a scent trail, marking their flight path with pheromones and, thus, queens of the same species will be attracted to the pheromones and follow the scent trail. The third mating behavior is where males wait at the entrance of a bumble bee nest for queens to leave ([35, 91, 92, 93, 94]).

3.2 General Description of the Algorithm

In the BBMOVRP algorithm, there are three kind of bumble bees in the colony, the queen bee, the worker bees and the drones (males). Initially, a number of bees are selected randomly. Each bee (a bee corresponds to an individual in the population) represents a candidate solution of the problem. Let n be the total number of variables. The bees are represented by vectors of dimension n. We use the path representation of the tour and a relative position indexing transformation (see section 3.3). Afterwards, the fitness of each bee is calculated (see section 3.4) and the best bee is selected as the queen. All the other bees in the initialization phase of the algorithm are the drones (males). The queen selects the drones that are used for mating by using the second mating behavior where it is assumed in the algorithm that the fittest drones let larger amount of pheromone in their flight paths and, thus, the queen selects the most promising paths. This procedure is realized by sorting of all drones based on their fitness function. Each time the queen successfully mates with a drone, the genotype of the drone is stored in her spermatheca until the maximum number of matings has been reached.

After the mating, the queen finds a place to hibernate and in the next year (a year corresponds to an external iteration) finds a place to create the hive and to begin to lay eggs. There are three kinds of bees that a queen lays: new queens, workers and drones. The first two kinds of bees are created by crossover of the genotype of the queen and the genotype of the drones using a specific crossover operator (see section 3.5).

The fittest of the broods are the candidate for becoming new queens while the rest are the workers. Initially, the new queens are fed from the queen and, afterwards, from the workers and the queen. The reason that we use this procedure is to improve the genotype (solution) of each new queen. This is achieved by using a local search phase where each new queen selects from the workers and the queen who is going to feed her by using the following equation:

$$nq_{ij} = nq_{ij} + (b_{max} - \frac{(b_{max} - b_{min}) * lsi}{lsi_{max}}) * (nq_{ij} - q_j) + \frac{1}{M} * \sum_{k=1}^{M} (b_{min} - \frac{(b_{min} - b_{max}) * lsi}{lsi_{max}}) * (nq_{ij} - w_{kj}) \quad (11)$$

where nq_{ij} is the solution of the new queen i, q_j is the solution of the old queen, w_{kj} is the solution of the worker, M is the number of the workers that each queen selects for feeding her and it is different for each queen, b_{max}, b_{min} are two parameters with values in the interval $(0, 1)$ that control if the new queen is fed from the old queen, from the workers or from both of them, lsi is the current local search iteration and lsi_{max} is the maximum number of local search iterations in each external iteration. Initially, the new queens are fed more from the old queen and as the local search iterations increase, then only the workers feed the new queen. The appropriate choice of the values of b_{max} and b_{min} controls the feeding process, i.e. in order to have the feeding process described previously, a large value for b_{max} and a value almost equal to zero for b_{min} are necessary. Afterwards, the new queens leave from the hive.

In the initially proposed Bumble Bees Mating Optimization algorithm [59] the drones are produced by mutate the queen's genotype or by mutate the fittest workers' genotype using a random mutation operator. In this mutation operator, the changes in the genotype of the queen or the workers are performed randomly. As this procedure is not suitable for the Vehicle Routing Problem we use instead of a random mutation operator a procedure called Expanding Neighborhood Search Strategy ([60]) (see section 3.6). The number of the drones per colony in each external generation is given by the equation:

$$number\ of\ drones\ per\ colony = \frac{number\ of\ drones}{number\ of\ queens} \quad (12)$$

The drones, then, leave from the hive and they are looking for new queens for mating. As the drones leave from the hive they are moving in a swarm in order to find the best places to wait for the new queens to find them by their marked flight paths. The movement of the drones away from the hive is calculated from the following equation:

$$d_{ij} = d_{ij} + \alpha * (d_{kj} - d_{lj}) \quad (13)$$

where d_{ij}, d_{kj} and d_{lj} are the solutions of the drones i, k, l respectively and α is a parameter that determines the percentage that the drone i is affected by the two other drones k and l. In the next external iteration, the best fertilized queens survive and all the other members of the population die. A pseudocode of the proposed algorithm is presented in Table 1.

Table 1. Bumble Bees Mating Optimization Algorithm

```
Algorithm Bumble Bees Mating Optimization Algorithm
Definition of parameters for the main phase of the algorithm
    Definition of the maximum number of external iterations
    Definition of the maximum number of matings
    Definition of the maximum number of queens
Initialization Phase
    Generate the initial population of the bumble bees
    Calculation of the fitness function of each bumble bee
    Selection of the bee with the best fitness function as the queen
    Selection of the rest bees as the drones
    Sorting the drones according to their fitness' functions
    Selection of the drones for mating by the queen
    Storing the drones' genotype to queen's spermatheca
Main Phase
do while the maximum number of external iterations has not been reached
    Creation of the broods by using a crossover operator
    Calculation of the fitness function of each brood
    Sorting the broods according to their fitness' functions
    Selection of the best broods as the new queens
    Selection of the rest broods as the workers
    do while the maximum number of local search iterations
            has not been reached
        Feeding of the new queens by the old queens and the workers
    enddo
    Creation of a percentage of the drones by mutating of
        the old queens' genotypes using ENS
    Creation of the rest of the drones by mutating of
        the workers' genotypes using ENS
    Calculation of the fitness function of each drone
    Calculation of the moving direction of the drones away from the hive
    Sorting the drones according to their fitness' functions
    do while the maximum number of matings for each new queen
            has not been reached
        Selection of the drones for mating by each new queen
        Storing the drones' genotypes to each new queen's spermatheca
    enddo
    Survival of the new queens for the next iteration
    Dying of all the other members (workers and drones) of the population
enddo
return The best queen (best solution found)
```

It should be noted that the Bumble Bees Mating Optimization (BBMO) algorithm, that is inspired from the mating behavior of the bumble bees, has a number of differences compared to another nature inspired algorithm, that is based on the mating behavior of honey bees, the Honey Bees Mating Optimization (HBMO) algorithm ([1, 57]). More precisely:

- In the BBMO the workers are different solutions while in the HBMO they are local search phases. This helps the exploration abilities of the population by searching in different places in the solution space.
- In the BBMO after the mating of the queen three kinds of bumble bees are produced, the new queens and the workers (by using a crossover operator) and the drones (by using a mutation operator). On the other hand, in the HBMO after the mating of the queen two kinds of honey bees are produced, the queen and the drones (both of them by using a crossover operator). By using in the proposed algorithm a mutation operator to produce new solutions we have the possibility to obtain completely different solutions.
- In the BBMO the fittest of the broods produced by the crossover operator are the new queens and all the others are the workers while in the HBMO the fittest of the broods is the new queen and all the others are the drones.
- In the BBMO the drones are produced by mutation of the queen or by mutation of the fittest workers. In the HBMO the drones are all the bees produced by the crossover operator except of the queen. By using in the proposed algorithm a mutation operator to produce new solutions we have the possibility to obtain completely different solutions.
- In the BBMO the drones are moving away of the hive and this affects their solutions.
- The feeding procedure in the BBMO is as described previously using the Equation (11) while in the HBMO the feeding procedure is local search phases that are applied independently in each brood.

3.3 Path Representation

All the solutions (bees) are represented with the path representation of the tour. For example if we have a bee with five nodes a possible path representation is the following:

1 3 5 2 4

As the calculation of the feeding equation of each new queen and the equation of the movement of the males away of the hive are performed by the equations (11) and (13) (see above), the above mentioned representation should be transformed appropriately. We transform each element of the solution into a floating point interval [0,1], calculate the velocities of all bees and then convert back into the integer domain using relative position indexing ([47]). Thus, initially we divide each element of the solution by the vector's largest element, and for the previous example the bee becomes:

0.2 0.6 1 0.4 0.8

In order to calculate the best new queens the elements of the vectors are transformed back into the integer domain by assigning the smallest floating value to the smallest integer, the next highest floating value to the next integer and so on. Thus, if the vector of a bee is

0.37 0.42 0.17 0.58 0.28

the backward transformation gives

3 4 1 5 2

3.4 Calculation of Fitness Function

In VRP, the fitness of each bee is related to the route length of each tour. For each bee i its fitness is calculated by the following equation

$$fitness_i = J_{max} - J_i + 1 \tag{14}$$

where the J_{max} is the objective function value of the bee in the population with the maximum cost and J_i (Eq. (1)) is the objective function value of the current bee. It should be noted that, since the probability of selecting a bee for mating is related to its fitness and since the bee with the worst cost has fitness equal to zero, it will never be selected for mating. Thus, the addition of 1 to the difference between J_{max} and J_i ensures that the worst solution is not totally excluded.

3.5 Crossover Operator

In this crossover operator, the points are selected randomly from the selected drones and from the queen. Thus, initially a crossover operator number is selected (Cr_1) that controls the fraction of the parameters that are selected for the drones and the queen. The Cr_1 value is compared with the output of a random number generator, $rand_i(0,1)$. If the random number is less or equal to the Cr_1 the corresponding value is inherited from the queen, otherwise it is selected, randomly, from the solutions of one of the drones' genotypes that are stored in spermatheca. Thus, if the solution of the brood i is denoted by $b_{ij}(t)$ (t is the external iteration number and j is the dimension of the problem ($j = 1, \cdots, n$)), the solution of the queen is denoted by $q_j(t)$ and the solution of the drone k is denoted by $d_{kj}(t)$:

$$b_{ij}(t) = \begin{cases} q_j(t), & \text{if } rand_i(0,1) \le Cr_1 \\ d_{kj}(t), & \text{otherwise.} \end{cases} \tag{15}$$

3.6 Expanding Neighborhood Search

The local search method that is used in this chapter is the Expanding Neighborhood Search ([60]). Expanding Neighborhood Search (ENS) is a metaheuristic algorithm ([55, 56, 60, 61, 62]) that can be used for the solution of a number of combinatorial optimization problems with remarkable results. The main features of this algorithm are:

- the use of the Circle Restricted Local Search Moves Strategy,
- the use of an expanding strategy.

- the ability of the algorithm to change between different local search strategies, and

These features are explained in detail in the following.

In the Circle Restricted Local Search Moves - CRLSM strategy, the computational time is decreased significantly compared to other heuristic and metaheuristic algorithms because all the edges that are not going to improve the solution are excluded from the search procedure. This happens by restricting the search space into circles around the candidate for deletion edges. It has been observed [56, 60, 61], for example, in the 2-opt local search algorithm that there is only one possibility for a trial move to reduce the cost of a solution, i.e. when at least one new (candidate for inclusion) edge has cost less than the cost of one of the two old edges (candidate for deletion edges) and the other edge has cost less than the sum of the costs of the two old edges. Thus, in the Circle Restricted Local Search Moves strategy, for all selected local search strategies, circles are created around the end nodes of the candidate for deletion edges and only the nodes that are inside these circles are used in the process of finding a better solution.

In Expanding Neighborhood Search strategy, the size of the neighborhood is **expanded** in each iteration. In order to decrease even more the computational time and because it is more possible to find a better solution near to the end-nodes of the candidate for deletion edge, we do not use from the begin the largest possible circle but the search for a better solution begins with a circle with a small radius. For example, in the 2-opt algorithm if the length of the candidate for deletion edge is equal to A, the initial circle has radius $A/2$, then, the local search strategies are applied as they are described in the following and if the solution can not be improved inside this circle, the circle is expanding by a percentage θ (θ is determined empirically) and the procedure continues until the circle reaches the maximum possible radius which is set equal to $A + B$, where B is the length of one of the other candidate for deletion edges.

The ENS algorithm has the ability to change between different local search strategies. The idea of using a larger neighborhood to escape from a local minimum to a better one, had been proposed initially by Garfinkel and Nemhauser [27] and recently by Hansen and Mladenovic [38]. Garfinkel and Nemhauser proposed a very simple way to use a larger neighborhood. In general, if with the use of one neighborhood a local optimum was found, then a larger neighborhood is used in an attempt to escape from the local optimum. Hansen and Mladenovic proposed a more systematical method to change between different neighborhoods, called Variable Neighborhood Search.

In the Expanding Neighborhood Search a number of local search strategies are applied inside the circle. The procedure works as follows: initially an edge of the current solution is selected (for example the edge with the worst length) and the first local search strategy is applied. If with this local search strategy a better solution is not achieved, another local search strategy is selected for the same edge. This procedure is continued until a better solution is found or all local search strategies have been used. In the first case the solution is updated, a new edge is selected and the new iteration of the Expanding Neighborhood Search strategy begins, while in the second case the circle is expanded and the local search strategies are applied in the new circle until a better solution is found or the circle reach the maximum possible radius. If the maximum possible radius have been reached, then a new candidate for deletion edge is selected.

The local search strategies for the Vehicle Routing Problem are distinguished between local search strategies for a single route and local search strategies for multiple routes. The local search strategies that are chosen and belong to the category of the single route interchange (strategies that try to improve the routing decisions) are the well known methods for the TSP, the 2-opt and the 3-opt ([10]). In the single route interchange all the routes have been created in the initial phase of the algorithm. The Local Search Strategies for Multiple Route Interchange try to improve the assignment decisions. This, of course, increases the complexity of the algorithms but gives the possibility to improve even more the solution. The multiple route interchange local search strategies that are used are the 1-0 relocate, 2-0 relocate, 1-1 exchange, 2-2 exchange and crossing ([84]).

4 Computational Results

The whole algorithmic approach was implemented in Fortran 90 and was compiled using the Lahey f95 compiler on a Centrino Mobile Intel Pentium M 750 at 1.86 GHz, running Suse Linux 9.1. The parameters of the proposed algorithm are selected after thorough testing. A number of different alternative values were tested and the ones selected are those that gave the best computational results concerning both the quality of the solution and the computational time needed to achieve this solution. The selected parameters are given in Table 2. After the selection of the final parameters, 50 different runs with the selected parameters were performed for each of the benchmark instances.

The quality is given in terms of the relative deviation from the best known solution, that is $\omega_{BBMOVRP} = \frac{(c_{BBMOVRP} - c_{BKS})}{c_{BKS}}\%$, where $c_{BBMOVRP}$ denotes the cost of the solution found by BBMOVRP and c_{BKS} is the cost of the best known solution.

The algorithm was tested on two sets of benchmark problems. The 14 benchmark problems proposed by Christofides ([12]) and the 20 large scale vehicle routing problems proposed by Golden ([34]). Each instance of the first set contains between 51 and 200 nodes including the depot. Each problem includes capacity constraints while the problems 6-10, 13 and 14 have, also, maximum route length restrictions and non zero service times. For the first ten problems, nodes are randomly located over a square, while for the remaining ones, nodes are distributed in clusters and the depot is not centered. The second set of instances contains between 200 and 483 nodes including the depot. Each problem instance includes capacity constraints while the first eight have, also, maximum route length restrictions but with zero service times. The efficiency of the BBMOVRP algorithm is measured by the quality of the produced solutions. The

Table 2. Parameter Values

Parameter	Value
Number of the total bees (workers - males - queens)	200
Number of external iterations	1000
Number of queens	10
Number of workers	90
Number of drones	100
lsi_{max}	100
b_{max}	0.99
b_{min}	0.01
α	0.8
θ	10%

quality is given in terms of the relative deviation from the best known solution, $\omega_{BBMOVRP}$.

In the first column of Tables 3 and 4 the number of nodes of each instance is presented, while in the second, third and fourth columns the most important characteristics of the instances, namely the maximum capacity of the vehicles (Cap. - column 2), the maximum tour length (m.t.l. - column 3) of each vehicle and the service time (s.t. - column 4) of each customer, are presented. In the last four columns, the results of the proposed algorithm (column 5), the best known solution (BKS - column 6), the quality of the solution of the proposed algorithm ($\omega_{BBMOVRP}$ - column 7) and the CPU time need to find the solution by the proposed algorithm for each instance (column 8) are presented, respectively. It can be seen from Table 3, that the algorithm, in eleven of the fourteen instances of the first set has reached the best known solution. For the other three instances the quality of the solutions is between 0.02% and 0.06% and the average quality for the fourteen instances is 0.009%. For the 20 large scale vehicle routing problems (Table 4) the algorithm has found the best known solution in three of them, for the rest the quality is between 0.07% and 0.58% and the average quality of the solutions is 0.26%. Also, in these Tables the computational time needed (in minutes) for finding the best solution by BBMOVRP is presented. The CPU time needed is significantly low for the first set of instances and only for two instances (instance 5 and 10) is somehow increased but still is very efficient. In the second set of instances, the problems are more complicated and, thus, the computational time is increased but is still less than 10 min in all instances. These results denote the efficiency of the proposed algorithm.

In Tables 5 and 6 comparisons of the proposed algorithm with four different implementations of the Honey Bees Mating Optimization Algorithm proposed in ([56]) are presented. In all implementations, the parameters were chosen in such a way that in all algorithms to have the same number of function evaluations. In Tables 5 and 6, the cost and the quality of the solutions given by the algorithms are presented. As it can be observed from these tables, the results are significantly improved with the use of the proposed algorithm. More precisely, the improvement in the quality of the results of the proposed method from the

Table 3. Results of BBMOVRP in Christofides benchmark instances

Nodes	Cap.	m.t.l.	s.t.	BBMOVRP	BKS	$\omega_{BBMOVRP}$ (%)	CPU (min)
51	160	∞	0	524.61	524.61 [76]	0.00	0.04
76	140	∞	0	835.26	835.26 [76]	0.00	0.22
101	200	∞	0	826.14	826.14 [76]	0.00	0.30
151	200	∞	0	1028.42	1028.42 [76]	0.00	1.05
200	200	∞	0	1291.54	1291.45 [76]	0.02	2.19
51	160	200	10	555.43	555.43 [76]	0.00	0.06
76	140	160	10	909.68	909.68 [76]	0.00	0.32
101	200	230	10	865.94	865.94 [76]	0.00	0.89
151	200	200	10	1163.12	1162.55 [76]	0.05	1.54
200	200	200	10	1395.85	1395.85 [76]	0.00	2.17
121	200	∞	0	1042.11	1042.11 [76]	0.00	0.55
101	200	∞	0	819.56	819.56 [76]	0.00	0.33
121	200	720	50	1542.10	1541.14 [76]	0.06	0.38
101	200	1040	90	866.37	866.37 [76]	0.00	0.37

Table 4. Results of BBMOVRP in the 20 benchmark Golden instances

Nodes	Cap.	m.t.l.	s.t.	BBMOVRP	BKS	$\omega_{BBMOVRP}$ (%)	CPU (min)
240	550	650	0	5643.27	5627.54 [64]	0.28	1.75
320	700	900	0	8455.12	8444.50 [71]	0.13	2.11
400	900	1200	0	11083.49	11036.22 [73]	0.43	6.15
480	1000	1600	0	13671.18	13624.52 [70]	0.34	7.13
200	900	1800	0	6460.98	6460.98 [79]	0.00	1.15
280	900	1500	0	8461.18	8412.8 [70]	0.58	1.51
360	900	1300	0	10198.25	10181.75 [69]	0.16	2.45
440	900	1200	0	11695.24	11643.90 [71]	0.44	6.14
255	1000	∞	0	583.39	583.39 [64]	0.00	1.58
323	1000	∞	0	743.19	741.56 [64]	0.22	2.57
399	1000	∞	0	922.17	918.45 [64]	0.41	3.15
483	1000	∞	0	1111.28	1107.19 [64]	0.37	8.09
252	1000	∞	0	860.17	859.11 [64]	0.12	3.11
320	1000	∞	0	1085.24	1081.31 [64]	0.36	2.35
396	1000	∞	0	1346.18	1345.23 [64]	0.07	7.81
480	1000	∞	0	1625.89	1622.69 [64]	0.20	9.34
240	200	∞	0	710.87	707.79 [64]	0.44	2.15
300	200	∞	0	1001.17	997.52 [64]	0.37	2.44
360	200	∞	0	1366.86	1366.86 [64]	0.00	3.12
420	200	∞	0	1824.14	1820.09 [64]	0.22	5.18

HBMO1 algorithm is between 0.00% to 1.60% in the Christofides benchmark instances with average improvement equal to 0.35% and is between 0.16% to 2.28% in the Golden benchmark instances with average improvement equal to 1.01%. The improvement in the quality of the results of the proposed method from the

HBMO2 algorithm is between 0.00% to 0.52% in the Christofides benchmark instances with average improvement equal to 0.13% and is between 0% to 0.98% in the Golden benchmark instances with average improvement equal to 0.39%. The improvement in the quality of the results of the proposed method from the HBMO3 algorithm is between 0.00% to 0.75% in the Christofides benchmark instances with average improvement equal to 0.12% and is between 0.00% to 0.89% in the Golden benchmark instances with average improvement equal to 0.37%. The improvement in the quality of the results of the proposed method from the HBMOVRP algorithm is between 0.00% to 0.16% in the Christofides benchmark instances with average improvement equal to 0.02% and is between 0.00% to 0.48% in the Golden benchmark instances with average improvement equal to 0.17%. As it can be observed from the previous results the use of the proposed method BBMOVRP improves the results obtained compared to the results of the honey bees mating optimization algorithm.

Table 5. Comparison of the proposed algorithm with implementations of Honey Bees Mating Optimization algorithm in the 14 Christofides benchmark instances

HBMO1		HBMO2		HBMO3		HBMOVRP		BBMOVRP	
cost	ω (%)	cost	ω (%)	cost	ω (%)	cost	ω (%)	cost	ω (%)
524.61	0.00	524.61	0.00	524.61	0.00	524.61	0.00	524.61	0.00
836.21	0.11	835.26	0.00	835.26	0.00	835.26	0.00	835.26	0.00
826.14	0.00	826.14	0.00	826.14	0.00	826.14	0.00	826.14	0.00
1031.18	0.27	1030.21	0.17	1029.28	0.08	1028.42	0.00	1028.42	0.00
1312.21	1.62	1298.27	0.54	1301.18	0.77	1292.57	0.10	1291.54	0.02
555.43	0.00	555.43	0.00	555.43	0.00	555.43	0.00	555.43	0.00
909.68	0.00	909.68	0.00	909.68	0.00	909.68	0.00	909.68	0.00
867.31	0.16	867.31	0.16	867.31	0.16	867.31	0.16	865.94	0.00
1172.12	0.82	1169.18	0.57	1167.21	0.40	1163.52	0.08	1163.12	0.05
1411.98	1.16	1399.21	0.24	1397.28	0.10	1395.85	0.00	1395.85	0.00
1042.11	0.00	1042.11	0.00	1042.11	0.00	1042.11	0.00	1042.11	0.00
820.98	0.17	819.71	0.02	820.76	0.15	819.56	0.00	819.56	0.00
1547.21	0.39	1544.21	0.20	1543.37	0.14	1542.21	0.07	1542.10	0.06
868.59	0.26	866.37	0.00	866.51	0.02	866.37	0.00	866.37	0.00

A statistical analysis based on the Mann-Whitney U-test for all algorithms based on a bee inspired method is presented in Table 7. In this Table, a value equal to 1 indicates a rejection of the null hypothesis at the 5% (or 10%) significance level, which means that the metric is statistically significant different from the other metrics. On the other hand, a value equal to 0 indicates a failure to reject the null hypothesis at the 5% (or 10%) significance level, meaning that no statistical significant difference exists between the two metrics. As it can be seen from this Table, at the 5% significance level the results with the Bumble Bees Mating Optimization are statistically significant different from the results of all implementations of Honey Bees Mating Optimization algorithm except of

Table 6. Comparison of the proposed algorithm with implementations of Honey Bees Mating Optimization algorithm in the 20 Golden instances

HBMO1		HBMO2		HBMO3		HBMOVRP		BBMOVRP	
cost	ω (%)	cost	ω (%)	cost	ω (%)	cost	ω (%)	cost	ω (%)
5708.21	1.43	5688.17	1.08	5675.71	0.86	5645.51	0.32	5643.27	0.28
8488.21	0.52	8472.51	0.33	8469.37	0.29	8458.72	0.17	8455.12	0.13
11136.59	0.91	11097.39	0.55	11087.18	0.46	11086.21	0.45	11083.49	0.43
13749.49	0.92	13688.29	0.47	13699.51	0.55	13675.23	0.37	13671.18	0.34
6471.29	0.16	6460.98	0.00	6461.21	0.01	6460.98	0.00	6460.98	0.00
8491.20	0.93	8471.17	0.69	8474.37	0.73	8467.57	0.65	8461.18	0.58
10271.29	0.88	10267.35	0.84	10254.71	0.72	10212.23	0.30	10198.25	0.16
11901.57	2.21	11809.57	1.42	11798.31	1.33	11710.53	0.57	11695.24	0.44
588.31	0.84	587.01	0.62	586.95	0.61	586.52	0.54	583.39	0.00
748.32	0.91	745.38	0.52	745.21	0.49	745.21	0.49	743.19	0.22
934.17	1.71	927.12	0.94	927.08	0.94	924.34	0.64	922.17	0.41
1136.54	2.65	1120.28	1.18	1119.12	1.08	1112.52	0.48	1111.28	0.37
875.05	1.86	863.81	0.55	863.72	0.54	863.79	0.54	860.17	0.12
1098.08	1.55	1089.07	0.72	1088.52	0.67	1088.27	0.64	1085.24	0.36
1368.72	1.75	1354.17	0.66	1358.21	0.96	1352.57	0.55	1346.18	0.07
1650.91	1.74	1631.01	0.51	1633.14	0.64	1629.28	0.41	1625.89	0.20
715.09	1.03	711.21	0.48	711.09	0.47	711.09	0.47	710.87	0.44
1013.17	1.57	1005.29	0.78	1006.12	0.86	1003.28	0.46	1001.17	0.37
1379.45	0.92	1366.98	0.01	1366.86	0.00	1366.86	0.00	1366.86	0.00
1835.01	0.82	1829.10	0.50	1827.34	0.40	1826.54	0.30	1824.14	0.22

Table 7. Results of Mann - Whitney test.

	5% significance level				
	HBMO1	HBMO2	HBMO3	HBMOVRP	BBMOVRP
HBMO1	-	1	1	1	1
HBMO2	1	-	0	1	1
HBMO3	1	0	-	0	1
HBMOVRP	1	1	0	-	0
BBMOVRP	1	1	1	0	-
	10% significance level				
	HBMO1	HBMO2	HBMO3	HBMOVRP	BBMOVRP
HBMO1	-	1	1	1	1
HBMO2	1	-	0	1	1
HBMO3	1	0	-	1	1
HBMOVRP	1	1	1	-	1
BBMOVRP	1	1	1	1	-

the HBMOVRP while at the 10% significance level the results with the Bumble Bees Mating Optimization are statistically significant different from the results of all algorithms.

Table 8. Comparison of other metaheuristics and nature inspired algorithms with BBMOVRP in Christofides benchmark instances

Rank	Algorithm	Quality (%)	n_{opt}	CPU (min)	Computer Used
		metaheuristic algorithms			
1	RT [76]	0.00	14	n/m*	Silicon Graphics 100MHz
2	**BBMOVRP**	0.009	11	0.743	Pentium M 750 at 1.86 GHz
3	HBMOVRP [57]	0.029	10	0.79	Pentium M 750 at 1.86 GHz
4	AGES best[64]	0.03	13	0.27	Pentium IV 2GHz
5	HybGENPSO [52]	0.046	10	0.95	Pentium III 667MHz
6	Taillard [77]	0.051	12	n/m	Silicon Graphics 100MHz
7	HybPSO [53]	0.084	7	0.80	Pentium M 750 at 1.86 GHz
8	best-Prins[70]	0.085	10	5.2	Pentium 1000 MHz
9	Best-SEPAS [78]	0.182	11	6.6	Pentium II 400 MHz
10	St-SEPAS [78]	0.195	9	5.6	Pentium II 400 MHz
		nature inspired methods			
1	**BBMOVRP**	0.009	11	0.743	Pentium M 750 at 1.86 GHz
2	HBMOVRP [57]	0.029	10	0.79	Pentium M 750 at 1.86 GHz
3	AGES best[64]	0.03	13	0.27	Pentium IV 2GHz
4	HybGENPSO [52]	0.046	10	0.95	Pentium III 667MHz
5	HybPSO [53]	0.084	7	0.80	Pentium M 750 at 1.86 GHz
6	best-Prins[70]	0.085	10	5.2	Pentium 1000 MHz
7	stand-Prins[70]	0.235	8	5.2	Pentium 1000 MHz
8	RSD[72]	0.383	6	7.7	Pentium 900 MHz
9	VRPBilevel [62]	0.479	7	0.76	Pentium III 667MHz
10	D-Ants[73]	0.481	5	3.28	Pentium 900 MHz

* not mentioned

The results obtained by the proposed algorithm are also compared to the results of the ten most efficient metaheuristic algorithms and the ten most efficient nature inspired algorithms that have ever been presented for the Vehicle Routing Problem. In Tables 8 and 9, the ranking of all algorithms is presented (in Golden instances we didn't find ten nature inspired algorithms and thus we present the seven algorithms used for the solution of these instances). The proposed algorithm is ranked in the second place among the ten most efficient metaheuristics and in the first place among the nature inspired methods used for the solution of the VRP in Christofides instances. In the set of the large scale vehicle routing instances the algorithm is ranked in the second place both in metaheuristics and nature inspired methods. In these Tables, the number of optimally solved instances for every method is, also, presented (n_{opt}). Finally, in the last two columns of these tables, the average CPU time (in minutes) and the computers used for the metaheuristic and nature inspired algorithms of the comparisons are presented. It should be noted that a fair comparison in terms of computational efficiency is difficult because the computational speed is affected, mainly, by the compiler and the hardware that are used.

Table 9. Comparison of other metaheuristics and nature inspired algorithms with BBMOVRP in Golden benchmark instances

Rank	Algorithm	Quality (%)	n_{opt}	CPU (min)	Computer Used
		metaheuristic algorithms			
1	AGES best [64]	0.00	20	0.63	Pentium IV 2GHz
2	**BBMOVRP**	0.26	3	3.96	Pentium M 750 at 1.86 GHz
3	HBMOVRP [57]	0.40	2	4.06	Pentium M 750 at 1.86 GHz
4	D-Ants [73]	0.59	4	49.33	Pentium 900 MHz
5	stand-SEPAS [78]	0.60	2	45.48	Pentium II 400 MHz
6	HybGENPSO [52]	0.60	1	4.19	Pentium III 667MHz
7	BoneRoute [79]	0.74	1	42.05	Pentium 400 MHz
8	best-Prins [70]	0.91	6	66.6	Pentium 1000 MHz
9	VRTR [46]	1.05	0	0.97	Athlon 1 GHz
10	VRPBilevel [62]	1.22	0	3.44	Pentium III 667MHz
		nature inspired methods			
1	AGES best [64]	0.00	20	0.63	Pentium IV 2GHz
2	**BBMOVRP**	0.26	3	3.96	Pentium M 750 at 1.86 GHz
3	HBMOVRP [57]	0.40	2	4.06	Pentium M 750 at 1.86 GHz
4	D-Ants [73]	0.59	4	49.33	Pentium 900 MHz
5	HybGENPSO [52]	0.60	1	4.19	Pentium III 667MHz
6	Prins [70]	0.91	6	66.6	Pentium 1000 MHz
7	VRPBilevel [62]	1.22	0	3.44	Pentium III 667MHz

5 Conclusions

In this chapter, a nature inspired approach was introduced for the effective handling of the Vehicle Routing Problem (VRP). More specifically the Bumble Bees Mating Optimization algorithm for the VRP (BBMOVRP) that gave remarkable results both to quality and computational efficiency. The algorithm was applied in two sets of benchmark instances and gave very satisfactory results. More specifically, in the set with the classic benchmark instances proposed by Christofides, the average quality is 0.009% while in the second set of instances proposed by Golden the average quality is 0.26%. A future work will be focused in the implementation of the algorithm to other variants of the vehicle routing problem, like vehicle routing problem with time windows, open vehicle routing problem and stochastic vehicle routing problem.

References

1. Abbass, H.A.: A monogenous MBO approach to satisfiability. In: International Conference on Computational Intelligence for Modelling, Control and Automation, CIMCA 2001, Las Vegas, NV, USA (2001)
2. Abbass, H.A.: Marriage in honey-bee optimization (MBO): a haplometrosis polygynous swarming approach. In: The Congress on Evolutionary Computation (CEC 2001), Seoul, Korea, pp. 207–214 (May 2001)

3. Afshar, A., Haddad, O.B., Marino, M.A., Adams, B.J.: Honey-bee mating optimization (HBMO) algorithm for optimal reservoir operation. J. Franklin Inst 344, 452–462 (2007)
4. Altinkemer, K., Gavish, B.: Altinkemer K., Gavish, B. Parallel savings based heuristics for the delivery problem. Oper. Res. 39(3), 456–469 (1991)
5. Baker, B.M., Ayechew, M.A.: A genetic algorithm for the vehicle routing problem. Comput. Oper. Res. 30(5), 787–800 (2003)
6. Barbarosoglu, G., Ozgur, D.: A tabu search algorithm for the vehicle routing problem. Comput. Oper. Res. 26, 255–270 (1999)
7. Baykasoglu, A., Ozbakir, L., Tapkan, P.: Artificial bee colony algorithm and its application to generalized assignment problem. In: Chan, F.T.S., Tiwari, M.K. (eds.) Swarm Intelligence, Focus on Ant and Particle Swarm Optimization, pp. 113–144. I-Tech Education and Publishing (2007)
8. Berger, J., Barkaoui, M.: A hybrid genetic algorithm for the capacitated vehicle routing problem. In: Proceedings of the Genetic and Evolutionary Computation Conference, Chicago, pp. 646–656 (2003)
9. Bodin, L., Golden, B.: Classification in vehicle routing and scheduling. Networks 11, 97–108 (1981)
10. Bodin, L., Golden, B., Assad, A., Ball, M.: The state of the art in the routing and scheduling of vehicles and crews. Comput. Oper. Res. 10, 63–212 (1983)
11. Bullnheimer, B., Hartl, P.F., Strauss, C.: An improved ant system algorithm for the vehicle routing problem. Ann. Oper. Res. 89, 319–328 (1999)
12. Christofides, N., Mingozzi, A., Toth, P.: The vehicle routing problem. In: Christofides, N., Mingozzi, A., Toth, P., Sandi, C. (eds.) Combinatorial Optimization, Wiley, Chichester (1979)
13. Clarke, G., Wright, J.: Scheduling of vehicles from a central depot to a number of delivery points. Oper. Res. 12, 568–581 (1964)
14. Clerc, M., Kennedy, J.: The particle swarm: explosion, stability and convergence in a multi-dimensional complex space. IEEE T Evolut. Comput. 6, 58–73 (2002)
15. Cordeau, J.F., Gendreau, M., Laporte, G., Potvin, J.Y., Semet, F.: A guide to vehicle routing heuristics. J. Oper. Res. Soc. 53, 512–522 (2002)
16. Cordeau, J.F., Gendreau, M., Hertz, A., Laporte, G., Sormany, J.S.: New heuristics for the vehicle routing problem. In: Langevine, A., Riopel, D. (eds.) Logistics Systems: Design and Optimization, pp. 279–298. Wiley and Sons, Chichester (2005)
17. Dantzig, G.B., Ramser, J.H.: The truck dispatching problem. Manage Sci. 6(1), 80–91 (1959)
18. Dasgupta, D. (ed.): Artificial immune systems and their application. Springer, Heidelberg (1998)
19. Desrochers, M., Verhoog, T.W.: A matching based savings algorithm for the vehicle routing problem. Les Cahiers du GERAD G-89-04, Ecole des Hautes Etudes Commerciales de Montreal (1989)
20. Dorigo, M., Stützle, T.: Ant colony optimization. A Bradford Book. The MIT Press, Cambridge (2004)
21. Drias, H., Sadeg, S., Yahi, S.: Cooperative bees swarm for solving the maximum weighted satisfiability problem. In: Cabestany, J., Prieto, A.G., Sandoval, F. (eds.) IWANN 2005. LNCS, vol. 3512, pp. 318–325. Springer, Heidelberg (2005)
22. Engelbrecht, A.P.: Computational intelligence: An introduction, 2nd edn. John Wiley and Sons, England (2007)
23. Fathian, M., Amiri, B., Maroosi, A.: Application of honey bee mating optimization algorithm on clustering. Appl. Math. Comput. 190, 1502–1513 (2007)

24. Fisher, M.L.: Vehicle routing. In: Ball, M.O., Magnanti, T.L., Momma, C.L., Nemhauser, G.L. (eds.) Network Routing, Handbooks in Operations Research and Management Science, vol. 8, pp. 1–33. North Holland, Amsterdam (1995)
25. Fisher, M.L., Jaikumar, R.: A generalized assignment heuristic for vehicle routing. Networks 11, 109–124 (1981)
26. Foster, B.A., Ryan, D.M.: An integer programming approach to the vehicle scheduling problem. Oper. Res. 27, 367–384 (1976)
27. Garfinkel, R., Nemhauser, G.: Integer Programming. John Wiley and Sons, New York (1972)
28. Gendreau, M., Hertz, A., Laporte, G.: A tabu search heuristic for the vehicle routing problem. Manage Sci. 40, 1276–1290 (1994)
29. Gendreau, M., Laporte, G., Potvin, J.Y.: Vehicle routing: modern heuristics. In: Aarts, E.H.L., Lenstra, J.K. (eds.) Local search in Combinatorial Optimization, pp. 311–336. Wiley, Chichester (1997)
30. Gendreau, M., Laporte, G., Potvin, J.Y.: Metaheuristics for the Capacitated VRP. In: Toth, P., Vigo, D. (eds.) The Vehicle Routing Problem, Monographs on Discrete Mathematics and Applications, pp. 129–154. SIAM, Philadelphia (2002)
31. Gillett, B.E., Miller, L.R.: A heuristic algorithm for the vehicle dispatch problem. Oper. Res. 22, 240–349 (1974)
32. Golden, B.L., Assad, A.A.: Vehicle Routing: Methods and Studies. North Holland, Amsterdam (1988)
33. Golden, B.L., Raghavan, S., Wasil, E.: The Vehicle Routing Problem: Latest Advances and New Challenges. Springer LLC, Heidelberg (2008)
34. Golden, B.L., Wasil, E.A., Kelly, J.P., Chao, I.M.: The impact of metaheuristics on solving the vehicle routing problem: algorithms, problem sets, and computational results. In: Crainic, T.G., Laporte, G. (eds.) Fleet management and logistics, pp. 33–56. Kluwer Academic Publishers, Boston (1998)
35. Goulson, D.: Bumblebees: Behaviour, Ecology, and Conservation. Oxford University Press, USA (2009)
36. Hackel, S., Dippold, P.: The bee colony-inspired algorithm (BCiA): a two stage approach for solving the vehicle routing problem with time windows. In: GECCO 2009: Proceedings of the 11th Annual Conference on Genetic and Evolutionary Computation, pp. 25–32 (2009)
37. Haddad, O.B., Afshar, A., Marino, M.A.: Honey-bees mating optimization (HBMO) algorithm: A new heuristic approach for water resources optimization. Water Resour Manag. 20, 661–680 (2006)
38. Hansen, P., Mladenovic, N.: Variable neighborhood search: Principles and applications. Eur. J. Oper. Res. 130, 449–467 (2001)
39. Holland, J.H.: Adaptation in natural and artificial systems. University of Michigan Press, Ann Arbor (1975)
40. Karaboga, D., Basturk, B.: A powerful and efficient algorithm for numerical function optimization: artificial bee colony (ABC) algorithm. J. Global. Optim. 39, 459–471 (2007)
41. Karaboga, D., Basturk, B.: On the performance of artificial bee colony (ABC) algorithm. Appl. Soft. Comput. 8, 687–697 (2008)
42. Karaboga, D., Akay, B.: A survey: algorithms simulating bee swarm intelligence. Artif. Intell. Rev. (2009), doi:10.1007/s10462-009-9127-4
43. Kennedy, J., Eberhart, R.: Particle swarm optimization. In: IEEE International Conference on Neural Networks, vol. 4, pp. 1942–1948 (1995)

44. Laporte, G., Semet, F.: Classical heuristics for the capacitated VRP. In: Toth, P., Vigo, D. (eds.) The Vehicle Routing Problem, Monographs on Discrete Mathematics and Applications, pp. 109–128. SIAM, Philadelphia (2002)
45. Laporte, G., Gendreau, M., Potvin, J.Y., Semet, F.: Classical and modern heuristics for the vehicle routing problem. Int. Trans. Oper. Res. 7, 285–300 (2000)
46. Li, F., Golden, B., Wasil, E.: Very large-scale vehicle routing: new test problems, algorithms and results. Comput. Oper. Res. 32(5), 1165–1179 (2005)
47. Lichtblau, D.: Discrete optimization using Mathematica. In: Callaos, N., Ebisuzaki, T., Starr, B., Abe, J.M., Lichtblau, D. (eds.) World Multi-Conference on Systemics, Cybernetics and Informatics (SCI 2002), International Institute of Informatics and Systemics, vol. 16, pp. 169–174 (2002)
48. Lin, S.: Computer solutions of the Traveling Salesman Problem. Bell. Syst. Tech. J. 44, 2245–2269 (1965)
49. Lin, S., Kernighan, B.W.: An Effective Heuristic Algorithm for the Traveling Salesman Problem. Oper. Res. 21, 498–516 (1973)
50. Marinaki, M., Marinakis, Y., Zopounidis, C.: Honey bees mating optimization algorithm for financial classification problems. Appl. Soft. Comput. (2009) (available on line – doi: 10.1016/j.asoc.2009.09.010)
51. Marinakis, Y., Marinaki, M.: A hybrid honey bees mating optimization algorithm for the probabilistic traveling salesman problem. In: IEEE Congress on Evolutionary Computation (CEC 2009), Trondheim, Norway (2009)
52. Marinakis, Y., Marinaki, M.: A Hybrid Genetic - Particle Swarm Algorithm for the Vehicle Routing Problem. Expert Syst. Appl. 37, 1446–1455 (2010)
53. Marinakis, Y., Marinaki, M., Dounias, G.: A Hybrid Particle Swarm Optimization Algorithm for the Vehicle Routing Problem. Eng. Appl. of Artif. Intell. (accepted 2010)
54. Marinakis, Y., Migdalas, A.: Heuristic solutions of vehicle routing problems in supply chain management. In: Pardalos, P.M., Migdalas, A., Burkard, R. (eds.) Combinatorial and Global Optimization, pp. 205–236. World Scientific Publishing Co., Singapore (2002)
55. Marinakis, Y., Marinaki, M., Dounias, G.: Honey bees mating optimization algorithm for the vehicle routing problem. In: Krasnogor, N., Nicosia, G., Pavone, M., Pelta, D. (eds.) Nature inspired cooperative strategies for optimization - NICSO 2007, Studies in Computational Intelligence, vol. 129, pp. 139–148. Springer, Berlin (2008)
56. Marinakis, Y., Marinaki, M., Dounias, G.: Honey bees mating optimization algorithm for large scale vehicle routing problems. Nat. Comput. (2009) (available on line - doi: 10.1007/s11047-009-9136-x)
57. Marinakis, Y., Marinaki, M., Matsatsinis, N.: A hybrid clustering algorithm based on Honey Bees Mating Optimization and Greedy Randomized Adaptive Search Procedure. In: Maniezzo, V., Battiti, R., Watson, J.-P. (eds.) LION 2007 II. LNCS, vol. 5313, pp. 138–152. Springer, Heidelberg (2008)
58. Marinakis, Y., Marinaki, M., Matsatsinis, N.: Honey bees mating optimization for the location routing problem. In: IEEE International Engineering Management Conference (IEMC - Europe 2008), Estoril, Portugal (2008)
59. Marinakis, Y., Marinaki, M., Matsatsinis, N.: A hybrid bumble bees mating optimization – GRASP algorithm for clustering. In: Corchado, E., Wu, X., Oja, E., Herrero, Á., Baruque, B. (eds.) HAIS 2009. LNCS, vol. 5572, pp. 549–556. Springer, Heidelberg (2009)
60. Marinakis, Y., Migdalas, A., Pardalos, P.M.: Expanding neighborhood GRASP for the traveling salesman problem. Comput. Optim. Appl. 32, 231–257 (2005)

61. Marinakis, Y., Migdalas, A., Pardalos, P.M.: A hybrid Genetic-GRASP algortihm using langrangean relaxation for the traveling salesman problem. J. Comb. Optim. 10, 311–326 (2005)
62. Marinakis, Y., Migdalas, A., Pardalos, P.M.: A new bilevel formulation for the vehicle routing problem and a solution method using a genetic algorithm. J. Global. Optim. 38, 555–580 (2007)
63. Mester, D., Braysy, O.: Active guided evolution strategies for the large scale vehicle routing problems with time windows. Comput. Oper. Res. 32, 1593–1614 (2005)
64. Mester, D., Braysy, O.: Active guided evolution strategies for large scale capacitated vehicle routing problems. Comput. Oper. Res. 34, 2964–2975 (2007)
65. Mole, R.H., Jameson, S.R.: A sequential route-building algorithm employing a generalized savings criterion. Oper. Res. Quart. 27, 503–511 (1976)
66. Osman, I.H.: Metastrategy simulated annealing and tabu search algorithms for combinatorial optimization problems. Ann. Oper. Res. 41, 421–451 (1993)
67. Pereira, F.B., Tavares, J.: Bio-inspired Algorithms for the Vehicle Routing Problem. Studies in Computational Intelligence, vol. 161. Springer, Heidelberg (2008)
68. Pham, D.T., Ghanbarzadeh, A., Koc, E., Otri, S., Rahim, S., Zaidi, M.: The bees algorithm - A novel tool for complex optimization problems. In: IPROMS 2006 Proceeding 2nd International Virtual Conference on Intelligent Production Machines and Systems. Elsevier, Oxford (2006)
69. Pisinger, D., Ropke, S.: A general heuristic for vehicle routing problems. Comput. Oper. Res. 34, 2403–2435 (2007)
70. Prins, C.: A simple and effective evolutionary algorithm for the vehicle routing problem. Comput. Oper. Res. 31, 1985–2002 (2004)
71. Prins, C.: A GRASP $\times$ Evolutionary Local Search Hybrid for the Vehicle Routing Problem. In: Pereira, F.B., Tavares, J. (eds.) Bio-inspired Algorithms for the Vehicle Routing Problem, SCI 161, pp. 35–53. Springer, Heidelberg (2008)
72. Reimann, M., Stummer, M., Doerner, K.: A savings based ant system for the vehicle routing problem. In: Proceedings of the Genetic and Evolutionary Computation Conference, New York, pp. 1317–1326 (2002)
73. Reimann, M., Doerner, K., Hartl, R.F.: D-Ants: savings based ants divide and conquer the vehicle routing problem. Comput. Oper. Res. 31, 563–591 (2004)
74. Rego, C.: A subpath ejection method for the vehicle routing problem. Manage Sci. 44, 1447–1459 (1998)
75. Rego, C.: Node-ejection chains for the vehicle routing problem: sequential and parallel algorithms. Parallel Comput. 27, 201–222 (2001)
76. Rochat, Y., Taillard, E.D.: Probabilistic diversification and intensification in local search for vehicle routing. J. Heuristics 1, 147–167 (1995)
77. Taillard, E.D.: Parallel iterative search methods for vehicle routing problems. Networks 23, 661–672 (1993)
78. Tarantilis, C.D.: Solving the vehicle routing problem with adaptive memory programming methodology. Comput. Oper. Res. 32, 2309–2327 (2005)
79. Tarantilis, C.D., Kiranoudis, C.T.: BoneRoute: an adaptive memory-based method for effective fleet management. Ann. Oper. Res. 115, 227–241 (2002)
80. Tarantilis, C.D., Kiranoudis, C.T., Vassiliadis, V.S.: A backtracking adaptive threshold accepting metaheuristic method for the Vehicle Routing Problem. System Analysis Modeling Simulation (SAMS) 42, 631–644 (2002)
81. Tarantilis, C.D., Kiranoudis, C.T., Vassiliadis, V.S.: A list based threshold accepting algorithm for the capacitated vehicle routing problem. Int. J. Comput. Math. 79, 537–553 (2002)

82. Teo, J., Abbass, H.A.: A true annealing approach to the marriage in honey bees optimization algorithm. Int. J. Comput. Intell. Appl. 3(2), 199–211 (2003)
83. Teodorovic, D., Dell'Orco, M.: Bee colony optimization - A cooperative learning approach to complex transportation problems. In: Advanced OR and AI Methods in Transportation. Proceedings of the 16th Mini - EURO Conference and 10th Meeting of EWGT, pp. 51–60 (2005)
84. Toth, P., Vigo, D.: The Vehicle Routing Problem, Monographs on Discrete Mathematics and Applications. SIAM, Philadelphia (2002)
85. Toth, P., Vigo, D.: The granular tabu search (and its application to the vehicle routing problem). INFORMS J. Comput. 15, 333–348 (2003)
86. Storn, R., Price, K.: Differential evolution - A simple and efficient heuristic for global optimization over continuous spaces. J. Global Optim. 11(4), 341–359 (1997)
87. Wark, P., Holt, J.: A repeated matching heuristic for the vehicle routing problem. J. Oper. Res. Soc. 45, 1156–1167 (1994)
88. Wedde, H.F., Farooq, M., Zhang, Y.: BeeHive: An Efficient Fault-Tolerant Routing Algorithm Inspired by Honey Bee Behavior. In: Dorigo, M., Birattari, M., Blum, C., Gambardella, L.M., Mondada, F., Stützle, T. (eds.) ANTS 2004. LNCS, vol. 3172, pp. 83–94. Springer, Heidelberg (2004)
89. Xu, J., Kelly, J.P.: A new network flow-based tabu search heuristic for the vehicle routing problem. Transport Sci. 30, 379–393 (1996)
90. Yang, X.-S.: Engineering Optimizations via Nature-Inspired Virtual Bee Algorithms. In: Mira, J., Álvarez, J.R. (eds.) IWINAC 2005. LNCS, vol. 3562, pp. 317–323. Springer, Heidelberg (2005)
91. http://www.bumblebee.org
92. http://www.everythingabout.net/articles/biology/animals/arthropods/insects/bees/bumble_bee
93. http://bumbleboosters.unl.edu/biology.shtml
94. http://www.colostate.edu/Depts/Entomology/courses/en570/papers_1998/walter.htm

Part C
Ant Colony Optimization

Ant Colony Optimization: Principle, Convergence and Application

Haibin Duan

School of Automation Science and Electrical Engineering, Beijing University of Aeronautics and Astronautics, Beijing 100191, P.R. China
hbduan@buaa.edu.cn

Abstract. Ant Colony Optimization (ACO) is a meta-heuristic algorithm for the approximate solution of combinatorial optimization problems that has been inspired by the foraging behaviour of real ant colonies. In this Chapter, we present a novel approach to the convergence proof that applies directly to the basic ACO model, and a kind of parameters tuning strategy for nonlinear PID(NLPID) controller using a grid-based ACO algorithm is also presented in detail. A series of simulation experimental results are provided to verify the performance the whole control system of the flight simulator with the grid-based ACO algorithm optimized NLPID.

1 Introduction

Ant Colony Optimization(ACO) is a meta-heuristic algorithm for the approximate solution of combinatorial optimization problems that has been inspired by the foraging behaviour of real ant colonies[1~5]. The principle of ACO is that ants deposit a chemical substance (called pheromone) on the ground, thus, they mark a path by the pheromone trail. In this process, a kind of positive feedback mechanism is adopted[6].

Proportional-Integral-Derivative (PID) control has been used successfully for regulating processes in industry for more than 60 years. Today, more than 95% of the control loops are of PID type, and PID controllers can be found in all areas where control is used. The controllers come in many different forms[7]. There are stand-alone systems in boxes for one or a few loops, which are manufactured by the hundred thousands yearly. PID control is also an important ingredient of complex control systems. The controllers are also embedded in many special-purpose control systems[8]. PID control is often combined with logic, sequential functions, selectors, and simple function blocks to build the complicated automation systems used for energy production, transportation, and manufacturing.

PID controllers often provide acceptable control even in the absence of tuning, but performance can generally be improved by careful tuning, and performance may be unacceptable with poor tuning. PID tuning is a difficult problem, even though there are only three parameters and in principal is simple to describe, because it must

B.K. Panigrahi, Y. Shi, and M.-H. Lim (Eds.): Handbook of Swarm Intelligence, ALO 8, pp. 373–388.
springerlink.com

satisfying complex criteria within the limitations of PID control. There are accordingly various methods for loop tuning, and more sophisticated techniques are the subject of patents.

In this Chapter, we present a novel approach to the convergence proof that applies directly to the basic ACO model, and a kind of parameters tuning strategy for Nonlinear PID(NLPID) controller using a grid-based ACO algorithm is also presented in detail. Series simulation experimental results are provided to verify the performance the whole control system of the flight simulator with the grid-based ACO algorithm optimized NLPID.

2 The Basic Principle of Ant Colony Optimization

Ant Colony Optimization has been inspired by the observation on real ant colony's foraging behavior, and on that ants can often find the shortest path between food source and their nest. In ACO algorithm, the computational resources are allocated to a set of relatively simple agents that exploit a form of indirect communication mediated by the environment to construct solutions to the finding the shortest path from ant nest to a considered problem[3, 4]. Real ants are capable of food source, because, while walking, ants deposit pheromone on the ground, and real ants have a probabilistic preference for paths with larger amount of pheromone. Fig. 1 shows the principle that ants exploit pheromone to establish shortest path from a nest to a food source and back.

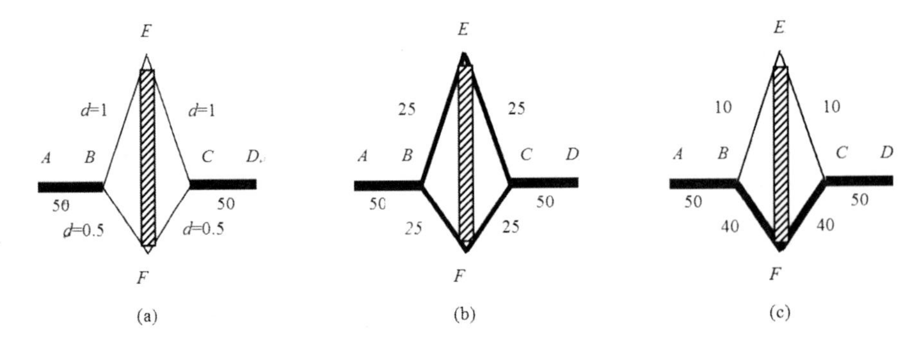

Fig. 1. Diagram of ACO algorithm principle

In Fig. 1, the point A is a nest, the point E is a food source, the point B or the point C is a path-crotch, the line EF is an obstacle, and d is the distance between two points. Because of barrier EF, the ants, which comes out of nest A or go back from the food source D, can reach their destination only via E or F. Suppose all ants go there and back between A and D at a speed of 1 unit length per unit time, and there are 50 ants per unit time from A and D to D and A to decide whether to turn left or right at B and C(shown as in Fig. 1a), respectively. At first, they choose randomly (25 left and 25 right separately, shown as in Fig. 1b) for there are not any pheromones on the ways. After 1 unit time, 50 ants have passed through the shorter path BFC and deposit pheromone on the ground, which only 25 ants have done through the longer path BEC (length double BFC). Therefore, the amount of pheromone on BFC is two times as much as on BEC. Then, there again are 50 ants to choose path at B and C respectively. Since the amount of pheromone on BFC is different from BEC, the path choice of the ants is influenced by it and is roughly proportional to it. Therefore, about 40 ants turn to direction F and the other 10 ants turn to E(shown as in Fig. 1c), and the outcome is that more pheromone deposits on shorter path BFC. As time goes on and the process repeats, the pheromone amount of the shorter path accumulates faster, and more and more ants choose this path. This process is thus characterized by a positive feedback loop, where the probability with which an ant chooses a path increases with the number of ants that previously chose the same path. With the above positive feedback mechanism, all ants will choose the shorter path in the end.

3 Ant Colony Optimization Mathematical Model

The Ant Colony Optimization mathematical model has first been applied to the Traveling Salesman Problem(TSP)[1, 6]. The aim of the TSP is to find the shortest path that traverses all cities in the problem exactly once, returning to the starting city.

We define the transition probability from city i to city j for the k-th ant as:

$$p_{ij}^{k}(t)=\begin{cases}\dfrac{[\tau_{ij}(t)]^{\alpha}[\eta_{ij}]^{\beta}}{\sum\limits_{k\in allowed_k}[\tau_{ik}(t)]^{\alpha}[\eta_{ik}]^{\beta}} & if \quad j\in allowed_k \\ 0 & otherwise\end{cases} \tag{1}$$

After the ants in the algorithm ended their tours, the pheromone trail τ_{ij} values of every edge (i, j) are updated according to the following formulas:

$$\tau_{ij}(t+n)=\rho\cdot\tau_{ij}(t)+\Delta\tau_{ij} \tag{2}$$

$$\Delta\tau_{ij}=\sum_{k=1}^{m}\Delta\tau_{ij}^{k} \tag{3}$$

$$\Delta\tau_{ij}^{k}=\begin{cases}\dfrac{Q}{L_k} & if \quad k-th \quad ant \quad uses \quad (i,j) \quad in \quad its \quad tour \\ 0 & otherwise\end{cases} \tag{4}$$

This iteration process goes on until a certain termination condition: a certain number of iterations have been achieved, a fixed amount of CPU time has elapsed, or solution quality has been achieved.

4 Convergence Proof of Ant Colony Optimization

Proposition 1[9]: Let $\tau_{\max}$ be the maximum quantity of pheromone of $\tau_{ij}(t)$, and the possible solution set is defined as $w^{*}=(w_0,w_1,\cdots,w_L)$, $w^{*}\subset L$ and $w^{*}\not\in\Phi$.

The maximum possible quantity of pheromone added to any edge l_{ij} after any iteration is $g(w_i)$. ρ is a constant, and $\rho \subset [0,1)$. Then for $\forall \tau_{ij}$, it holds that

$$\lim_{t\to\infty} \tau_{ij}(t) \le \tau_{\max} = \frac{1}{\rho} \cdot g(w_i) \tag{5}$$

Proposition 2[9]**:** Let $\tau_{\min}$ be the minimum quantity of pheromone $\tau_{ij}(t)$. Once an optimal solution $\tau_{ij}^{\ *}(t)$ has been found, as $\tau_{ij}^{\ *}(t)$ is monotonically increasing and $\tau_{ij}^{*}(t) \ge \tau_{\min}$, then for $\forall (i, j) \subset w^*$, it holds that

$$\lim_{t\to\infty} \tau_{ij}^{*}(t) = \tau_{\max} = \frac{1}{\rho} \cdot g(w_i) \tag{6}$$

Theorem 1: Let *M* be the number of the connections for the optimal solution and let $\lambda = \min\{[\eta_{ij}(t)]^\beta \mid (i,j) \in w^*\}$, $p_{A_i}(t)$ be the probability of ant A_i in the *t*-th iteration, $\hat{A}$ be the event that at one ant find the optimal path, $\tau_{ij}(1)$ be the pheromone amount at the initialized stage, and let $p_{\hat{A}}$ be the probability that at least one ant find the optimal path in finite iterations. Then, it holds that

$$p_{\hat{A}} \ge 1-(1-\lambda^M \rho^{M(t-1)} \prod_{(i,j)\in w^*} \tau_{ij}(1))^m \tag{7}$$

Proof: Suppose *t* is the iteration of ant. With $\Delta\tau_{ij} \ge 0$, $\rho > 0$, and Formula (2), we have

$$\tau_{ij}(t+1) \ge \rho \tau_{ij}(t) \tag{8}$$

Starting from 1st iteration, it holds that

$$\tau_{ij}(t) \ge \rho^{t-1} \tau_{ij}(1) \tag{9}$$

Because of $0 < \eta_{ij}(t) \le 1$, we obtain

$$0 < [\eta_{ij}(t)]^\beta \le 1 \tag{10}$$

Hence

$$\sum_{j \subset allowed_k} [\tau_{ij}(t)]^\alpha \cdot [\eta_{ij}(t)]^\beta \le \sum_{j \subset allowed_k} [\tau_{ij}(t)]^\alpha \le 1$$

Meanwhile, it holds that

$$p_{ij}(t) = \frac{[\tau_{ij}(t)]^\alpha \cdot [\eta_{ik}(t)]^\beta}{\sum_{s \subset allowed_k} [\tau_{is}(t)]^\alpha \cdot [\eta_{is}(t)]^\beta} \ge [\tau_{ij}(t)]^\alpha \cdot [\eta_{ik}(t)]^\beta$$

Let $\lambda = \min\{[\eta_{ij}(t)]^{\beta} \mid (i,j) \in w^*\}$. Then by Formula (9) and (10), we have

$$p_{A_i}(t) = \prod_{i=0}^{M-1} p_{w_i,w_i-1}(t,(w_0,w_1,\cdots,w_i)) \geq \prod_{i=0}^{M-1} [\tau_{w_i,w_i-1}(t)]^{\alpha}[\eta_{w_i,w_i-1}]^{\beta}$$
$$\geq \lambda^M \prod_{i=0}^{M-1} [\tau_{w_i,w_i-1}(t)]^{\alpha} \geq \lambda^M \prod_{i=0}^{M-1} \rho^{t-1}\tau_{w_i,w_i-1}(1)$$
$$= \lambda^M \rho^{M(t-1)} \prod_{(i,j)\in w^*} \tau_{ij}(1)$$

The m ants are independent from each other. Hence, in t iterations, the probability that no ants transverse the optimal path is as follows:

$$\overline{p}_{\hat{A}} \leq (1-\lambda^M \rho^{M(t-1)} \prod_{(i,j)\in w^*} \tau_{ij}(1))^m \tag{11}$$

Therefore, the probability that at least one ant find the optimal path in finite iterations is given as follows

$$p_{\hat{A}} \geq 1-(1-\lambda^M \rho^{M(t-1)} \prod_{(i,j)\in w^*} \tau_{ij}(1))^m$$

Theorem 2: Let $p_{\hat{A}}^{(m)}$ be the probability that m ants find the optimal path in finite iterations. Then, for $m \to \infty$, it holds that

$$\lim_{m\to\infty} p_{\hat{A}}^{(m)} = 1 \tag{12}$$

Proof: From Proposition 1 and Proposition 2, the pheromone quantity $\tau_{ij}(t)$ is bounded with $\tau_{\max}$ and $\tau_{\min}$. Let $\hat{p}_{\min}(t)$ be the lower-bounded probability in t iterations. Then, it holds that

$$\hat{p}_{\min}(t) \geq \frac{\tau_{\min}^{\alpha} \cdot [\eta_{ik}(t)]^{\beta}}{(t-1)\cdot\tau_{\max}^{\alpha} \cdot [\eta_{ik}(t)]^{\beta} + \tau_{\min}^{\alpha} \cdot [\eta_{ik}(t)]^{\beta}} \tag{13}$$

Where t is the iteration number. Obviously, for the probability $p_{\hat{A}}(t)$ after t-th iterations, it holds that

$$p_{\hat{A}}^{(m)}(t) \geq p_{\min}(t) > 0$$

From Theorem 1, it is immediate to see that, for finite iterations of m ants, the probability $p_{\hat{A}}^{(m)}(t)$ that at least one ant find the optimal path becomes

$$p_{\hat{A}}^{(m)} \geq 1-(1-\lambda^M \rho^{M(t-1)} \prod_{(i,j)\in w^*} \tau_{ij}(1))^m$$

when $m \to \infty$, we have

$$\lim_{m\to\infty}(1-\lambda^M \rho^{M(t-1)} \prod_{(i,j)\in w^*} \tau_{ij}(1))^m = 0$$

Therefore $\lim_{m\to\infty} p_{\hat{A}}^{(m)} = 1$

5 NLPID Controller

The conventional PID controller is a widely used industrial controller that uses a combination of proportional, integral and derivative action on the control error to form the output of the controller. Due to its simple structure, easy tuning and effectiveness, this technology has been being the mainstay for so long among practicing engineers.

It is known that the linear combination these components can at most achieve a compromised performance in terms of system response speed and stability. Although its popularity, it is difficult to obtain satisfactory control results to employ the conventional PID controller for the systems or processes which are nonlinearity, long time-delaying, time-varying and strong cross-coupling. In order to improve the performance of conventional PID, Han developed a kind of NLPID controller with the use of nonlinear characteristics[8]. A nonlinear combination can provide additional degree of freedom to achieve a much improved system performance. However, this improvement can only be achieved at the expense of higher complexity in the controller. Artificial intelligence approaches can alleviate some of the difficulties by fusing a priori information into the control design.

The control law of the common NLPID controller is presented as follows[9].

$$u = K\left(\left| \frac{\dot{e}}{T_i} \right|^{\lambda} sign(\dot{e}) + |e|^{\lambda} sign(e) + \left| \frac{\int edt}{T_d} \right|^{\lambda} sign(\int edt) \right) \tag{14}$$

Where λ is a constant, K is the proportional gain, T_i is the integral time constant, T_d is the derivation time constant, u is the combination of error, differentor of error and integrator of error. In order to improve the control quality, a novel type NLPID controller can be constructed as in Fig. 2.

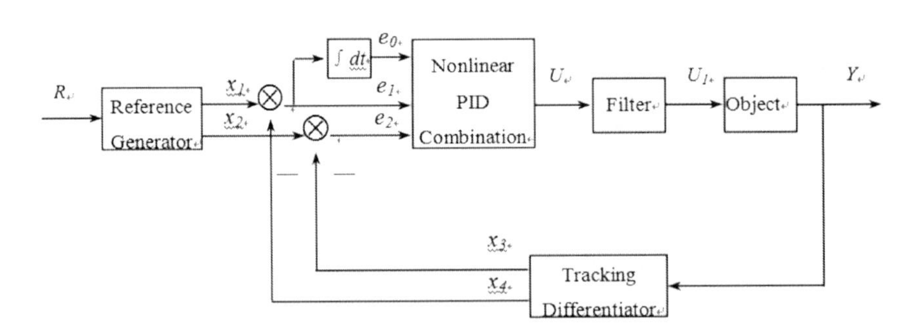

Fig. 2. NLPID controller

The differential tracker is used to track output position and its differential[1, 9]. The reference generator is used to generate reference input signals[5]. From Fig. 2, it is clearly that the input error signals for the NLPID combination can be expressed as follows.

$$\begin{cases} e_1 = x_1 - x_4 \\ e_2 = x_2 - x_3 \\ e_0 = \int_0 e_1(\tau)d\tau \end{cases} \tag{15}$$

Where e_1 denotes angular position tracking error, e_2 denotes angular velocity tracking error.

The proposed reference generator is expressed as follows.

$$\begin{cases} \dot{x}_1 = x_2 \\ A_1 = x_1 - R + \dfrac{x_2|x_2|}{2R_1} \\ \dot{x}_2 = -R_1 Sat(A_1, \delta_1) \end{cases} \tag{16}$$

Similarly, for tracking differentiator:

$$\begin{cases} \dot{x}_3 = x_4 \\ A_2 = x_3 - Y + \dfrac{x_4|x_4|}{2R_2} \\ \dot{x}_4 = -R_2 Sat(A_2, \delta_2) \end{cases} \tag{17}$$

Where Y denotes the feedback position of object.

$$Sat(A,\delta) = \begin{cases} sgn(A), & |A| \geq \delta \\ \dfrac{A}{\delta}, & |A| < \delta \end{cases} \tag{18}$$

Where δ is constant, and empirically, $\delta = 0.1\sim0.2$.

Therefore, we can get the NLPID combination as follows.

$$\begin{cases} \dot{x}_5 = x_1 - x_4 \\ e_0 = x_5 \end{cases} \tag{19}$$

$$\begin{cases} U = k_i fal(e_0, \alpha, \delta) + k_p fal(e_1, \alpha, \delta) \\ \qquad + k_d fal(e_2, \alpha, \delta) \\ \dot{x}_6 = -\rho_0(x_6 - u_1) \end{cases} \tag{20}$$

Where α is constant, and empirically, $\alpha = 0.5\sim1.0$. k_i denotes integral coefficient, k_p denotes proportional coefficient, k_d denotes differential coefficient. The proportional term plays the essential role of making the controller respond to the error while the integral and derivative helps to eliminate steady state error and prevent overshoot respectively. $fal(e,\alpha,\delta)$ is a selected nonlinear function. Fig. 3 shows the characteristics of the $fal(e,\alpha,\delta)$ function. A linear relationship is thus effectively used when $e < |\delta|$. Therefore, the $fal(e,\alpha,\delta)$ function can provide a smoother control action when e is near zero.

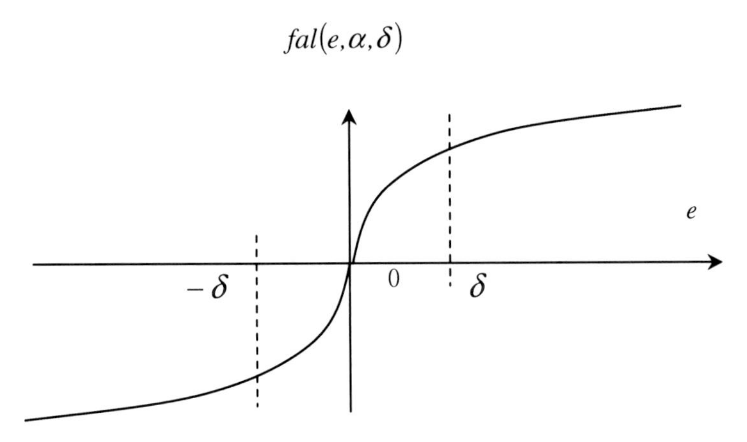

Fig. 3. Diagram of $fal(e,\alpha,\delta)$ function characteristics

Suppose the system error $\varepsilon = R - Y$, subsequently, Let the objective function be as follows.

$$J = \int_0^t \varepsilon^2(t)dt \tag{21}$$

Eq. (21) is also named cost function.

6 Grid-Based ACO Algorithm

The flight simulator architecture of NLPID controller with the grid-based ACO algorithm parameters tuning is constructed as in Fig. 4.

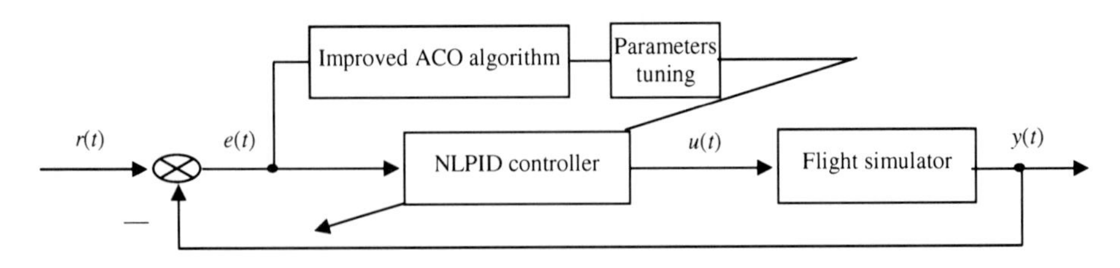

Fig. 4. Flight simulator architecture of NLPID controller with the grid-based ACO algorithm

The parameters tuning of NLPID can be summed up as the typical continual spatial optimization problem[10], and the general method of solving continual space optimization question using the grid-based ACO algorithm is generalized as follows.

Firstly, we may estimate the optimal solution the scope according to the problem nature, the estimated value scope of the variable can be described as $x_{jlower} \le x_j \le x_{jupper} (j = 1,2,3,\cdots,n)$. Secondly, we can divide the grids in the variable region, and the spatial grid corresponds to a condition. The ants move between different grid points, and leave different pheromone amount according to various mesh points goal function value, which can influence next batch of ants' travel direction[8, 9]. After periods of time, the pheromone amount corresponding to the neighboring point objective function value is quite big. The mesh point with biggest pheromone amount could be found according to the pheromone amount. The variable scope is minimized, and the ant colony move around this point. This iteration process goes on until a certain termination condition: a certain number of iterations have been achieved; a fixed amount of CPU time has elapsed; or solution quality has been achieved.

Suppose the variable is decomposed into n parts, therefore the n variables composed the decision problem with n grades. There are N+1 mesh points in each grade, and there are $(N+1)\times n$ mesh points from grade one to grade n. By this way, a solution in the solution space is formulated. The above process can be illustrated in Fig. 5.

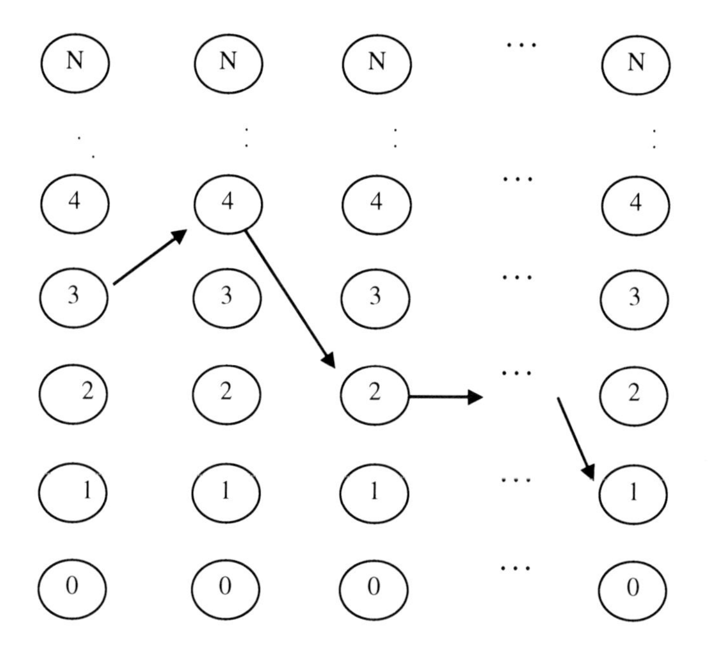

Fig. 5. Schematic figure of state space solution

In Fig 5, the state space shows the condition is (3, 4, 2, …, 1), and the corresponding solution is as follows.

$$(x_1, x_2, x_3, \cdots, x_n) = (x_{1lower} + \frac{x_{1upper} - x_{1lower}}{N} \times 3, x_{2lower} + \frac{x_{2upper} - x_{2lower}}{N} \times 4, \\ x_{3lower} + \frac{x_{3upper} - x_{3lower}}{N} \times 2, \cdots, x_{nlower} + \frac{x_{nupper} - x_{nlower}}{N} \times 1) \tag{22}$$

The performance index of smallest Integrated Time Absolute Error (ITAE) was first proposed by Graham D and Lathrop L C. The ITAE seldom considers the big initial error, and emphasizes the overshoot and the control time. ITAE reflects the rapidity and accuracy of the control system, and is widely used in the field of control technology[10]. The formula of ITAE is as follows.

$$J_{\mathrm{ITAE}} = \int t|e(t)|dt = \min \tag{23}$$

Discretize the above formula, can result in the difference equation as follows.

$$J_{\mathrm{ITAE}} = \sum_{k=0}^{N} |e(kT)| kT = \min \tag{24}$$

Where T denotes the step of simulation, N denotes the nodes number of real-time simulation.

Let m is the number of the whole ant colony, and m ants are scattered randomly on the mesh nodes. In discrete time steps, all ants select their next mesh node then simultaneously move to their next node[11~13]. For each ant l, we can define the appraisal function value as the difference between J_i and J_j, which can be shown as follows.

$$\Delta J_{ij} = J_i - J_j, \quad \forall i, j \tag{25}$$

Ants deposit pheromone on each edge they visit to indicate the utility of these edges. The accumulated strength of pheromone on edge (i, j) is denoted by τ_{ij}. We assume that an real ant l located at node i chooses its next node j by applying Eq. (26), which is also called state transition rule.

$$P_{ij} = \begin{cases} \dfrac{[\tau_{ij}]^{\alpha}[\Delta J_{ij}]^{\beta}}{\sum_{r \in allowed_l} [\tau_{ir}]^{\alpha}[\Delta J_{ir}]^{\beta}}, & \text{if } j \in allowed_l \\ 0, & \text{otherwise} \end{cases} \tag{26}$$

Where $allowed_l$ denotes the set of mesh points available for ant l, α is the pheromone heuristic parameter, β weighs the relative importance of the heuristic value. If $\beta = 0$ only pheromone amplification will occur and the distance between nodes has no direct influence on the choice.

We may spread the ants in the spatial mesh grids according to the stochastic principle, and record the elitist ant which has the best objective function value. Then, we can move each ant according to Eq. (26). Neighbor search mechanism is adopted in the search process, *e. g.* when $\Delta J_{ij} > 0$, the ant l migrate from neighborhood i to neighborhood j according to Eq(26). When $\Delta J_{ij} \le 0$, the ant l will carry on neighborhood search. In this way, a better solution can be found.

Once all ants have constructed a tour, global updating of the pheromone takes place. Here, we use elitist strategy: edges that compose the best solution are rewarded with a relatively large increase in their pheromone level. This is expressed in Eq. (27).

$$\begin{cases} \tau_{ijNew} = \rho\tau_{ij} + \Delta\tau_{ij} \\ \Delta\tau_{ij} = \sum_{l=1}^{m} \Delta\tau_{ij}^{l} \end{cases} \tag{27}$$

Where ρ is the pheromone decay parameter, and $\rho \subset (0,1)$. $\Delta\tau_{ij}^{l}$ is used to reinforce the pheromone on the edge (i, j), which can be calculated by using Eq. (28) [14].

$$\Delta\tau_{ij}^{l} = \begin{cases} \frac{Q}{J_l}, & \text{if ant } l \text{ transverse edge } (i,j) \\ 0, & \text{otherwise} \end{cases} \tag{28}$$

Where Q is a constant, J_l is the objective function value of ant l in this cycle.

The bigger the parameter ρ, the bigger the iteration time N_C, and the convergence speed of the grid-based ACO algorithm will become slow. Otherwise, although the convergence speed become fast, but the calculation result is easy to fall into local best. Therefore, in order to enhance the resolution efficiency of the grid-based ACO algorithm, self-adaptive control strategy for the parameter ρ is adopted [14], this strategy can be illustrated by Eq. (29).

$$\rho(t+1) = \begin{cases} 0.9 \cdot \rho(t), & \text{if } 0.9 \cdot \rho(t) > \rho_{\min} \\ \rho_{\min}, & \text{otherwise} \end{cases} \tag{29}$$

Where 0.9 is the pre-determined evaporation restraint coefficient, $\rho_{\min}$ denotes lower limit of the pheromone decay parameter.

In order to enhance the searching speed, pair ants which search from two different mesh points at the same time are used [15]. This parallel processing strategy may effectively enhance the global convergence speed of the grid-based ACO algorithm.

7 Experimental Results

The NLPID parameters tuning strategy proposed in the previous section can be appraised via typical simulation tests. Here, we take a novel kind of flight simulator as an example, which has high requirements for the tracking property, *i. e.* high precision, fast tracking response and high reliability. Flight simulator is an important device and a typical high performance servo system used in the hardware-in-the-loop simulation of flight control system. It can be used to simulate the dynamic characteristics and various flying posture. Flight simulator is usually driven by using three

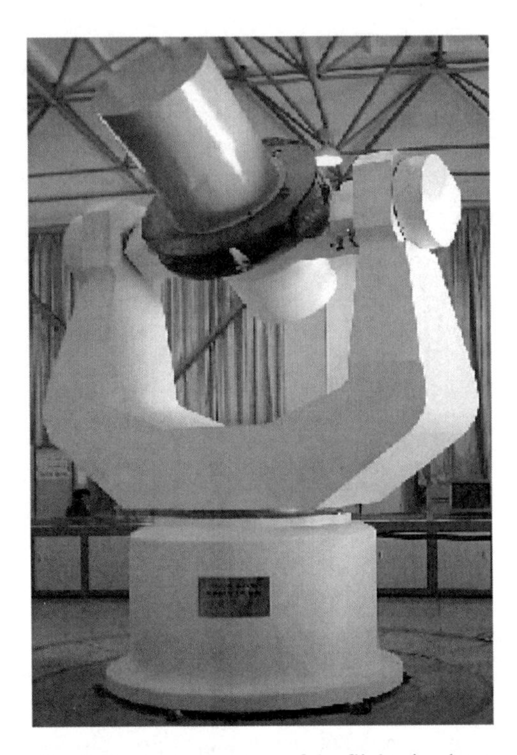

Fig. 6. The body structure of the flight simulator

Fig. 7. The control system of the flight simulator

D. C. motors, which have three rotating parts, namely roll axis including simulation object; pitch axis including roll motor; and yaw axis including pitch motor [16]. Fig. 6 shows the body structure of a type of flight simulator considered in this chapter, the

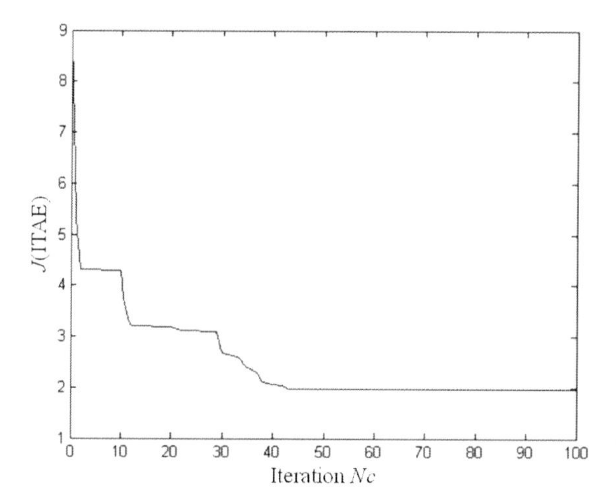

Fig. 8. Evolution curve of objective function with iteration number

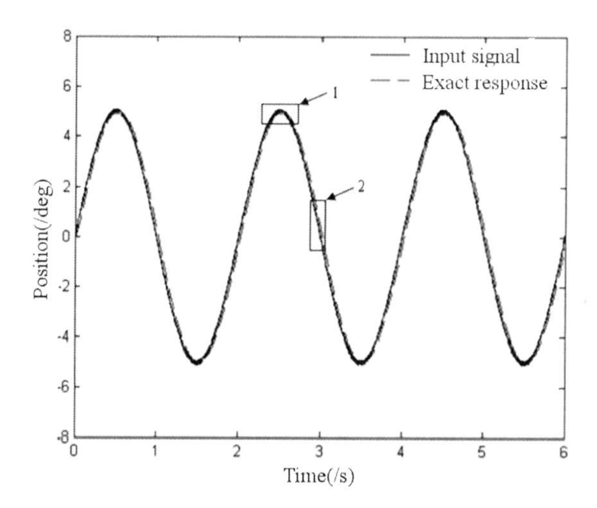

Fig. 9. The standard sine wave response of flight simulator with grid-based ACO algorithm optimized NLPID

whole structure is designed to be axis-symmetric. Fig. 7 is a photograph of the control system of the type of flight simulator.

In the experiments, the sampling time of the flight simulator is 0.0008s. The grid-based ACO algorithm parameters were set to the following values: α=1.5, $\beta = 4.5$, $\rho_{\min} = 0.2$, $\tau_{ij} = 1$, $\Delta\tau_{ij}(0) = 0$, $m = 20$, $\rho_0 = 0.15$, R_1=150,R_2=100, $\delta_1 = 0.002$, $\delta_2 = 0.0013$, $\delta = 0.01$, $\alpha' = 0.6$, N_{Cmax}=100, Q=300. After trying several times, the scope of the control coefficients could be determined as $\beta_0 \subset [0.01, 0.09]$,

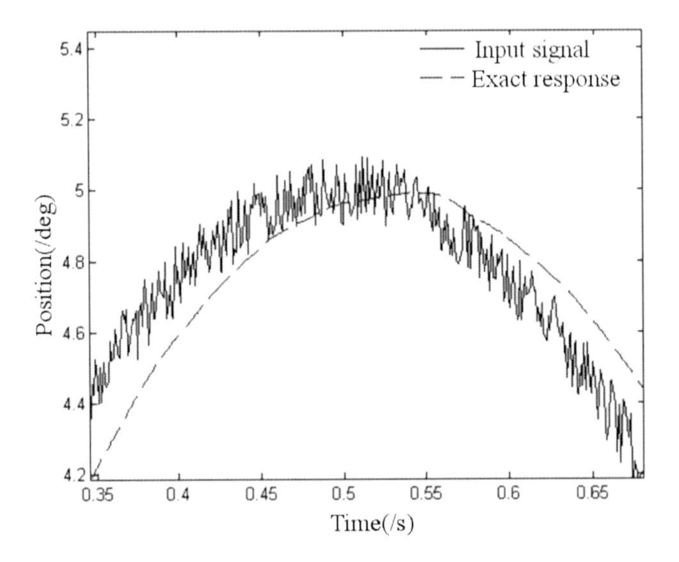

Fig. 10. No. 1 partial enlarged figure of Fig 6

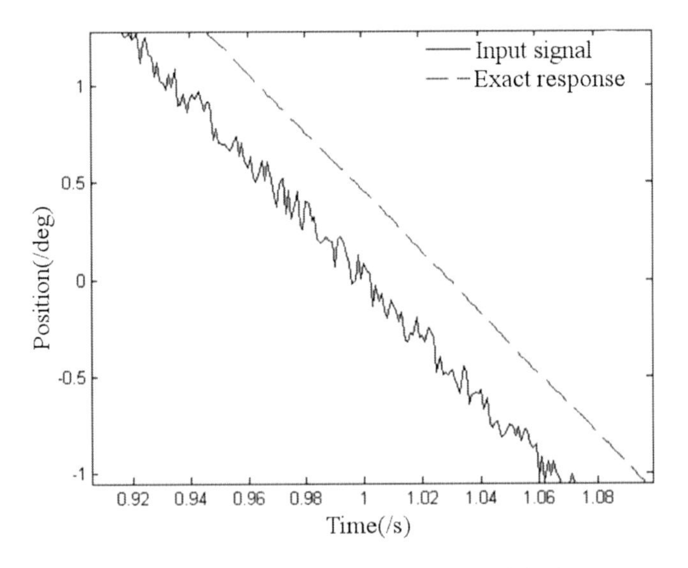

Fig. 11. No. 2 partial enlarged figure of Fig 6

$\beta_1 \subset [770,790]$, $\beta_2 \subset [0.1,0.7]$. The simulation experiments have been conducted 10 times, and the obtained average optimal solution values are $\overline{\beta}_0^* = 0.03$, $\overline{\beta}_1^* = 776.39$, $\overline{\beta}_3^* = 0.54$. The evolution process of objective function J with iteration number can be shown in Fig 8.

From the evolution curve in Fig. 8, it is obvious that although the optimal objective function is diverge in the initial period, but still could converge with relatively quick speed. This process has good convergence characteristic and strong robustness.

To assess the effectiveness of the grid-based ACO algorithm optimized nonlinear PID parameters, we also use test the tracking quality of the whole system of flight simulator. The standard sine wave response is shown in Fig. 9, and the corresponding partial enlarged figures are shown in Fig. 10 and Fig. 11. The random wave response with white noise is shown in Fig. 12.

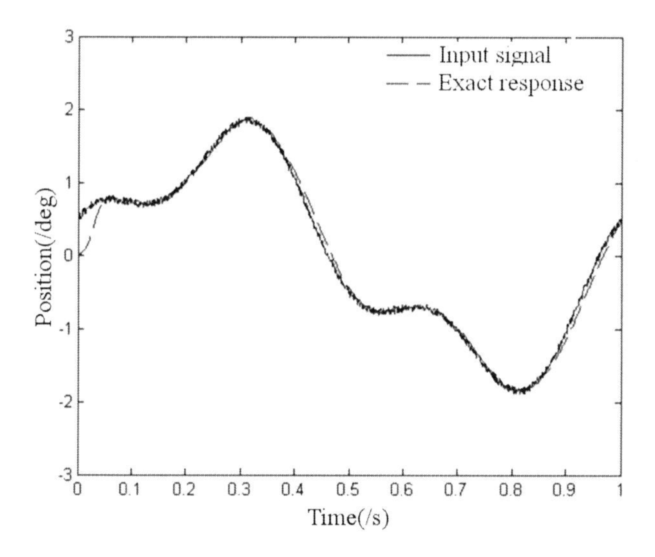

Fig. 12. The random input signal response of flight simulator with grid-based ACO algorithm optimized NLPID

It is clear that the tracking error is very small for the standard and random input signals, and the whole control system of the flight simulator with grid-based ACO algorithm optimized NLPID has strong robustness and high precision.

8 Conclusions

ACO algorithm has shown great advantages in solving combination optimization problems recently, because it is realized easily, and its adaptability is extraordinary good. It is significant to enlarge the application for ACO algorithm further.

In this chapter, we have introduced a grid-based ACO algorithm approach for tuning the parameters of NLPID controller. From the results of simulation experiments, it is concluded that this proposed NLPID controller using the grid-based ACO algorithm is very effective. It possesses good control and robust performance, and can be widely used to control different kind of objects and processes.

References

1. Colorni, A., Dorigo, M., Maniezzo, V., et al.: Distributed optimization by ant colonies. In: Proceedings of the 1st European Conference on Artificial Life, Paris, pp. 134–142 (1991)
2. Bonabeau, E., Dorigo, M., Theraulaz, G.: Inspiration for optimization from social insect behavior. Nature 406(6), 39–42 (2000)
3. Geis, M., Middendorf, M.: Creating melodies and baroque harmonies with ant colony optimization. International Journal of Intelligent Computing and Cybernetics 1(2), 213–238 (2009)
4. Duan, H.B.: Ant colony algorithms: theory and applications. Science Press, Beijing (2005)
5. Duan, H.B., Wang, D.B., Zhu, J.Q., et al.: Development on Ant Colony Optimization theory and its application. Control and Decision 19(12), 1321–1326, 1340 (2004)
6. Dorigo, M., Maniezzo, V., Colorni, A.: The ant system: optimization by a colony of cooperating agents. IEEE Transactions on Systems, Man, and Cybernetics-Part B 26(1), 29–41 (1996)
7. Tan, K.K., Lee, T.H., Zhou, H.X.: Micro-positioning of linear-piezoelectric motors based on a learning nonlinear PID controller. IEEE/ASME Trans. on Mechatronics 6(4), 428–436 (2001)
8. Han, J.Q.: Nonlinear PID controller. Acta Automatica Sinica 20(4), 487–490 (1994)
9. Stüezle, T., Dorigo, M.: A short convergence proof for a class of ant colony optimization algorithms. IEEE Transactions on Evolutionary Computation 6(4), 358–365 (2002)
10. Xiong, W.Q., Wei, P.: A kind of ant colony algorithm for function optimization. In: Proceedings of the 1st International Conference on Machine Learning and Cybernetics, Beijing, pp. 552–555 (2002)
11. Duan, H.B., Wang, D.B.: A novel improved ant colony algorithm with fast global optimization and its simulation. Information & Control 33(2), 193–197 (2004)
12. Duan, H.B., Wang, D.B., Yu, X.F.: Research on the optimum configuration strategy for the adjustable parameters in ant colony algorithm. Journal of Communication and Computer 2(9), 32–35 (2005)
13. Ma, L., Yao, J., Fan, B.Q.: The application of ant algorithm in traffic assignment. Bulletin of Science and Technology 19(5), 377–380 (2003)
14. Wang, Y., Xie, J.Y.: An adaptive ant colony optimization algorithm and simulation. Journal of System Simulation 14(1), 31–33 (2002)
15. Duan, H.B., Wand, D.B., Yu, X.F.: Grid-based ACO algorithm for parameters tuning of NLPID controller and its application in flight simulator. International Journal of Computatio-nal Methods 3(2), 163–175 (2006)
16. Duan, H.B., Wand, D.B., Yu, X.F.: A novel approach to the convergence of ant colony algorithm and its Matlab GUI-based realization. International Journal of Plant Engineering and Management 11(2), 124–128 (2006)

Optimization of Fuzzy Logic Controllers for Robotic Autonomous Systems with PSO and ACO

Oscar Castillo, Patricia Melin, Fevrier Valdez, and Ricardo Martínez-Marroquín

Tijuana Institute of Technology, Tijuana México
ocastillo@hafsamx.org

Abstract. In this chapter we describe the application of Ant Colony Optimization (ACO) and Particle Swarm Optimization (PSO) to the optimization of the membership functions' parameters of a Fuzzy Logic Controller (FLC) in order to find the optimal intelligent controller for an Autonomous Wheeled Mobile Robot. The results obtained by the simulations performed are statistically compared among them and with the results obtained with genetic algorithms in previous work in order to find the best optimization technique for this particular robotics problem.

1 Introduction

The complexity in developing fuzzy systems can be found at the time of deciding which are the best parameters of the membership functions, the number of rules or even the best granularity that could give us the best solution for the problem that we want to solve [3].

A solution to the above mentioned problem is the application of bio-inspired algorithms for the optimization of fuzzy systems. Optimization algorithms are a useful tool for this problem due to their capabilities in solving nonlinear problems, well-constrained or even NP-hard problems [5, 6]. Among the most used optimization methods we can find [11]: Genetic Algorithms (GA), Ant Colony Optimization (ACO), Particle Swarm Optimization (PSO), etc.

This paper describes the application of ACO and PSO as optimization methods for the membership functions' parameters of the FLC in order to find the best possible intelligent controller for an autonomous wheeled mobile robot and making a statistical comparison of the techniques. Recently, PSO has been applied to the solution of diverse real-world problems, like in [4, 12, 17, 25]. Also, ACO has been successfully applied in complex optimization problems, like in [18, 19].

This chapter is organized as follows: Section 2 shows the basic concepts of ACO and PSO and a description of the S-ACO and gbest algorithms respectively, which are the techniques that were applied for optimization. Section 3 presents the

B.K. Panigrahi, Y. Shi, and M.-H. Lim (Eds.): Handbook of Swarm Intelligence, ALO 8, pp. 389–417.
springerlink.com

problem statement and the dynamic and kinematic model of the unicycle mobile robot. Section 4 shows the proposed fuzzy logic controller and in Section 5 the development of the optimization methods is described. In Section 6 the simulation results are presented. Section 7 describes the statistical results for the comparison of the optimization methods Finally, Section 8 shows the Conclusions.

2 S-ACO and *gbest* PSO Algorithms

This section describes the theoretical basis of the algorithms applied in this work. The algorithms we used for FLC optimization were the simplest versions of ACO and PSO respectively.

2.1 S-ACO Algorithm

ACO is a probabilistic technique that can be used for solving problems that can be reduced to finding good paths along graphs. This method is inspired on the behavior presented by ants in finding paths from the nest or colony to the food source.

The S-ACO is an algorithmic implementation that adapts the behavior of real ants to solutions of minimum cost path problems on graphs [6]. A number of artificial ants build solutions for a certain optimization problem and exchange information about the quality of these solutions making allusion to the communication system of real ants [7, 8].

Let us define the graph $G=(V,E)$, where V is the set of nodes and E is the matrix of the links between nodes. G has $n_G=|V|$ nodes. Let us define L^K as the number of hops in the path built by the ant k from the origin node to the destiny node. Therefore, it is necessary to find:

$$Q=\{q_a,\dots,q_f \mid q_1\in C\} \tag{1}$$

where Q is the set of nodes representing a continuous path with no obstacles; $q_a,\dots,q_f$ are former nodes of the path and C is the set of possible configurations of the free space. If $x^k(t)$ denotes a Q solution in time t, $f(x^k(t))$ expresses the quality of the solution. The S-ACO algorithm is based on Equations (2), (3) and (4):

$$p_{ij}^k(t)=\begin{cases}\dfrac{\tau_{ij}^k}{\sum_{j\in N_{ij}^k}\tau_{ij}^{\alpha}(t)} & if \;\; j\in N_i^k\\ 0 & if \;\; j\notin N_i^k\end{cases} \tag{2}$$

$$\tau_{ij}(t)\leftarrow(1-\rho)\tau_{ij}(t) \tag{3}$$

$$\tau_{ij}(t+1)=\tau_{ij}(t)+\sum_{k=1}^{n_k}\Delta\tau_{ij}^{k}(t) \tag{4}$$

Equation (2) represents the probability for an ant k located on a node i selects the next node denoted by j, where, N_i^k is the set of feasible nodes (in a neighborhood) connected to node i with respect to ant k, τ_{ij} is the total pheromone concentration of link ij, and α is a positive constant used as a gain for the pheromone influence.

Equation (3) represents the evaporation pheromone update, where $\rho \in [0,1]$ is the evaporation rate value of the pheromone trail. The evaporation is added to the algorithm in order to force the exploration of the ants, and avoid premature convergence to sub-optimal solutions [18]. For $\rho = 1$ the search becomes completely random [8].

Equation (4), represents the concentration pheromone update, where $\Delta\tau_{ij}^{k}$ is the amount of pheromone that an ant k deposits in a link ij in a time t.

The general steps of S-ACO are the following:

1. Set a pheromone concentration τ_{ij} to each link (i,j).
2. Place a number k=1, 2,..., n_k in the nest.
3. Iteratively build a path to the food source (destiny node), using Equation (2) for every ant.

- Remove cycles and compute each route weight $f\left(x^k(t)\right)$. A cycle could be generated when there are no feasible candidates nodes, that is, for any i and any k, $N_i^k = \varnothing$; then the predecessor of that node is included as a former node of the path.

4. Apply evaporation using Equation (3).
5. Update of the pheromone concentration using equation (4)
6. Finally, finish the algorithm in any of the three different ways:
   - When a maximum number of epochs has been reached.
   - When it has been found an acceptable solution, with $f(x_k(t)) < \varepsilon$.
   - When all ants follow the same path.

2.2 *gbest* PSO Algorithm

PSO is a stochastic optimization technique based on population inspired in the social behavior of big masses of birds, fish or bees.

In PSO, the potential solutions (called particles), fly trough the space problem, where the less optimum particles fly to optimum particles, doing this iteratively until all the particles converge at the same point (solution). To achieve convergence, PSO applies two types of knowledge, personal experience or *cognitive component*, which is the experience that every particle gets along optimization process; and social experience or *social component*, which is the experience that all the swarm gets during the optimization process.

In simple terms, the particles are "flown" through a multidimensional search space where the position of each particle is adjusted according to its own experience and that of its neighbors [11].

Let $x_i(t)$ denote the position of a particle i in search space at time step t. The position of the particle is changed by adding a velocity $v_i(t)$, to the current position

$$x_i(t) = x_i(t) + v_i(t+1) \tag{5}$$

with $x_i(0) \sim U(x_{min}, x_{max})$.

Originally two PSO algorithms have been developed which differ in the size of their neighborhood [10]. These two algorithms are called *lbest* and *gbest*. In the *lbest* algorithm the swarm is divided in neighborhoods of size n while in the *gbest* the neighborhood for each particle is the entire swarm, or in other words, the social component of the particle velocity update reflects information obtained from all the particles in the swarm [23, 24]. For the *gbest* PSO, the velocity of particle i is calculated as

$$v_{ij}(t+1) = v_{ij}(t) + c_1 r_{1j}(t)\left[y_{ij}(t) - x_{ij}(t)\right] + c_2 r_{2j}(t)\left[\hat{y}_j(t) - x_{ij}(t)\right] \tag{6}$$

where $v_{ij}(t)$ is the velocity of particle i in dimension $j = 1,2,3,\dots,n_x$ at time step t, $x_{ij}(t)$ is the position of particle i in dimension j at time step t, c_1 and c_2 are positive acceleration constants used to scale the contribution of the cognitive and social components respectively, $r_1(t)$ and $r_2(t) \sim U(0,1)$ are random values in the range [0, 1], sampled for uniform distribution. The random values introduce a stochastic element to the algorithm [11]. y_i is the best position that a particle i since the first time step and $\hat{y}$ is the best position found all the particles in the swarm.

Considering minimization problems, the personal best y_i is updated by equation (7)

$$y_i(t+1) = \begin{cases} y_i(t) & \text{if } f(x_i(t+1)) \geq y_i(t) \\ x_i(t+1) & \text{if } f(x_i(t+1)) < y_i(t) \end{cases} \tag{7}$$

where $f: \mathbb{R}^{n_x} \rightarrow \mathbb{R}$ is the fitness function.

The global best position, $\hat{y}(t)$, at time step t is defined by equation (8)

$$\hat{y}(t) \in \left\{ y_0(t), \dots, y_{n_s}(t) \middle| f(\hat{y}(t)) = \mathbf{min}\left\{ f(y_0(t)), \dots, f(y_{n_s}(t)) \right\} \right\} \tag{8}$$

where n_s is the size of the swarm. Table 1 shows the *gbest* PSO algorithm.

Table 1. *gbest* PSO

```
Create an initialize an n_x-dimensional swarm, S;
repeat
    for each particle I = 1,2,...,S.n_s do
        //set the personal best position If f(x_i(t+1)) < y_i(t) then
            S.y_i = S.x_i;
        end
        //set the global best position If S.y_i(t) < f(ŷ_i(t)) then
            S.ŷ_i = S.y_i;
        end
    end
    for each particle I = 1,2,...,S.n_s do
        update the velocity using equation (2)
        update the position using equation (1)
    end
    until stopping condition is true;
```

3 Problem Statement

The model of the robot considered in this chapter is that of a unicycle mobile robot (see Figure 1), that consists of two driving wheels mounted of the same axis and a front free wheel.

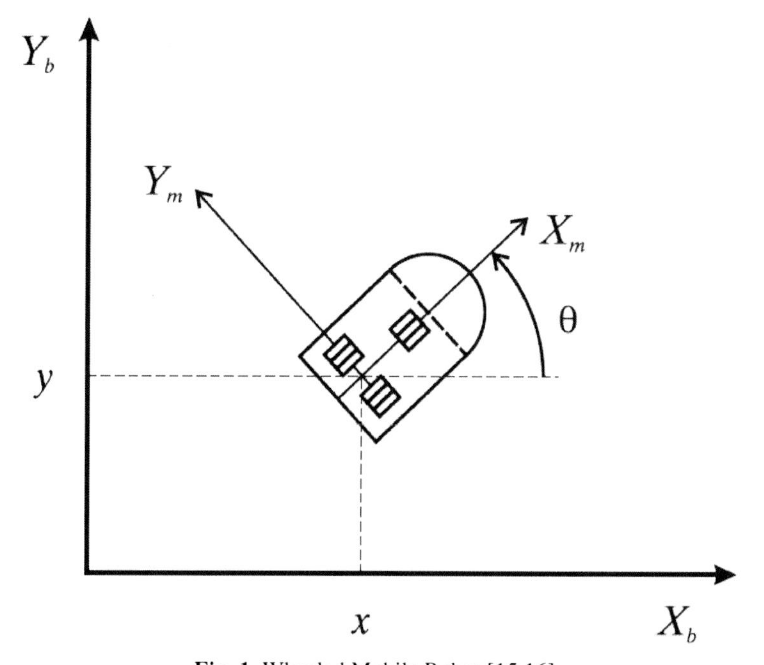

Fig. 1. Wheeled Mobile Robot [15,16]

A unicycle mobile robot is an autonomous, wheeled vehicle capable of performing missions in fixed or uncertain environments [13]. The robot body is symmetrical around the perpendicular axis and the center of mass is at the geometrical center of the body. It has two driving wheels that are fixed to the axis that passes through C and one passive wheel prevents the robot from tipping over as it moves on a plane. In what follows, it´s assumed that motion of the passive wheel can be ignored in the dynamics of the mobile robot presented by the following set of equations [8]:

$$M(q)\dot{\vartheta}+C(q,\dot{q})\vartheta+D\vartheta=\tau+F_{ext}(t) \tag{9}$$

$$\dot{q}=\underbrace{\begin{bmatrix}\cos\theta & 0\\ \sin\theta & 0\\ 0 & 1\end{bmatrix}}_{J(q)}\underbrace{\begin{bmatrix}v\\ w\end{bmatrix}}_{\vartheta} \tag{10}$$

where $q=(x,y,\theta)^T$ is the vector of the configuration coordinates; $\vartheta=(v,w)^T$ is the vector of linear and angular velocities; $\tau=(\tau_1,\tau_2)$ is the vector of torques applied to the wheels of the robot where τ_1 and τ_2 denote the torques of the right and left wheel respectively (Figure 1); $F_{ext}\in\mathbb{R}^2$ uniformly bounded disturbance vector; $M(q)\in\mathbb{R}^{2x2}$ is the positive-definite inertia matrix; $C(q,\dot{q})\vartheta$ is the vector of centripetal and Coriolis forces; and $D\in\mathbb{R}$ is a diagonal positive-definite damping matrix. Equation (10) represents the kinematics of the system, where (x, y) is the position of the mobile robot in the X-Y (world) reference frame, θ is the angle between heading direction and the x-axis v and w are the angular and angular velocities, respectively.

Furthermore, the system (9)-(10) has the following non-holonomic constraint [1]:

$$\dot{y}\cos\theta-\dot{x}\sin\theta=0 \tag{11}$$

which corresponds to a no-slip wheel condition preventing the robot from moving sideways[14]. The system (10) fails to meet Brockett's necessary condition for feedback stabilization [2], which implies that non-continuous static state-feedback controller exists that stabilizes the close-loop system around the equilibrium point.

The control objective is to design a fuzzy logic controller of τ that ensures:

$$\lim_{t\to\infty}\|q_d(t)-q(t)\|=0 \tag{12}$$

for any continuously, differentiable, bounded desired trajectory $q_d\in\mathbb{R}^3$ while attenuating external disturbances.

A more detailed description can be found on references [15, 16].

4 Fuzzy Logic Control Design

In order to satisfy the control objective it is necessary to design a fuzzy logic controller for the real velocities of the mobile robot. To do that, a Takagi-Sugeno fuzzy logic controller was designed, using linguistic variables in the input and

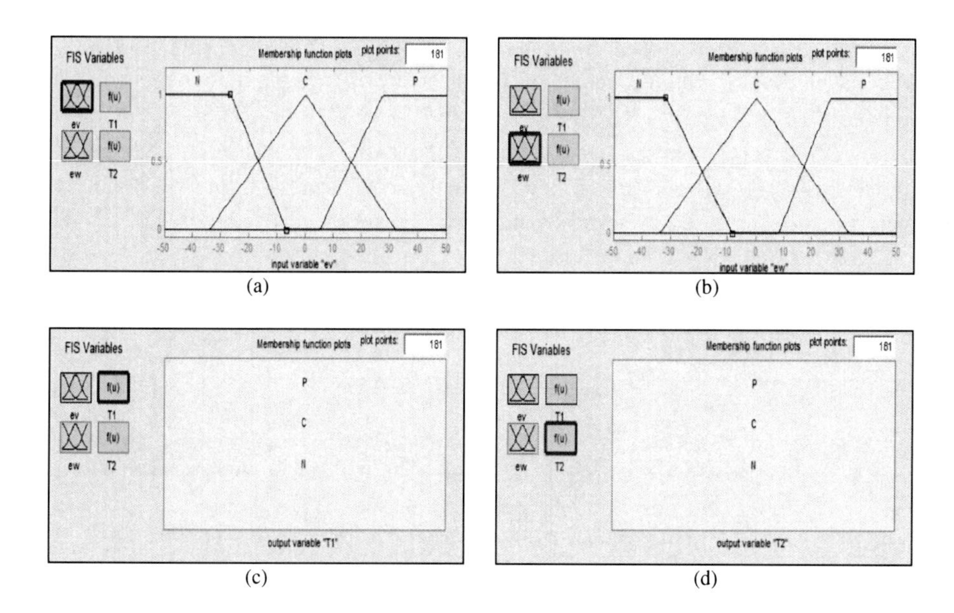

Fig. 2. (a) Linear velocity error. (b) Angular velocity error. (c) Right output (τ_1). (d) Left output (τ_2).

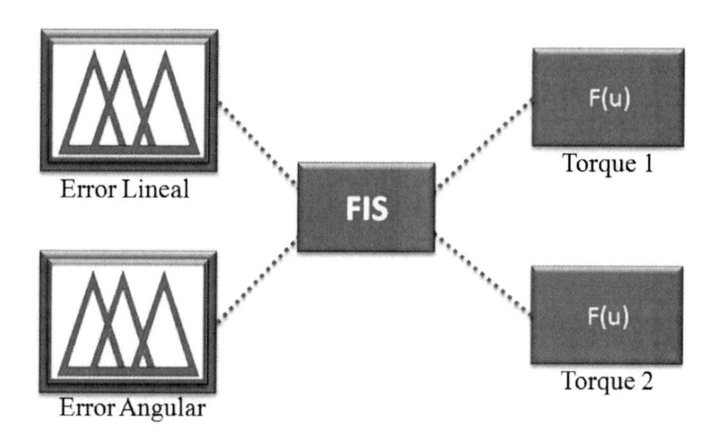

Fig. 3. Fuzzy Logic Controller Architecture.

mathematical functions in the output. The error of the linear and angular velocities (v_d, w_d respectively), were taken as inputs variables, while the right (τ_1) and left (τ_2) torques as outputs. The membership functions used on the input are trapezoidal for the negative (*N*) and positive (*P*), and a triangular was used for the zero (*C*) linguistic terms. The interval used for this fuzzy controller is [-50 50] [15,16]. Figure 2 shows the input and output variables, and figure 3 shows the general FLC architecture.

The rule set of the FLC contains 9 rules, which governs the input-output relationship of the FLC and this adopts the Takagi-Sugeno style inference engine

[15, 16], and it is used with a single point in the outputs, this means that the outputs are constant values, obtained using weighted average defuzzification procedure. In Table 2 we present the rule set whose format is established as follows:

$$\text{Rule } i: \text{if } e_v \text{ is } G_1 \text{ and } e_w \text{ is } G_2 \text{ then } F \text{ is } G_3 \text{ and } N \text{ is } G_4$$

where $G_1..G_4$ are the fuzzy set associated to each variable i=1,2,…,9.

Table 2. Fuzzy rules set

e_v / e_w	N	Z	P
N	N / N	N / Z	N / P
Z	Z / N	Z / Z	Z / P
P	P / N	P / Z	P / P

To find the best FLC, we used a *gbest* PSO to find the parameters of the membership functions. Table 3 shows the parameters of the membership functions, the minimal and maximum values in the search range for the *gbest* PSO algorithm to find the best fuzzy logic controller.

Table 3. Parameters of the membership Functions.

MF TYPE	POINT	MINIMAL VALUE	MAXIMAL VALUE
Trapezoidal	a	-50	-50
	b	-50	-50
	c	-15	-5.1
	d	-1.5	-0.5
Triangular	a	-5	-1.8
	b	0	0
	c	1.8	5
Trapezoidal	a	0.5	1.5
	b	5.1	15
	c	50	50
	d	50	50
Constant (N)	a	-50	-50
Constant (C)	a	0	0
Constant (P)	a	50	50

It is important to remark that the values shown in Table 2 are applied to both inputs and outputs of the fuzzy logic controller. As we can see on Table 3 there are several values that will not change with the optimization process because their minimal and maximum values are the same; because of that we decided to ignore the constant

Table 4. Parameters of the MFs Included in optimization Process.

MF TYPE	POINT	MINIMAL VALUE	MAXIMAL VALUE
Trapezoidal	c	-15	-5.1
	d	-1.5	-0.5
Triangular	a	-5	-1.8
	c	1.8	5
Trapezoidal	a	0.5	1.5
	b	5.1	15

values and let them out of the optimization process. Table 4 shows the parameters include in optimization process.

5 Optimization of the Fuzzy Controllers

5.1 ACO Architecture for the FLC Optimization

A S-ACO algorithm was applied for the optimization of the membership functions for the type-1 fuzzy logic controller. For developing the architecture of the algorithm it was necessary to follow the next steps:

1. Representing the architecture of the FLC as a graph that artificial ants could traverse.
2. Achieving an adequate handling of the pheromone but permitting the algorithm to evolve by itself.

5.1.1 Limiting the Problem and Graph Representation

One of problems found on the development of the S-ACO algorithm was to make a good representation of the FLC. First step was to represent those parameters shown in Table 4; to do that it was necessary to discretize the parameters in a range of possible values in order to represent every possible value as a node in the search graph. The level of discretization between minimal and maximal value was of 0.1 (by example: -1.5, -1.4, -1.3,…, -0.5).

Table 5 shows the number of possible values that each parameter can take.

Table 5. Number of Possible Values of the Parameters of MFs of the type-1 FLC

Mf Type	Point	Combinations
Trapezoidal	c	100
	d	15
Triangular	a	33
	c	33
Trapezoidal	a	15
	b	100

Figure 4 shows the search graph for the proposed S-ACO algorithm, the graph can be viewed as a tree where the root is the nest and the last node is the food source.

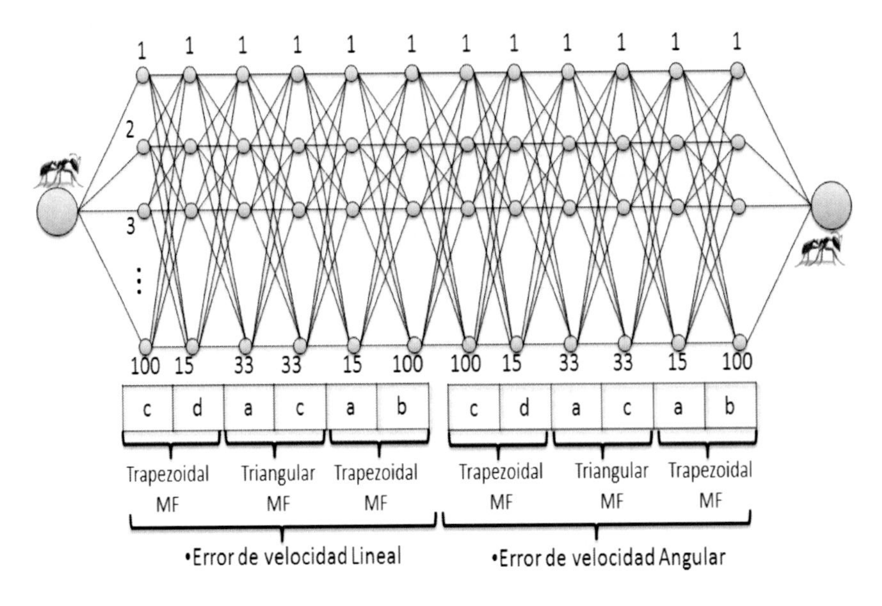

Fig. 4. S-ACO Architecture

5.1.2 References Updating Pheromone Trail

An important issue is that the update of pheromone trail has to be applied in the best way possible. In this sense we need to handle the evaporation (Equation (3)), and increase or deposit of pheromone (Equation (4)), where the key parameter in evaporation is denoted by ρ that represents the rate of evaporation and in deposit of pheromone is denoted by $\Delta\tau$ that represents the amount of pheromone that an ant *k* deposits in a link *ij* in a time *t*. For ρ we assign a random value and Equation (13) shows the way of how the increase of pheromone is calculated.

$$\Delta\tau = \frac{(e_{\max} - e_k)}{e_{\max}} \tag{13}$$

where $e_{\max} = 10$ is the maximum error of control permitted and e_k is error of control generated by a complete path of an ant *k*. We decided to allocate $e_{\max} = 10$ in order to stand $\Delta\tau \in [0,1]$. We also defined $\alpha = 0.2$ (see equation (2)) for all experiments presented in this paper because of the results obtained after the tests. And finally we use a random value for the evaporation coefficient ρ = [0,1] (see equation (3)) in each iteration in order to simulate the uncertain environment conditions which affects the real pheromone trails deposited by real ants.

5.2 PSO Architecture (*gbest*)

This section describes the PSO architecture used for the optimization of the FLC for the autonomous mobile robot. An important issue was achieving an adequate handling of the parameters c_1, c_2 and velocity v_{ij} that ensures a good performance of the algorithm.

The First step was coding the parameters shown in Table 4 into a multi-dimensional space of search. Figure 4 shows the representation of a particle in PSO.

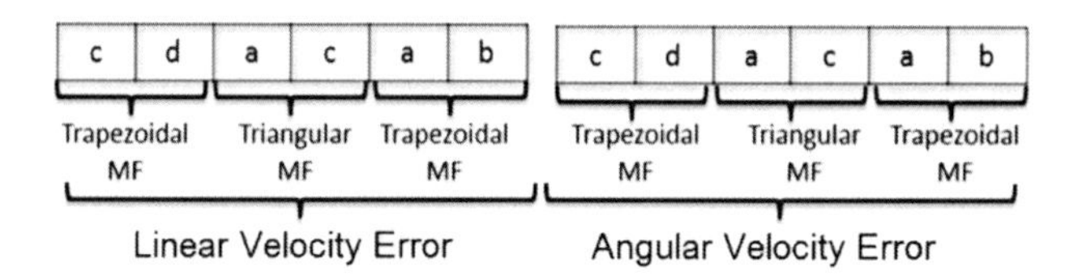

Fig. 5. *gbest* PSO Particle.

5.3 PSO Parameters Handling

Before the application of the PSO, we needed to initialize the first swarm. Equation (14) shows the initializing method we used

$$x(0) = x_{\min,j} + r_j\left(x_{\max,j} - x_{\min,j}\right), \forall j = 1,\dots,n_x, \forall i = 1,\dots,n_s \tag{14}$$

where $x(0)$ is the matrix of dimensions, $x_{min,j}$ and $x_{max,j}$ are the minimal an maximum value respectively that a dimension j can take, and $r_j \sim U(0,1)$ is a random value.

Equation (15) shows the way we calculate the velocity and Equation (16) shows the maximum velocity

$$v_{ij}(t+1) = \begin{cases} v'_{ij}(t+1) & \text{if } v'_{ij}(t+1) < V_{\max,j} \\ V_{\max,j} & \text{if } v'_{ij}(t+1) \geq V_{\max,j} \end{cases} \tag{15}$$

$$V_{\max,j} = \delta\left(x_{\max,j} - x_{\min,j}\right) \quad \delta \in (0,1] \tag{16}$$

where $v'_{ij}(t+1)$ is the velocity calculated by Equation (6), δ is a gain that is problem–dependent. For this parameter we used the heurist value of $\delta = 0.5$.

As we didn't know the adequate values for the constants c_1 and c_2 we decided to apply the method of Ratnaweera, who proposed that c_1 decreases linearly over time, while c_2 increases linearly [20]. This strategy focuses on exploration in the early stages of the optimization, while encourage convergence to a good optimum near the end to the optimization process by attracting particles more towards the neighborhood best (or global best) positions [21]. The values of $c_1(t)$ and $c_2(t)$ at time step t are calculated by Equation (17).

$$c_1(t) = (c_{1,\min} - c_{1,\max})\frac{t}{n_t} + c_{1,\max}$$
$$c_2(t) = (c_{2,\max} - c_{1,\min})\frac{t}{n_t} + c_{1,\min} \tag{17}$$

where $c_{1,max}$=$c_{2,max}$=2.5 and $c_{1,min}$=$c_{2,min}$ = 0.5.

Further improvements in performance have been obtained using a fuzzy self-adaptive acceleration scheme [21].

6 Simulation Results

In this section we present the results of the proposed controller to stabilize the unicycle mobile robot, defined by Equation (9) and Equation (10), where the matrix values

$$M(q) = \begin{bmatrix} 0.3749 & -0.0202 \\ -0.0202 & 0.3739 \end{bmatrix},$$
$$C(q,\dot{q}) = \begin{bmatrix} 0 & 0.1350\dot{\theta} \\ -0.150\dot{\theta} & 0 \end{bmatrix},$$

and $D = \begin{bmatrix} 10 & 0 \\ 0 & 10 \end{bmatrix}$

were taken from [9]. The evaluation was made through a computer simulation performed in MATLAB® and SIMULINK®.

The desired trajectory is the following one:

$$\vartheta_d(t) = \begin{cases} v_d(t) = 0.2(1 - \exp(-t)) \\ w_d(t) = 0.4\sin(0.5t) \end{cases} \tag{18}$$

and was chosen in terms of its corresponding desired linear v_d and angular w_d velocities, subject to the initial conditions

$$q(0) = (0.1, 0.1, 0)^T \text{ and } \vartheta(0) = 0 \in \mathbb{R}^2$$

The gains γ_i, i=1, 2, 3 of the kinematic model (see [15]) are $\gamma_1 = 5$, $\gamma_2 = 24$ and $\gamma_3 = 3$ were taken from [15, 16].

6.1 S-ACO Results for the Optimization of the Type-1 FLC

Table 6 shows the results of the type-1 FLC, obtained varying the values of maximum iterations and number of artificial ants, where the highlighted row shows the best result obtained with the method. Figure 6 shows the optimization behavior of the method.

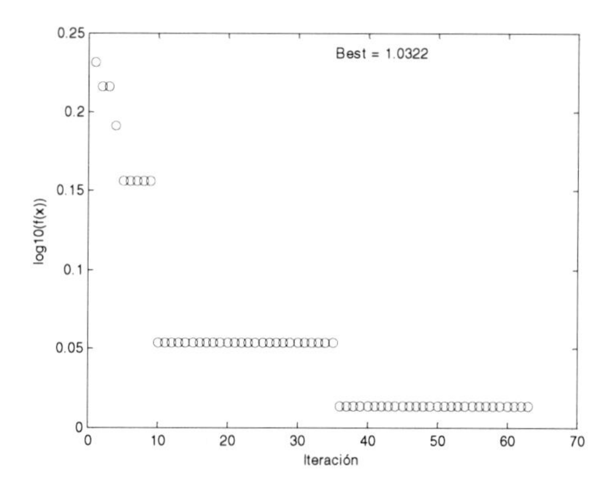

Fig. 6. Optimization Behavior for the S-ACO on Type-1FLC Optimization

Table 6. S-ACO results for type-1 FLC optimization.

Experiment	Iterations	Ants	Average Error	Time
01	25	80	0.1263	00:14:55
02	25	80	0.1432	00:15:00
03	25	80	0.1069	00:14:52
04	25	80	0.1119	00:15:41
05	50	80	0.1100	00:28:43
06	50	80	0.0982	00:29:22
07	100	18	0.1078	00:13:59
08	100	18	0.1095	00:13:16
09	100	18	0.1197	00:12:09
10	15	20	0.1343	00:02:19
12	15	20	0.1308	00:02:17
13	15	20	0.1353	00:02:18
14	50	20	0.1222	00:07:30
15	50	20	0.1233	00:07:29
16	50	20	0.1098	00:07:30
17	50	20	0.1215	00:08:09
18	50	20	0.1142	00:07:14
19	50	20	0.1164	00:07:39
20	50	20	0.1217	00:07:22
21	100	200	0.1087	02:15:46
22	50	80	0.1086	00:29:56
23	60	90	0.1338	00:41:56
24	60	90	0.1091	00:39:48
25	60	90	0.1044	00:40:00
26	17	28	0.1211	00:04:44
27	17	28	0.1211	00:05:51
28	40	50	0.1211	00:17:51
29	40	50	0.1211	00:18:14

Figure 7 shows the membership functions of the FLC obtained by the S-ACO algorithm.

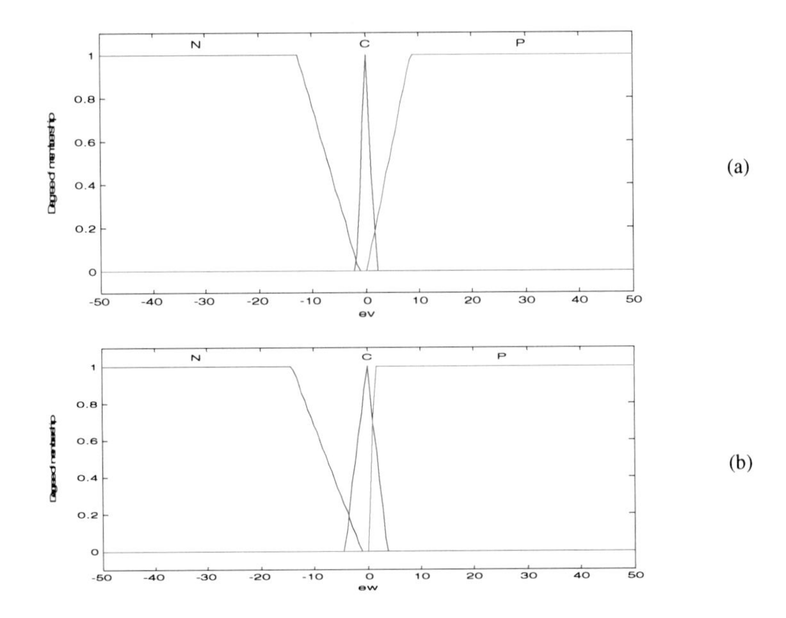

Fig. 7. Membership Functions: (a)Linear Velocity Error and (b) Angular Velocity Error optimized by S-ACO Algorithm.

The block diagram used for the FLC that obtained the best results can be seeing in [15]. Figure 8 shows the desired trajectory and obtained trajectory.

6.2 S-ACO Results for the Optimization of the Type-2 FLC

Table 7 shows the results of the type-2 FLC, obtained varying the values of maximum iterations and number of artificial ants, where the highlighted row shows the best result obtained with the method. Figure 9 shows the optimization behavior of the method.

Table 7. S-ACO results for type-2 FLC optimization.

Experiment	Iterations	Ants	Average Error	Time
01	10	10	0.0866	00:30:43
02	10	10	0.0756	00:57:06
03	10	10	0.0730	00:32:07
04	10	10	0.0690	00:31:15
05	18	13	0.0659	01:14:49

Table 7. *(continued)*

06	18	13	0.0666	01:15:53
07	100	14	0.0663	05:27:42
08	45	14	0.0679	02:25:38
09	45	14	0.0675	02:26:55
10	45	14	0.0717	02:18:58
11	45	14	0.0691	02:16:57
12	25	15	0.0745	01:26:11
13	25	15	0.0710	01:28:04
14	25	15	0.0765	01:30:24
15	25	15	0.0666	01:30:25
16	27	17	0.0724	01:49:05
17	27	17	0.0657	01:30:53
18	27	17	0.0748	01:40:17
19	20	20	0.0741	02:11:05
20	20	20	0.0666	04:11:49
21	20	20	0.0733	02:08:06
22	20	20	0.0656	02:59:22
23	12	23	0.0663	01:32:07
24	12	23	0.0663	01:28:43
25	12	23	0.0781	01:29:23

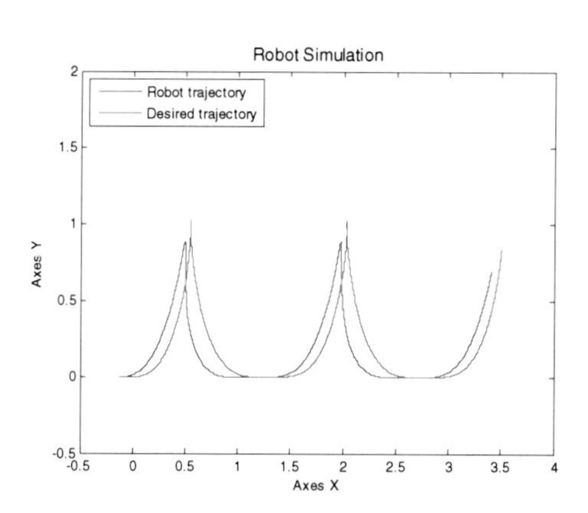

Fig. 8. Obtained Trajectory with Type-1 optimization.

Figure 10 shows the membership functions of the FLC obtained by S-ACO algorithm. Figure 11 shows the desired trajectory and obtained trajectory.

6.3 *gbest* PSO Algorithm Results for Type-1 FLC Optimization

Table 8 shows some results chosen randomly of the FLC, obtained varying the values of maximum iterations and number of particles, where the highlighted row shows the

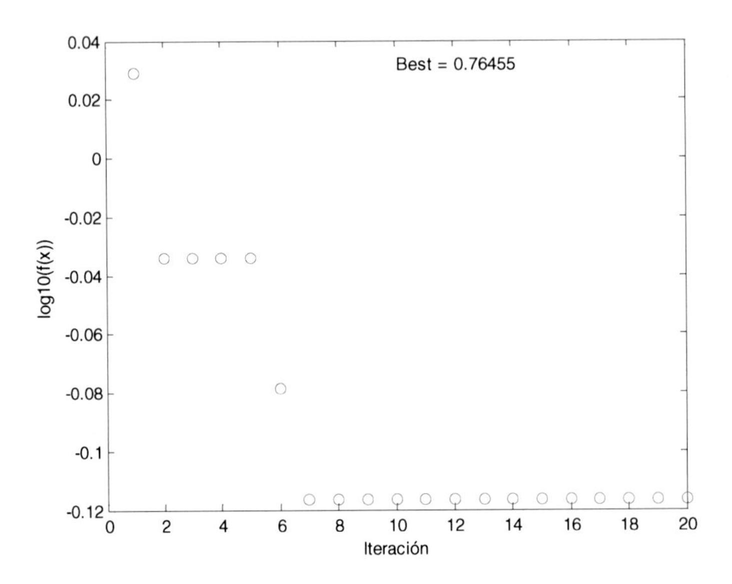

Fig. 9. Optimization behavior for the S-ACO on Type-2 FLC optimization

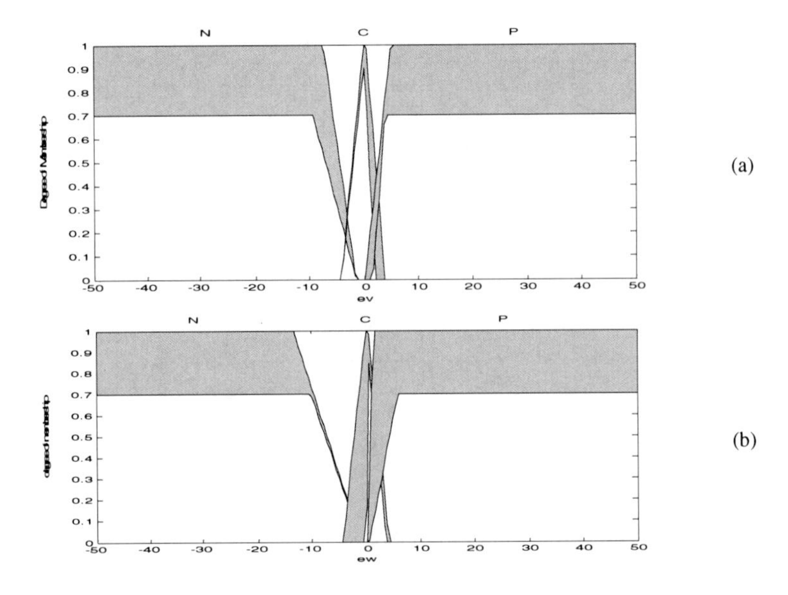

Fig. 10. Membership Functions: (a) Linear velocity error, (b) angular velocity error optimized by S-ACO algorithm.

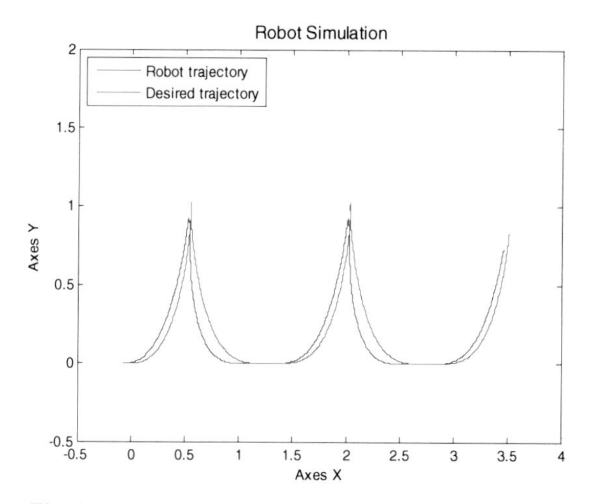

Fig. 11. Obtained trajectory with type-2 FLC optimization

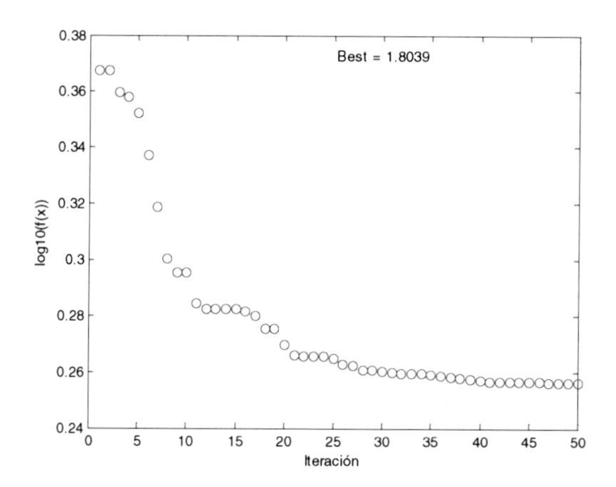

Fig. 12. Optimization Behavior of *gbest* PSO for Type-1 FLC.

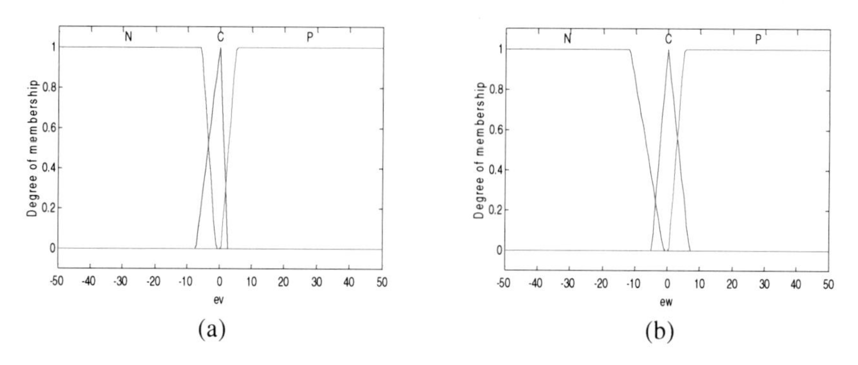

Fig. 13. (a) Linear velocity error, and (b) angular velocity error optimized by *gbest* PSO algorithm.

best result obtained with the method. Figure 12 shows the optimization behavior of the method for a type-1 FLC optimization.

Figure 13 shows the membership functions of the type-1 FLC obtained by the *gbest* PSO algorithm.

Figure 14 shows the desired trajectory and the obtained trajectory, where we can appreciate that they are very close.

Table 8. *gbest* PSO Results of Simulations for Type-1 FLC Optimization.

Experiment	Iterations	Swarm	Average Error	Time
1	50	100	0.1523	00:09:08
2	15	150	0.1512	00:32:06
3	15	150	0.1516	00:33:03
4	15	25	0.1642	00:06:29
5	35	25	1.8981	00:04:18
6	**50**	**150**	**0.1509**	**01:00:50**
7	71	34	0.1515	00:02:35
8	71	34	0.1510	00:17:44
9	73	38	0.1520	00:03:19
10	250	50	0.1511	00:57:32
11	95	51	0.1912	01:16:20
12	51	57	0.1615	00:02:06
13	67	65	0.1728	01:29:28
14	67	65	0.1728	01:29:28
15	80	67	0.1994	01:46:27
16	65	68	0.1830	01:05:56
17	50	80	0.1591	00:18:24
18	78	82	0.1510	00:43:13
19	81	82	0.1510	00:40:09
20	17	91	0.1619	00:22:25
21	85	97	0.1510	01:31:51

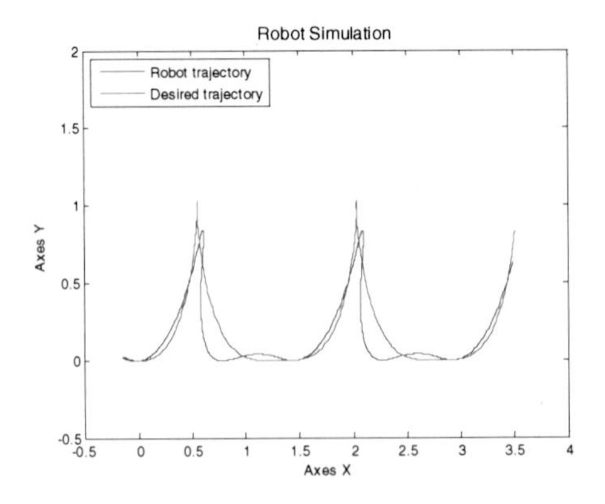

Fig. 14. Obtained trajectory with type-1 optimization

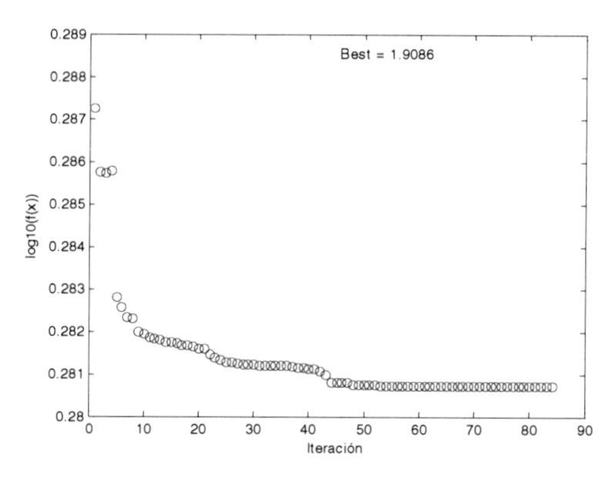

Fig. 15. Optimization Behavior of *gbest* PSO forType-2 FLC.

Table 9. *gbest* PSO Results of Simulations for Type-2 FLC Optimization.

Experiment	Iterations	Size of Swarm	Average Error	Time
1	10	10	0.1608	00:09:41
2	16	17	0.1607	00:16:32
3	21	17	0.1606	00:26:38
4	16	19	0.1614	00:18:06
5	28	20	0.1605	00:40:22
6	22	21	0.1607	00:25:25
7	28	22	0.1606	00:57:47
8	17	26	0.1606	00:24:22
9	19	26	0.1607	00:37:47
10	22	28	0.1606	00:42:26
11	27	28	0.1606	00:52:08
12	27	28	0.1606	00:51:56
13	28	28	0.1606	00:53:00
14	23	29	0.1608	00:39:09
15	71	34	0.1605	03:56:16
16	**84**	**91**	**0.1601**	**07:46:58**

6.4 *gbest* PSO Algorithm Results for Type-2 FLC Optimization

Table 9 shows some results chosen randomly of the type-2 FLC, obtained varying the values of maximum iterations and number of particles, where the highlighted row shows the best result obtained with the method. Figure 15 shows the optimization behavior of the method for type-2 FLC optimization. Figure 16 shows the membership functions of the type-2 FLC obtained by the *gbest* PSO algorithm.

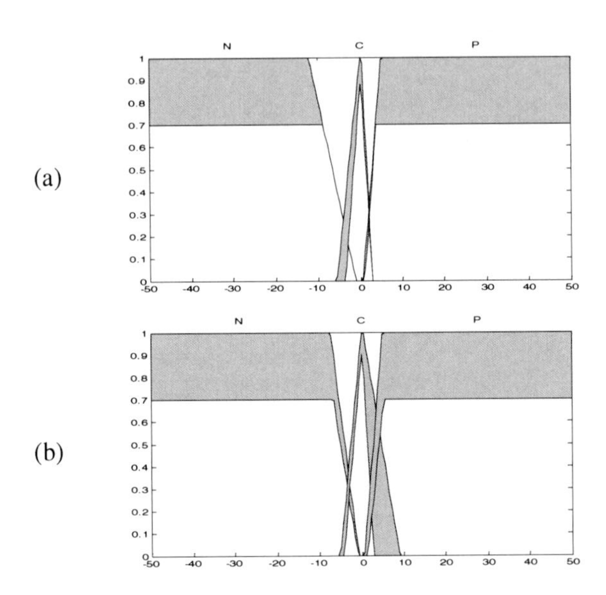

Fig. 16. (a) Linear velocity error, and (b) angular velocity error optimized by *gbest* PSO algorithm.

Figure 17 shows the desired trajectory and the obtained trajectory, where we can appreciate that they are very close.

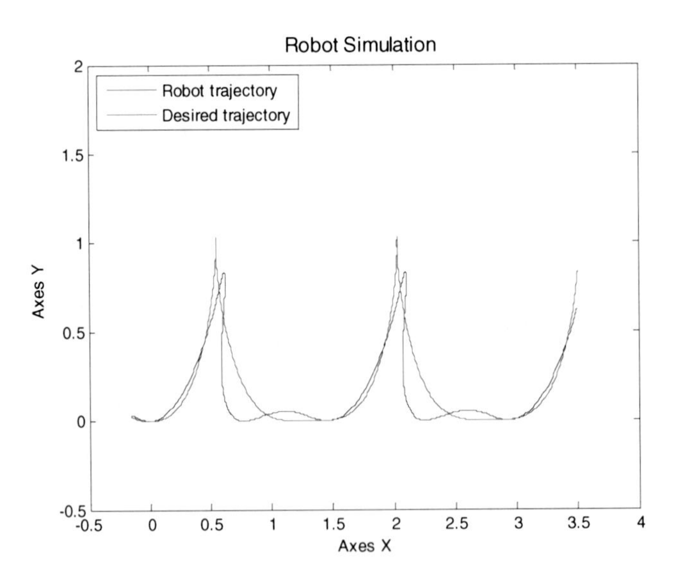

Fig. 17. Obtained trajectory with type-2 optimization

7 Statistical Comparison

This section describes the comparisons among the different results obtained with S-ACO, PSO and GA. The GA results were taken from [15, 16]. Table 10 shows the

Table 10. GAs Results For Type-2 FLC Optimization

Experiment	Ind.	Gen.	Replace	Cross.	Mutation.	Average Error	Time
01	50	30	0.7	0.7	0.2	0.4122618	07:18
02	100	25	0.7	0.8	0.3	0.4212924	12:11
03	20	15	0.7	0.8	0.2	0.5524043	01:26
04	10	20	0.7	0.8	0.2	0.4899811	01:21
05	80	25	0.7	0.7	0.4	0.4126189	09:57
06	150	50	0.7	0.6	0.3	0.4094381	43:15
07	90	60	0.7	0.9	0.4	0.4087614	44:43
08	10	25	0.7	0.8	0.2	0.5703853	02:09
09	65	40	0.7	0.8	0.2	0.4099531	22:52
10	30	25	0.7	0.9	0.5	0.4086178	06:21
11	70	50	0.7	0.8	0.3	0.4086729	29:17
12	80	50	0.7	0.9	0.3	0.4099137	33:32
13	200	100	0.7	0.4	0.1	0.4085207	2:43:28
14	15	10	0.7	0.8	0.5	0.5669795	01:14
15	15	25	0.7	0.9	0.2	0.4789307	03:03
16	30	40	0.7	0.7	0.2	0.4108032	10:28
17	50	60	0.7	0.6	0.4	0.4111103	1:05:14
18	20	50	0.7	0.6	0.2	0.4339689	09:13
19	80	20	0.7	0.8	0.6	0.4490967	13:13
20	100	80	0.7	0.8	0.3	0.4083982	06:16
21	30	60	0.7	0.8	0.5	0.4943807	14:43
22	25	40	0.7	0.8	0.6	0.4247892	08:10
23	70	60	0.7	0.8	0.4	0.4084446	34:44
24	35	40	0.7	0.7	0.3	0.4099876	11:30
25	45	50	0.7	0.7	0.3	0.4128472	18:19
26	60	40	0.9	0.8	0.5	0.4106830	20:38
27	26	30	0.9	0.6	0.3	0.4082359	06:40
28	40	30	0.9	0.9	0.4	0.4102437	10:07
29	100	35	0.9	0.7	0.4	0.4094340	29:01

Table 11. GAs Results For Type-2 FLC Optimization

No.	Ind.	Gen.	Replace	Cross.	Mutation	Average Error	Time
1	50	20	0.7	0.8	0.4	0.3993130	4:52:08
2	20	15	0.7	0.8	0.5	0.4008340	1:13:03
3	23	20	0.7	0.8	0.4	0.3994720	02:56:23
4	40	25	0.7	0.8	0.5	0.3993860	6:37:16
5	30	19	0.7	0.9	0.5	0.3994950	3:02:35
6	35	10	0.7	0.8	0.5	0.4111980	1:15:03
7	45	25	0.7	0.9	0.5	0.4008810	7:22:52
8	38	18	0.7	0.7	0.3	0.3991930	3:40:29
9	60	20	0.7	0.8	0.6	0.3989860	6:40:59
10	45	20	0.7	0.8	0.6	0.4007900	5:56:20
11	45	15	0.7	0.7	0.5	0.4068480	3:22:18
12	58	25	0.9	0.6	0.4	0.3995240	7:49:24
13	40	18	0.9	0.9	0.6	0.3990670	3:29:21
14	58	45	0.9	0.8	0.6	0.3989470	15:20:34
15	26	18	0.9	0.9	0.5	0.4021550	3:48:35
16	10	15	0.9	0.8	0.5	0.4028900	1:43:28
17	15	15	0.9	0.7	0.4	0.4006630	1:45:03
18	25	22	0.9	0.8	0.5	0.3995900	2:11:58
19	60	25	0.9	0.9	0.4	0.4002830	9:38:31
20	30	15	0.9	0.8	0.4	0.4110670	2:23:33
21	20	18	0.9	0.6	0.4	0.3989810	1:28:53
22	46	28	0.9	0.6	0.4	0.3991000	6:19:01
23	70	25	0.9	0.7	0.5	0.3989980	10:36:57
24	54	20	0.9	0.8	0.6	0.3992210	6:19:40
25	66	30	0.9	1	0.6	0.3989810	12:54:21
26	42	35	0.9	0.8	0.6	0.3989410	7:11:52
27	26	10	0.9	0.6	0.4	0.4027990	2:29:51
28	40	20	0.9	0.6	0.4	0.3990530	4:24:41
29	50	15	0.9	0.9	0.5	0.4005340	3:33:39
30	80	12	0.9	0.9	0.6	0.3997710	6:32:30

results Type-1 FLC GA optimization and Table 11 Shows the Type-2 FLC GA optimizations. To perform the statistical comparison we applied hypothesis testing with the t-Student statistic or distribution *t*.

Equation (19) describes the t-student statistic definition.

$$T_0^* = \frac{\overline{X}_1 - \overline{X}_2}{\sqrt{\frac{S_1^2}{n_1} + \frac{S_2^2}{n_2}}} \tag{19}$$

where $\overline{X}_1$ and $\overline{X}_2$ are the means of the samples X_1 and X_2 respectively, S_1^2 and S_2^2 are the variance of sample 1 and sample 2 respectively and finally n_1 is the size of the sample 1 and n_2 is the size of the sample 2.

The hypothesis testing was performed in order to determine which of the bio-inspired algorithms for optimization (ACO, PSO and GA) was the most adequate for solving the particular problem of control.

The samples taken for the test we have size 30 and represent the best 30 results of every bio-inspired optimization method.

Statdisk ® 9.5.2 software was employed for computing the t-Student statistic.

7.1 Statistical Comparison among Bio-inspired Methods for Type-1 FLCs

Table 12 shows the results of statistical comparison between ACO and GA. Figure 18 shows the t distribution. The value t=-64.9980 gives us sufficient statistical information to say that ACO outperforms GA along the realized simulations.

Table 12. ACO vs. GA

$X_1 = \text{ACO}$	$X_2 = GA$
Null Hypothesis	$\overline{X}_1 \neq \overline{X}_2$
Test Statistical t	-64.9980
Critical values t	±2.039352
Value P	0.00000
Grades of Freedom	31.0599
Interval of Confidence	95%: -0.3294214< $\mu_1 - \mu_2$ <-0.3093786
Reject Null Hypothesis.	YES

Table 13 shows the results of statistical comparison between ACO and PSO. Figure 19 shows the t distribution. The value t=-50.374 gives us sufficient statistical information to say that ACO outperforms PSO along the realized simulations.

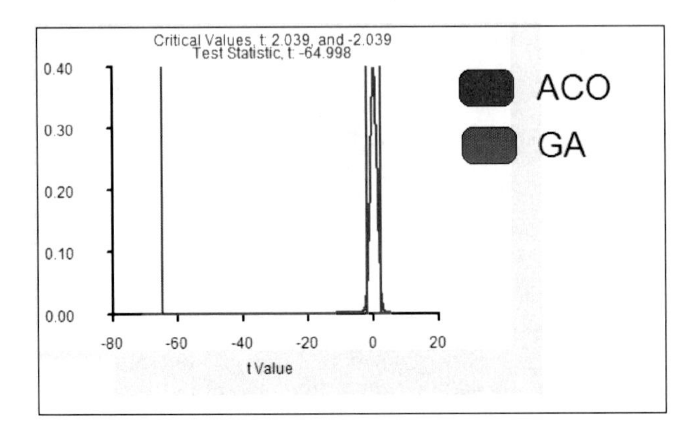

Fig. 18. ACO vs. GA

Table 13. ACO vs. PSO

X_1 = ACO	$X_2 = PSO$
Null Hypothesis	$\bar{X}_1 \neq \bar{X}_2$
Test Statistical t	-50.374000
Critical values t	±2.037876
Value P	0.000000
Grades of Freedom	58.000000
Interval of Confidence	95%: -0.0487973< $\mu_1 - \mu_2$ <-0.0450027
Reject Null Hypothesis.	Yes

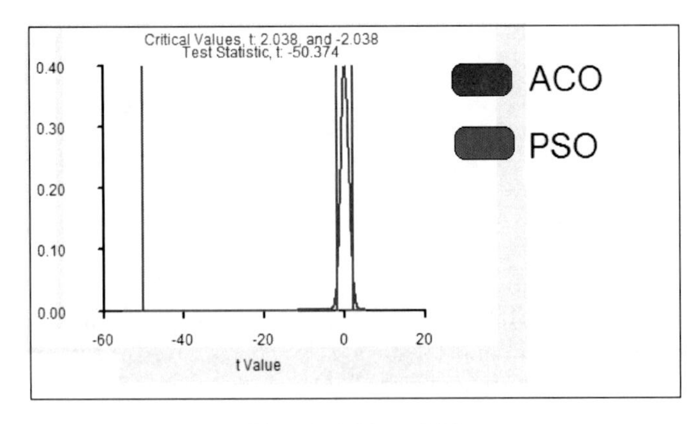

Fig. 19. ACO vs. PSO

Table 14 shows the results of statistical comparison between PSO and GA. The value *t*= 56.3858 gives us sufficient statistical information to say that PSO outperforms GA along the realized simulations.

Table 14. PSO vs. GA

$X_2 = PSO$	$X_2 = GA$
Null Hypothesis	$\overline{X}_1 \neq \overline{X}_2$
Test Statistical t	56.385800
Critical values t	±2.044942
Value P	0.000000
Grades of Freedom	29.0936
Interval of Confidence	95%: -0.2626172< $\mu_1 - \mu_2$ <-0.2823828
Reject Null Hypothesis.	yes

7.2 Statistical Comparison among Bio-inspired Methods for Type-2 FLCs

Table 15 shows the results of statistical comparison between ACO and GA. Figure 20 shows the distribution, the value *t*= -524.1978 gives us sufficient statistical information to say that ACO outperforms GA along the realized simulations.

Table 15. ACO vs. GA

$X_1 = \text{ACO}$	$X_2 = GA$
Null Hypothesis	$\overline{X}_1 \neq \overline{X}_2$
Test Statistical t	-524.1978
Critical values t	±2.025097
Value P	0.000000
Grades of Freedom	37.6013
Interval of Confidence	95%: -0.3350715< $\mu_1 - \mu_2$ <-0.3324925
Reject Null Hypothesis.	YES

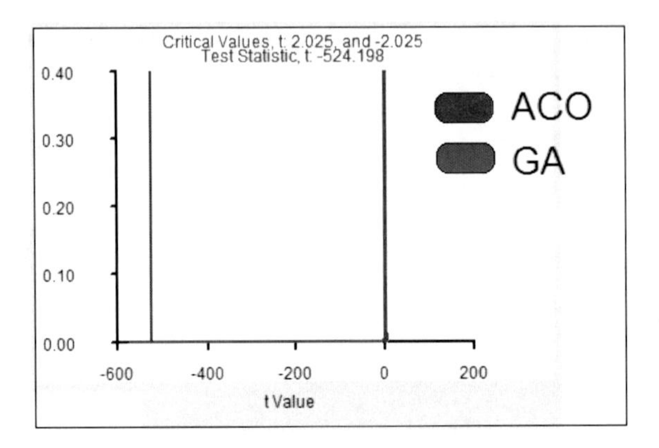

Fig. 20. ACO vs.GA

Table 16. ACO vs. PSO

$X_1 = \text{ACO}$	$X_2 = PSO$
Null Hypothesis	$\bar{X}_1 \neq \bar{X}_2$
Test Statistical t	-295.9147
Critical values t	±2.001876
Value P	0.000000
Grades of Freedom	58.782000
Interval of Confidence	95%: -0.0945084< $\mu_1 - \mu_2$ <-0.0932382
Reject Null Hypothesis.	SI

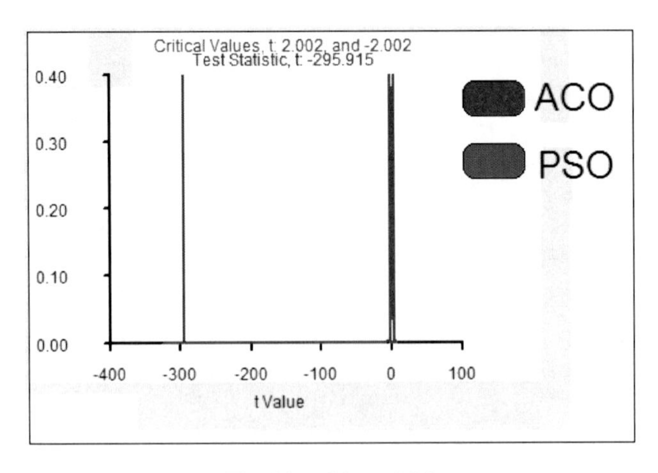

Fig. 21. ACO vs. PSO

Table 16 shows the results of statistical comparison between ACO and PSO. Figure 21 shows the distribution, the value *t*=-295.9147 gives us sufficient statistical information to say that ACO outperforms PSO along the realized simulations.

Table 17 shows the results of statistical comparison between GA and PSO. Figure 22 shows the distribution, the value of *t*=379.8209 gives sufficient statistical information to say that PSO outperforms GA along the realized simulations.

Table 17. GA vs. PSO

$X_1 = GA$	$X_2 = PSO$
Null Hypothesis	$\overline{X}_1 \neq \overline{X}_2$
Test Statistical t	379.8209
Critical values t	±2.026859
Value P	0.000000
Grades of Freedom	36.6433
Interval of Confidence	95%: 0.238719< $\mu_1 - \mu_2$ <0.2412804
Reject Null Hypothesis.	YES

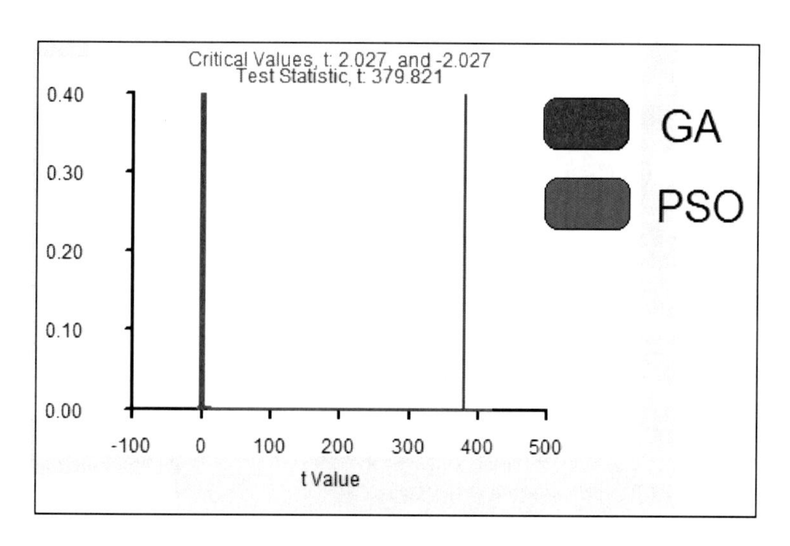

Fig. 22. GA vs. PSO

In view of the presented results, we can say that ACO outperforms both PSO and GAs for this problem of intelligent control of autonomous robots. The advantage of ACO is even better in the case of designing the type-2 fuzzy controller due to the higher number of parameters involved in this case.

8 Conclusions

A trajectory tracking controller has been designed based on the dynamics and kinematics of the autonomous mobile robot through the application of ACO and PSO for the optimization of membership functions for the type-1 and type-2 FLCs with good results obtained after simulations. Previous works results of GAs for FLC optimization were taken for comparison. Statistical results give us sufficient statistical evidence to say that ACO outperforms PSO and GAs based on the simulations. This conclusion is based on the types of ACO and PSO algorithms that were used for optimization. Future work will consist in applying different versions of ACO and PSO for optimizing the fuzzy controllers. Also, a more general theoretical analysis and computational complexity of the algorithms will be part of the future research work.

References

[1] Bloch, A.M.: Nonholonomic mechanics and control. Springer, New York (2003)

[2] Brockett, R.W.: Asymptotic stability and feedback stabilization. In: Millman, R.S., Susman, H.J. (eds.) Differential Geometric Control Theory, p. 181. Birkhauser, Boston (1983)

[3] Castillo, O., Melin, P.: Soft Computing for Control of Non-Linear Dynamical Systems. Springer, Heidelberg (2001)

[4] Chen, D., Zhao, C.X.: Particle swarm optimization with adaptive population size and its application. Applied Soft Computing Journal 9(1), 39–48 (2009)

[5] Clerc, M.: Particle Swarm Optimization. ISTE Ltd., London (2006)

[6] Dorigo, M., Birattari, M., Blum, C., Gambardella, L.M., Mondada, F., Stützle, T. (eds.): ANTS 2004. LNCS, vol. 3172. Springer, Heidelberg (2004)

[7] Dorigo, M., Birattari, M., Stützle, T.: Ant Colony Optimization. IEEE Computational Intelligence Magazine, 28–39 (November 2006)

[8] Dorigo, M., Stützle, T.: Ant Colony Optimization, Bradford, Cambridge, Massachusetts (2004)

[9] Duc Do, K., Zhong-Ping, Pan, J.: A global output-feedback controller for simultaneous tracking and stabilizations of unicycle-type mobile robots. IEEE Trans. Automat. Contr. 20(3), 589–594 (2004)

[10] Eberhart, R.C., Kennedy, J.: A new optimizer using particle swarm theory. In: Proceedings of the sixth international symposium on micro machine and human science, pp. 39–43. IEEE service center, Piscataway (1995)

[11] Engelbrecht, A.P.: Fundamentals of Computational Swarm Intelligence. Wiley, J., England (2005)

[12] Hsu, C.-C., Shieh, W.-Y., Gao, C.-H.: Digital redesign of uncertain interval systems based on extremal gain/phase margins via a hybrid particle swarm optimizer. Applied Soft Computing Journal 10(2), 602–612 (2010)

[13] Lee, T.-C., Song, K.-T., Lee, C.-H., Teng, C.-C.: Tracking control of unicycle-modeled mobile robot using a saturation feedback controller. IEEE trans. Contr. Syst. Technol. 9(2), 305–318 (2001)

[14] Liberzon, D.: Switching in Systems and control. Bikhauser (2003)

[15] Martínez, R., Castillo, O., Aguilar, L.: Intelligent control for a perturbed autonomous wheeled mobile robot using type-2 fuzzy logic and genetic algorithms. JAMRIS 2(1), 1–11 (2008)

[16] Martínez, R., Castillo, O., Aguilar, L.: Optimization of interval type-2 fuzzy logic controllers for a perturbed autonomous wheeled mobile robot using genetic algorithms. Information Sciences 179, 2158–2174 (2009)

[17] Pedersen, M.E.H., Chipperfield, A.J.: Simplifying Particle Swarm Optimization. Applied Soft Computing Journal 10(2), 618–628 (2010)

[18] Porta-García, M., Montiel, O., Sepulveda, R.: An ACO path planner using a FIS for a Path Selection Adjusted with a Simple Tuning Algorithm. JAMRIS 2(1), 1–11 (2008)

[19] Porta Garcia, M.A., Montiel, O., Castillo, O., Sepúlveda, R., Melin, P.: Path planning for autonomous mobile robot navigation with ant colony optimization and fuzzy cost function evaluation. Applied Soft Computing Journal 9(3), 1102–1110 (2009)

[20] Ratnaweera, A., Halgamuge, S.K., Watson, H.C.: Particles Swarm Optimizer with Time Varying Acceleration Coefficients. In: Proceedings of the International Conference on Soft Computing and Intelligent Systems, pp. 240–255 (2002)

[21] Ratnaweera, A., Halgamuge, S.K., Watson, H.C.: Particles Swarm Optimizer with Self-Adaptive Acceleration Coefficients. In: Proceedings of the First International Conference on Fuzzy Systems and Knowledge Discovery, pp. 264–268 (2003)

[22] Sepúlveda, R., Castillo, O., Melin, P., Montiel, O.: An Efficient Computational Method to Implement Type-2 Fuzzy Logic In control Applications, Analysis and Design of Intelligent Systems Using Soft Computing Techniques. Advances in Soft Computing 41, 45–52 (2007)

[23] Shi, Y., Eberhart, R.C.: Parameter selection in particle swarm optimization. In: Porto, V.W., Waagen, D. (eds.) EP 1998. LNCS, vol. 1447, pp. 591–600. Springer, Heidelberg (1998)

[24] Shi, Y., Eberhart, R.C.: A modified particle swarm optimizer. In: Proceedings of the IEEE International Conference on Evolutionary Computation, pp. 69–73. IEEE Press, Piscataway (1998)

[25] van den Bergh, F., Engelbrecht, A.P.: A study of particle swarm optimization particle trajectories. Information Sciences 176(8), 937–971 (2006)

Part D
Other Swarm Techniques

A New Framework for Optimization Based-On Hybrid Swarm Intelligence

Pei-Wei Tsai[1], Jeng-Shyang Pan[1], Peng Shi[2], and Bin-Yih Liao[1]

[1] Department of Electronic Engineering, National Kaohsiung University of Applied Sciences, Taiwan
No.415, Chien-Kung Rd., Kaohsiung City 80778, Taiwan
pwtsai@bit.kuas.edu.tw, {jspan,byliao}@cc.kuas.edu.tw

[2] School of Technology, University of Glamorgan, UK
Pontypridd, CF37 1DL, UK
Peng.Shi@vu.edu.au

Summary. A hybrid optimization algorithm based on Cat Swarm Optimization (CSO) and Artificial Bee Colony (ABC) is proposed in this chapter. CSO is an optimization algorithm designed to solve numerical optimization problems, and ABC is an optimization algorithm generated by simulating the behavior of bees finding foods. By hybridizing these two algorithms, the hybrid algorithm called Hybrid PCSOABC is presented. Five benchmark functions are used to evaluate the accuracy, convergence, the speed, and the stabilization of the Hybrid PCSOABC. In this chapter, the literature review regarding CSO, AS, ACS, BF, PSO, ABC, and the parallel version of CSO are given at the beginning. The proposed hybrid framework combining different algorithms is given in the fourth section. And the experimental results are presented at the end of the chapter with the conclusions.

1 Introduction

In recent years, swarm intelligence becomes more and more attractive for the researchers. It is one of the branches in evolutionary computing. The algorithms in swarm intelligence are often applied to solve problems of optimization. Several algorithms for optimization issues related to swarm intelligence are proposed one after another. Soft computing includes several research topics, e.g. swarm intelligence, evolutionary computing, fuzzy set theory, artificial neural network, and so forth. Each of them includes various practical theories, algorithms, and applications. For example, fuzzy logic is applied to the control system of the high speed rail with a big success; artificial neural network is applied to real-time control system [30], and evolutionary computing is applied to design and to control the power systems [20].

The algorithms, which we are going to review, under evolutionary computing can be used to solve the optimization problems. A series of algorithms regarding this subject are presented in the third section; the proposed hybrid framework

B.K. Panigrahi, Y. Shi, and M.-H. Lim (Eds.): Handbook of Swarm Intelligence, ALO 8, pp. 421–449.
springerlink.com

is presented in the fourth section; and the experimental results, the discussion, and the conclusions are presented in the last section.

2 Optimization in Swarm Intelligence

Swarm intelligence can be defined as the measure, which introduces the collective behavior of social insect colonies, other animal societies, or the relationship description of unsophisticated agents interacting with their environment, to design algorithms or distributed problem-solving devices. By collecting the characteristics and the behaviors of creatures, several algorithms of the optimization issues related to swarm intelligence are proposed one after another. In addition, several applications of optimization algorithms based on computational intelligence or swarm intelligence are also presented continuously [14][21][23].

Shi et al. proposed Particle Swarm Optimization (PSO)[3][4][17][26] by observing the flock of birds and the school of fishes in 1995, and his work is refined by inserting a weighting factor into the formula of the particle's movement in 1999. Dorigo et al. proposed Ant Colony System (ACS) [5][6][12]by simulating the process of the ants finding the shortest path from the nest to the food source by exploiting pheromone information in 1991. Passino proposed Bacterial Foraging (BF)[25][22][28] by simulating the movement of the bacterium, which is called E. coli, in 2002. Karaboga proposed Artificial Bee Colony (ABC)[18][19] optimization by simulating the information sharing between the bees and the process of the bees finding nectar sources in 2005. In 2006, Chu et al. proposed Cat Swarm Optimization (CSO)[7][8] by employing the particular behaviors of cats to construct the movement for solving problems regarding the optimization issue, and the parallel version of CSO is proposed by Tsai et al. in 2008 [29]. In this section, we are going to review these algorithms and discuss the experimental results.

2.1 Particle Swarm Optimization

Kennedy and Eberthart found that an optimization problem can be formulated as that a flock of birds fly across an area seeking for spot with abundant food. After modeling it into the mathematical structure, Particle Swarm Optimization (PSO) is proposed. PSO is a methodology that is based on social behavior of evolution, which means it is naturally not like those methodologies that use natural evolution as the weeding-out process[10][15][16]. It imitates the behavior of the flock of birds or a school of fishes, and takes the intelligence of these creatures to solve the problems of optimization. In addition, PSO usually can find the nearly best solution in much less evolution than the others [2][13].

In 1999, Shi et al. [26]presented a new kind of PSO with weighted factor to control the movement velocity; this factor implants the phenomenon of the first theory of physics into the movement of particles in PSO and refines the convergence of the original PSO. The traditional particle swarm optimization (PSO) can be described as follows and presented in Figure 1. It is composed of four

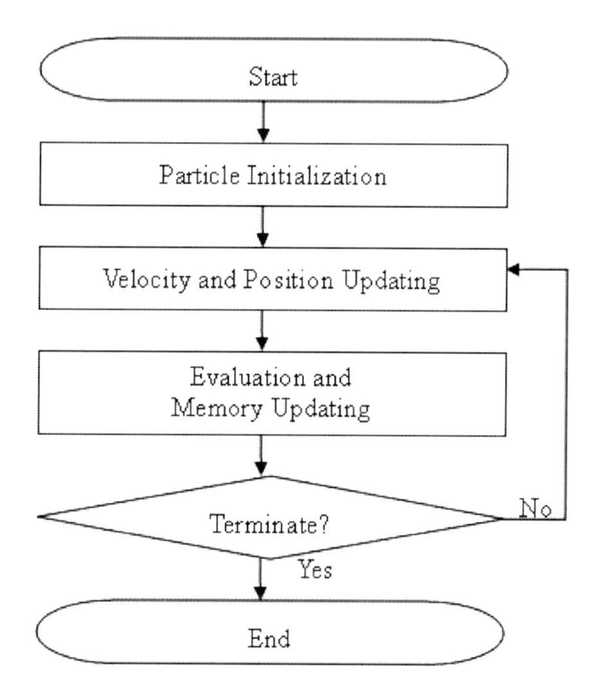

Fig. 1. The diagram of PSO.

steps, namely, particle initialization, velocity and position update, evaluation and memory update, and termination checking. The operations of these steps are listed described in Figure 1.

- The Coding Scheme and Parameter Settings
  In the processes of PSO, the numerical coding scheme is the most popular coding method due to it provides a quite convenient way to combine the problems of optimization with PSO. In addition, the potential solutions in PSO are called particles, and the position of the particles in the solution space can be described in equation (1), where X is an $M-$dimension vector. Furthermore, the velocity and the best solutions of them can be represented by equation (2) and equation (3).

$$X = \{x_0, x_1, x_2, \ldots, x_{M-1}\} \tag{1}$$

$$V = \{v_0, v_1, v_2, \ldots, v_{M-1}\} \tag{2}$$

$$B = \{b_0, b_1, b_2, \ldots, b_{M-1}\} \tag{3}$$

To employ PSO to solve problems of optimization, some parameters need to be set before the search process starts. Tehy are the object function, population size, the maximum velocity, and two constant number, which are called c_1 and c_2.

- Object Function:
  The object function, which is also called the fitness function, is the key point combining the optimization algorithm with the problems of optimization to solve, and it is defined by the user. This results in the widely adaptive ability for optimization algorithms to solve almost all the problems of optimization in every fields and applications.

  The design of the fitness function should obviously represent the characteristics of the problem to solve. A well designed fitness function gives much help for locating the global optimum; on the other hand, a clumsily designed fitness function raises the computation cost, and in worse cases, it may not be able to find the global optimum.

- Population Size:
  The population size denotes the number of particles that take part in the search process. As the same as in other optimization algorithms, the larger the population size is, the larger the dispersal of particles is. Simultaneously, it also results in the larger calculation cost.

- The Maximum Velocity:
  In order to prevent the velocities from becoming too large, there are constrains for the velocities. The maximum velocities limit the moving variances of the particles, in other words, the moving length of the particles in each generation is limited.

  Like the population size, the setting of the maximum velocity is also related to the application. If the maximum velocity is too small, it results in slow convergence; if the maximum velocity is too large, the global optimum will not easily be located.

- The Constant Numbers, c_1 and c_2:
  According to the definition in equation (4), c_1 and c_2 are pre-defined constants. One of them controls the maximum search step of the direction to the current near global best solution, and the other one affects the search step of the direction to the particle's memorized best solution.

- Particle Initialization
  In the process of initialization, all the particles are randomly spread into the solution space, and so are the memorized best solutions of all particles. According to the constraints of the maximum velocity, all the velocities for the particles are randomly generated.

- Velocity Updating
  In this step, the process of velocity update is processed by equation (4), where c_1 and c_2 are constants, r_1 and r_2 are random variables in the range

from 0 to 1, $b_i(t)$ is the best solution of the ith particle for the iteration number up to the tth iteration and the $G(t)$ is the found near best solution over all particles. To prevent the velocity from becoming too large, we set a maximum value to limit the range of velocity by equation (5).

$$v_i(t+1) = v_i(t) + c_1 \cdot r_1 \cdot [b_i(t) - x_i(t)] + c_2 \cdot r_2 \cdot [G(t) - x_i(t)] \quad (4)$$

$$-V_{MAX} \leq V \leq V_{MAX} \quad (5)$$

Moreover, the difference between the weighted PSO and the original PSO is that the weighted PSO has a weighting factor, which decreases the velocity of particles according to the increase of evolution time. The velocity function of the weighted PSO is presented in equation (6), where $w(t)$ is the inertia weight at iteration t.

$$v_i(t+1) = w(t) \cdot v_i(t) + c_1 \cdot r_1 \cdot [b_i(t) - x_i(t)] + c_2 \cdot r_2 \cdot [G(t) - x_i(t)] \quad (6)$$

In the experiment of Shi, he finds that if the constants, c_1 and c_2, are set equally to 2.0, the experimental result reports a better solution than those set these constant to others. In equation (4) and equation (6), r_1 and r_2 present random variables in the range from 0 to 1.

- Particle Position Updating and Memory Updating
  After updating the velocity of the particles, the movement process is applied to the particles by equation (7), and this is what we called the particle position updating.

$$x_i(t+1) = x_i(t) + v_i(t)\,\textbf{, where } 0 \leq i < M \quad (7)$$

  If there exists a better solution than $G(t)$ in $G(t+1)$, $G(t)$ will be replaced by $G(t+1)$ after the evaluation. Otherwise, $G(t+1)$ will keep the same content as $G(t)$.

- Termination Checking
  The velocity updating, particle position updating and memory updating processes keep repeating until certain termination condition is met, such as when a predefined number of iteration is reached or when it failed to have progress for certain number of iterations. Once terminated, PSO reports the global best solution, $G(t)$, and its fitness value as the solution.

2.2 Bacterial Foraging

In 2002, Kevin M. Passino proposed the idea of BF [25] for solving optimization problems. The framework of BF was based on parroting the behaviors of bacterium, i.e., the way they search nutrients, evade noxious environments, and the moving circumstance. By way of imitating the existence of the bacterium, the optimization problem can be solved.

To apply BF to solve optimization problems, like most of the other methods, a fitness function for describing the solution space is required. Assume that we are going to find the minimum of a fitness function. In equation (8), θ is a coordinate in the solution space, and $J(\theta)$ is an attractant-repellant profile, which represents where nutrients and noxious substances are located.

$$J(\theta), \theta \in \mathbf{R}^p \tag{8}$$

$$P(j,k,l) = \{\theta^i(j,k,l) | i = 1,2,\ldots,S\} \tag{9}$$

Equation (9) denotes the positions of each bacterium in the population. For the ith bacteria, it represents the bacteria at the jth chemo-tactic step, the kth reproduction step, and the lth elimination-dispersal event.

When evaluating the fitness value of each bacterial, the users can choose whether to consider the cell-to-cell signaling via attractant and repellent coefficients. These coefficients are calculated by equation (10), and then combined with the fitness value in equation (11), where J_{cc} represents the combined cell-to-cell attraction and repelling effects, and $\theta = [\theta_1, \theta_2, \ldots, \theta_p]^T$, where d, h and w are all the attractant and repellent coefficients, which we mentioned above.

$$\begin{aligned} J_{cc}(\theta, P(j,k,l)) &= \textstyle\sum_{m=1}^{S} J_{cc}^i(\theta, \theta^i(j,k,l)) \\ &= \textstyle\sum_{m=1}^{S}[-d_{attract} \cdot \exp(-w_{attract} \sum_{m=1}^{p}(\theta_m - \theta_m^i)^2)] \\ &+ \textstyle\sum_{m=1}^{S}[-h_{repellnt} \cdot \exp(-w_{repellant} \sum_{m=1}^{p}(\theta_m - \theta_m^i)^2)] \end{aligned} \tag{10}$$

$$J(\theta) = J(i,j,k,l) + J_{cc}(\theta, P) \tag{11}$$

The movement of each bacterium in BF is applied after the evaluation process. Every bacterial in the population will tumble once, and the bacterial, which have better fitness values after the tumble, will go on the swim step under the swim length limit Ns. The movement is described in equation (12), where $C(i)$ denotes the step size, and $\phi(j)$ is a random variable in $[0,1]$.$C(i)\phi(j)$ is the moving length that the ith bacterium takes.

$$\theta^i(j+1,k,l) = \theta^i(j,k,l) + C(i)\phi(j) \tag{12}$$

The evaluation and movement processes take turns to execute until the lifetime of the bacteria reaches the limit, and then the reproduction process executes. In the process of reproduction, the whole population will only keep half of the bacteria, which perform better healthy values, and directly reproduce them to replace the bacteria with worse healthy values. Once the reproduction is done, the process goes back into the cycles of evaluation and movement for the reproduced bacteria. The diagram of the BF is listed in Figure 2.

When the reproduction times exceed the user defined limit, the elimination-dispersal even takes part in the process. This process eliminates the bacteria with probability, if a bacterium is eliminated, a bacterium, which fits in with the initial conditions, is generated to replace the eliminated one. Then the whole processes described above repeat until all the process times are exceeded.

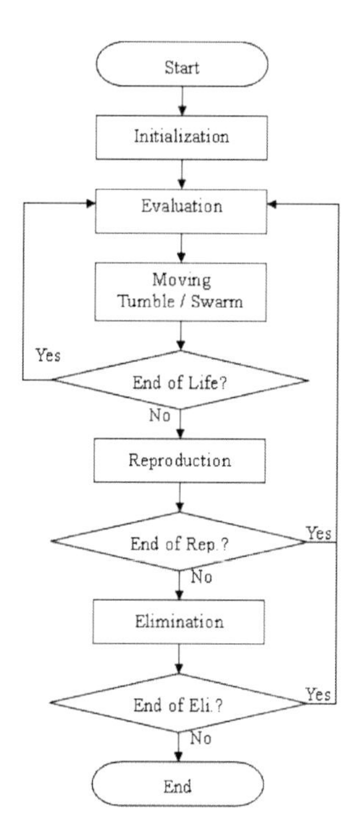

Fig. 2. The diagram of BF.

2.3 Ant System and Ant Colony System

In the real world, the ant can naturally find the shortest path between its nest to the food source by observing the concentration of an attracting substance called pheromone, which is dropped on the paths when it moves. Two paths with different distances from the ant's nest to the food source are represented in Figure 3. When ants arrive at a decision point, they firstly choose the path randomly. After a period of time, several ants have passed the decision point and the pheromone on both paths is getting stronger, but the amount of the passing ants should be larger on shorter path than the other path. Gradually, all the ants will choose the shortest path due to the pheromone level is dense.

Dorigo et al. were the first to apply this idea to the traveling salesman problem [9][11], and the algorithm is referred to as Ant System (AS). ACS is developed based on the AS with a new global update strategy. The diagram of ACS is represented in Figure 4, and the process of ACS is listed as follows:

Step 1. Initialization:
Initialize the pheromone intensity to be τ_0 on all edges. The trace pheromone density is distributed in the whole environment since all the living creatures secrete the pheromone and leave it on the things they contacted.

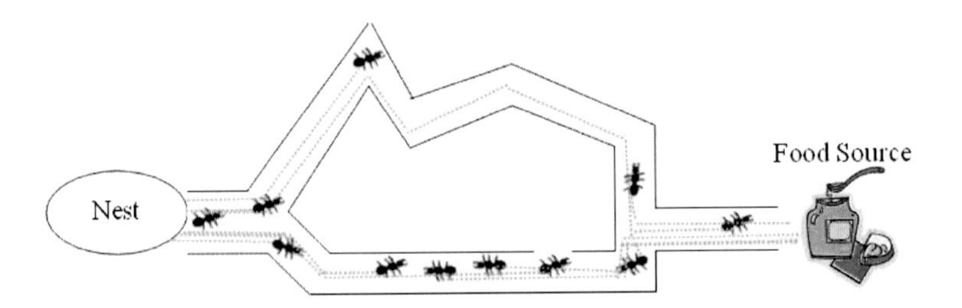

Fig. 3. How real ants find the shortest path.

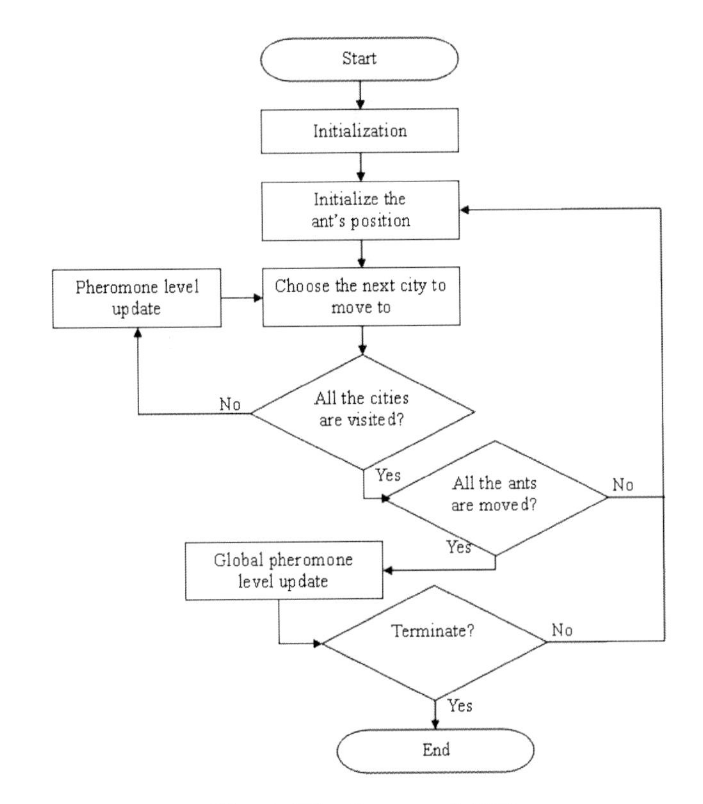

Fig. 4. The diagram of ACS.

Step 2. Initialize the beginning city of the ants:
In every cycle of ACS, each ant starts its travel from a randomly chosen city and is required to visit all the cities only once.

Step 3. Move the ants:
Calculate the transition probability from city r to city s for the kth ant by equation (13), where r is the current city, s denotes the next city, $\tau(r, s)$ represents

the pheromone level between the city r and the city s, $J_k(r)$ is the set of cities, which remain to be visited by the kth ant, and β represents a parameter determining the relative importance of the pheromone level versus the distance. Select the next target city s for the kth ant according to the probability $P_k(r,s)$.

$$P_k(r,s) = \begin{cases} \frac{[\tau(r,s)]\cdot[\eta(r,s)]^\beta}{\sum_{u\in J_k(r)}[\tau(r,u)]\cdot[\eta(r,u)]^\beta} & \textbf{, if } s \in J_k(r) \\ 0 & \textbf{, otherwise} \end{cases} \tag{13}$$

Generate a random number q to make the decision of where to move the kth ant to by equation (14), where q is a random number distributed uniformly in the range of $[0,1]$, q_0 is a predetermined parameter in the range of $[0,1]$, and S denotes the random variable selected according to the transition probability.

$$P_k(r,s) = \begin{cases} \textbf{arg } max_{u\in J_k(r)}\{[\tau(r,u)]\cdot[\eta(r,u)]^\beta\} & \textbf{, if } q \leq q_0 \\ S & \textbf{, otherwise} \end{cases} \tag{14}$$

Step 4. Local pheromone level update:
Update the pheromone level between cities by equation (15) and (16). $\Delta\tau(r,s)$ represents the pheromone level, which is laid down by the kth ant, between the city r and the city s. L_k denotes the route length crept by the kth ant, m is the number of ants, and α is the pheromone decay parameter in the range of $[0,1]$. Repeat the process of step 3 to step 4 till all the ants finish their trip.

$$\tau(r,s) \leftarrow (1-\alpha)\cdot\tau(r,s) + \sum_{k=1}^{m}\Delta\tau_k(r,s) \tag{15}$$

$$\Delta\tau(r,s) = \begin{cases} \frac{1}{L_k} & \textbf{, if } (r,s) \in \textbf{route passed by the } k^{th} \textbf{ ant} \\ 0 & \textbf{, otherwise} \end{cases} \tag{16}$$

Step 5. Global pheromone level update:
The pheromone level on the shortest path, which is found by the ant in the current cycle, is updated again by equation (17), where L_{gb} represents the length of the shortest path and α is the pheromone decay parameter.

$$\tau(r,s) \leftarrow (1-\alpha)\cdot\tau(r,s) + \alpha\cdot\Delta\tau(r,s) \tag{17}$$

$$\textbf{where } \Delta\tau(r,s) = \begin{cases} \frac{1}{L_{gb}} & \textbf{, if}(r,s) \in \textbf{the global best route} \\ 0 & \textbf{, otherwise} \end{cases} \tag{18}$$

Step 6. Termination checking:
If the termination condition is satisfied, end the program and output the shortest path, otherwise go back to step 2.

The difference between AS and ACS is the global pheromone level update only executed in ACS, the effect of equation (14) is not considered in AS, and the ants in AS move only by equation (13). The diagram of AS is presented in Figure 5.

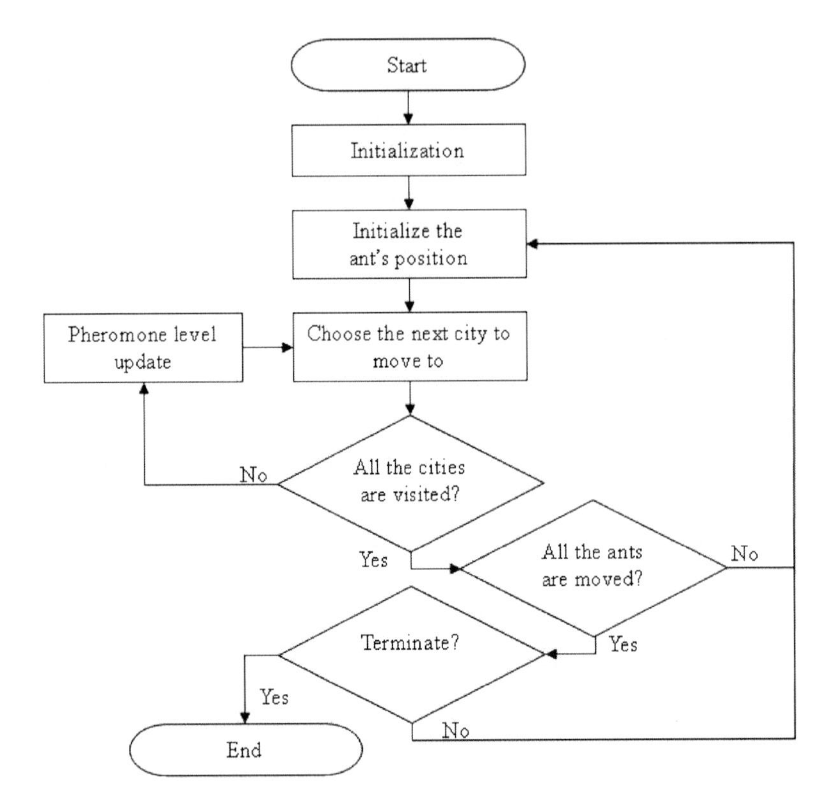

Fig. 5. The diagram of AS.

2.4 Artificial Bee Colony Optimization

Artificial Bee Colony (ABC) algorithm is proposed by Karaboga [18] in 2005, and the performance of ABC is analyzed in 2007 [19]. It is developed based on inspecting the behaviors of real bees on finding nectar and sharing the information of food sources to the bees in the nest.

The behavior of finding nectar sources of the real bees in the world can be generally described in three steps: Parts of the bees fly away from their nest searching for the nectar sources. When a bee finds a nectar source, it flies back to the nest and pass the information of the food location by dancing. The other bees in the nest watch the dance to determine how far and how plentiful the food source is. By tasting the pollen stuck on the dancing bee, the breed of the food source is recognized. In a period of time, there may be more than one bee fly back to the nest and bring back the information regarding the food source. Subsequently, the bee in the nest will choose one of the food sources, of which it is interested in, to feed itself and collect the nectar.

Three kinds of bees, namely, the employed bee, the onlooker, and the scout, are defined in ABC. Each of them has its own duty: the employed bee stays on a food source and provides the neighborhood of the source in its memory;

the onlooker gets the information of food sources from the employed bees in the hive and select one of the food source to gathers the nectar; and the scout is responsible for finding new food, the new nectar, sources. To implement ABC optimization, the procedures are given as follows:

Step 1. Initialization:
Spray n_e percentage of the populations into the solution space randomly, and then calculate their fitness values, called the nectar amounts, where n_e represents the ratio of employed bees to the total population. Once these populations are positioned into the solution space, they are called the employed bees.

Step 2. Move the Onlookers:
Calculate the probability of selecting a food source by equation (19), select a food source to move to by roulette wheel selection for every onlooker bees and then determine the nectar amounts of them. The onlookers move by equation (20).

Step 3. Move the Scouts:
If the fitness values of the employed bees do not be improved by a continuous predetermined number of iterations, which is called "*Limit*", those food sources are abandoned, and these employed bees become the scouts. To move the scouts, the equation (21) is applied.

Step 4. Update the Best Food Source Found So Far:
Memorize the best fitness value and the position, which are found by the bees.

Step 5. Termination Checking:
Check if the amount of the iterations satisfies the termination condition. If the termination condition is satisfied, terminate the program and output the results; otherwise go back to Step 2.

$$P_i = \frac{F(\theta_i)}{\sum_{k=1}^{S} F(\theta_k)} \tag{19}$$

where θ_i denotes the position of the ith employed bee, S represents the number of employed bees, and P_i is the probability of selecting the ith employed bee.

$$x_{ij}(t+1) = \theta_{ij}(t) + \phi(\theta_{ij}(t) - \theta_{kj}(t)) \tag{20}$$

where x_i denotes the position of the ith onlooker bee, t denotes the iteration number, θ_k is the randomly chosen employed bee, j represents the dimension of the solution and $\phi(\bullet)$ produces a series of random variable in the range $[-1, 1]$.

$$\theta_{ij} = \theta_{j_{min}} + r \cdot (\theta_{j_{max}} - \theta_{j_{min}}) \tag{21}$$

where r is a random number and $r \in [0, 1]$.

2.5 Cat Swarm Optimization

CSO is proposed by Chu et al. in 2006 [8] by employing the particular behaviors of animals, and the performance analysis is given in 2007 [7]. Two sub-modes, namely the seeking mode and the tracing mode, are proposed in CSO to simulate the motions of cats. There are about thirty-two different species of creatures in cat, e.g. lion, tiger, leopard etc. according to the classification of biology. Though they live in different environments, most of their behaviors are still similar. The instinct of hunting ensures the survival of outdoor cats, which live in the wild. Nevertheless, this instinct of the indoor cats behaves on the strongly curious about any moving things. Contrary to the instinct of hunting, the cats are usually inactive when they are awake. The alertness of cats is very high. They always stay alert even if they are resting. Thus, you can simply see that the cats usually look lazy, but stare their eyes hugely looking around to observe the environment. Chu et al. utilize the conduct of the cats to construct CSO and find that the outcomes of employing CSO to solve problems of optimization present high convergence.

In different algorithms of evolutionary computing, the solutions are presented in different forms. The solutions are presented by chromosomes in Genetic Algorithm (GA) [16][24]; the solutions are presented by particles in PSO; and CSO forms the solutions with the coordinates of the cats. Each cat in CSO carries 4 kinds of different information, namely, the coordinate of the current position, the velocities in each dimension, the fitness value, and a flag that presents the motion of it-self. Suppose the population size is chosen to be N, and the problem to be solved is an M-dimensional problem. Figure 2 presents the structure of CSO. The diagram of CSO is presented in Figure 6, and the operation of CSO can be described in 5 steps and is listed as follows:

1. Create N cats and randomly sprinkle the cats into the M-dimensional solution space within the constrain ranges of the initial value and the velocities for each dimension are also generated. Set the motion flag of the cats to make them move into the tracing mode or the seeking mode according to the ratio MR, where MR is a ratio indicates how many motion flags of the cats should be set to move in the seeking mode.
2. Evaluate the fitness value of the cats by taking the coordinates into the fitness function, which represents the benchmark and the characteristics of the problem you want to solve. After calculating the fitness values one by one, record the coordinate (x_{best}) and the fitness value of the cat, which owns the best fitness value we find so far.
3. Move the cats by taking the operations in seeking mode or tracing mode according to the information form the motion flag.
4. Reset the motion flag of all cats and separate them into statuses that indicating seeking or tracing according to MR.
5. Check if the process satisfies the termination condition. If the process is terminated, output the coordinate, which presents the best solution we find, otherwise go back to step 2 and repeat the process.

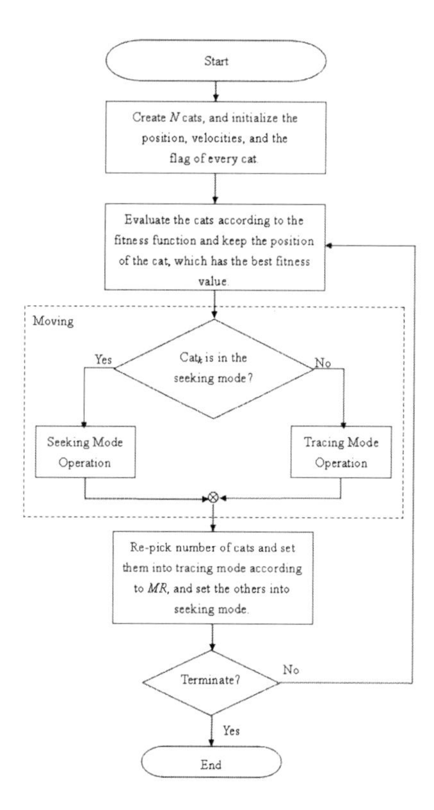

Fig. 6. The flowchart of the Cat Swarm Optimization.

- The Seeking Mode Process

  In the seeking mode, the cat moves slowly and conservatively. It observes the environment before it moves. Four essential factors are defined in the seeking mode:

  Seeking Memory Pool (SMP)
  Seeking Range of the selected Dimension (SRD)
  Counts of Dimension to Change (CDC)
  Self-Position Considering (SPC)

SMP is used to define the size of the seeking memory for each cat to indicate the points sought by the cat. The cat would pick a point from the memory pool according to the rules described later. SRD declares the mutative ratio for the selected dimensions. CDC discloses how many dimensions will be varied. In the seeking mode, if a dimension is selected to mutate, the difference between the new value and the old one cannot be out of the range of which is defined by SRD. SPC is a Boolean variable, which decides whether the point, where the cat is already standing, will be one of the candidates to move to. No matter the value of SPC is true or false; the value of SMP will not be influenced. These factors are all playing important roles in the seeking mode. How the seeking mode works can be described as follows:

1. Generate j copies of cat_k, where $j = SMP$. If SPC is true, let $j = SMP - 1$ and retain the present position as one of the candidates.
2. According to CDC, plus/minus SRD percents of the current value randomly and replace the old one for all copies by equation (22) – (24).
3. Calculate the fitness value (FS) of all candidate points.
4. If all FS are not exactly equal, calculate the selecting probability of each candidate point by equation (25), otherwise set all the selecting probability of each candidate point be equal. If the goal of the fitness function is to find the minimum solution, let $FS_b = FS_{max}$, otherwise $FS_b = FS_{min}$. FS_{max} denotes the largest FS in the candidates, and FS_{min} represents the smallest one.
5. Pick the point to move to randomly from the candidate points, and replace the position of cat_k.

$$M = Modify \cup (1 - Modify) \tag{22}$$

$$|Modify| = CDC \times M \tag{23}$$

$$x_{jd} = \left\{ \begin{array}{ll} x_{jd} & , d \notin Modify \\ (1 + rand \times SRD) \times x_{jd} & , d \in Modify \end{array} \right|_{d=1,2,\ldots,M} \quad \forall j \tag{24}$$

$$P_i = \frac{|FS_i - FS_b|}{FS_{max} - FS_{min}}, \textbf{ where } 0 < i < j \tag{25}$$

- The Tracing Mode Process:
The tracing mode models the case of the cat in tracing targets. Once a cat goes into the tracing mode, it moves according to its own velocities for every dimension. The action of the tracing mode can be described as follows:

  1. Update the velocity by equation (26). The new velocity should satisfy the constraint of the range.
  2. Move the cat according to equation (27).

$$v_{k,d}^{t+1} = v_{k,d}^{t} + r_1 \times c_1 \times (x_{best,d} - x_{k,d}^{t}), \textbf{ where } d = 1, 2, \ldots, M \tag{26}$$

$$x_{k,d}^{t+1} = x_{k,d}^{t} + v_{k,d}^{t+1} \tag{27}$$

where $v_{k,d}$ is the velocity for cat_k, t denotes the rounds, r_1 is a random variable satisfies $r_1 \in [0, 1]$, c_1 is a constant, x_{best} represents the best solution found so far, $x_{k,d}$ is the current position of the cat, and M is the dimension of the solution.

2.6 Parallel Cat Swarm Optimization

The parallel version of CSO is proposed by Tsai et al. in 2008 [29]. According to the experience and the knowledge from our life, things will be done in the best way when the consideration is complete and is observed from different angles. Sometimes, an expert may get a new idea or a breakthrough by the opinions from a novice. Hence, the power and the affection of the cooperation have the potential to be stronger than the power of solo.

The precedents of splitting the amount of the population into several sub-populations to construct the parallel structure can be found in several algorithms such as Island-model Genetic Algorithm, Parallel Genetic Algorithm [1], Ant colony system with communication strategies [5], and Parallel Particle Swarm Optimization Algorithm with Communication Strategies [4]. Each of the sub-populations evolves independently and shares the information they have occasionally. Although it results in the reducing of the population size for each sub-population, the benefit of cooperation is achieved.

By inspecting the structure of CSO, it is clear that the individuals work independently when they stay in the seeking mode. Oppositely, they share the identical information, the global best solution, as they know according to their knowledge to the present in the tracing mode. In PCSO, it still follows the structure of this framework.

Based on the structure of CSO, most of the individuals work as a stand along system. To parallelize these individuals in CSO, The procedure in the tracing mode is modified to make it becomes more cooperatively. In PCSO, the individuals are separated into several sub-populations at the beginning. Hence, the individuals in the tracing mode do not move forward to the global best solution directly, but they move forward to the local best solution of its own group in general. Only if the predetermined iteration is achieved, the sub-populations pop the local best solutions at present and randomly pick a sub-population to replace the worst individual in the selected sub-population.

Basically, the main framework of the PCSO is similar to the CSO. Assume that we set G equals to 1; then the PCSO becomes the original CSO since there is only one group overall populations. The difference between the PCSO and the CSO is that at the beginning of the PCSO algorithm, N individuals are created and separate into G groups, the calculation in the tracing mode is different, and there exists a information exchanging process. The Parallel tracing mode process and the information exchanging process are described as follows, and the diagram of the PCSO is represented in Figure 7.

- Parallel Tracing Mode Process
  The parallel tracing mode process can be described as follows:
  1. Update the velocities for every dimension $v_{k,d}(t)$ for the cat_k at the current iteration according to equation (28), where t indicates the iteration number, $x_{l_{best},d}(t-1)$ is the position of the cat, which has the best fitness value, at the previous iteration in the group that cat_k belongs to.
  2. Check if the velocities are in the range of maximum velocity. In case the new velocity is over-range, it is set equal to the limit.
  3. Update the position of cat_k according to equation (29).

$$v_{k,d}^{t+1} = v_{k,d}^{t} + r_1 \times c_1 \times \left(x_{l_{best},d} - x_{k,d}^{t}\right), \textbf{ where } d = 1, 2, \ldots, M \quad (28)$$

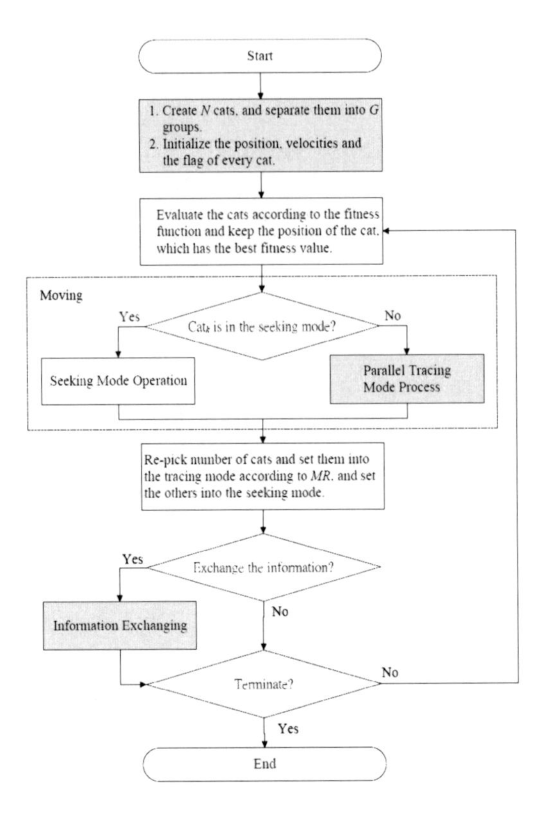

Fig. 7. The flowchart of the Parallel Cat Swarm Optimization.

$$x_{k,d}^{t+1} = x_{k,d}^{t} + v_{k,d}^{t+1} \tag{29}$$

- Information Exchanging
  The procedure forces the sub-populations exchange their information, and achieve somehow the cooperation. We define a parameter ECH to control the exchanging of the information between sub-populations. The information exchanging is applied once per ECH iterations. The information exchanging consists of 4 steps:
  1. Pick up a group of the sub-populations sequentially and sort the individuals in this group according to their fitness value.
  2. Randomly select a local best solution from an unrepeatable group.
  3. The individual, whose fitness value is the worst in the group, is replaced by the selected local best solution.
  4. Repeat step 1 to 3 until all the groups exchanged information to someone else.

3 The Hybrid Swarm Intelligence Based on PCSO and ABC

In this chapter, Hybrid PCSOABC, the hybrid framework, based on PCSO and ABC algorithm is proposed. The precedents of splitting the amount of the population into several sub-populations to construct the parallel structure can be found in several algorithms such as Island-model Genetic Algorithm, Parallel Genetic Algorithm [1], Ant colony system with communication strategies [5], and Parallel Particle Swarm Optimization Algorithm with Communication Strategies [4]. Each of the sub-populations evolves independently and shares the information they have occasionally. Since PCSO presents high convergence when the population size is small and the iteration is less [29], it is useful for reducing the computational time. The proposed hybrid framework starts with the parallel system, PCSO operation, with only a small population size. After a few iterations, the populations are cloned randomly to increase the population size to the specific number and every population is ensured to be remained. Afterward, ABC takes over the process and output the result when the termination condition is satisfied. The diagram of Hybrid PCSOABC is represented in Figure 8.

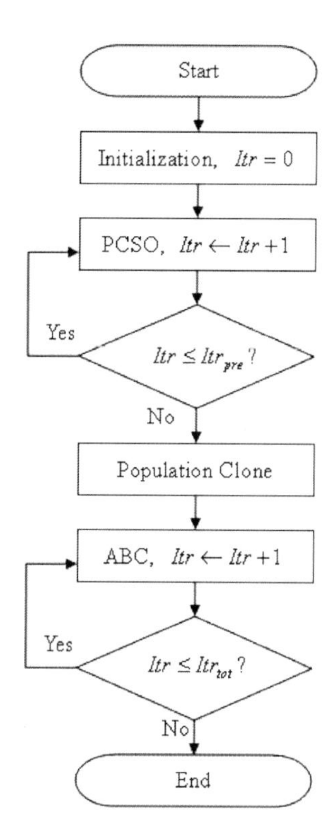

Fig. 8. The Hybrid PCSOABC

The process of the hybrid framework can be described as follows:

Step 1. Initialization:
Create N_{pre} cats and separate them into G groups. Set the iteration counter (Itr) to be 0, define the iteration number (Itr_{pre}) for executing the PCSO process, and define the total iteration number (Itr_{tot}) by equation (30).

$$\textbf{Let } Itr_{pre} < Itr_{tot}, \textbf{ where } Itr_{pre} \in \mathbf{C} \textbf{ and } Itr_{tot} \in \mathbf{C} \tag{30}$$

Step 2. The PCSO Process:
Executing the PCSO process with the N_{pre} populations until the Itr grows larger than the predefined Itr_{pre}. In other words, execute PCSO algorithm, which is described in the section 2.6, with N_{pre} iterations by equation (22)-(25) and (28)-(29), and the information exchanging process accordingly.

Step 3. Population Clone:
Randomly clone the populations from the N_{pre} cats to N copies, and make sure that each population is retained in the N copies. In this step, the population size is extended by equation (31) from N_{pre} to N.

$$|N_{pre}| < |N| \textbf{ and } N \subset N_{pre} \tag{31}$$

Step 4. The ABC Process:
Executing step 2 to step 5 listed in the ABC process, which is described in the section 2.4, with the N populations by equation (19)-(21).

Step 5. Termination Checking:
If the Itr is still less than the predefined total iteration number (Itr_{tot}), go back to step 4, otherwise output the final result and terminate the program.

4 Experimental Results and Conclusions

According to the experience obtained from PCSO [29], the ECH is set to be 20, the population size N_{pre} is set to 16 for the PCSO process, the group number G is set to be 4, and the population size N for executing the ABC process is set to 160. The parameter setting for ABC algorithm follows the values listed in ABC algorithm [19], and the rest of the parameter setting is listed in Table 1.

Table 1. The parameter setting

Parameter	Value or Range	Parameter	Value or Range
MP	5	c_1	2.0
SRD	20%	r_1	[0 1]
CDC	80%	Itr_{pre}	500
MR	2%	Itr_{tot}	5000

Table 2. The test functions

Function	Initial Range
$f_1(\overrightarrow{x}) = \frac{1}{4000}\left[\sum_{i=1}^{n}(x_i - 100)^2\right] - \left[\prod_{i=1}^{n}\cos\frac{x_i - 100}{\sqrt{i}}\right] + 1$	$0 \leq x_i \leq 600$
$f_2(\overrightarrow{x}) = \sum_{i=1}^{n}[x_i^2 - 10cos(2\pi x_i) + 10]$	$-5 \leq x_i \leq 5$
$f_3(\overrightarrow{x}) = \sum_{i=1}^{n-1} 100(x_{i+1} - x_i^2)^2 + (x_i - 1)^2$	$-100 \leq x_i \leq 100$
$f_4(\overrightarrow{x}) = -20\exp^{-0.2\sqrt{\frac{1}{n}\sum_{i=1}^{n}x_i^2}} - \exp^{\frac{1}{n}\sum_{i=1}^{n}cos(2\pi x_i)} + 20 + \exp^1$	$-32 \leq x_i \leq 32$
$f_5(\overrightarrow{x}) = \sum_{i=1}^{n}\left(\sum_{j=1}^{i}x_j\right)^2$	$-100 \leq x_j \leq 100$

To analysis the convergence and the performance of Hybrid PCSOABC, 5 well-known test functions, which are adopted from the IEEE CEC competitions [27], are applied in the experiments. The test functions and the initial ranges are listed in Table 2. The results are compared with ABC algorithm and PCSO algorithm simultaneously. The population size is set to be 160 for ABC algorithm and 16 for PCSO algorithm. The maximum velocity for Hybrid PCSOABC and PCSO are set to the upper limit of the initial bounds. For all the algorithms, 50 runs are taken in average to present the final results, and the dimension of the solution space is set to 30 and 100 to test the searching ability of the algorithms in different scale optimizations. The termination condition is set to be executing the program for 5000 iterations. The movement of the artificial agents are constrained under the range of the feasible solution, which are indicated by the initial range. The experimental results are listed in Table 3 and Figure 9 to Figure 18. To test the reaction of the artificial agents under the unconstrained condition, the initial range of the first three benchmark functions are shifted to the space, which does not content the global optimum solution respectively. The shifted initial range for the benchmark functions are listed in Table 4. Since the global optimum is not inside the initial range, there is no bound for the artificial agents during the optimization process. The result of the unconstrained experiments are listed in Figure 19 to Figure 21.

According to the experimental results, Hybrid PCSOABC presents satisfactory searching ability with in average less computation time than single-handed employing PCSO and ABC to solve optimization problems. The solution obtained by Hybrid PCSOABC, in some conditions, is worse than PCSO since the hybrid framework in this case is a simple structure without any feed-back to adjust the parameter or the structure of the process. However, it provides a trade-off between the accuracy and the computational speed. The computational speed of Hybrid PCSOABC can save hundreds times of the computational cost than PCSO in some cases. In the future work, creating a hybrid framework with a feed-back parameter for adjusting the key parameters in the algorithm is a feasible way for the research and the applications.

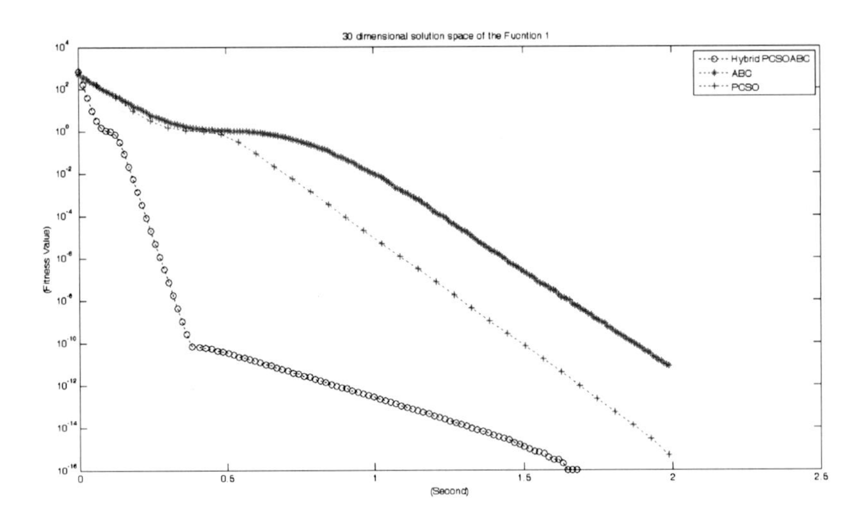

Fig. 9. The experimental result of $f_1(X)$ with 30 dimensions.

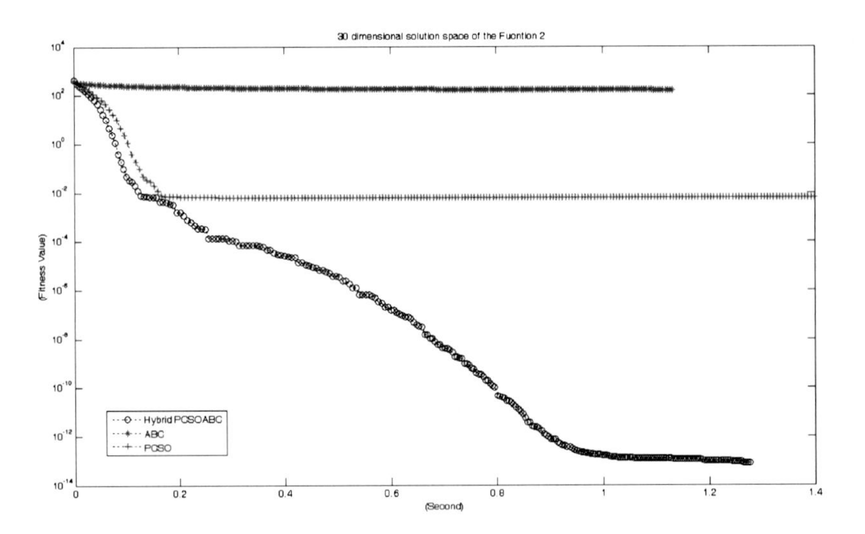

Fig. 10. The experimental result of $f_2(X)$ with 30 dimensions.

Table 3. The experimental results

	The 30 dimensional solution space of $f_1(\overrightarrow{x})$		
	Cost (in mSec.)	*Fitness in average*	STD
Hybrid PCSOABC	1.709×10^3	0	7.193×10^{-11}
ABC	1.996×10^3	1.70×10^{-14}	9.70×10^{-14}
PCSO	1.900×10^3	0	0
	The 30 dimensional solution space of $f_2(\overrightarrow{x})$		
	Cost (in mSec.)	*Fitness in average*	STD
Hybrid PCSOABC	1.281×10^3	8.130×10^{-14}	7.445×10^{-3}
ABC	1.159×10^3	1.598×10^2	9.859×10^0
PCSO	1.417×10^3	6.417×10^{-3}	3.176×10^{-2}
	The 30 dimensional solution space of $f_3(\overrightarrow{x})$		
	Cost (in mSec.)	*Fitness in average*	STD
Hybrid PCSOABC	8.694×10^2	2.374×10^1	4.684×10^0
ABC	6.267×10^2	2.667×10^2	4.330×10^2
PCSO	1.048×10^4	2.821×10^1	8.151×10^{-1}
	The 30 dimensional solution space of $f_4(\overrightarrow{x})$		
	Cost (in mSec.)	*Fitness in average*	STD
Hybrid PCSOABC	1.658×10^3	8.000×10^{-15}	1.243×10^{-6}
ABC	1.548×10^3	2.797×10^{-4}	1.033×10^{-3}
PCSO	1.497×10^3	4.00×10^{-16}	0
	The 30 dimensional solution space of $f_5(\overrightarrow{x})$		
	Cost (in mSec.)	*Fitness in average*	STD
Hybrid PCSOABC	8.388×10^2	4.627×10^{-2}	8.759×10^{-2}
ABC	8.620×10^2	2.596×10^5	8.445×10^4
PCSO	1.350×10^3	0	0
	The 100 dimensional solution space of $f_1(\overrightarrow{x})$		
	Cost (in mSec.)	*Fitness in average*	STD
Hybrid PCSOABC	4.920×10^3	1.616×10^{-5}	2.437×10^{-5}
ABC	4.910×10^3	7.558×10^2	6.040×10^1
PCSO	5.464×10^3	0	0
	The 100 dimensional solution space of $f_2(\overrightarrow{x})$		
	Cost (in mSec.)	*Fitness in average*	STD
Hybrid PCSOABC	3.372×10^3	1.779×10^{-2}	9.356×10^{-2}
ABC	3.315×10^3	1.408×10^3	4.135×10^1
PCSO	3.768×10^3	0	0
	The 100 dimensional solution space of $f_3(\overrightarrow{x})$		
	Cost (in mSec.)	*Fitness in average*	STD
Hybrid PCSOABC	1.489×10^3	9.726×10^1	1.633×10^1
ABC	1.331×10^3	1.111×10^{11}	1.002×10^{10}
PCSO	2.543×10^3	9.850×10^1	3.373×10^{-1}
	The 100 dimensional solution space of $f_4(\overrightarrow{x})$		
	Cost (in mSec.)	*Fitness in average*	STD
Hybrid PCSOABC	3.395×10^3	7.301×10^{-4}	3.999×10^{-4}
ABC	3.273×10^3	2.061×10^1	1.007×10^{-1}
PCSO	3.932×10^3	4.00×10^{-16}	0
	The 100 dimensional solution space of $f_5(\overrightarrow{x})$		
	Cost (in mSec.)	*Fitness in average*	STD
Hybrid PCSOABC	4.137×10^3	4.983×10^3	6.034×10^3
ABC	4.138×10^3	4.427×10^7	1.562×10^7
PCSO	5.280×10^3	0	0

Table 4. The test functions

Function	Initial Range
$f_1(\overrightarrow{x}) = \frac{1}{4000}\left[\sum_{i=1}^{n}(x_i-100)^2\right] - \left[\prod_{i=1}^{n}\cos\frac{x_i-100}{\sqrt{i}}\right]+1$	$300 \leq x_i \leq 600$
$f_2(\overrightarrow{x}) = \sum_{i=1}^{n}[x_i^2 - 10cos(2\pi x_i)+10]$	$2.56 \leq x_i \leq 5.12$
$f_3(\overrightarrow{x}) = \sum_{i=1}^{n-1} 100(x_{i+1}-x_i^2)^2 + (x_i-1)^2$	$300 \leq x_i \leq 600$

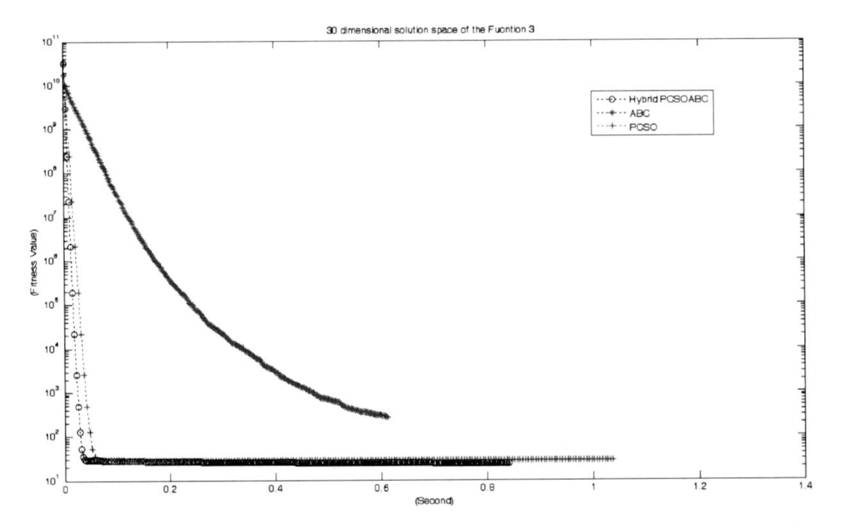

Fig. 11. The experimental result of $f_3(X)$ with 30 dimensions.

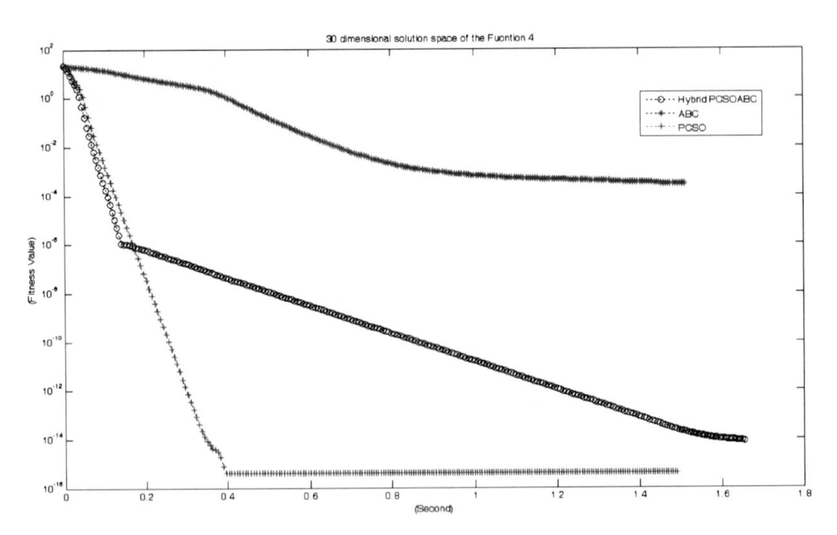

Fig. 12. The experimental result of $f_4(X)$ with 30 dimensions.

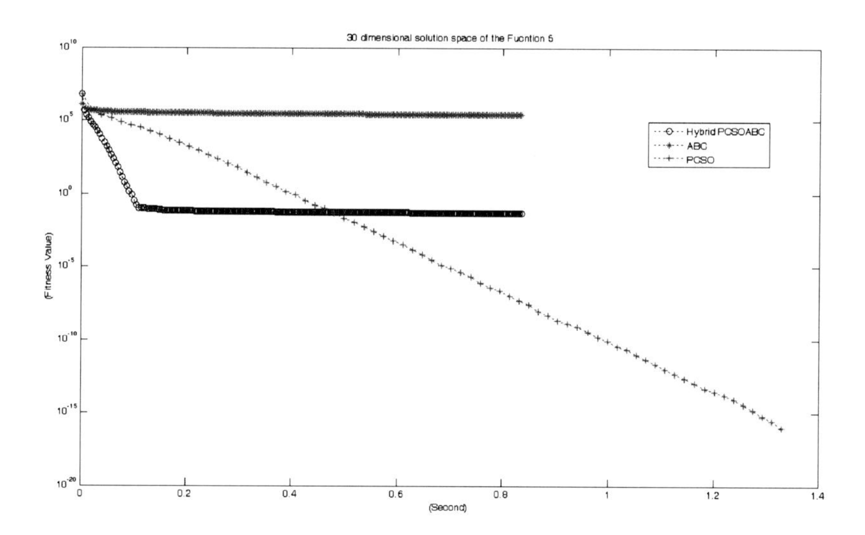

Fig. 13. The experimental result of $f_5(X)$ with 30 dimensions.

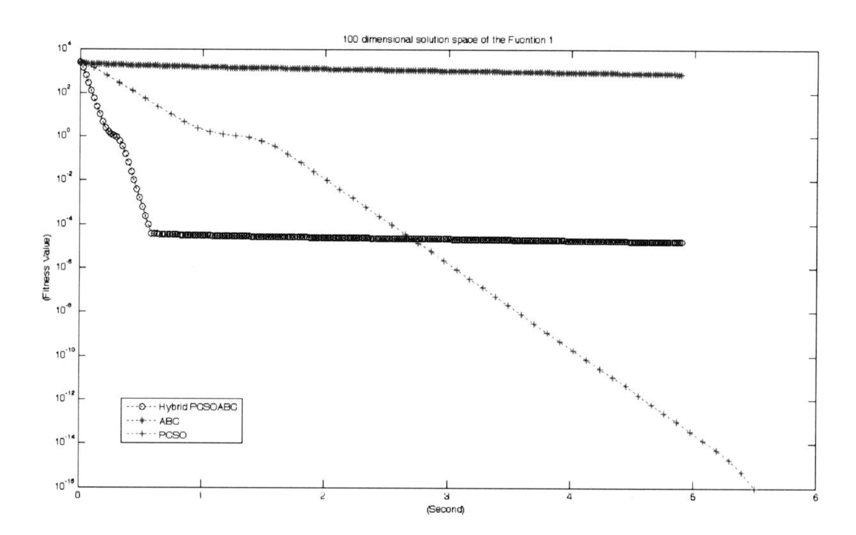

Fig. 14. The experimental result of $f_1(X)$ with 100 dimensions.

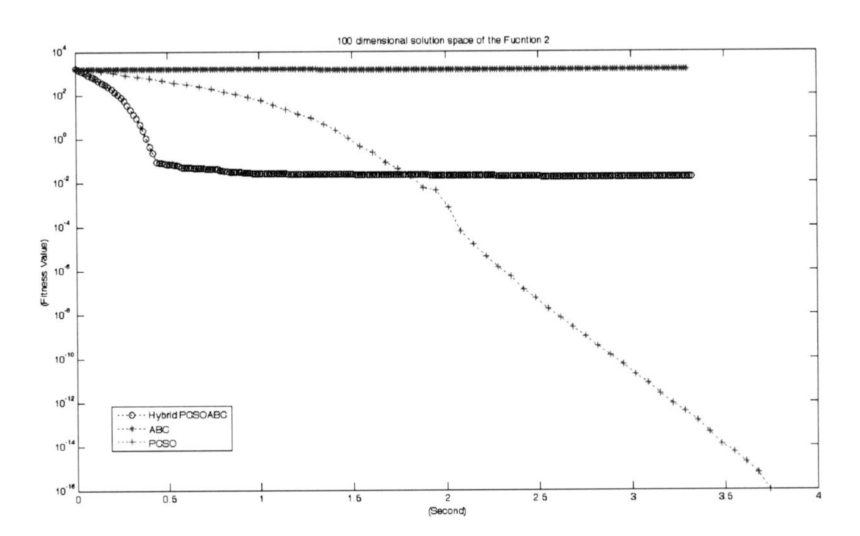

Fig. 15. The experimental result of $f_2(X)$ with 100 dimensions.

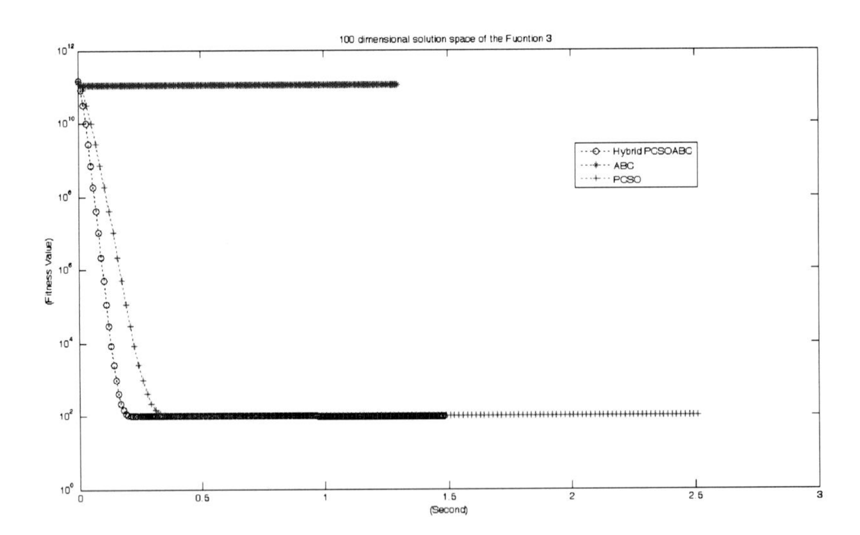

Fig. 16. The experimental result of $f_3(X)$ with 100 dimensions.

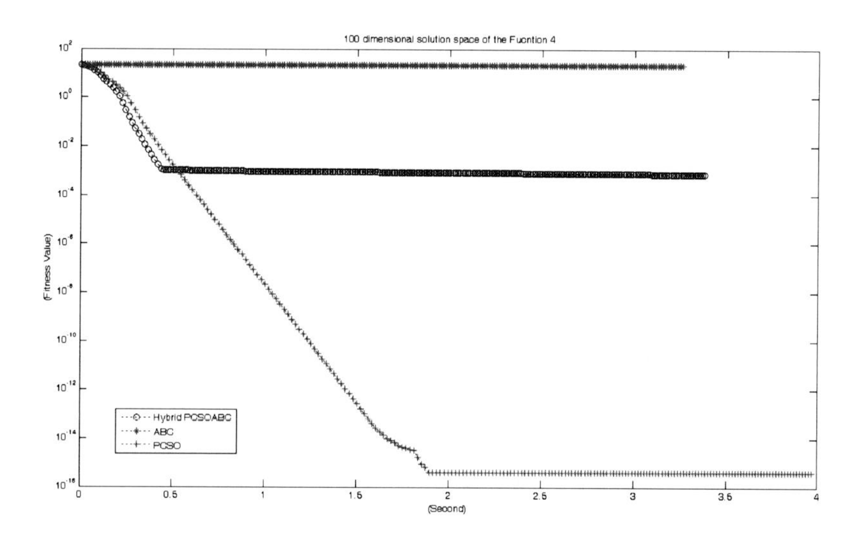

Fig. 17. The experimental result of $f_4(X)$ with 100 dimensions.

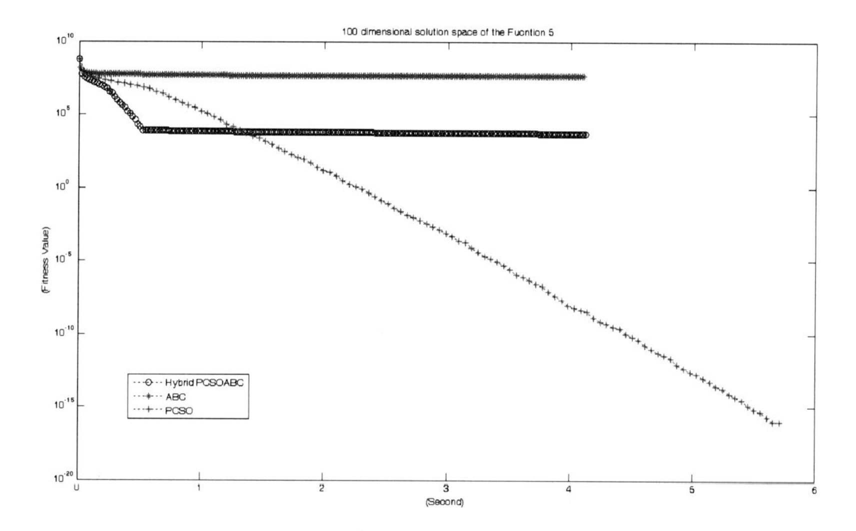

Fig. 18. The experimental result of $f_5(X)$ with 100 dimensions.

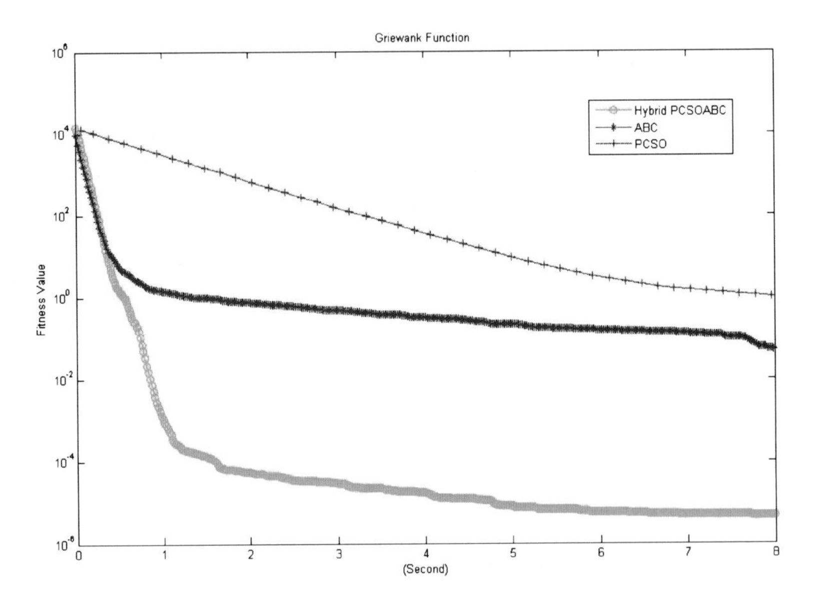

Fig. 19. The experimental result of the unconstrained $f_1(X)$.

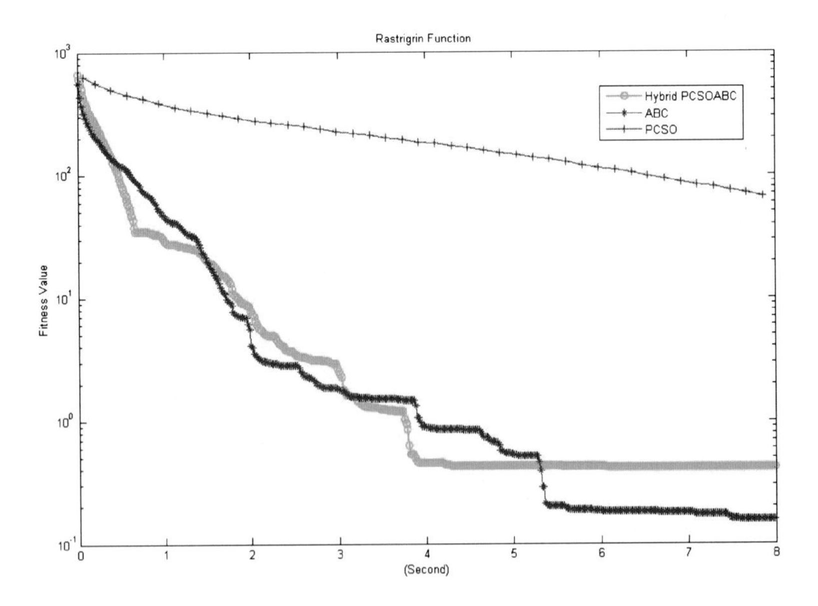

Fig. 20. The experimental result of the unconstrained $f_2(X)$.

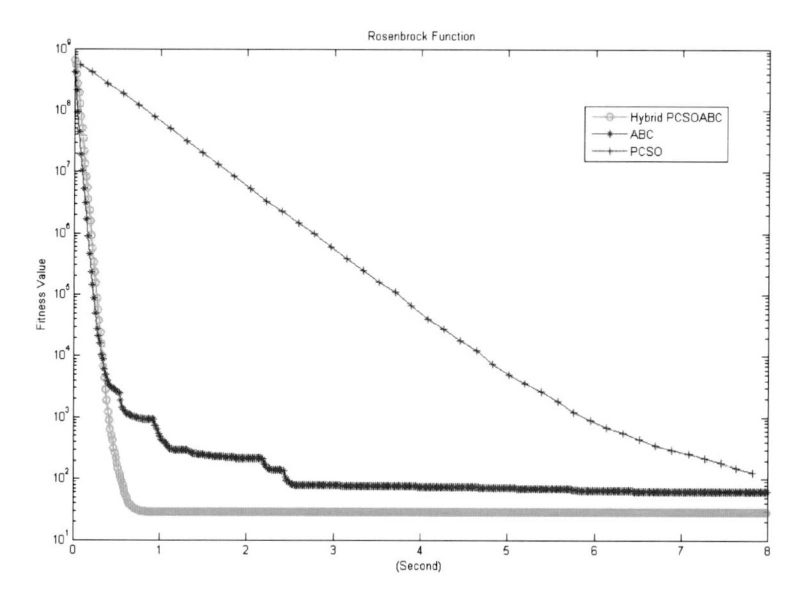

Fig. 21. The experimental result of the unconstrained $f_3(X)$.

References

1. Abramson, D., Abela, J.: A Parallel Genetic Algorithm for Solving the School Timetabling Problem. Technical Report, Division of Information Technology, CSIRO (1991)
2. Angeline, P.J.: Evolutionary Optimization Versus Particle Swarm Optimization. Philosophy and performance differences. In: Porto, V.W., Saravanan, N., Waagen, D., Eiben, A.E. (eds.) EP 1998. LNCS, vol. 1447, Springer, Heidelberg (1998)
3. Cai, X., Cui, Z., Zeng, J., Tan, Y.: Particle Swarm Optimization with Self-adjusting Cognitive Selection Strategy. International Journal of Innovative Computing, Information and Control 4(4), 943–952 (2008)
4. Chang, J.F., Chu, S.C., Roddick, J.F., Pan, J.S.: A parallel particle swarm optimization algorithm with communication strategies. Journal of Information Science and Engineering 21(4), 809–818 (2005)
5. Chu, S.C., Roddick, J.F., Pan, J.S.: Ant colony system with communication strategies. Information Sciences 167, 63–76 (2004)
6. Chu, S.C., Roddick, J.F., Su, C.-J., Pan, J.-S.: Constrained Ant Colony Optimization for Data Clustering. In: Zhang, C., W. Guesgen, H., Yeap, W.-K. (eds.) PRICAI 2004. LNCS (LNAI), vol. 3157, pp. 534–543. Springer, Heidelberg (2004)
7. Chu, S.C., Tsai, P.W.: Computational Intelligence Based on the Behavior of Cats. International Journal of Innovative Computing, Information and Control 3(1), 163–173 (2007)
8. Chu, S.C., Tsai, P.W., Pan, J.S.: Cat Swarm Optimization. In: Yang, Q., Webb, G. (eds.) PRICAI 2006. LNCS (LNAI), vol. 4099, pp. 854–858. Springer, Heidelberg (2006)

9. Colorni, A., Dorigo, M., Maniezzo, V.: Distributed Optimization by Ant Colonies. In: Varela, F., Bourgine, P. (eds.) First Eur. Conference Artificial Life, pp. 134–142 (1991)
10. Davis, L.: Handbook of Genetic Algorithms. Van Nostrand Reinhold, New York (1991)
11. Dorigo, M., Maniezzo, V., Colorni, A.: Ant System: Optimization by a Colony of Cooperating Agents. IEEE Transactions on Systems, jernetics–Part B: Cybernetics 26(1), 29–41 (1996)
12. Dorigo, M., Gambardella, L.M.: Ant colony system: a cooperative learning approach to the traveling salesman problem. IEEE Trans. on Evolutionary Computation 26(1), 53–66 (1997)
13. Eberhart, R., Shi, Y.: Comparison Between Genetic Algorithms and Particle Swarm Optimization. In: Proc. Seventh Annual Conference on Evolutionary Programming, pp. 611–619 (1998)
14. Gao, X.Z., Ovaska, S.J., Wang, X.: A GA-based Negative Selection Algorithm. International Journal of Innovative Computing, Information and Control 4(4), 971–979 (2008)
15. Gen, M., Cheng, R.: Genetic Algorithm and Engineering Design. John Wiley and Sons, New York (1997)
16. Goldberg, D.E.: Genetic algorithm in search. In: Optimization and Machine Learning. Addison-Wesley Publishing Company, Reading (1989)
17. Guo, Y., Gao, X., Yin, H., Tang, Z.: Coevolutionary Optimization Algorithm with Dynamic Sub-population Size. International Journal of Innovative Computing, Information and Control 3(2), 435–448 (2007)
18. Karaboga, D.: An Idea Based On Honey Bee Swarm For Numerical Optimization. Technical Report-TR06, Erciyes University, Engineering Faculty, Computer Engineering Department (2005)
19. Karaboga, D., Basturk, B.: On the performance of artificial bee colony (ABC) algorithm. Applied Soft Computing 8, 687–697 (2007)
20. Kassabalidis, I.N., El-Sharkawi, M.A., Marks II, R.J., Moulin, L.S., Alves da Silva, A.P.: Dynamic Security Border Identification Using Enhanced Particle Swarm Optimization. IEEE Trans. on Power Systems 17(3) (2002)
21. Kim, S.S., Kim, I.H., Mani, V., Kim, H.J.: Ant Colony Optimization for SONET Ring Loading Problem. International Journal of Innovative Computing, Information and Control 4(7), 1617–1626 (2008)
22. Mishra, S., Bhende, C.N.: Bacterial Foraging Technique-Based Optimized Active Power Filter for Load Compensation. IEEE Transactions on Power Delivery 22(1), 457–465 (2007)
23. Nakajima, S., Arimoto, H., Rensha, H., Toriu, T.: Measurement of a Translation and a Rotation of a Tooth after an Orthodontic Treatment Using GA. International Journal of Innovative Computing, Information and Control 3(6)(A),1399–1406 (2007)
24. Pan, J.S., McInnes, F.R., Jack, M.A.: Application of parallel genetic algorithm and property of multiple global optima to VQ codevector Index assignment. Electronics Letters 32(4), 296–297 (1996)
25. Passino, K.M.: Biomimicry of Bacterial Foraging for Distributed Optimization and Control. IEEE Control Systems Magazine, 52–67 (2002)
26. Shi, Y., Eberhart, R.: Empirical Study of Particle Swarm Optimization. Congress on Evolutionary Computation, 1945–1950 (1999)

27. Suganthan, P.N., Hansen, N., Liang, J.J., Deb, K., Chen, Y.P., Auger, A., Tiwari, S.: Problem Definitions and Evaluation Criteria for the CEC, Special Session on Real-Parameter Optimization (2005)
28. Tang, W.J., Wu, Q.H., Saunders, J.R.: Bacterial Foraging Algorithm for Dynamic Environments. In: IEEE Congress on Evolutionary Computation, pp. 1324–1330 (2007)
29. Tsai, P.W., Pan, J.S., Chen, S.M., Liao, B.Y., Hao, S.P.: Parallel Cat Swarm Optimization. In: 7th International Conference on Machine Learning and Cybernetics, pp. 3328–3333 (2008)
30. Xia, F., Tian, Y.C., Sun, Y., Dong, J.: Neural Feedback Scheduling of Real-time Control Tasks. International Journal on Innovative Computing, Information & Control 4(11) (2008)

Glowworm Swarm Optimization for Multimodal Search Spaces

K.N. Krishnanand[1] and D. Ghose[2]

[1] Department of Computer Science, University of Vermont, Burlington, VT, USA
kkrishna@uvm.edu, krishna.iisc@gmail.com

[2] Guidance, Control and Decision Systems Laboratoty (GCDSL) Department of Aerospace Engineering Indian Institute of Science, Bangalore, India
dghose@aero.iisc.ernet.in

Summary. This chapter presents glowworm swarm optimization (GSO), a novel swarm intelligence algorithm, which was recently proposed for simultaneous capture of multiple optima of multimodal functions. In particular, GSO prescribes individual-level rules that cause a swarm of agents deployed in a signal medium to automatically partition into subswarms that converge on the multiple sources of the signal profile. The sources could represent multiple optima in a numerical optimization problem or physical quantities like sound, light, or heat in a realistic robotic source localization task. We present the basic GSO model and use a numerical example to characterize the group-level phases of the algorithm that gives an insight into how GSO explicitly addresses the issue of achievement/maintenance of swarm diversity. We briefly summarize the results from the application of GSO to the following three problems–multimodal function optimization, signal source localization, and pursuit of mobile signal sources.

1 Introduction

Nature abounds in examples of swarming, a form of collective behavior found in insect and animal societies. Ants use simple trail-laying and trail-following behaviors to self-organize into complex foraging patterns [1]; ants gathered in groups can carry prey that are so large that if they were fragmented, the original members of the group would be unable to carry all the fragments [2]. Honeybee swarms use group decision making to find a future nest: the scouts find potential sites in all directions and advertise a dozen or more of them to the recruits, but eventually they reach a consensus about a single site [3]. Schooling in fish serves to confuse a predator [4]. Birds that gather in large flocks for migrations achieve aerodynamic efficiency higher than that of a single bird leading to reduced fatigue and higher chances of survival at the end of the migration [5].

The decentralized decision-making mechanisms found in the above examples, and others in the natural world, offer an insight into the basis to devise distributed algorithms that solve complex problems related to diverse fields such as optimization, multi-agent decision making, and collective robotics. Examples of such nature-inspired algorithms include ant colony optimization (ACO)

B.K. Panigrahi, Y. Shi, and M.-H. Lim (Eds.): Handbook of Swarm Intelligence, ALO 8, pp. 451–467.
springerlink.com

techniques [7], particle swarm optimization (PSO) algorithms [8], bacterial chemotaxis based optimization [12], social foraging swarms that respond to environmental stimuli and perform gradient climbing [13], and swarm based collective robotic algorithms [14].

This chapter presents glowworm swarm optimization (GSO), a novel algorithm for the simultaneous capture of multiple optima of multimodal functions [6]. In particular, GSO prescribes individual-level rules that cause a swarm of agents deployed in a signal medium to automatically partition into subswarms that converge on the multiple sources of the signal profile. GSO is based on behavior of glowworms (also known as fireflies or lightning bugs). The behavior pattern of glowworms, which is used in this algorithm, is the apparent capability of the glowworms to change the intensity of the luciferin emission and thus appear to glow at different intensities. The GSO algorithm makes the agents glow at intensities approximately proportional to the function value being optimized. It is assumed that glowworms of brighter intensities attract glowworms that have lower intensity. The algorithm incorporates a dynamic decision range by which the effect of distant glowworms are discounted when a glowworm has sufficient number of neighbors or the range goes beyond the range of perception of the glowworms. This algorithm allows swarms of glowworms to split into sub-groups and converge to high function value points. This property of the algorithm allows it to be used to identify multiple peaks of a multi-modal function. Later, other researchers have proposed the firefly algorithm [26], which essentially follows the same logic as GSO with some minor variations. It has been recognized that GSO's approach of explicitly addressing the issue of partitioning a swarm required by multiple source localization is very effective [27]. We also believe that GSO is the first algorithm to do this and prior methods provided only indirect solutions to the problem.

In this chapter, we present the basic working principle of the GSO algorithm, describe the various algorithmic phases, and present the equations that represent the basic GSO model. We show through simulations that the implementation of the GSO algorithm at the individual agent-level gives rise to two major phases at the group level: splitting of the agent-swarm into subswarms and local convergence of agents in each subswarm to the peak locations. We show how GSO can be applied to problems ranging from numerical multimodal optimization to realistic collective robotics tasks of localizing static signal sources and pursuing mobile signal sources.

The chapter is organized as follows. Related work on multimodal function optimization is described in Section 2. The basic principle of GSO and the equations describing the GSO model are presented in Section 3. A characterization of the group-level phases of GSO and a numerical example to illustrate GSO's efficacy in capturing multiple peaks of multimodal functions are given in Section 4. GSO applications are briefly summarized in Section 5. Conclusions are given in Section 6.

2 Multimodal Function Optimization

Seeking multiple optima by maintaining population diversity has received some attention in the domain of genetic algorithms [15]–[18] and particle swarm optimization algorithms [8]–[11]. Niche-preserving techniques that allow a GA to identify multiple optima of a multimodal function include sharing [15], clearing [16], crowding [17], and restricted tournament selection [18].

Parsopoulos et al. [9] studied altering the fitness value via fitness function stretching to adapt PSO to sequentially find peaks in a multimodal environment. In particular, a potentially good solution is isolated once it is found (if its fitness is below a threshold value) and then the fitness landscape is stretched to keep other particles away from this area of the search space. The isolated particle is checked to see if it is a global optimum, and if it is below the desired accuracy, a small population is generated around this particle to allow a finer search in this area. The main swarm continues its search for the rest of the search space for other potential global optima. With this modification, the algorithm was able to locate all the global optima of the test functions successfully.

Brits et al. [10] proposed a niching particle swarm optimization (NichePSO) algorithm, which has some improvements to the Parsopoulos and Vrahitis's model. The NichePSO is a parallel niching algorithm that locates and tracks multiple solutions simultaneously. In NichePSO, the subswarm approach [19] is adapted to maintain and optimize niches in the objective function space. Particles in the main swarm do not share knowledge about the best solution and use only their own knowledge. If a particles fitness shows very little change over a small number of iterations, a subswarm is created with the particle and its closest topological neighbor. A subswarm radius is defined as the maximum distance between the global best particle and any other particle within the subswarm. A particle that is outside of the swarm radius of a subswarm is merged into it when the particle moves into the swarm, that is, when its distance to the global best particle becomes less than the swarm radius. In this manner, population diversity is preserved in the form of different niches (subswarms) and each subswarm converges to a different optimum solution. The algorithm was reported to be successful at detecting global maxima and sometimes local maxima.

Li [11] proposed a species-based PSO (SPSO) that incorporates the idea of species into PSO for solving multimodal optimization problems. Initially, a population of particles is generated randomly. At each iteration step, different species seeds are identified for multiple species and then used as the *lbest* (particle that has the best fitness among the members of the same topological neighborhood) for different species accordingly. For this purpose, all the particles are evaluated and sorted in descending order of their fitness values. The particle with the best fitness is set as the initial species seed. All particles that are within a radius r_s of the species seed's position, along with the seed, form one species. The next best particle that falls outside the r_s range of the first seed is set as the next species seed. The above process is repeated until all the particles are checked against the species seeds. These multiple adaptively formed species are then used to optimize towards multiple optima in parallel, without interference across different species.

3 Basic Principle of GSO

The agents in GSO are modeled after glowworms that carry a luminescence quantity called luciferin along with them. Natural glowworms primarily use the bioluminescent light to signal other individuals of the same species for reproduction [20]. This natural light is also used to attract prey. The general idea in GSO is similar in these aspects in the sense that glowworm agents are assumed to be attracted to move toward other glowworm agents that have brighter luminescence (higher luciferin value). Hereafter, we refer to the agents in GSO as glowworms. However, they are endowed with other behavioral mechanisms (not found in their natural counterparts) that enable them to selectively interact with their neighbors and decide their movements at each iteration. The brightness of a natural glowworm's glow as perceived by its neighbor reduces with increase in the distance between the two glowworms. However, in GSO, we assume that luciferin value of a glowworm as perceived by its neighbor does not reduce due to distance. The glowworms encode the function profile values at their current locations into a luciferin value and broadcast the same to other glowworms in their neighborhood. Each glowworm i regards only those incoming luciferin data as useful that are broadcast by other glowworms located within a adaptive local-decision domain whose range r_d^i is bounded by a hard-limited sensor range r_s^i $(0 < r_d^i \leq r_s^i)$. Each glowworm selects, using a probabilistic mechanism, a neighbor that has a luciferin value higher than its own and moves toward it. These movements-based only on local information and selective neighbor interactions-enable the swarm of glowworms to partition into disjoint subgroups that converge on multiple optima of a given multimodal function. The significant difference between our work and most earlier approaches to multimodal function optimization problems is the adaptive local-decision domain which we use effectively to locate multiple peaks. Figure 1 sets out the difference between PSO and GSO. In both algorithms a group of particles/agents are initially deployed in the objective function space and each agent selects, and moves in, a direction based on respective position update formulae. The particle movement directions in original PSO are adjusted according to its own and global best previous positions. However, movement directions are aligned along the line-of-sight between neighbors in GSO. In PSO, The net improvement in the objective function at the iteration t is stored in $P_g(t)$. However, in GSO, a glowworm with the highest luciferin value in a neighborhood indicates its potential proximity to an optimum location. Figure 1(a) shows the trajectories of six agents, their current positions, and the positions at which they encountered their personal best and the global best. Figure 1(b) shows the GSO decision for an agent which has four neighbors in its dynamic range of which two have higher luciferin value and thus only two probabilities are calculated. Figure 1(c)

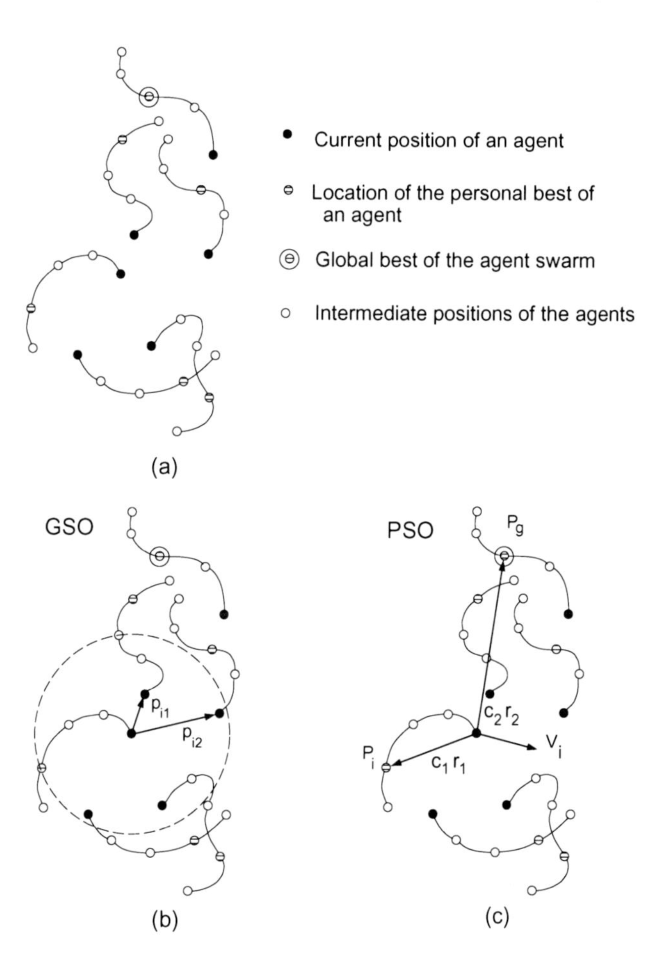

Fig. 1. (a) Traces of six agents, their current positions, and the positions at which they encountered their personal best and the global best. (b) GSO decision for an agent which has four neighbors in its dynamic range of which two have higher luciferin value and thus only two probabilities are calculated. (c) PSO decision which is a random combination between the current velocity of the agent, the vector to the personal best location, and the vector to the global best location.

shows the PSO decision, which is a random combination between the current velocity of the agent, the vector to the personal best location, and the vector to the global best location. Experiments based comparison with PSO is available in [6].

The algorithm is presented here for maximization problems. However, it can be easily modified and used to find multiple minima of multimodal functions. The algorithm starts by placing the glowworms randomly, according to a uniform distribution, in the objective function space. Each iteration consists of a

GLOWWORM SWARM OPTIMIZATION (GSO) ALGORITHM

Set number of dimensions = m
Set number of glowworms = n
Let s be the step size
Let $x_i(t)$ be the location of glowworm i at time t
deploy_agents_randomly;
for $i = 1$ to n do $\ell_i(0) = \ell_0$
for $i = 1$ to n do $r_d^i(0) = r_0$
Set maximum iteration number $= iter\_max$;
Set $t = 1$;
while ($t \leq iter\_max$) do:
{
  for each glowworm i do: % Luciferin-update phase
    $\ell_i(t) = (1-\rho)\ell_i(t-1) + \gamma J(x_i(t))$;

  for each glowworm i do: % Movement-phase
  {
    $N_i(t) = \{j : d_{ij}(t) < r_d^i(t); \ell_i(t) < \ell_j(t)\}$;
    for each glowworm $j \in N_i(t)$ do:
      $p_{ij}(t) = \frac{\ell_j(t)-\ell_i(t)}{\sum_{k \in N_i(t)} \ell_k(t)-\ell_i(t)}$;
    $j = select\_glowworm(\mathbf{p})$;
    $x_i(t+1) = x_i(t) + s\left(\frac{x_j(t)-x_i(t)}{\|x_j(t)-x_i(t)\|}\right)$
    $r_d^i(t+1) = \min\{r_s, \max\{0, r_d^i(t) + \beta(n_t - |N_i(t)|)\}\}$;
  }
  $t \leftarrow t+1$;
}

luciferin-update phase, movement phase based on a transition rule, and a local-decision range update phase. The equations describing the GSO model are given in the inset box.

We have addressed the parameter selection problem by using extensive numerical experiments in [6] to show that the quantities ρ, γ, s, β, and n_t are algorithmic constants (Table 1) and only n and r_s are the parameters that influence the algorithm's behavior, and need to be selected. Even for these parameters, the selection is straightforward.

Table 1. Values of algorithmic parameters that are kept fixed for all the experiments

ρ	γ	β	n_t	s	ℓ_0
0.4	0.6	0.08	5	0.03	5

4 Group Level Phases of GSO Algorithm

Implementation of GSO at the individual agent-level gives rise to two important phases at the group-level that are described below.

4.1 Splitting of the Agent-Swarm into Subgroups

The local decision-domain update rule enables each glowworm to select its neighbors in such a way that its movements get biased toward the nearest peak. The above individual agent behavior leads to a collective behavior of agents that constitutes the autonomous splitting of the whole group into subgroups whose number is equal to the number of peak locations and where each subgroup of agents gets allocated to a nearby peak (that is a peak to which the agent-distance, averaged over the subgroup, is minimum among those distances to all the peaks in the environment). We present here a simulation example in order to clearly characterize the splitting behavior of the glowworm swarm. We consider the $J(x, y)$ function (1) for this set of experiments (Figure 2):

$$J(x,y) = \sum_{i=1}^{Q} a_i \exp(-b_i((x - x_i)^2 + (y - y_i)^2)) \tag{1}$$

where, Q represents the number of peaks and (x_i, y_i) represents the location of each peak. The function $J(x, y)$ represents a linear sum of two dimensional exponential functions centered at the peak-locations. The constant b_i determines how the function profile changes slope in the vicinity of the peak i. Initially, we keep b_i equal for all the individual exponentials by choosing $b_i = 3$, $i = 1, ...Q$. A workspace of $(-5, 5) \times (-5, 5)$ and a set of ten peaks ($Q = 10$) are considered for the purpose. The values of a_i, x_i, and y_i are generated according to the following equations:

$$\begin{aligned} a_i &= 1 + 2\vartheta \\ x_i &= -5 + 10\vartheta \\ y_i &= -5 + 10\vartheta \end{aligned}$$

where, ϑ is uniformly distributed within the interval [0, 1].

Table 2. Values of a_i, x_i, and y_i used to generate the function profile $J(x, y)$

	1	2	3	4	5	6	7	8	9	10
a_i	2.616	2.427	1.223	2.366	1.338	2.285	2.325	1.002	1.709	2.985
x_i	2.933	-4.084	3.459	2.066	-4.814	4.381	-1.111	-1.711	-1.728	-3.233
y_i	-0.559	-2.146	-2.803	2.710	4.789	3.282	-1.496	-3.977	0.244	-3.166

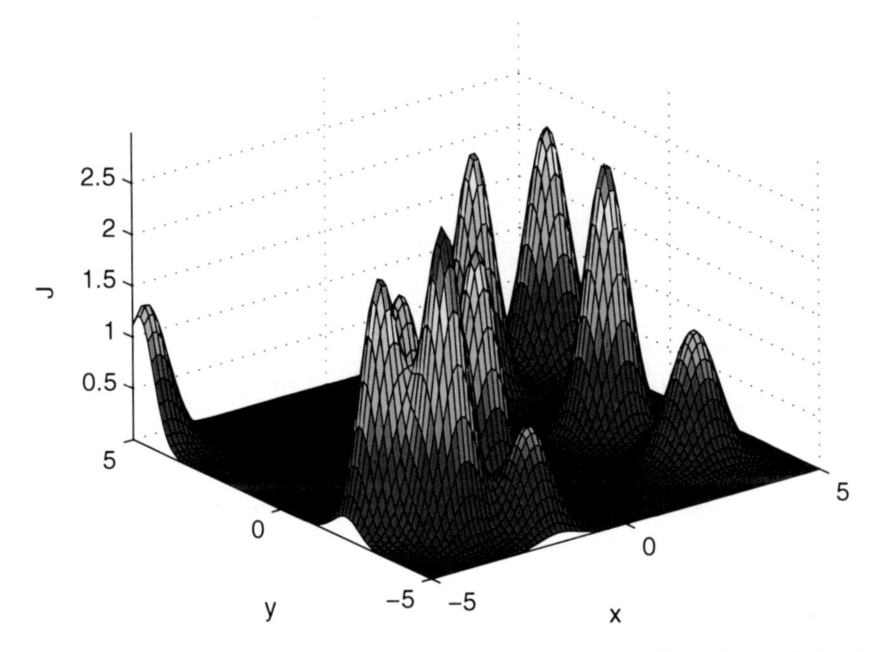

Fig. 2. Multimodal function profile used to demonstrate the splitting behavior of agents in GSO

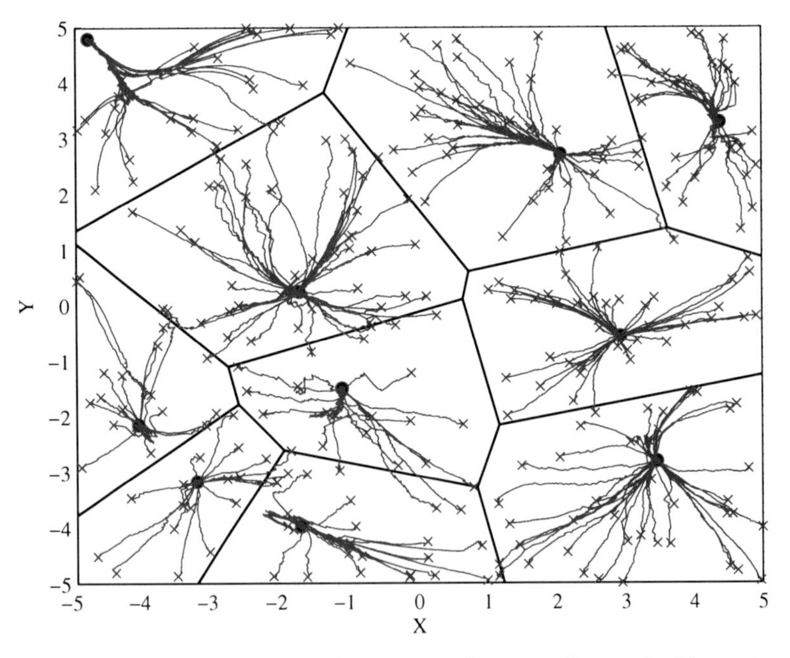

Fig. 3. Trace of agent-movements: the swarm splits according to the Voronoi-partition of the peak locations.

The function profile $J(x, y)$ is representative of a varied set of problem scenarios in the sense that it has peaks at different random points giving rise to equal/unequal peaks, closely spaced peaks, distant peaks, and peaks on the edges of the workspace. The values of $\{a_i,\ x_i,\ y_i,\ i\ =\ 1, ..., 10\}$ are shown in Table 2. Figure 2 shows the multimodal function profile $J(x, y)$. Figure 3 shows the trace of agent movements. Note that when the slopes of the peaks are equal, the swarm splits into subgroups according to the Voronoi-partition of the peak locations (Figure 3). However, a few agents located near the border region of a Voronoi-partition migrate into adjacent partitions and eventually get co-located at the respective peaks. The locations of agents at different time instants ($t\ =\ 0,\ 20,\ 40,\ 60$) are plotted in Figures 4(a) – 4(d), which show the formation of subgroups as a function of time.

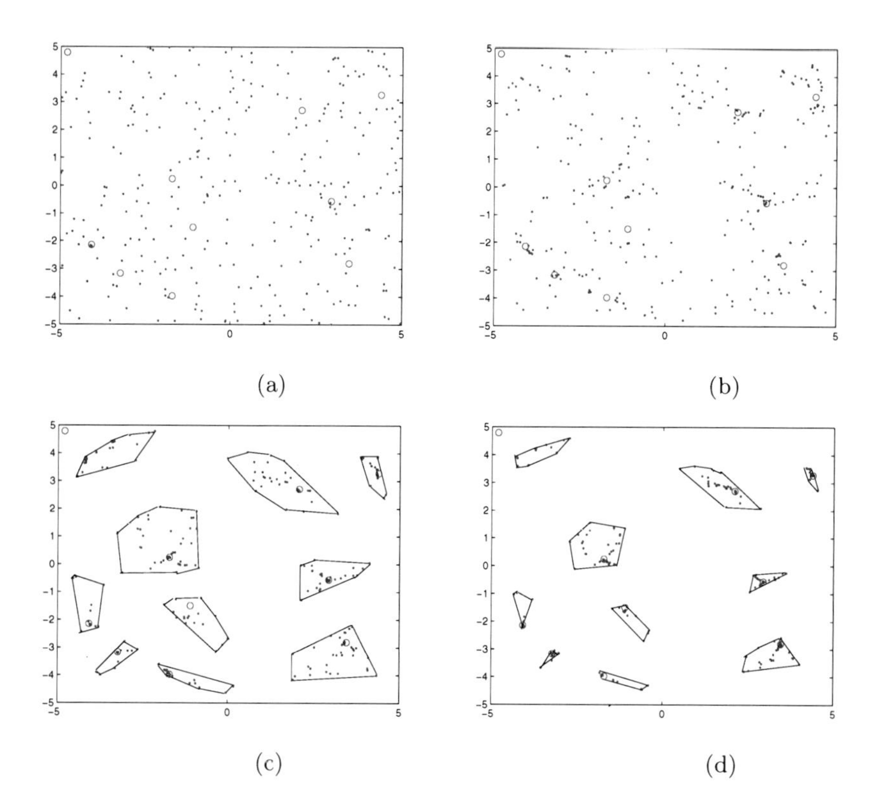

Fig. 4. Location of agents at different time instants (corresponding to Figure 3(a)) (a) $t = 0$ (b) $t = 20$ (c) $t = 40$ (d) $t = 60$

Next, we characterize the splitting of the agents in the case where different slope profiles (b_i) are used for the individual exponentials in the $J(x, y)$ function. For this purpose, we consider two peaks at $(-3, 0)$ and $(3, 0)$, with $a_1\ =\ 3, b_1\ =\ 0.8$ and $a_2\ =\ 2,\ b_2\ =\ 0.1$, respectively. Figure 5 shows the trace of agent movements when 100 agents are randomly deployed in the workspace. The line L, that passes through each equi-valued contour at a point where the gradient shifts direction from one peak to the other, divides the workspace into two attraction-regions. Note that all agents on the concave side of line L converge to the left peak and a majority (86%) of the agents on the convex side of the line L converge to the right peak. When there are more than two peaks, we can obtain the attraction-regions for each peak in a similar manner as described above and show that the swarm splits into subgroups according to the attraction-regions and eventually converge to the respective peak locations. It is easy to see that the partitioning of the region obtained based on the attraction-regions of the peaks coincides with the Voronoi-partition of the peaks when their slopes become equal.

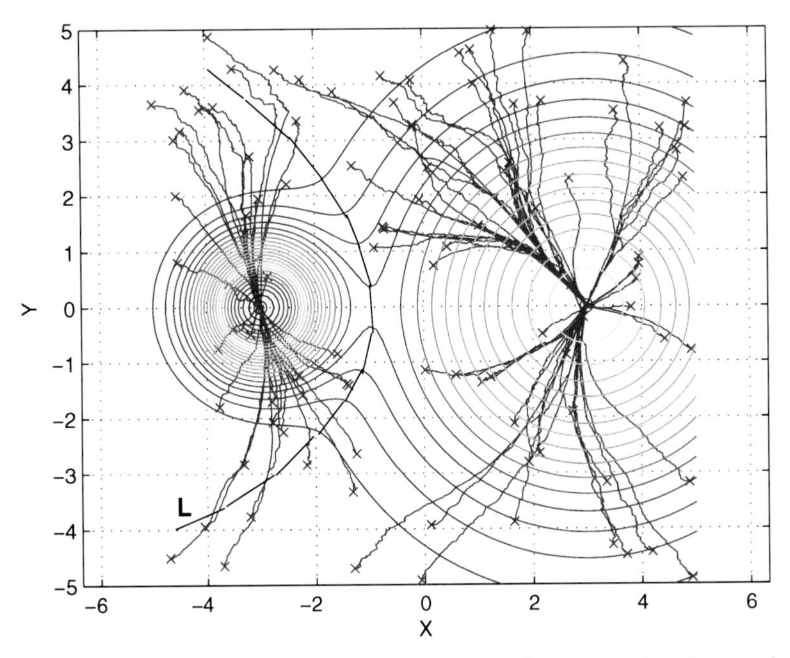

Fig. 5. Trace of agent-movements when two unequal peaks with different slopes are considered. The line L divides the region into two attraction-regions.

4.2 Local Convergence of Agents in Each Subgroup to the Peak Locations

The relative initial placement of the agents with respect to various peaks in the environment gives rise to different subgroup-peak configurations. Accordingly, the respective local convergence behaviors are different from each other. Among these, we consider two major configurations that occur frequently:

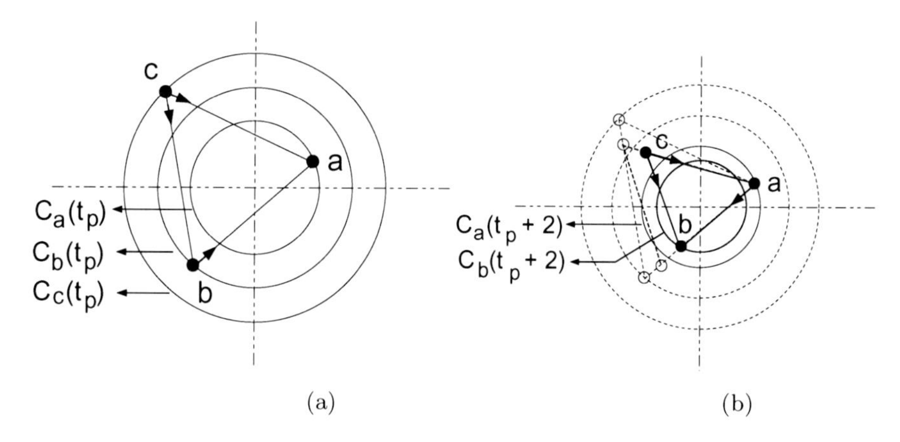

Fig. 6. (a) State of agents in configuration 1 at time t_p. (b) State of agents in configuration 1 at time $t_p + 2$.

Configuration 1: The peak is located within the convex-hull of the initial positions of agents in the subgroup. The convergence behavior of this configuration can be explained in the following way. For simplicity, we consider a radially symmetric function-profile with a single peak at the center and an initial placement of three agents a, b, and c as shown in Figure 6(a). At time instant t_p, agent a remains stationary, agent b makes a deterministic movement toward a, and agent c moves either toward a or b (since, $\ell_a > \ell_b > \ell_c$). Note from Figure 6(b) that the agent movements at any time instant are within the convex hull of all the current positions of agents. Agent a does not move until after two time steps (i.e., at $t_p + 2$), when b would have crossed the equi-valued contour $C_a(t_p + 2)$, leading to the condition $\ell_b > \ell_a$. Now, b remains stationary and a starts moving toward b. This cycle repeats, leading to the asymptotic convergence of agents to the peak.

Configuration 2: The peak is located outside the convex-hull of initial agent positions and all the agents are situated on one side of the peak. In order to describe the convergence behavior, we consider the same radially symmetric function-profile with a single peak at the center. However, we consider an initial placement of agents a, b, and c such that b and c are located on one side of the tangent line T at agent a's location as shown in Figure 7(a). At time instant t_p, agent a remains stationary, agent b makes a deterministic movement toward a, and agent c moves either toward a or b (since, $\ell_a > \ell_b > \ell_c$). Agent a does not move until after two time steps (i.e., at $t_p + 2$), when b leapfrogs over a, leading to the condition $\ell_b > \ell_a$ (Figure 7(b)). Unlike in the previous case, agents movements at any instant are not necessarily within the convex-hull of the current agent positions. For example, at $t_p + 2$, agent a moves outside the convex-hull when it leapfrogs over b. However, the leapfrogging mechanism ensures that the agents eventually converge to the peak.

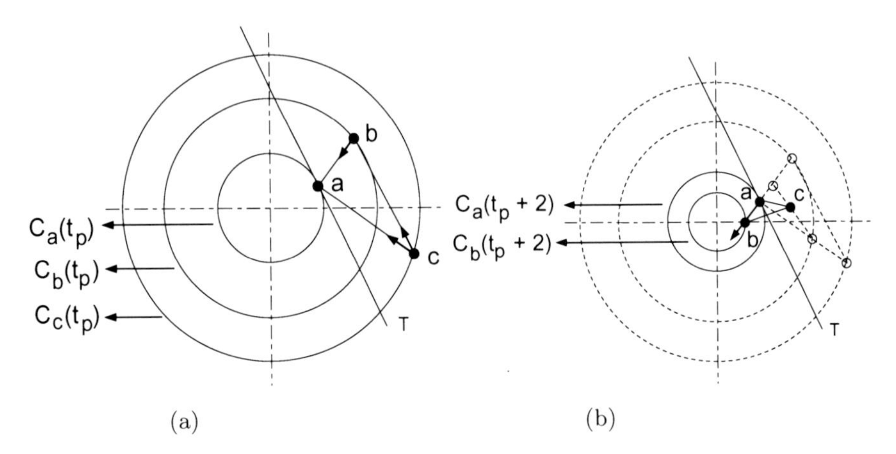

Fig. 7. (a) State of agents in configuration 2 at time t_p. (b) State of agents in configuration 2 at time $t_p + 2$.

5 GSO Applications

The GSO algorithm has been applied to problems ranging from multimodal function optimization to multiple signal source localization and pursuit of mobile signal sources. A brief summary of these experiments is given below.

5.1 Multimodal Function Optimization

GSO was shown to be effective in capturing multiple peaks of several benchmark multimodal test functions [6] that pose different cases of complexity: unequal peaks, equal peaks, peaks of concentric circles, peak-regions involving step-discontinuities, and plateaus of equal heights. GSO was also used to address the problem of searching higher dimensional spaces [21]. Results reported from tests conducted up to a maximum of eight dimensions show the efficacy of GSO in capturing multiple peaks in high dimensions. With an ability to search for local peaks of a function (which is the measure of fitness) in high dimensions, GSO can be applied to identification of multiple data clusters, satisfying some measure of fitness defined on the data, in high dimensional databases.

5.2 Multiple Signal Source Localization

Embodied simulations and real-robot-experiments were used in [22] to demonstrate the potential of GSO for signal source localization applications. The problem involves the deployment of a group of mobile robots that use their sensory perception of signal-signatures at distances that are potentially far from a source and interaction with their neighbors as cues to guide their movements toward, and eventually co-locate at, the signal-emitting source. Certain algorithmic aspects need modifications while implementing in a robotic network mainly because of the point-agent model of the basic GSO algorithm and the physical

Fig. 8. Multi-robot system built for source localization experiments

dimensions and dynamics of a real robot. We used a set of four wheeled robots, called Kinbots (See Figure 8), that were originally built for experiments related to robot formations [23]. By making necessary modifications to the Kinbot hardware, the robots are endowed with the capabilities required to implement the various behavioral primitives of GSO. The various hardware modules of each robot that are used to achieve the above tasks are shown in Figure 9. We presented the results from an experiment where two Kinbots use GSO to localize a light source. The paths traced by the robots as they execute the GSO is shown in Figure 10. Kinbots implementing GSO to localize a sound source is demonstrated in [24].

5.3 Pursuit of Mobile Signal Sources

In [25], we investigated the behavior of agents that implement GSO when mobile sources are considered. In particular, we use GSO to develop a coordination scheme that enables a swarm of mobile pursuers, with hard-limited sensing ranges, to split into subgroups, exhibit simultaneous taxis toward, and eventually pursue a group of mobile signal sources. Examples of such sources include hostile mobile targets in a battlefield and moving fire-fronts that are created in forest fires. We assume that the mobile source radiates a signal whose intensity peaks at the source location and decreases monotonically with distance from the source. For the case where the positions of the pursuers and the moving source are collinear, we presented a theoretical result that provides an upper bound on the relative speed of the mobile source below which the pursuers succeed in chasing the source. We used simulations to demonstrate the efficacy of the

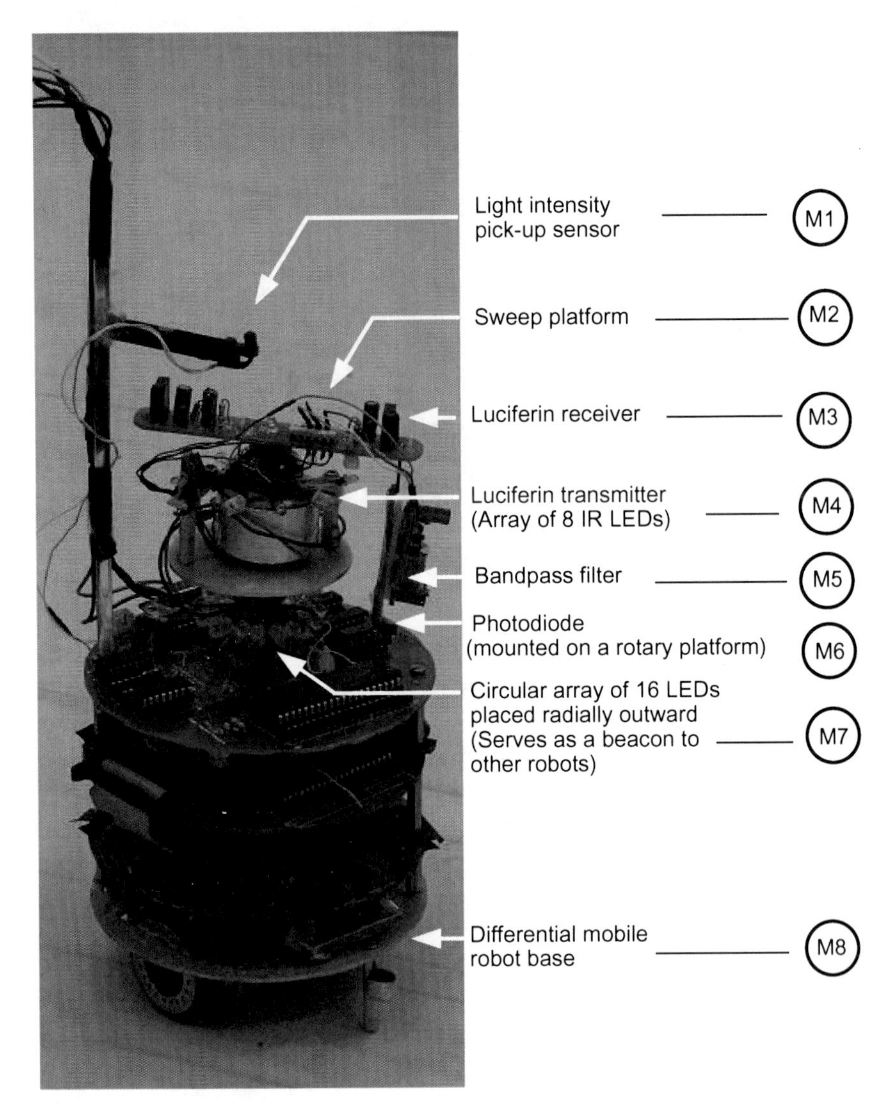

Fig. 9. Various hardware modules used by the Kinbots to implement the GSO behaviors

algorithm in addressing these pursuit problems. In particular, we presented simulation results for single and two source cases, respectively, where each source moves in a circular trajectory and at constant angular speed. In the case where the positions of the pursuers and the moving source are non-collinear, we used numerical experiments to determine an upper bound on the relative speed of the mobile source below which the pursuers succeed in chasing the source.

Fig. 10. Paths traced by the robots K_A and K_B as they taxis toward, and co-locate at, the light source

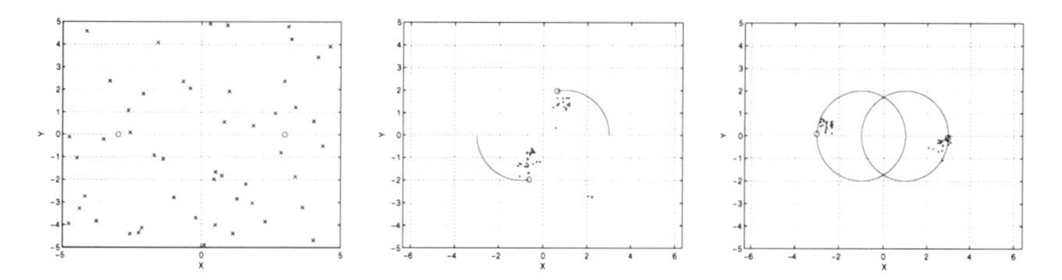

Fig. 11. Snapshots of positions of pursuers and sources at t = 0, t = 40, and t = 140.

For illustration purpose, we consider the following simulation example: two sources move at constant angular speed in circular trajectories of equal radii ($r = 2$) and centered at (-1, 0) and (1, 0), respectively. A set of 50 pursuers is randomly deployed in a workspace of (-5, 5)×(-5, 5) units (Figure 11(a)). The mobile sources are deployed at locations (3, 0) and (-3, 0), respectively. Snapshots of the pursuer and source positions at $t = 0$, $t = 40$, and $t = 140$ are shown in Figure 11(a) to 11(c). Note that at $t = 40$, the swarm splits into two subgroups and each one of them pursues one of the two sources.

6 Conclusions

GSO is a new swarm intelligence method for simultaneous capture of multiple optima of multimodal functions. We presented the basic GSO model and provided a characterization of the group-level phases of the algorithm that gives an insight into how GSO explicitly addresses the issue of autonomous formation and maintenance of subswarms. Finally, we briefly summarized results from application of GSO to the three problems of multimodal function optimization, signal source localization, and pursuit of mobile signal sources.

References

1. Deneubourg, J.L., Aron, S., Goss, S., Pasteels, J.M.: Journal of Insect Behavior 3(2), 159–168 (1990)
2. Franks, N.R.: Behavioral Ecology and Sociobiology 18, 425–429 (1986)
3. Seeley, T.D., Buhrman, S.C.: Behavioral Ecology and Sociobiology 45, 19–31 (1999)
4. Fuiman, L.A., Magurran, A.E.: Reviews in Fish Biology and Fisheries 4, 145–183 (1994)
5. Lissaman, P.B.S., Shollenberger, C.: Science 168, 1003–1005 (1970)
6. Krishnanand, K.N., Ghose, D.: Swarm Intelligence 3(2), 87–124 (2009)
7. Dorigo, M., Stützle, T.: Ant Colony Optimization. MIT Press, Cambridge (2004)
8. Clerc, M.: Particle Swarm Optimization. Hermes Science Publications (2006)
9. Parsopoulos, K.E., Plagianakos, V.P., Magoulas, G.D., Vrahatis, M.N.: Proceedings of the Particle Swarm Oprimimtion Workshop, pp. 22–29 (2001)
10. Brits, R., Engelbrecht, A.P., van den Bergh, F.: Proceedings of the Fourth Asia-Pacific Conference on Simulated Evolution and Learning (SEAL 2002), pp. 692–696 (2002)
11. Li, X.: Proceedings of the Genetic and Evolutionary Computation Conference, pp. 105–116 (2004)
12. Muller, S.D., Marchetto, J., Airaghi, S., Koumoutsakos, P.: IEEE Transactions on Evolutionary Computation 6(6), 16–29 (2002)
13. Liu, Y., Passino, K.M.: IEEE Transactions on Automtic Control 49(1), 30–43 (2004)
14. Hayes, A.T., Martinoli, A., Goodman, R.M.: Robotica 21, 427–441 (2003)
15. Goldberg, D., Richardson, J.: Genetic Algorithms and Their Applications: Proceedings of the Second International Conference on Genetic Algorithms, pp. 44–49 (1987)
16. Petrowski, A.: Proceedings of Third IEEE International Conference on Evolutionary Computation, pp. 798–803 (1996)
17. Mahfoud, S.: Parallel Problem Solving. Nature 2, 27–37 (1992)
18. Harick, G.: Proceedings of the Sixth International Conference on Genetic Algorithms, pp. 24–31 (1997)
19. LØvbjerg, M., Rasmussen, T.K., Krink, T.: Proceedings of the Genetic and Evolutionary Computation Conference (2001)
20. Blair, K.G.: Luminous insects. Nature 96, 411–415 (1915)
21. Krishnanand, K.N., Ghose, D.: Glowworm swarm optimization for searching higher dimensional spaces. In: Lim, C.P., Jain, L.C., Dehuri, S. (eds.) Innovations in Swarm Intelligence. Springer, Heidelberg (2009)
22. Krishnanand, K.N.: Glowworm swarm optimization: A multimodal function optimization paradigm with applications to multiple signal source localization tasks. PhD Thesis, Indian Institute of Science, Bangalore (2007)
23. Krishnanand, K.N., Ghose, D.: Robotics and Autonomous Systems 53, 194–213 (2005)

24. Krishnanand, K.N., Amruth, P., Guruprasad, M.H., Bidargaddi, S.V., Ghose, D.: IEEE International Conference on Robotics and Automation (ICRA 2006), pp. 958–963 (2006)
25. Krishnanand, K.N., Ghose, D.: Chasing multiple mobile signal sources: A glowworm swarm optimization approach. In: Third Indian International Conference on Artificial Intelligence, Pune (2007)
26. Yang, X.S.: Firefly algorithms for multimodal optimization. In: Watanabe, O., Zeugmann, T. (eds.) SAGA 2009. LNCS, vol. 5792, pp. 169–178. Springer, Heidelberg (2009)
27. McGill, K., Taylor, S.: Robot locolization of multiple sources, ACM Computing Surveys (to appear, 2010)

Direct and Inverse Modeling of Plants Using Cat Swarm Optimization

Ganapati Panda[1], Pyari Mohan Pradhan[1], and Babita Majhi[2]

[1] Indian Institute of Technology Bhubaneswar, India
ganapati.panda@gmail.com, pyarimohan.pradhan@gmail.com
[2] Institute of Technical Education and Research, Bhubaneswar, India
babita.majhi@gmail.com

Abstract. Derivative based learning rule poses stability problem when used in adaptive plant modeling. In addition the performance of these techniques deteriorates when used for non-linear plant modeling. In this chapter, the plant modeling task is formulated as an optimization problem. A recently introduced evolutionary algorithm, cat swarm optimization (CSO), is used to develop a new population based learning rule for the model. Adaptive modeling of a benchmarked plant is carried out through simulation study. The performance of the CSO in presence of nonlinearity in the plant is also studied. The results demonstrate superior performance of the CSO compared to that achieved by genetic algorithm (GA) and particle swarm optimization (PSO) based approaches for adaptive modeling.

Keywords: Cat swarm optimization, direct modeling, inverse modeling.

1 Introduction

Direct modeling and inverse modeling are very important applications of adaptive filters. The direct modeling finds applications in control system engineering including robotics [1], intelligent sensor design [2], process control [3], power system engineering [4], image and speech processing [4], geophysics [5], acoustic noise and vibration control [6] and biomedical engineering [7]. Similarly inverse modeling technique is used in digital data reconstruction [8], channel equalization in digital communication [9], digital magnetic data recording [10], intelligent sensor [2], deconvolution of seismic data [11]. The direct modeling mainly refers to adaptive identification of unknown plants in presence of additive White Gaussian noise(AWGN). Inverse modeling of a telecommunication and magnetic medium channels is also important for reducing the effect of inter symbol interference (ISI) and achieving faithful reconstruction of original data. Adaptive inverse model of sensors is required to extend their linearities for direct digital readout and enhancement of dynamic range.

In recent past, population based optimization techniques have been reported which fall under the category of evolutionary computing [12] or computational intelligence [13]. These are also called bio-inspired technique which includes genetic algorithm (GA) [14], particle swarm optimization (PSO) [15], bacterial foraging optimization (BFO) [16] and artificial immune system (AIS) [17]. These techniques are suitably

B.K. Panigrahi, Y. Shi, and M.-H. Lim (Eds.): Handbook of Swarm Intelligence, ALO 8, pp. 469–485.
springerlink.com

employed to obtain efficient iterative learning algorithms for developing adaptive direct and inverse models of complex plants and channels.

Recently Majhi [18] has proposed an efficient scheme of pole-zero system identification using PSO technique. In 2008, an adaptive nonlinear system identification using Comprehensive Learning PSO is proposed by Katari [19]. The performance of evolutionary algorithms is improved by combining the stochastic gradient techniques with the evolutionary techniques [20].

Development of direct and inverse adaptive models essentially consists of two components The first component is an adaptive network which may be linear or nonlinear in nature and the second component is the learning algorithm which is used to train the parameters associated with the network. The linear networks chosen for modeling are adaptive linear combiner or all-zero or FIR structure and pole-zero or IIR structure [7]. Under nonlinear category low complexity single layer function link artificial neural network (FLANN) [21] and multilayer perceptron network (MLP) [22] are used. The second component is the training or learning algorithm used to train the parameters of the model. Depending upon the complexity and nature of the plants to be identified, proper combination of the model and the learning rule is employed so that the combination yields the best possible performance in direct and inverse modeling tasks.

In this chapter, a new evolutionary algorithm, cat swarm optimization (CSO) [23, 24, 25], is introduced. This algorithm is applied to direct and inverse modeling problems using two sets of standard plant coefficients. The performance of CSO based models is compared with those obtained by GA and PSO based counterparts and the results are analyzed qualitatively as well as quantitatively.

2 System Modeling

A system is described using a difference equation as follows

$$y_0(n) = H(z)x(n) \tag{1}$$

where $x(n)$ is the input signal and $y_0(n)$ is the output of the plant. $H(z)$ is the transfer function of the unknown plant.

$$H(z) = \sum_{i=0}^{L} a_i z^{-i} \tag{2}$$

The overall output of the plant is given as

$$y(n) = y_0(n) + v(n) \tag{3}$$

where $v(n)$ is an additive white Gaussian noise. Combining (1) and (3), we get

$$y(n) = H(z)x(n) + v(n) \tag{4}$$

2.1 Direct Modeling

The block diagram of direct modeling of a plant or system is shown in Fig. 1. The input signal acts as input to both the adaptive filter as well as the unknown system. In order to

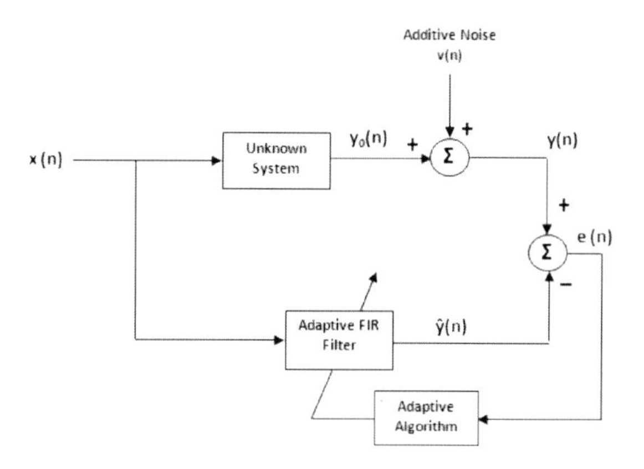

Fig. 1. Block diagram of direct modeling

reduce the error, the adaptive filter tries to emulate the systems transfer characteristic. After adaptation the system is modeled in the sense that its transfer function is same as that of the adaptive filter.

The adaptive filter is characterized by the difference equation:

$$\hat{y}(n) = \hat{H}(z)x(n) \tag{5}$$

where $\hat{H}(z)$ is the transfer function of the model and is given as

$$\hat{H}(z) = \sum_{i=0}^{L} \hat{a}_i z^{-i} \tag{6}$$

The adaptive algorithm estimates the filter coefficients in such a way that its input/output relationship matches closely to that of the unknown system. This identification task is formulated as an optimization problem where mean square error(MSE) is used as the cost function and is given by

$$J = E[e^2(n)] \approx \frac{1}{N} \sum_{n=1}^{N} e^2(n) \tag{7}$$

where $e(n) = y(n) - \hat{y}(n)$ is the error between the plant and model outputs and N is the number of input samples and $E(.)$ represents the statistical expectation operator.

2.2 Inverse Modeling

The block diagram to build an inverse model of a system is shown in Fig. 2. The inverse modeling configuration employs an adaptive filter for recovering the delayed version of a signal, which is altered by the unknown system and corrupted with additive noise. The delay is used to compensate for the propagation delay between the input of the unknown system and output of the inverse filter.

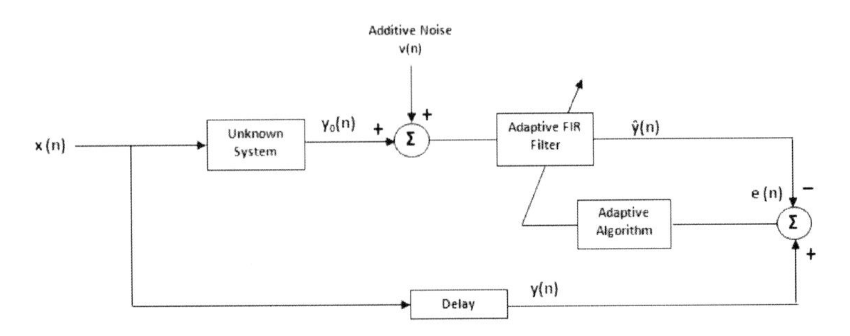

Fig. 2. Block diagram of inverse modeling

The adaptive filter is governed by the difference equation (5). In this case, the adaptive algorithm searches for the adaptive filter coefficients such that its input/output relationship matches closely to unity. The plant is to be modeled using the transfer function $\hat{H}(z)$ of the adaptive filter such that $H(z)\hat{H}(z) = 1$. This shows that $\hat{H}$ needs to be an IIR filter whose order is equal to the order of the plant. An IIR filter can also be approximated using a FIR filter of infinite order. Since a FIR filter with order infinity is not practically realizable, a FIR filter of large order is used. Hence $\hat{H}(z)$ can be expressed as follows

$$\hat{H}(z) = \sum_{i=0}^{L} \hat{a}_i z^{-i} \tag{8}$$

where L is the order of the FIR filter, which is more than the order of the unknown plant. Mean square error(MSE) is used as the cost function for this optimization problem and is given by (7). In this case the error signal is $e(n) = d(n) - \hat{d}(n)$, where $d(n)$ is the desired output obtained by using α number of delays for the input signal $x(n)$. Hence $d(n)$ is given as $d(n) = z^{-\alpha}x(n)$. Inverse modeling can be used to alleviate the inter symbol interference effects of a communication channel. It can also be used for inverse model of complex plants.

3 Cat Swarm Optimization

In 2006, Chu et al. [23] have proposed a new optimization algorithm which imitates the natural behavior of cats. Cats have a strong curiosity towards moving objects and possess good hunting skill. Even though cats spend most of their time in resting, they always remain alert and move very slowly. When the presence of a prey is sensed, they chase it very quickly spending large amount of energy. These two characteristics of resting with slow movement and chasing with high speed are represented by seeking and tracing respectively. In CSO these two modes have been mathematically modeled for solving complex optimization problems.

3.1 Seeking Mode

The seeking mode corresponds to a global search technique in the search space of the optimization problem. Some of the terms related to this mode are:

- Seeking Memory Pool(SMP): It is the number of copies of a cat produced in seeking mode.
- Seeking Range of selected Dimension(SRD): It is the maximum possible change in the position of a cat.
- Counts of Dimension to Change(CDC): It is the number of dimensions to be mutated.
- Self Position Consideration(SPC): It decides if the current position of the cat can have the chance to be retained in this generation. The SPC is a predefined flag with the value "true" or "false".

The steps involved in this mode are:

1. If SPC="false", create T copies of ith cat i.e. $U_j = X_i$ where $j = 1,2,.....,T$. Otherwise create $T-1$ copies of ith cat because the original i^{th} cat will take one place in the candidates.
2. Based on CDC, update the position of each copy by randomly adding or subtracting SRD percents the present position value.

$$U_{jd} = U_{jd} * (1 + k * SRD) \tag{9}$$

   where d represents the dimension selected for updation and k is a random binary number i.e. [-1,1].
3. Evaluate the fitness of all copies and store the best candidate as U_{best}
4. Replace the ith cat with the best candidate U_{best} i.e. $X_i = U_{best}$.

3.2 Tracing Mode

In this mode, the rapid chase of the cat is mathematically modelled as a large change in its position. Define position and velocity of ith cat in the D-dimensional space as $X_i = (X_{i1}, X_{i2}, ..., X_{iD})$ and $V_i = (V_{i1}, V_{i2}, ..., V_{iD})$ where $d(1 \le d \le D)$ represents the dimension. The global best position of the cat swarm is represented as $P_g = (P_{g1}, P_{g2}, ..., P_{gD})$. The update equations [24] are:

$$V_{id} = w * V_{id} + c * r * \left(P_{gd} - X_{id}\right) \tag{10}$$

$$X_{id} = X_{id} + V_{id} \tag{11}$$

where w is the inertia weight, c is the acceleration constant and r is a random number uniformly distributed in the range [0, 1].

3.3 Algorithm

The CSO algorithm reaches its optimal solution using two groups of cats i.e. one group containing cats in seeking mode and other group containing cats in tracing mode. The two groups combine to solve the optimization problem. A mixture ratio(MR) is used which defines the ratio of number of cats in tracing mode to that of number of cats in seeking mode. The CSO algorithm is described as follows:

1. Randomly initialize the position of cats in D-dimensional space for the population i.e. X_{id} representing position of the ith cat in the dth dimension.
2. Randomly initialize the velocity for cats i.e. V_{id}
3. According to MR, cats are randomly picked from the population and their flag is set to seeking mode, and for others the flag is set to tracing mode.
4. Evaluate the fitness of each cat and store the position of the cat with best fitness P_{gm} where $m = 1, 2, \ldots D$.
5. If the ith cat is in seeking mode, apply the cat to the seeking mode process, otherwise apply it to the tracing mode process. The process steps are presented above.
6. Cats are again randomly picked from the population according to MR and their flag is set to seeking mode, and for others the flag is set to tracing mode.
7. Check the termination condition, if satisfied, terminate the program. Otherwise repeat steps 4 to 7.

The flowchart of the algorithm is shown in Fig. 3.

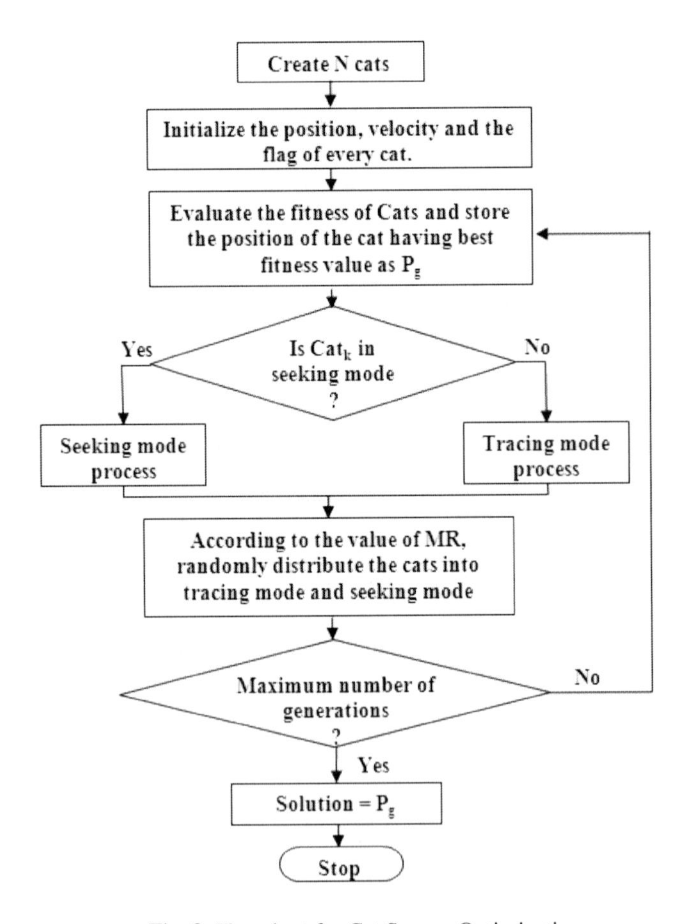

Fig. 3. Flowchart for Cat Swarm Optimization

4 Simulation Results

In order to demonstrate the potentiality of CSO algorithm, simulation study is carried out in MATLAB environment and the results of CSO are compared with those of PSO and GA. The input is a white signal with zero mean, unit variance and uniform distribution. The additive noise is a Gaussian white signal with variance 10^{-3}. The number of input samples used for training is 100. The number of samples used for testing is 50. To compare the performance of the new method, the results of GA and PSO methods are obtained. The initial population chosen for all the three algorithms is 50. The simulation parameters used for the three algorithms are as follows:

- **CSO:** SMP = 10, SRD = 20%, SPC="false", CDC = 80%, MR = 50%, C = 2, inertia weight is linearly decreased from 0.9 to 0.4 and r in the range [0 1]
- **PSO:** inertia weight is linearly decreased from 0.9 to 0.4, both the acceleration constants are taken as 2 and the random numbers are chosen in the range [0 1]
- **GA:** single point crossover with probability 0.8, probability of mutation is 0.1 and number of bits per dimension are 10

Two performance measures i.e. Residual Mean Square Error (RMSE) and Mean Square Deviation (MSD) are used to compare the performance of CSO, PSO and GA based approaches. The RMSE is defined as the steady state MSE value and MSD is defined as

$$MSD = \frac{1}{Q}\sum_{i=0}^{Q-1}[\Phi(i) - \hat{\Phi}(i)]^2 \tag{12}$$

where Φ is the desired parameter vector, $\hat{\Phi}$ is the estimated parameter vector and Q represents the total number of parameters to be estimated.

Different tests are carried out for modeling of a benchmark system. In each case the results obtained from 30 independent experiments are reported. Four different experiments are carried out:

1. Direct modeling of the plant
2. Inverse modeling of the plant
3. Direct modeling of the plant with nonlinearity
4. Inverse modeling of the plant with nonlinearity

The transfer function of the standard plant is,

$$H(z) = 0.26 + 0.93z^{-1} + 0.26z^{-2} \tag{13}$$

The nonlinearity introduced in the plant is hyperbolic tangent function.

4.1 Case-1

The transfer function of the model is,

$$\hat{H}(z) = a_0 + a_1z^{-1} + a_2z^{-2} \tag{14}$$

The convergence characteristic shown in Fig. 4 exhibits superior performance of CSO in comparison to GA and PSO. CSO reaches the global optimal solution in 30 generations whereas PSO takes 75 generations to reach the same value. In this case also, GA falls into a local solution and hence the chromosomes become stagnant. Figure 5, 6 and 7 shows that CSO locates the global best solution much faster than PSO. We can see in Fig. 5 that CSO reaches the global value of a_0=0.26 in 3 generations whereas PSO takes 9 generations to reach the same value. Similarly for a_1 and a_2, CSO takes approximately 3 and 4 generations respectively to reach the global value. In both cases PSO takes about 10 generations to reach the final solution. GA does not perform well with respect to any parameter. As shown in Fig. 8, 9 and 10, the estimated output obtained using CSO and PSO perfectly matches with the desired signal but there is mismatch between the desired signal and the estimated signal obtained using GA.

Table 1 shows that GA is not able to reach the global solution. CSO performs better than PSO as indicated by the smaller value of standard deviation for the three coefficients a_0, a_1 and a_2. The average results with respect to RMSE and MSD are shown in Table 2. It can be noted that CSO performs best with respect to both the metrics.

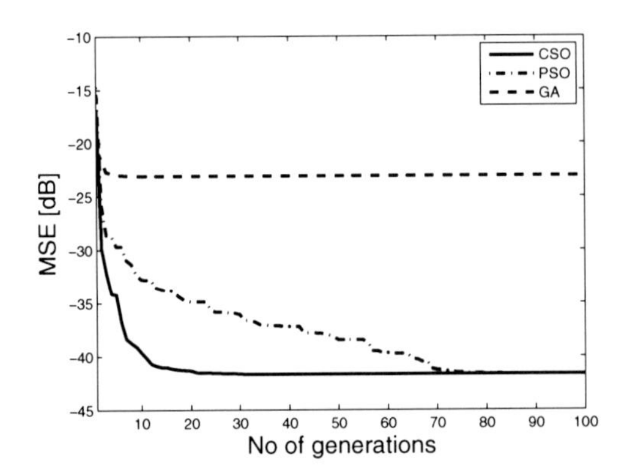

Fig. 4. Convergence characteristic for case-1

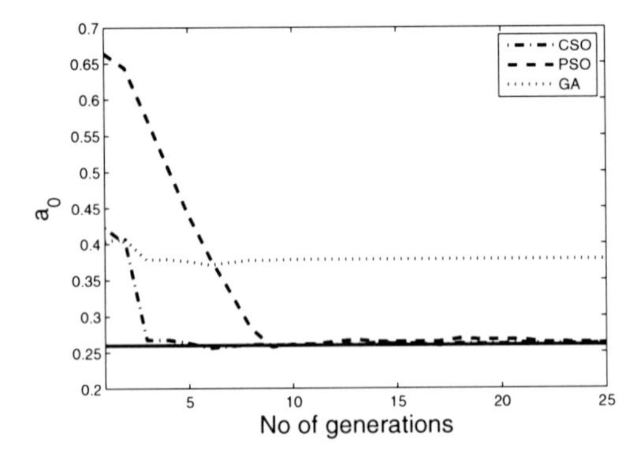

Fig. 5. Evolution of the parameter a_0 in case-1

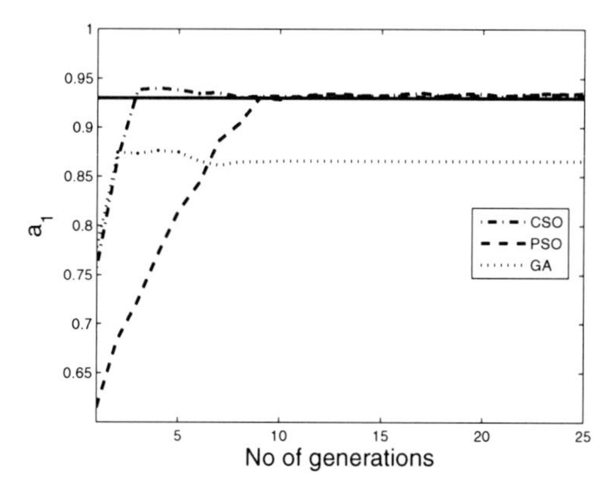

Fig. 6. Evolution of the parameter a_1 in case-1

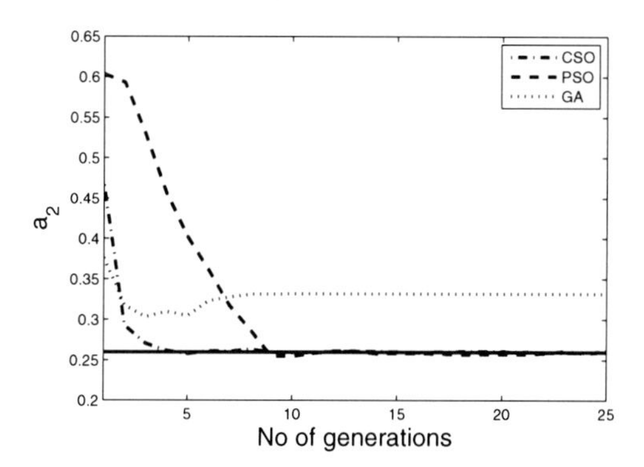

Fig. 7. Evolution of the parameter a_2 in case-1

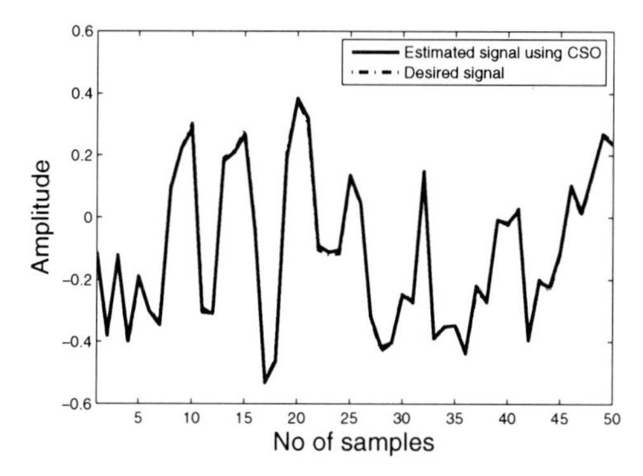

Fig. 8. Response matching for case-1 using CSO

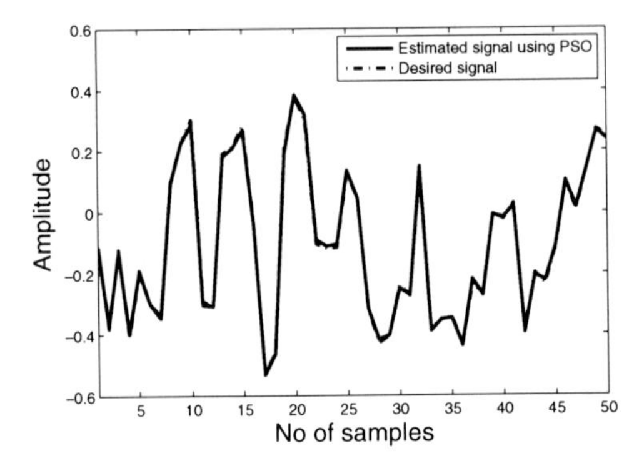

Fig. 9. Response matching for case-1 using PSO

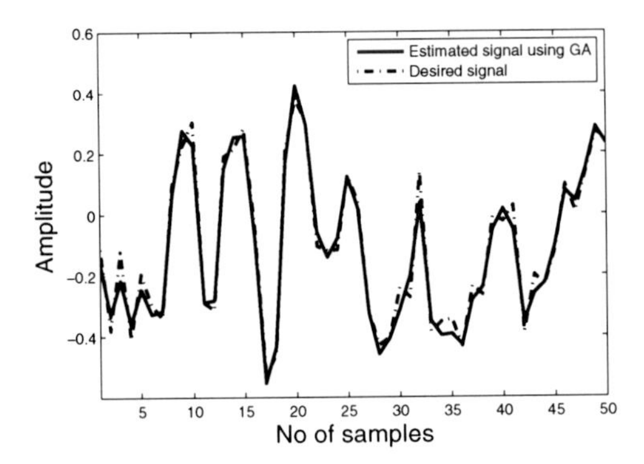

Fig. 10. Response matching for case-1 using GA

Table 1. Parameter estimation for case-1

Parameter		Estimated Value		
		CSO	PSO	GA
a_0	Average	0.2626	0.2626	0.3782
	Std. Dev.	0.1048×10^{-8}	0.14×10^{-4}	0.1257
a_1	Average	0.9341	0.9341	0.8658
	Std. Dev.	0.1871×10^{-8}	0.1977×10^{-4}	0.0926
a_2	Average	0.2626	0.2626	0.3321
	Std. Dev.	0.1536×10^{-8}	0.2596×10^{-4}	0.1011

Table 2. Performance measures for case-1

		CSO	PSO	GA
RMSE	Average	6.7892×10^{-5}	6.7894×10^{-5}	0.0049
	Std. Dev.	2.4934×10^{-19}	3.232×10^{-9}	0.0057
MSD	Average	2.0701×10^{-5}	2.085×10^{-5}	0.0221
	Std. Dev.	4.6971×10^{-12}	8.1854×10^{-7}	0.0256

4.2 Case-2

In this case, a 8th order FIR filter is used for simulation exercise. The transfer function of the model is,

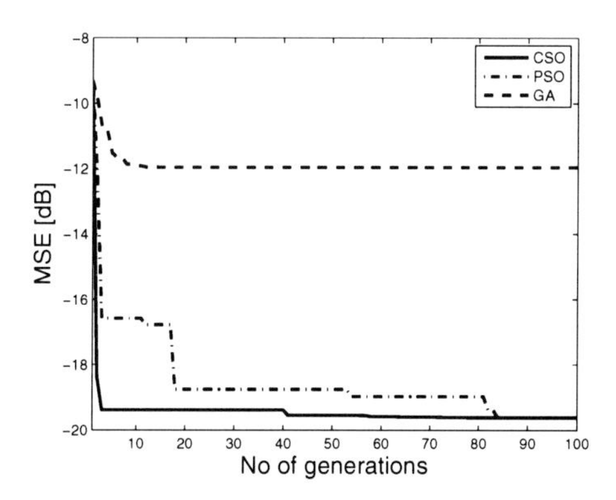

Fig. 11. Convergence characteristic for case-2

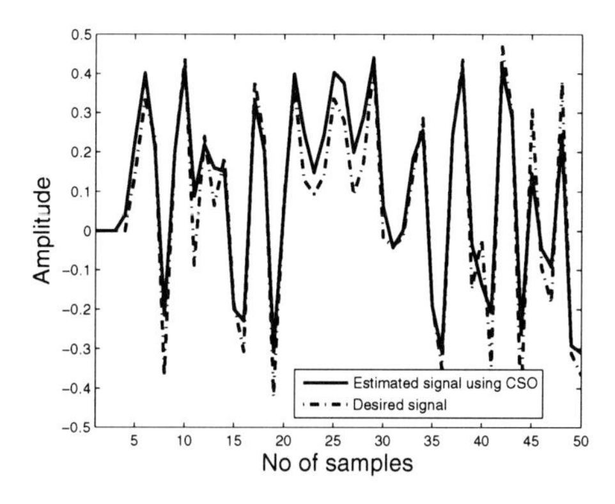

Fig. 12. Response matching for case-2 using CSO

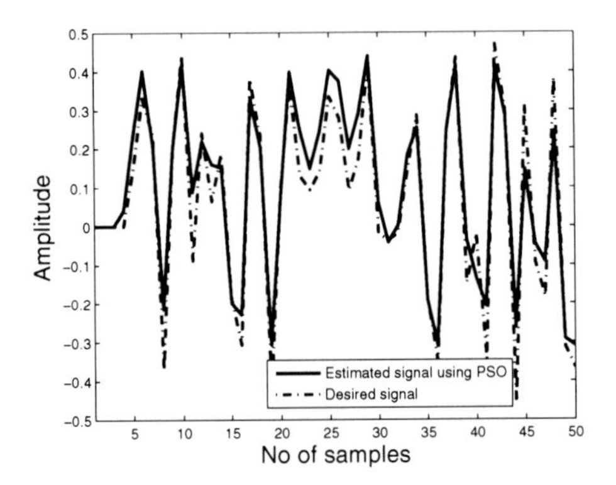

Fig. 13. Response matching for case-2 using PSO

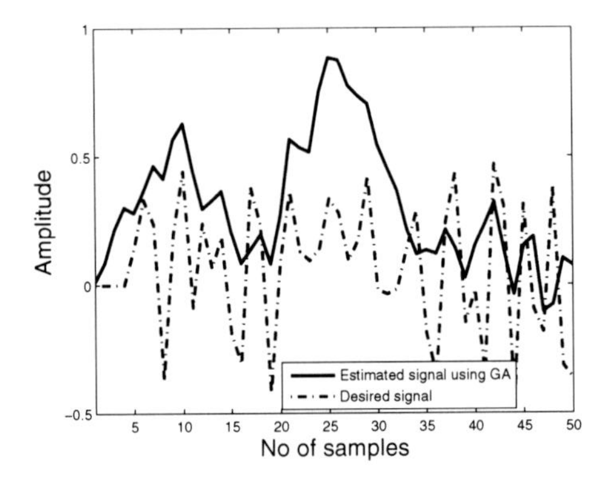

Fig. 14. Response matching for case-2 using GA

Table 3. Performance measures for case-2

		CSO	PSO	GA
RMSE	Average	0.0109	0.0109	0.0634
	Std. Dev.	5.2308×10^{-12}	6.831×10^{-7}	0.0203

$$\hat{H}(z) = a_0 + a_1 z^{-1} + a_2 z^{-2} + a_3 z^{-3} + a_4 z^{-4} + a_5 z^{-5} + a_6 z^{-6} + a_7 z^{-7} + a_8 z^{-8} \quad (15)$$

The convergence characteristic shown in Fig. 11 shows that CSO outperforms PSO with respect to convergence speed. In this case GA is getting trapped into a local solution in the early stage of training process, hence the fitness function for GA becomes stagnant at a value of 0.0634. CSO and PSO perform equally with respect to average value of RMSE but CSO . The response matching for CSO, PSO and GA shown in Fig. 12, 13 and 14 provides a qualitative assessment of the performance of the three algorithms. In this case, CSO as well as PSO is not able to perform efficiently in obtaining an exact inverse model, but the extent of mismatch between the desired signal and the estimated signal using GA is very high in comparison to CSO and PSO.

4.3 Case-3

The transfer function of the model is,

$$\hat{H}(z) = a_0 + a_1 z^{-1} + a_2 z^{-2} \tag{16}$$

Figure 15 shows that the rate of convergence of CSO is much higher than that of the PSO and the GA. Table 4 shows that the CSO and the PSO perform equally with respect to RMSE but the standard deviation achieved by the CSO is much smaller than that of the PSO and the GA. This is also demonstrated in terms of response matching shown in Fig. 16, 17 and 18.

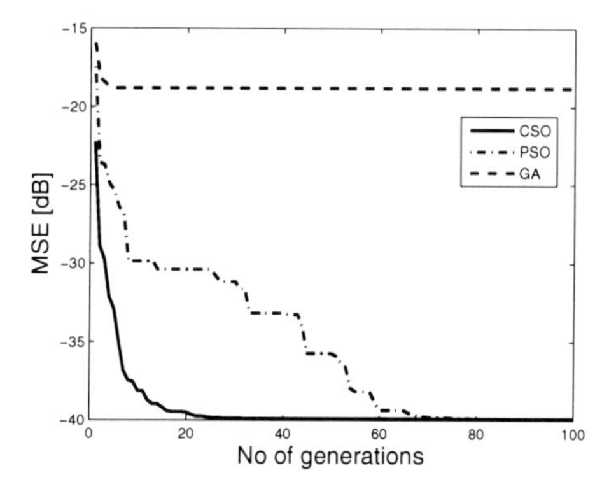

Fig. 15. Convergence characteristic for case-3

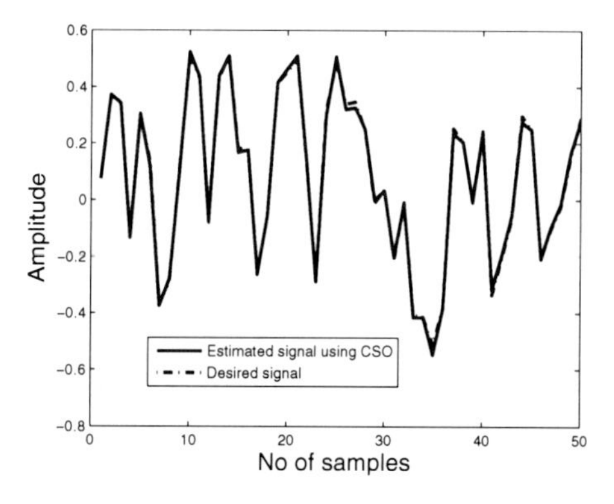

Fig. 16. Response matching for case-3 using CSO

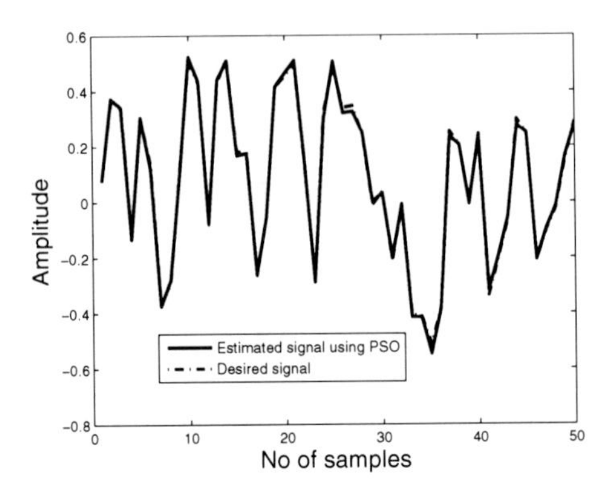

Fig. 17. Response matching for case-3 using PSO

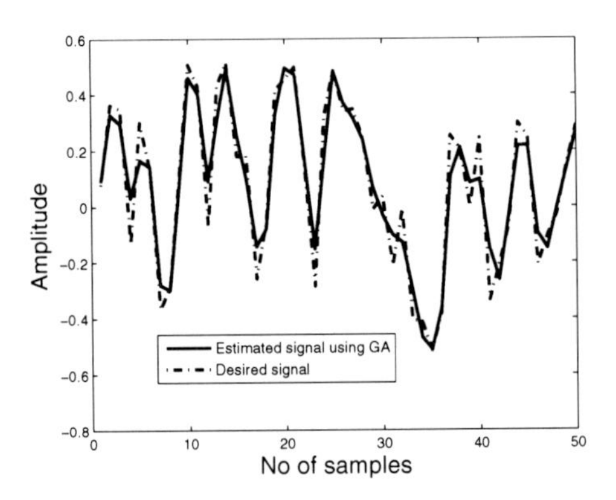

Fig. 18. Response matching for case-3 using GA

Table 4. Performance measures for case-3

		CSO	PSO	GA
RMSE	Average	1.0259×10^{-4}	1.0259×10^{-4}	0.0132
	Std. Dev.	1.4117×10^{-19}	1.2933×10^{-9}	0.0022

4.4 Case-4

In this case, a 8th order FIR filter is used for simulation exercise. The transfer function of the model is,

$$\hat{H}(z) = a_0 + a_1 z^{-1} + a_2 z^{-2} + a_3 z^{-3} + a_4 z^{-4} + a_5 z^{-5} + a_6 z^{-6} + a_7 z^{-7} + a_8 z^{-8} \quad (17)$$

As shown in Fig. 19, the convergence speed of the CSO is much better than that of the PSO and GA. Table 5 shows that the CSO and PSO provide equal RMSE value but the

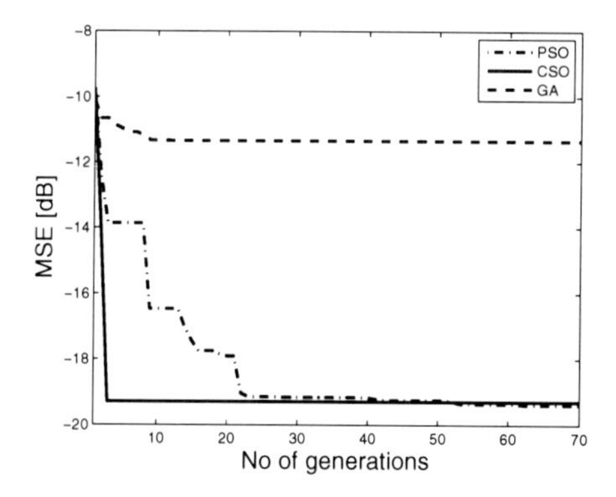

Fig. 19. Convergence characteristic for case-4

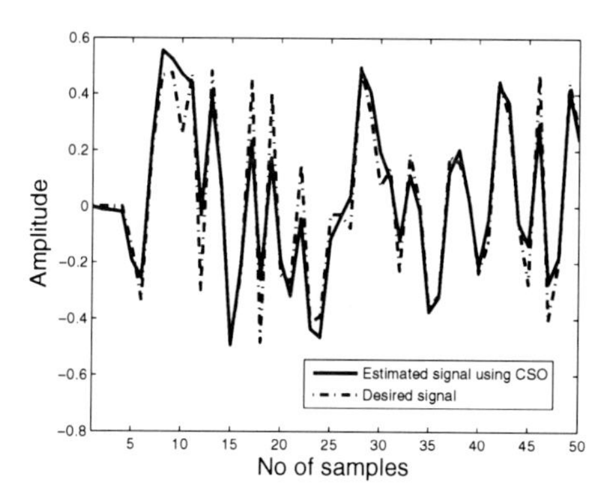

Fig. 20. Response matching for case-4 using CSO

CSO provides a better performance in terms of lower standard deviation. The response matching shown in Fig. 20 and 21 shows that the estimated output obtained using the CSO and the PSO match very well with the actual outputs. But the GA provides a very poor response matching as shown in Fig. 22.

Table 5. Performance measures for case-4

		CSO	PSO	GA
RMSE	Average	0.0115	0.0115	0.0647
	Std. Dev.	3.506×10^{-12}	1.6635×10^{-7}	0.0097

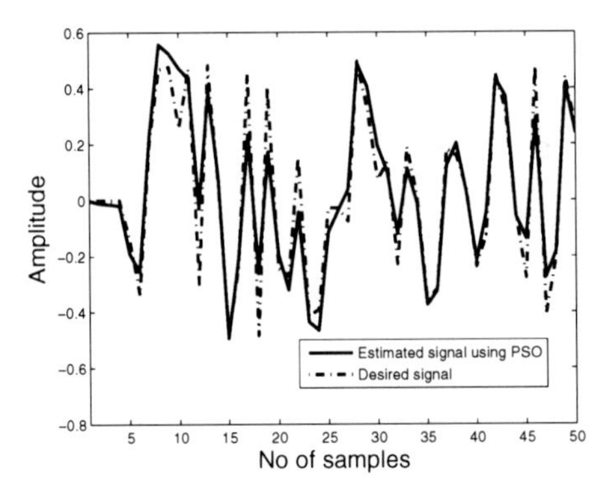

Fig. 21. Response matching for case-4 using PSO

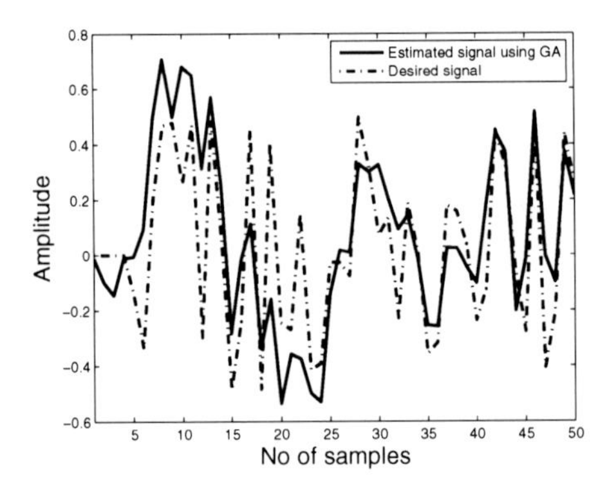

Fig. 22. Response matching for case-4 using GA

5 Conclusion

The CSO has been introduced as an optimization tool for carrying out direct and inverse modeling of linear and nonlinear plants. The simulation study is carried out to obtain the direct and inverse models of some standard plants. Comparison of results with those obtained by GA and PSO based approach, exhibits improved performance of the proposed method.

References

1. Narendra, K.S., Parthasarathy, K.: Identification and control of dynamical systems using neural networks. IEEE Trans. Neural Networks 1, 4–26 (1990)
2. Patra, J.C., Kot, A.C., Panda, G.: An intelligent pressure sensor using neural networks. IEEE Trans. Instrumentation and Measurement 49, 829–834 (2000)
3. Pachter, M., Reynolds, O.R.: Identification of a discrete time dynamical system. IEEE Trans. Aerospace Electronic System 36, 212–225 (2000)

4. Giannakis, G.B., Serpedin, E.: A bibliography on nonlinear system identification. Signal Processing 83(3), 533–580 (2001)
5. Robinson, E.A., Durrani, T.: Geophysical Signal Processing. Prentice-Hall, Englewood Cliffs (1986)
6. Das, D.P., Panda, G.: Active mitigation of nonlinear noise processes using a novel filtered-s lms algorithm. IEEE Trans. Speech and Audio Processing 12, 313–322 (2004)
7. Widrow, B., Strearns, S.D.: Adaptive Signal Processing. Prentice-Hall, Englewood Cliffs (1985)
8. Gibson, G.J., Siu, S., Cowan, C.F.N.: The application of nonlinear structures to the reconstruction of binary signals. IEEE Trans. signal processing 39(8), 1877–1884 (1991)
9. Lucky, R.W.: Techniques for adaptive equalization of digital communication systems. Bell Sys. Tech. J. 45, 255–286 (1966)
10. Sun, H., Mathew, G., Farhang-Boroujeny, B.: Detection techniques for high density magnetic recording. IEEE Trans. Magnetics 41(3), 1193–1199 (2005)
11. Griffiths, L.J., Smolka, F.R., Trenbly, L.D.: Adaptive deconvolution: a new technique for processing time varying seismic data. Geophysics (1977)
12. Eiben, A.E., Smith, J.E.: Introduction to Evolutionary Computing. Springer, Heidelberg (2003)
13. Engelbrecht, A.: Computational Intelligence: An introduction. Wiley & Sons, Chichester (2002)
14. Goldberg, D.E.: Genetic algorithms in search, optimization and machine learning. Addison-Wesley, Reading (1989)
15. Kennedy, J., Eberhart, R.C., Shi, Y.: Swarm intelligence. Morgan Kaufmann Publishers, San Francisco (2001)
16. Passino, K.M.: Biomimicry of bacterial foraging for distributed optimization and control. IEEE control system magazine 22, 52–67 (2002)
17. Dasgupta, D.: Artificial Immune Systems and their Applications. Springer, Heidelberg (1999)
18. Majhi, B., Panda, G., Choubey, A.: Efficient scheme of pole-zero system identification using particle swarm optimization technique. In: IEEE Congress on Evolutionary Computation, pp. 446–451 (2008)
19. Katari, V., Malireddi, S., Bendapudi, S.K.S., Panda, G.: Adaptive nonlinear system identification using Comprehensive Learning PSO. In: 3rd International Symposium on Communications, Control and Signal Processing, vol. 2008, pp. 434–439 (2008)
20. Theofilatos, K., Beligiannis, G., Likothanassis, S.: Combining evolutionary and stochastic gradient techniques for system identification. Journal of Computational and Applied Mathematics 227(1), 147–160 (2009)
21. Pao, Y.H.: Adaptive Pattern Recognition and Neural Networks. Addison Wesley, Reading (1989)
22. Haykin, S.: Neural Networks: A comprehensive foundation, 2nd edn. Pearson Education Asia (2002)
23. Chu, S., Tsai, P., Pan, J.: Cat swarm optimization. In: Yang, Q., Webb, G. (eds.) PRICAI 2006. LNCS (LNAI), vol. 4099, pp. 854–858. Springer, Heidelberg (2006)
24. Chu, S., Tsai, P.: Computational intelligence based on the behavior of cats. International Journal of Innovative Computing, Information and Control 3(1), 163–173 (2007)
25. Tsai, P., Pan, J., Chen, S., Liao, B., Hao, S.: Parallel cat swarm optimization. In: Proc. of the Seventh International Conference on Machine Learning and Cybernetics, Kunming, pp. 3328–3333 (2008)

Parallel Bacterial Foraging Optimization

S.S. Pattnaik[1], K.M. Bakwad[1], S. Devi[1], B.K. Panigrahi[2], and Sanjoy Das[3]

[1] National Institute of Technical Teachers' Training and Research, Chandigrah, India
[2] Indian Institute of Technology, New Delhi, India
[3] Kansas State University, Manhattan, KS 66506, USA

Abstract. This chapter focuses on concept of new variant of Bacterial Foraging Optimization (BFO) named as Parallel Bacterial Foraging Optimization (PBFO). The key issues on implementation of PBFO in parallel architecture are also addressed. PBFO and its fusions with Particle Swarm Optimization (PSO) and its variants to optimize multimodal functions with high dimensions are discussed. Fusion of PBFO with parameter free Particle Swarm Optimization (pf-PSO) is validated on unimodal and multimodal benchmark functions with high dimensions. The PBFO attains good quality solution as compared to BFO on mutimodal functions.

1 Introduction

Optimization is associated with almost every problem of engineering. The underlying principle in optimization is to enforce constraints that must be satisfied while exploring as many options as possible within tradeoff space. There exists numerous optimization techniques. Bio-inspired or nature inspired optimization techniques are class of random search techniques suitable for linear and nonlinear process. Hence, nature based computing or nature computing is an attractive area of research. Like nature inspired computing, their applications areas are also numerous. To list a few, the nature computing applications include optimization, data analysis, data mining, computer graphics and vision, prediction and diagnosis, design, intelligent control, and traffic and transportation systems. Most of the real life problem occurring in the field of science and engineering may be modeled as nonlinear optimization problems, which may be unimodal or multimodal. Multimodal problems are generally considered more difficult to solve because of the presence of several local and global optima.

Bacterial Foraging Optimization proposed in 2002 by K.M. Passino is based on the foraging behavior of Escherichia Coli (E. coli) bacteria present in the human intestine. BFO has been successfully applied to varied engineering problems, such as PID controller(Kim *et al.* 2005), harmonic estimation (Mishra 2005), transmission loss reduction (Tripathy *et al.* 2006), antenna parameter calculation (Pattnaik *et al.* 2008), image denoising (Pattnaik *et al.* 2009), economic load dispatch (Panigrahi and Pandi 2008) and machine learning (Kim and Cho 2005) etc. The quality performance of BFO due to its fixed step size decreases heavily with the growth of dimensions of the search and multimodal functions. The basic BFO i.e. the algorithm proposed by Passino in

B.K. Panigrahi, Y. Shi, and M.-H. Lim (Eds.): Handbook of Swarm Intelligence, ALO 8, pp. 487–502.
springerlink.com

the year 2002 is valid only for unimodal function optimization. Hence, there have been continuous attempts to improve upon the basic BFO algorithm. The adaptive run-length and hybridization are some of the approaches used to improve the performance of BFO. In a recent communication, the convergence and stability analysis of the BFO technique is obtained based on which adaptive schemes are proposed for the run-length unit (Dasgupta *et al.* 2009). However, these proposed adaptive schemes rely on well defined fitness landscapes that are not always available in more general search scenario. Eslamian *et al.* proposed new integer-code algorithm based on foraging behavior of *E-coli* bacteria which he applied for unit commitment (UC) problem. The mathematical model for the chemotactic movements of an artificial bacterium living in continuous time is derived and used to analyze the stability and convergence-behavior of the said dynamics (Das *et al.* 2009).

BFO hybridized with PSO is known as BSO, which is used for multimodal and high dimensional functions optimization (Biswas *et al.* 2007). Velocity Modulated Bacterial Foraging Optimization (VMBFO) is used to reduce the convergence time while maintaining high accuracy (Sastry *et al.* 2009). GA and BFO hybridization have been validated on various numerical benchmarks and on practical PID tuner design problem (Kim *et al.* 2007). Besides BFO, the other successful nature-inspired optimization techniques such as Genetic algorithm and Particle Swarm Optimization (Eberhart and Shi 1998) due to their stochastic nature are also sensitive to the increase of the problem complexity and dimensionality. Das Sanjoy *et al.* in their work addressed the key issues for applying nature-inspired algorithms such as evolutionary algorithms, Particle Swarm Optimization and artificial immune system to multiobjective problem. Different hybridization schemes such as sequential, parallel and implicit for multiobjective optimization have also been proposed.

2 Bacterial Foraging Optimization

The Bacterial Foraging Optimization (Passino 2002) is based on foraging strategy of *E. coli* bacteria. The foraging theory is based on the assumption that animals obtain maximum energy nutrients 'E' in a suppose to be a small time 'T'. The basic Bacterial Foraging Optimization consists of three principal mechanisms; namely chemotaxis, reproduction and elimination-dispersal. The brief descriptions of these steps involved in Bacterial Foraging are presented below.

2.1 Chemotaxis

During chemotaxis, the bacteria climb the nutrient concentration, avoid noxious substances, and search for a way out the of neutral media. This process is achieved through swimming and tumbling. Bacterial chemotaxis is a complex combination of swimming and tumbling that keeps bacteria in places of higher concentration of nutrients. Bacterial chemotaxis (Passino 2002) can also be considered as the optimization process of the exploitation. In chemotaxis, the flagellum is a left-handed helix configured so that, as the base of the flagellum rotates counterclockwise, it produces force against the bacterium and pushes the cell. Otherwise, each flagellum operates relatively independent of the others; rotates clockwise. Depending upon the rotation of the flagella in each bacterium, it decides whether it should move in a predefined direction (swimming) or in a different direction (tumbling). Therefore, an E. coli bacterium can move in two different ways; it can swim for a period or it can tumble. The bacteria usually take a tumble followed by a tumble, tumble followed by a run, or swim (Passino 2002). This movement of bacteria in each chemotaxis step can be expressed by Eq. (1)

$$\theta^i(j+1,k,l) = \theta^i(j,k,l) + C(i)\frac{\Delta(i)}{\sqrt{\Delta^T(i)\Delta(i)}} \tag{1}$$

Where $\theta^i(j,k,l)$ represents position vector of i-th bacterium, in j-th chemotaxis step, in k-th reproduction step and in l-th elimination and dispersal step. $C(i)$ shows the step size taken in the random direction specified by the tumble. $\Delta(i)$ depict the direction vector of the j-th chemotaxis step. When the bacterial movement is run or swim, $\Delta(i)$ is taken as same that was available in the last chemotaxis step; otherwise, $\Delta(i)$ is a random vector whose elements lie in [-1, 1].

2.2 Reproduction

Using Eq.(2), the health/fitness of the bacteria is calculated.

$$J^i_{health} = \sum_{j=1}^{Nc+1} J(i,j,k,l) \tag{2}$$

Where, Nc is the maximum step in a chemotaxis step. During reproduction, all bacteria are sorted in reverse order according to fitness values. The least healthy bacteria die and the rest healthiest bacteria each splits into two bacteria, which are placed in the same location in the search space. This makes the population of bacteria remains constant .The reproduction process of bacterial foraging aims to speed up the convergence suitable in static problems, but not in dynamic environment (Tang 2006).

2.3 Elimination and Dispersal

The elimination and dispersal events assist chemotaxis progress by placing the bacteria to the nearest required values. In BFO, the dispersion event happens after a certain number of reproduction processes. Each bacterium according to a fixed probability dispersed from their original position and move to best position within the search space. These events may prevent the local optima trapping but lead to disturb the optimization process. Elimination and dispersal helps to avoid premature convergence or, being trapped in local optima.

Step-by-Step Algorithm of Bacterial Foraging
Initialize Parameters *p, S, Nc, Ns, Nre, Ned, Ped and C (i), i= 1, 2... S*
Where,

p = Number of variables.
S = Number of bacteria in the population.
Nc = Number of chemotaxis steps.

Ns = Number of swimming steps
Nre = Number of reproduction steps.
Ned = Number of elimination –dispersal steps.
Ped = Elimination-dispersal with probability.
$C(i)$ = Step size taken in the random direction specified by the tumble.
$\theta^i(j,k,l)$ = Position vector of *i*-th bacterium, in *j*-th chemotaxis step, in *k*-th reproduction step and in *l*-th elimination and dispersal step.

[Step 1]: Elimination-dispersal loop: $l = l+1$
[Step 2]: Reproduction loop: $k = k+1$
[Step 3]: Chemotaxis loop: $j = j+1$

a) For $i = 1, 2, 3.., S$ take a chemotaxis step for bacterial i as follows
b) Compute fitness function $J(i, j, k, l)$.
c) Let $Jlast = J(i, j, k, l)$ to save this value since we may find a better cost via a run.
d) Tumble: Generate a random vector $\Delta(i) \in \Re^p$ with each element $\Delta_m(i)$, $m = 1, 2..,p$, a random number on [-1 1]
e) Move: Let

$$\theta^i(j+1,k,l) = \theta^i(j,k,l) + C(i)\frac{\Delta(i)}{\sqrt{\Delta^T(i)\Delta(i)}}$$

f) Compute $J(i,j+1,k,l)$.
g) Swim
   i) Let $m = 0$ (counter for swim length)
   ii) While $m < Ns$ (if have not climbed down too long)
   - Let $m = m+1$
   - If $J(i, j+1, k, l) < Jlast$ (if doing better),
     Let $Jlast = J(i, j+1, k, l)$ and let

$$\theta^i(j+1,k,l) = \theta^i(j,k,l) + C(i)\frac{\Delta(i)}{\sqrt{\Delta^T(i)\Delta(i)}}$$

     and use this $\theta^i(j+1,k,l)$ to compute the new $J(j+1, k, l)$ as we did in (f).
   - Else, let $m = Ns$. This is the end of the while statement.

h) Go to next bacteria if $i \neq S$

[Step 4]: If $j < Nc$, go to step 3. In this case, continue chemotaxis since the life of bacteria is not over.

[Step 5]: Reproductions:

a) For the given k and l, and for each $i = 1, 2, \ldots S$, let

$$J^i_{health} = \sum_{j=1}^{Nc+1} J(i,j,k,l)$$

be the health of bacterium i. Sort bacteria and chemotaxis parameter $C(i)$ in order of ascending cost J_{health} (higher cost means lower health).

b) The Sr bacteria with the highest J_{health} values die and other Sr bacteria with the best values split.

[Step 6]: If $k < Nre$, go to *step 2*. We have not reached the specified number of reproduction steps. So we have to start the next generation in the chemotaxis loop.

[Step 7]: Elimination-dispersal. For $i = 1, 2...S$, with probability p_{ed}, eliminate and dispersal each bacterium (This keeps the number of bacterium in the population constant). If you have eliminated a bacterium, simply disperse one to a random location on the optimization domain.

[Step 8]: If $l < Ned$, the go to step 1, otherwise end.

3 Parallel Bacterial Foraging Optimization (Pbfo)

In every optimization technique, generation, populations, step size of individual accomplish an exploration (global search) and exploitation (local search). The following steps perform exploration and exploitation in Bacterial Foraging Optimization.

1. Chemotaxis provides a basis for local search.
2. Reproduction process speeds up the convergence.
3. Elimination and dispersal (mutation) helps to avoid premature convergence and progresses search towards global optima by eliminating and dispersing the bacteria.

In BFO, step size of each generation is the main determining factor for accuracy as well as convergence of global best optima (Datta and Mishra 2008). Bacterial Foraging Optimization with fixed step size suffers from following major disadvantages.

1. If step size is very small then it requires many generations to reach optimum solution. It may not achieve global optima with less number of iterations.
2. If the step size is very high then the bacterium reach to optimum value quickly but accuracy of optimum value becomes low. To a large extent, only chemotaxis and reproduction are not enough for global optima searching. Bacteria may be stucked around the initial positions or local optima as dispersion event happens after a certain number of reproduction processes.
3. In Bacterial Foraging Optimization, it is possible either to change gradually or suddenly to eliminate the accident of being trapped into the local optima by introducing mutation operator. The mutation operator brings diversity in the population to avoid premature convergence or getting trapped in some local optima.

The authors suggest two modifications in Bacterial Foraging Optimization to improve convergence, accuracy and precision of optimal solution and to avail the possibility to execute the presented technique on parallel architecture. This is because no relationship exists among all the bacteria during same generation. In the first step, all bacteria positions are updated after all fitness evaluations in same generation. In second step, diversity for changing the bacteria positions gradually or suddenly and fine-tuning are achieved by mutation using PSO variants. In the suggested technique, a local search is accomplished by chemotaxis events and while reproduction and mutation by PSO variants help the global search in entire search space. The chemotaxis, mutation and reproduction events in Parallel Bacterial Foraging are explained below.

3.1 Chemotaxis

During chemotaxis, the bacterium swims or tumbles. All bacteria positions and directions are updated after all fitness evaluations instead of each fitness evaluations in chemotaxis step. During swimming or tumbling, the step size of PBFO is same as BFO. Bacterium decides whether it should move in a predefined direction (swimming) or altogether in a different direction (tumbling.). PBFO can easily be decomposed on parallel processors as shown in Fig 1.

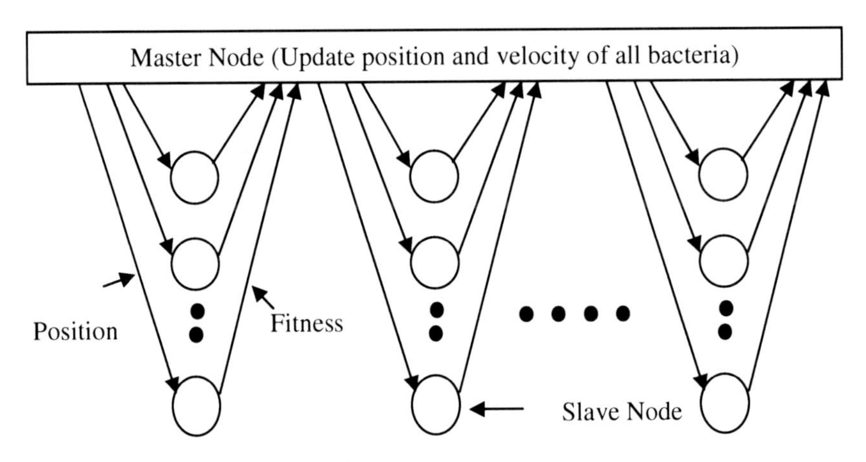

Fig. 1. Fitness Evaluations by Parallel Computers

During chemotaxis, fitness function evaluations can be performed on parallel computer architecture as shown in Fig. 1. Parallel computers consist of one master and other slave nodes. The number of slave node is equal to number of bacteria. The slave nodes must report fitness evaluations to the master node before starting next chemotaxis step. The synchronization is required between master and slave nodes for evaluating fitness value and updating the positions before starting the next iteration.

3.2 Mutation

After completing chemotaxis step, in order to accelerate the global performance of PBFO the bacteria positions are mutated by parameter free PSO (Ramana *et al.* 2009). The authors suggest that by using pfPSO, the PBFO does not require any additional parameter and velocity equation for fine-tuning as bacteria positions are updated directly by local and global best positions. The $\theta(i, j+1, k)$ can be updated as follows.

$$\theta(i, j+1, k) = \left(1 - \frac{\theta_{global}}{\theta(i, j, k)}\right) * r_1 * \theta_{global} + \left(\frac{\theta_{global}}{\theta(i, j, k)}\right) * r_2 * \theta_{pbest}(j, k) \quad (3)$$

Where,
$\theta(i, j, k)$= Position vector of i-th bacterium in j-th chemotaxis step and k-th reproduction steps.

$\theta_{pbest}(j,k)$ = Best position in *j*-th chemotaxis and *k*-th reproduction steps.

θ_{global} = Best position in the entire search space.

r_1 and r_2 = Random values [0 1].

At the initial stage, ratio of θ_{global} and $\theta(i, j, k)$ *is* very small resulting in large step size and during later stages, the step size is decreased because $\theta(i, j, k)$ becomes almost equal to θ_{global}. As the number of generation is increased, bacteria get attracted towards global optimum.

The authors also suggest the mutation by other PSO variants. The details of suggested PSO variants are given in section 4 and the names are listed below.

1. Standard Particle Swarm Optimization (Kennedy and Eberhart 1995).
2. Modified Particle Swarm Optimization (Shi and Eberhart 1998).
3. Linearly Decreasing Weight PSO (LDW-PSO) (Liu *et al.* 2004).
4. Supervisor Student Model in Particle Swarm Optimization (SSM-PSO) (Liu *et al.* 2004).
5. Particle Swarm Optimization with Time Varying Acceleration Coefficients (PSOTVAC) (Ratnaweera *et al.* 2004).
6. Global Local Best Particle Swarm Optimization (GLBest PSO) (Arumugam *et al.* 2008).
7. Parameter Free PSO algorithm (pf-PSO) (Ramana *et al.* 2009).
8. Particle swarm optimization with extrapolation technique (e-PSO) (Arumugam *et al.* 2009).
9. An Adaptive Particle Swarm Optimization (Panigrahi *et al.* 2008)

3.3 Reproduction

The reproduction step of PBFO is the same as that of BFO. During reproduction, the health/fitness of the bacteria is calculated and all bacteria are sorted in reverse order according to fitness values. The least healthy bacteria die and the other healthiest bacteria each split into two bacteria, which are placed on the same location in the search space. This makes the population of bacteria remains constant.

Step-by-Step algorithm of Parallel Bacterial Foraging Optimization

Initialize Parameters p, S, Nc, Ns, Nre and $C(i)$, $i= 1, 2 \ldots S$
Where, p = Number of variables.

S = Number of bacteria in the population.
Nc = Number of chemotaxis steps.
Ns= Number of swimming steps.
Nre = Number of reproduction steps.
$C(i)$ = Step size taken in the random direction specified by the tumble
$J(i, j, k)$ = Fitness value or cost of *i*-th bacteria in the *j*-th chemotaxis and *k*-th reproduction steps.
$\theta(i, j, k)$= Position vector of *i*-th bacterium in *j*-th chemotaxis step and *k*-th reproduction steps.

$\theta_{pbest}(j,k)$ = Best position in the *j*-th chemotaxis and *k*-th reproduction steps.
Jbest (j, k) = Fitness value or cost of best position in the *j*-th chemotaxis and *k*-th reproduction steps.
Jglobal= Fitness value or cost of the global best position in the entire search space.
θglobal = Position vector of the global best position in the entire search space.

[Step 1]: Update the following parameters
$J_{best}(j, k)$
$J_{global} = J_{best}(j, k)$
[Step 2]: Reproduction Loop: $k = k+1$
[Step 3]: Chemotaxis loop: $j = j+1$
a) Compute fitness function $J(i, j, k)$ for $i = 1, 2, 3... S$ ***(Parallel Fitness Evaluations)***

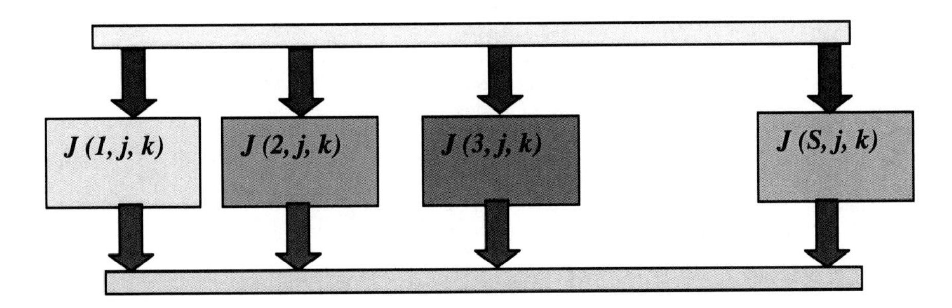

Update $J_{best}(j, k)$ and. $\theta_{pbest}(j,k)$

b) *Tumble:* Generate a random vector $\Delta(i) \in \Re^p$ with each element $\Delta_m(i)$ $m = 1, 2... p$, a random number on [-1 1]

c) Compute θ for $i = 1, 2...S$

$$\theta(i, j+1, k) = \theta(i, j, k) + C(i)\frac{\Delta(i)}{\sqrt{\Delta^T(i)\Delta(i)}}$$

d) *Swim*
i) Let $m = 0$ (counter for swim length)
ii) While $m < Ns$ (if have not climbed down too long
Let $m = m+1$
Compute fitness function $J(i, j+1, k)$ for $i = 1, 2, 3... S$ ***(Parallel Fitness Evaluations)***

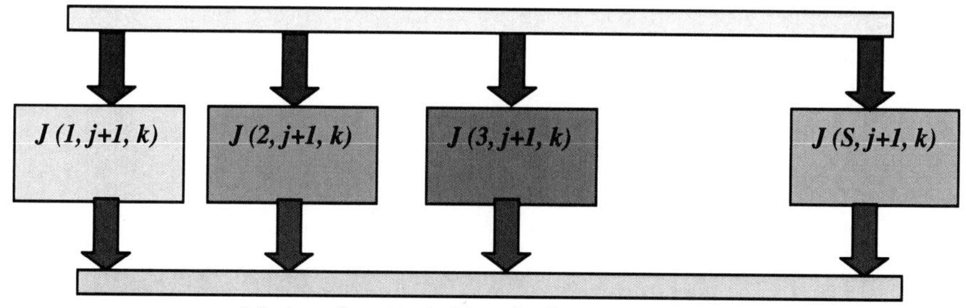

Update $J_{best}(j+1, k)$ and. $\theta_{pbest}(j+1,k)$

- If $J_{best}(j+1,k) < J_{best}(j, k)$ (if doing better),
  $J_{best}(j, k) = J_{best}(j+1, k)$
  $\theta_{pbest}(j,k) = \theta_{pbest}(j+1,k)$
  Compute θ for $i = 1, 2... S$

$$\theta(i, j+1, k) = \theta(i, j, k) + C(i)\frac{\Delta(i)}{\sqrt{\Delta^T(i)\Delta(i)}}$$

  Use this $\theta(i, j+1, k)$ to compute the new $J(i, j+1, k)$.
- Else, let $m = Ns$. This is the end of the while.

[Substep] e): Mutation**:** Change the position of bacteria by mutation operator.

$$\theta(i, j+1, k) = \left(1 - \frac{\theta_{global}}{\theta(i, j, k)}\right) * r_1 * \theta_{global} + \left(\frac{\theta_{global}}{\theta(i, j, k)}\right) * r_2 * \theta_{pbest}(j, k)$$

Mutation by other PSO variants (please see the sec.4)

[Step 4]: If $j < Nc$, go to step 3. In this case, continue chemotaxis, since the life of bacteria is not over.

[Step 5]: The $Sr=S/2$ bacteria with the highest cost function values die and other Sr bacteria with the best values split. Update J_{global} and θ_{global}

[Step 6]: If $k < Nre$, go to step2, otherwise end.

4 Mutation by PSO Variants

The mutation operator brings diversity in the population to avoid premature convergence or getting trapped in some local optima. The mutation changes the bacteria positions with different step size depending upon types of PSO variants used for mutation. The author suggests the following mutation operators.

4.1 Standard Particle Swarm Optimization

Particle swarm optimization (Kennedy and Eberhart 1995) is a population-based, self-adaptive search optimization technique. Kennedy and Eberhart design the original framework of PSO in 1995. Therefore it is known as standard PSO. The two updating fundamental equations in a PSO are velocity and position equations, which are expressed as Eq (4) and (5) respectively

$$V_{id}(t+1) = V_{id}(t) + c_1 * r_{1d}(t) * (p_{id}(t) - X_{id}(t)) + c_2 * r_{2d}(t) * (p_{gd}(t) - X_{id}(t)) \quad (4)$$

$$X_{id}(t+1) = X_{id}(t) + V_{id}(t+1) \quad (5)$$

Where,
t= Current iteration or generation.
i = Particle number.
d= Dimensions.
$V_{id}(t)$ = Velocity of i-th particle for d-dimension at iteration number t.
$X(t)$ = Position of i-th particle for d-dimension at iteration number t.
c_1 and c_2 = Acceleration constants.
$r_{1d}(t)$ and $r_{2d}(t)$ = Random values [0 1] for d- dimension at iteration number t.
$p_{id}(t)$ = Personal or local best i-th particle for d-dimension at iteration number t.
$p_{gd}(t)$ = Global best for d-dimension at iteration number t
Referring to the Eq. (4), the right side of which consists of three parts. The first part of Eq. (4) is the previous velocity of the particle. The second part is the cognition (self-knowledge) or memory, which represents the particle, is attracted by its own previous best position and move toward to it. The third part is the social (social knowledge) or cooperation, which represents that the particle is attracted by the best position so far in population and moves toward to it. There are restrictions among these three parts and can be used to determine the major performance of the algorithm.

4.2 Modified Particle Swarm Optimization

In modified particle swarm optimization (Shi and Eberhart 1998), inertia weight (w) brought into velocity equation of standard PSO for balancing global and local search. The position equation remains same. The two updating equations in a modified PSO are velocity and position equations, which are expressed as Eq. (6) and (7) respectively.

$$V_{id}(t+1) = w * V_{id}(t) + c_1 * r_{1d}(t) * (p_{id}(t) - X_{id}(t)) + c_2 * r_{2d}(t) * (p_{gd}(t) - X_{id}(t)) \quad (6)$$

$$X_{id}(t+1) = X_{id}(t) + V_{id}(t+1) \quad (7)$$

4.3 Linearly Decreasing Weight PSO (LDW-PSO)

By linearly decreasing the inertia weight (Liu *et al.* 2004) from a large value to small value, the PSO tends to have more global search ability at the beginning of the run while having more local search ability near the end of the run. The time varying inertia weight (TVIW) is given by Eq. (8).

$$w = (w_1 - w_2)\left(\frac{\max iter - iter}{\max iter}\right) + w_2 \quad (8)$$

Where w_1 and w_2 are higher and lower inertia weight. Current generation is denoted by "*iter*" and "*maxiter*" is the total number of generations.

4.4 Supervisor Student Model in Particle Swarm Optimization (SSM-PSO)

SSM-PSO (Liu *et al.* 2004) can prevent particles from flying out of defined region without checking the validity of positions at every iteration. In SSM-PSO, particles velocity and positions are manipulated according to the following Eq (9) and (10) respectively.

$$V_{id}(t+1) = V_{id}(t) + c_1 * r_{1d}(t) * (p_{id}(t) - X_{id}(t)) + c_2 * r_{2d}(t) * (p_{gd}(t) - X_{id}(t)) \tag{9}$$

$$X_{id}(t+1) = (1 - mc) * X_{id}(t) + mc * V_{id}(t+1) \tag{10}$$

Where mc is momentum factor ($0 < m < 1$)

4.5 Particle Swarm Optimization with Time Varying Acceleration Coefficients (PSOTVAC)

In PSOTVAC (Ratnaweera *et al.* 2004), time varying acceleration coefficients are used to control the local search and convergence to the global optimum solution. The time varying acceleration coefficients c_1 and c_2 reduce the cognitive component and increase social component. The time varying acceleration coefficients (TVAC) are given by Eq. (11) and (12).

$$c_1 = (c_{1i} - c_{1f})\left(\frac{\max iter - iter}{\max iter}\right) + c_{1f} \tag{11}$$

$$c_2 = (c_{2i} - c_{2f})\left(\frac{\max iter - iter}{\max iter}\right) + c_{2f} \tag{12}$$

Where, c_{1i} and c_{2i} are the initial values of the acceleration coefficients c_1 and c_2. c_{1f} and c_{2f} are the final values of the acceleration coefficients c_1 and c_2.

4.6 Global Local Best Particle Swarm Optimization (GLBest PSO)

In GLBestPSO (Arumugam *et al.* 2008), the inertia weight (w) and acceleration co-efficient (c_1 and c_2) are neither set to a constant value nor set as a linearly decreasing time varying function. Instead, they proposed in terms of global best (*gbest*) and local best (*pbest*) position of the particles. The GLBest inertia weight (GLBestTW) and GLBest acceleration co-efficient (GLBest AC) are given by Eq. (13) and (14).

$$GLBestIW = w_i = \left(1.1 - \frac{gbest_i}{(pbest_i)_{average}}\right) \tag{13}$$

$$GLBestAC = c_i = \left(1 + \frac{gbest_i}{(pbest_i)}\right) \tag{14}$$

The modified velocity equation for the GLBest PSO is given by Eq. (15). The position equation of GLBest PSO is same as that of other PSO methods.

$$V_i(t) = w_i * V_i(t-1) + c_i * r(t) * (pbest_i + gbest_i - 2X_i(t)) \tag{15}$$

4.7 Parameter Free PSO Algorithm (pf-PSO)

In pf-PSO (Ramana *et al.* 2009), local best (*pbest*) and global best (*gbest*) positions are used to update the position of particles. The position of particles is updated by following Eq. (16).

$$X_i(t+1) = [1 - \frac{gbest}{X_i(t)}] * Rnd_1 * gbest + [\frac{gbest}{X_i(t)}] * Rnd_2 * pbest_i \tag{16}$$

Where,
gbest= Global best.
pbest= Local best.
Rnd_1 and Rnd_2 = Random functions lays between [0 1]

4.8 Particle Swarm Optimization with Extrapolation Technique (e-PSO)

This method uses extrapolation technique to perform the optimization; hence, it is named as extrapolated PSO or shortly ePSO algorithm (Arumugam *et al.* 2009). The updated position equation for each particle is given in Eq. (17). The position of particles is updated by global best (*gbest*) position, local best position (*pbest*) and the current position of the particle.

$$X_i(t+1) = [gbest] + [e_1 * Rnd * gbest] + [e_1 * (gbest - X_i(t)) * \exp(e_2 * ((f(gbest) - f(X_i(t))) / f(gbest)))] \tag{17}$$

Where,
e_1 and e_2 = Extrapolation co-efficient.
$e_1 = e_2 = exp$ *(-current generation/max. no. of generation).*
gbest = Particle position where the best fitness solution is found.
Rnd = Random function that varies between (0, 1).
$X_i(t)$ = Current particle position in *t*-th generations.
f (gbest) = Fitness value at *gbest* position
f (Xi(t)) =Current particle's fitness value

5 Experimental Results and Discussions

The widely used and well-known three benchmark functions (Biswas *et al.* 2007) as listed in Table 1 are used to validate the performance of the PBFO. For the sake of comparison, search range and asymmetric initialization range (Biswas *et al.* 2007) shown in Table 2 are used in proposed method. The authors considered both unimodal and multimodal functions to test efficiency of the PBFO. The function f_1 is unimodal while f_2 and f_3 are multimodal.

Using PBFO, Rosenbrock, Rastrigin and Griewank functions with different number of dimensions and fitness evaluations are simuliated for 25 independent runs are presented in Table 3. In PBFO, position of bacteria are updated after all finteness evaluations i.e. at the same time. The mean and standard deviation for 25 independent runs obtained by BFO (Biswas *et al.* 2007) and PBFO are presented in Table 3. The mean best fitness value is measured as a function of number of function evalutions. During experimentation of three benchmark functions, the parameters of PBFO are selected as S=40, p=15,30,45,60, Nre=1, Nc=50, Ns=2. The PBFO improves the quality of optimum solution of unimodal and multimodal functions as compared

Table 1. Benchmark Functions for simulations, where d is the dimension of the function

Function Name	**Mathematical Description**
Rosenbrock $f_1(x)$	$\sum_{i=1}^{d}(100(x_{i+1}-x_i^2)^2+(x_i-1)^2$
Rastrigin $f_2(x)$	$\sum_{i=1}^{d}(x_i^2-10COS(2\pi x_i)+10)$
Griewank $f_3(x)$	$\frac{1}{4000}\sum_{i=1}^{d}x_i^2-\prod_{i=1}^{d}COS(\frac{x_i}{\sqrt{i}})+1$

Table 2. Search range and initialization range for the benchmark functions

Function	**Search Range** $[L_1, U_1]$ x.....x $[L_d, U_d]$	**Initialization Range** $[L'_1, U'_1]$ x.....x $[L'_d, U'_d]$
f_1	$[-100, 100]^d$	$[15, 30]^d$
f_2	$[-10, 10]^d$	$[2.56, 5.12]^d$
f_3	$[-600, 600]^d$	$[300, 600]^d$

Table 3. Mean and Standard Deviation over three Benchmarks

Function	**Dimension** d	**Function Evaluations (FEs)**	**BFO**	**PBFO**
f_1 Rosenbrock	15	50,000	26.705 (2.162)	**14.368 (0.985)**
	30	1,00000	58.216 (14.32)	**34.589 (3.589)**
	45	5,00000	96.873 (26.136)	**56.789 (10.547)**
	60	10,00000	154.705 (40.162)	**67.3254 (9.578)**

Table 3. (*continued*)

f_2 Rastrigin	15	50,000	6.9285 (2.092)	**0.0968 (0.008)**
	30	1,00000	17.0388 (4.821)	**2.8592 (0.058)**
	45	5,00000	30.9925 (7.829)	**5.8962 (0.256)**
	60	10,00000	45.8234 (9.621)	**10.3256 (1.568)**
f_3 Griewank	15	50,000	0.2812 (0.0216)	**0.0056 (0.003)**
	30	1,00000	0.3729 (0.046)	**0.0096 (0.006)**
	45	5,00000	0.6351 (0.052)	**0.0685 (0.008)**
	60	10,00000	0.8324 (0.076)	**0.3256 (0.012)**

to BFO. The overall performance of PBFO is better than the BFO for multimodal functions with high dimensions.

6 Issues for Implementation of Parallel Bacterial Foraging Optimization

The following important issues for implementation of PBFO on parallel processor architecture are taken into account.

1. The processor in parallel architecture should have same speed otherwise all the processor has to wait for next operation.
2. The synchronization is required between master and slave nodes for evaluating fitness values and updation of positions of all bacteria. The slave node reports fitness values before starting up the next operation. The synchronization signals are required between master and slave nodes.
3. Parallel fitness evaluations should not affect the convergence and accuracy of global optima.
4. The number of processors that can easily be added to parallel architecture depend upon complexity of problem. It should not place a limit on the amount of computational nodes that can be utilized.
5. If the memory is occupied by large jobs, one or more of the CPUs might be idle even though there are other jobs to run
6. Parallel programming languages and parallel computers must have a consistency model The consistency model defines rules for how operations on computer memory occur and how results are produced

7 Conclusion

Concept of Parallel Bacterial Foraging to improve convergence with less computational time requirement of multimodal functions with high dimensions is introduced. In addition, several approaches for fusion of Parallel Bacterial Foraging with PSO variants are also addressed by changing the mutation operator. Finally, issues related to implementation of the proposed technique on parallel architecture are also presented. Fusion of PBFO with pf-PSO improves the quality of global best value as number of fitness evalutions progress. Further research focuses on implementation of PBFO on parallel architecture to observe the performance on convergence and accuracy of global optima.

References

Arumugam, M.S., Rao, M.V.C., Chandramohan, A.: A new and improved version of particle swarm optimization algorithm with global local best parameters. Journal of Knowledge and Information System (KAIS) 16, 324–350 (2008)

Arumugam, M.S., Rao, M.V.C., Alan, W.C.T.: A new novel and effective particle swarm optimization like algorithm with extrapolation technique International. Journal of Applied Soft Computing 9, 308–320 (2009)

Biswas, A., Dasgupta, S., Das, S., Abraham, A.: Synergy of PSO and Bacterial Foraging Optimization-A Comparative Study on Numerical Benchmarks Innovations in Hybrid Intelligent Systems ASC, vol. 44, pp. 255–263. Springer, Heidelberg (2007)

Datta, T., Misra, I.S.: Improved Adaptive Bacteria Foraging Algorithm in Optimization of Antenna Array for Faster Convergence. Electromagnetic Research C 1, 143–157 (2008)

Das, S., Panigrahi, B.K., Pattnaik, S.S.: Nature-Inspired Algorithms for Multi-objective Optimization Handbook of Research on Machine Learning Applications and Trends: Algorithms Methods and Techniques Hershey New York, vol. 1, pp. 95–108 (2009)

Das, S., Panigrahi, B.K.: Multi-objective Evolutionary Algorithms, vol. 3, pp. 1145–1151. Encyclopedia of Artificial Intelligence Idea Group Publishing (2008)

Dasgupta, S., Das, S., Abraham, A., Biswas, A.: Adaptive computational chemotaxis in Bacterial foraging optimization: an analysis. IEEE Transactions on Evolutionary Computing 13, 919–941 (2009)

Das, S., Dasgupta, S., Biswas, A., Abraham, A., Konar, A.: On Stability of the Chemotactic Dynamics in Bacterial-Foraging Optimization Algorithm. IEEE Transactions on System, Man and Cybernetics 39, 670–679 (2009)

Eberhart, R.C., Shi, Y.: Comparison between genetic algorithm and particle swarm optimization. In: IEEE Int. Conf. Computt., Anchorage AK, pp. 611–616 (1998)

Eslamian, M., Hosseinian, S.H., Vahidi, B.: Bacterial foraging-based So-lution to the unit-commitment problem. IEEE Transactions Power System 24, 1478–1488 (2009)

Kim, D.H., Cho, J.H.: Adaptive Tuning of PID Controller for Multivariable System Using Bacterial Foraging Based Optimization. In: Szczepaniak, P.S., Kacprzyk, J., Niewiadomski, A. (eds.) AWIC 2005. LNCS (LNAI), vol. 3528, pp. 231–235. Springer, Heidelberg (2005)

Kim, D.H., Cho, C.H.: Bacterial Foraging Based Neural Network. In: Fuzzy Learning Indian International Conference on Artificial Intelligence, pp. 2030–2036 (2005)

Kim, D.H., Abraham, A., Cho, J.H.: A hybrid genetic algorithm and bacterial foraging approach for global optimization. Information Sciences 177, 3918–3937 (2007)

Kennedy, J., Eberhart, R.: Particle swarm optimization. In: IEEE International Conference on Neural Networks, pp. 1942–1948 (1995)
Yu, L., Qin, Z., He, X.: Supervisor-Student Model in Particle Swarm Optimization. In: IEEE Congress on Evolutionary Computation, vol. 1, pp. 542–547 (2004)
Mishra, S.: Hybrid least-square Fuzzy bacterial foraging strategy for harmonic estimation. IEEE Transactions on Evolutionary Computation 9, 61–73 (2005)
Mishra, S.: Hybrid least-square adaptive bacterial foraging strategy for harmonic estimation. IEEE Proceedings-Generation Transmission Distribution 152, 379–389 (2005)
Passino, K.M.: Biomimicry of bacterial foraging for distributed optimization and control. IEEE Control System Magazine 22, 52–67 (2002)
Panigrahi, B.K., Pandi, V.R.: Bacterial foraging optimization: NelderMead hybrid algorithm for economic load dispatch Generation Transmission & Distribution. IET 2, 556–565 (2008)
Panigrahi, B.K., Pandi, V.R., Das, S.: An Adaptive Particle Swarm Optimization Approach for Static and Dynamic Economic Load Dispatch. International Journal on Energy Conversion and Management 49, 1407–1415 (2008)
Pattnaik, S.S., Bakwad, K.M., Sohi, B.S., Devi, S., Panigrahi, B.K., Sastry, G.V.R.S.: Bacterial Foraging Optimization Technique Cascaded with Adaptive Filter to Enhance PSNR from a Single Image. IETE Journal of Research 55, 173–179 (2009)
Pattnaik, S.S., Sastry, G.V.R.S., Bajpai, O.P., Devi, S., Chintakindi, V.S., Patra, P.K., Bakwad, K.M.: Bacterial Foraging Optimization technique to calculate resonant frequency of rectangular microstrip antenna International. Journal of RF and Microwave Computer Aided Engineering 18, 383–388 (2008)
Ratnaweera, A., Halgamuge, S., Watson, H.: Self Organizing Hierarchical Particle Swarm Optimization with time varying acceleration coefficients. IEEE transactions on Evolutionary Transactions 8, 240–255 (2004)
Ramana, M.G., Arumugam, M.S., Loo, C.K.: Hybrid Particle Swarm Optimization Algorithm with fine tuning operators International. Journal of Bio-Inspired Computation 1, 14–31 (2009)
Shi, Y., Eberhart, R.C.: A modified particle swarm optimizer. In: IEEE Congress on Evolutionary Computation Piscataway NI, pp. 69–73 (1998)
Sastry, G.V.R.S., Pattnaik, S.S., Bajpai, O.P., Devi, S., Bakwad, K.M.: Velocity Modulated Bacterial Foraging Optimization Technique (VMBFO) Applied Soft Computing. Elsevier, Amsterdam (2009), doi:10.1016/j.asoc.2009.11.006
Tripathy, M., Mishra Lai, L.L., Zhang, Q.P.: Transmission Loss Reduction Based on FACTS and Bacteria Foraging Algorithm. In: Proceedings of the 2006 Parallel Problem Solving from Nature, pp. 222–231 (2006)
Tang, W.J., Wu, Q.H.: Bacterial Foraging Algorithm for Dynamic Environments. In: IEEE Congress on Evolutionary Computation, Sheraton Vancouver Wall Centre Hotel, Vancouver BC, Canada (2006)

Reliability-Redundancy Optimization Using a Chaotic Differential Harmony Search Algorithm

Leandro dos Santos Coelho[1], Diego L. de A. Bernert[1], and Viviana Cocco Mariani[2]

[1] Industrial and Systems Engineering Graduate Program (PPGEPS), Pontifical Catholic University of Parana, Curitiba, Parana, Brazil
leandro.coelho@pucpr.br, dbernert@gmail.com
[2] Department of Mechanical Engineering (PPGEM), Pontifical Catholic University of Parana, Curitiba, Parana, Brazil
viviana.mariani@pucpr.br

Abstract. In many industrial systems, reliability has been considered as an important design measure. In this context, the system reliability maximization subject to performance and cost constraints is well known as reliability optimization problem. The diversity of system structures, resource constraints, and options for reliability improvement has led to the construction and analysis of several optimization methods, such as dynamic programming, Lagrangian multiplier, and heuristic approaches. On the other hand, a broad class of metaheuristics has been developed for reliability-redundancy optimization. Metaheuristics can overcome many limitations of classical optimization methods and offer a practical way to solve complex optimization problems in reliability engineering. Our study considers one of the latest metaheuristics, harmony search (HS), to solve the reliability-redundancy optimization problems. The HS algorithm was originally inspired by the improvisation process of Jazz musicians. The HS algorithm uses a random search, which is based on random selection, memory consideration, and pitch adjusting. The purpose of this study is to introduce a modified HS approach combined with differential evolution and chaotic sequences to solve optimization problems in reliability engineering. The validity and efficiency of the proposed HS approach are evaluated in two benchmark problems (an overspeed protection system for a gas turbine and a series-parallel system) of mixed integer programming in reliability-redundancy design. Simulation results show that the proposed HS produces promising results in comparison to other optimization methods available in the recent literature for mentioned benchmark problems.

Keywords: Harmony search, evolutionary algorithm, reliability-redundancy optimization, differential evolution, chaotic sequences.

1 Introduction

The primary goal of reliability engineering is to improve the reliability system. The need to design structural systems with adequate levels of reliability and redundancy is widely acknowledged. It is as crucial that these desired levels are maintained above target levels throughout the life of the system. Many designers are devoted to improving the reliability of manufacturing systems or product components to be more

B.K. Panigrahi, Y. Shi, and M.-H. Lim (Eds.): Handbook of Swarm Intelligence, ALO 8, pp. 503–516.
springerlink.com

competitive in the market. Typical approaches to achieve higher system reliability are: (i) increasing the reliability of system components, and (ii) using redundant components in various subsystems in the system [1].

The optimal reliability design is to determine a system structure which has the highest reliability under several resource constraints. Redundancy strategy has been widely used to improve the reliability of a system by incorporating redundant components or subsystems. Design-reliability experts have focused a great deal of their efforts on the allocation of reliability and redundancy of components for maximizing system reliability. In this context, the well-known problem of reliability-redundancy optimization has been addressed in a number of studies [1,2].

Metaheuristics are general procedures that coordinate simple heuristics and rules to find good approximate solutions to computationally difficult optimization problems. Metaheuristics, in particular, offer flexibility and a practical way to solve complex optimization problems in reliability engineering. Metaheuristic approaches such as simulated annealing, tabu search, ant colony, particle swarm optimization, evolutionary algorithms have been used to handle system reliability problems (see examples in [1-4]).

One of the recent additions to metaheuristics is the harmony search algorithm (HS). In spite of it being a relatively new approach, it has been used in many applications [5-10] and it is attracting more attentions day by day. HS uses a stochastic random search instead of a gradient search so that derivative information is unnecessary. In addition, HS has ability to deal with discrete and continuous mathematics problem.

HS algorithm proposed in [5] has been recently developed in an analogy with music improvisation process where musicians in an ensemble continue to polish their pitches in order to obtain better harmony. Jazz improvisation seeks to find musically pleasing harmony similar to the optimum design process which seeks to find optimum solution. The pitch of each musical instrument determines the aesthetic quality, just as the objective function value is determined by the set of values assigned to each decision variable. The musician then tunes some of these notes to achieve a better harmony. Similarly it is then checked whether this candidate solution improves the objective function or not, much like finding out if the notes are euphonic. This candidate solution is then checked whether it satisfies the objective function or not, similar to the process of finding out whether euphonic music is obtained or not [6].

Recently, metaheuristic approaches based on evolutionary algorithms (EAs) are receiving increasing attention, because of their potential to effectively explore a wide range of complex optimization problems. EAs operate on a population of potential solutions, applying the principle of survival of the fittest to produce successively better approximations to a solution. A number of EA based approaches have been proposed during last decades.

Differential evolution (DE) as developed by Storn and Price [11] is one of the best EAs, and has proven to be a promising candidate to solve real valued optimization problems. DE is capable of handling non-differentiable, nonlinear and multi-modal objective functions [12-13]. The basic idea of DE is to use vector differences for perturbing the vector population. DE is a simple population based optimization technique which is at the same time very powerful and robust.

On other hand, chaos is a nonlinear system with deterministic dynamic behavior. It has ergodicity, stochastic and regularity properties, and is very sensitive to its initial conditions and parameters. Small differences in values in the initial solutions will result in great differences after many iterations. These characteristics of a chaotic system can enhance the search ability of classical HS in terms of global search.

The purpose of the present paper is to provide a HS approach with an operator inspired on DE and utilization of chaotic sequences, called CHSDE, to solve optimization problems in reliability engineering.

In general terms, the utilization of an operator based on DE can accelerate the convergence rate of the CHSDE, however, it also accelerates the premature convergence of the CHSDE and make it get into the local optimum. To overcome these disadvantages, chaotic sequences is introduced in the CHSDE design.

To illustrate the power of the proposed CHSDE, two benchmark problems of reliability-redundancy optimization [14-18] have been considered. Simulation results of HS and the proposed CHSDE approaches are compared with other optimization techniques presented in literature for the evaluated benchmark problems of reliability-redundancy optimization.

The remainder of the paper is organized as follows. In Section 2, the reliability-redundancy optimization problem is introduced, while the characteristics of HS and CHSDE approaches are detailed in Section 3. The problem formulation for reliability-redundancy optimization and its assumptions are given in Section 4. Moreover, Section 4 also presents the simulation results for two benchmark problems of reliability-redundancy optimization. Finally, the conclusion and further research are discussed in Section 5.

2 Fundamentals of Reliability-Redundancy Optimization

In many system design situations, the overall system is partitioned into a specific number of subsystems, according to the function requirements of the system. For each subsystem, there are different component types available with varying reliability, costs, weight, volume and other characteristics. The system reliability depends on the reliability of each subsystem. To maximize system reliability, the following approach can be considered: (i) using more reliable components, (ii) using redundant configuration with active or stand-by components in parallel, or (iii) a combination of (i) and (ii) [19].

In this work, the reliability-redundancy allocation problem of maximizing the system reliability subject to constraints can be formulated as

$$\text{Maximize } R_s = f(\boldsymbol{r},\boldsymbol{n}), \tag{1}$$

$$\text{subject to } g(\boldsymbol{r},\boldsymbol{n}) \leq l \tag{2}$$

$$0 \leq r_i \leq 1, \quad r_i \in \Re, \ n_i \in Z^+, \ 1 \leq i \leq m.$$

where R_s is the reliability of system, g is the set of constraint functions usually associated with system weight, volume and cost, $\boldsymbol{r}= (r_1, r_2, r_3,, \ldots, r_m)$ is the vector of the component reliabilities for the system, $\boldsymbol{n} = (n_1, n_2, n_3,, \ldots, n_m)$ is the vector of the redundancy allocation for the system; r_i and n_i are the reliability and the number of components in the ith subsystem respectively; $f(\cdot)$ is the objective function for the overall system reliability; and l is the resource limitation; and m is the number of subsystems in the system. Our goal is to determine the number of components, and the components' reliability in each system so as to maximize the overall system reliability. The problem belongs to the category of constrained nonlinear mixed-integer optimization problems. The two examples of reliability-redundancy optimization evaluated in this paper are formulated, which are outlined below.

2.1 Example 1: Overspeed Protection System for a Gas Turbine

This benchmark problem is related to an overspeed protection system for a gas turbine [14-16]. Overspeed detection is continuously provided by the electrical and mechanical systems. When an overspeed occurs, it is necessary to cut off the fuel supply [14]. For this purpose, four control valves (V_1-V_4) must close. The control system is modeled as a 4-stage series system (see diagram in Fig. 1).

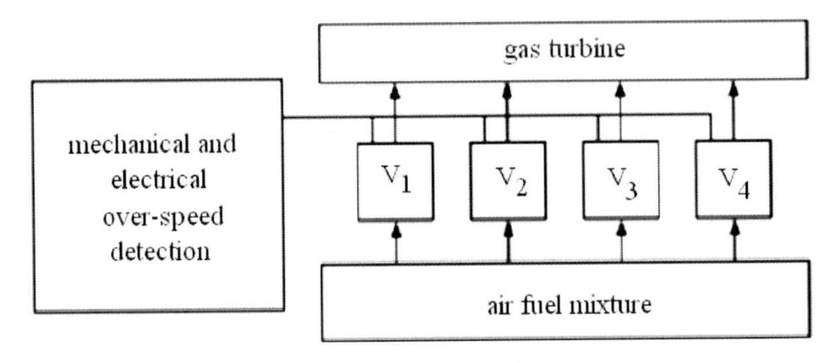

Fig. 1. Diagram for the overspeed protection system of a gas turbine.

This problem is formulated as the following mixed-integer nonlinear programming problem, i.e., the problem can be stated as

$$\text{Maximize } f(\boldsymbol{r},\boldsymbol{n}) = \prod_{i=1}^{m}\left[1-(1-r_i)^{n_i}\right], \tag{3}$$

subject to

$$g_1(\boldsymbol{r},\boldsymbol{n}) = \sum_{i=1}^{m} v_i \cdot n_i^2 \leq V \tag{4}$$

$$g_2(\boldsymbol{r},\boldsymbol{n}) = \sum_{i=1}^{m} C(r_i)\cdot\left[n_i + e^{0.25\cdot n_i}\right] \leq C \tag{5}$$

$$g_3(\boldsymbol{r},\boldsymbol{n}) = \sum_{i=1}^{m} w_i \cdot n_i \cdot e^{0.25\cdot n_i} \leq W \tag{6}$$

where $1 \leq n_i \leq 10$, $n_i \in Z^+$, where Z^+ is the discrete space of positive integers, $0.5 \leq r_i \leq 1\text{-}10^{-6}$, $r_i \in \Re$; v_i is the volume of each component in subsystem i; V is the upper limit on the sum of the subsystems' products of volume and weight; C is the upper limit on the system cost; $C(r_i) = \alpha_i \cdot [-T / \ln(r_i)]^{\beta_i}$ is the cost of each component with reliability r_i at subsystem i; T is the operating time during which the component must not fail; and W is the upper limit on the weight of the system. The input parameters of the overspeed protection system for a gas turbine are shown in Table 1.

Table 1. Data of overspeed protection system.

Stage	$10^5 \cdot \alpha_i$	β_i	v_i	w_i	V	C	W	T
1	1.0	1.5	1	6	250	400	500	1000 h
2	2.3	1.5	2	6				
3	0.3	1.5	3	8				
4	2.3	1.5	2	7				

2.2 Example 2: Series-Parallel System

The second example problem was evaluated in [14],[17]. Fig. 2 shows the series-parallel system analyzed in this paper.

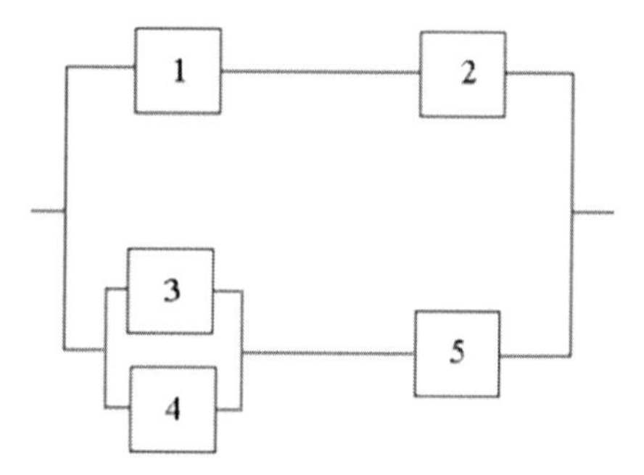

Fig. 2. Series-parallel system.

The optimization problem of series-parallel system can be stated as

$$\text{Maximize } f(\boldsymbol{r},\boldsymbol{n}) = 1 - (1 - R_1 R_2)(1 - (1 - R_3)(1 - R_4) R_5) \tag{7}$$

subject to

$$g_1(\boldsymbol{r},\boldsymbol{n}) = \sum_{i=1}^{m} w_i \cdot v_i^2 \cdot n_i^2 \leq V \tag{8}$$

$$g_2(\boldsymbol{r},\boldsymbol{n}) = \sum_{i=1}^{m} \alpha_i \cdot \left(-\frac{1000}{\ln r_i} \right)^{\beta_i} \cdot \left[n_i + e^{0.25 n_i} \right] \leq C \tag{9}$$

$$g_3(\boldsymbol{r},\boldsymbol{n}) = \sum_{i=1}^{m} w_i \cdot n_i \cdot e^{0.25 n_i} \leq W \tag{10}$$

Equations (7), (8), and (9) are constraints about the reliability, system, cost and weight, respectively. The constraint given by (7) is a combination of weight, redundancy allocation, and volume; (8) is a cost constraint; and (9) is a weight constraint. In this context, V is the upper limit on the sum of the subsystems' products of volume, C is the upper limit on the cost of the system, and W is the upper limit on the weight of the system, w_i is a factor with constant value for the ith system component; $n_i \in Z^+$, where Z^+ is the discrete space of positive integers, and $0 \le r_i \le 1$, $r_i \in \Re$, $1 \le i \le m$. The parameters β_i and α_i are physical features of ith system component. The input parameters of the series-parallel system are given in Table 2. The data shown in Table 2 are also available in [14],[16].

Table 2. Data used in series-parallel system.

Stage	$10^5 \cdot \alpha_i$	β_i	$w_i \cdot v_i^2$	w_i	V	C	W
1	2.500	1.5	2	3.5	180	175	100
2	1.450	1.5	4	4.0			
3	0.541	1.5	5	4.0			
4	0.541	1.5	8	3.5			
5	2.100	1.5	4	4.5			

3 Optimization Algorithms

This section describes the proposed CHSDE algorithm. First, a brief overview of the HS is provided, and finally the modification procedures of the proposed CHSDE algorithm are presented.

3.1 Harmony Search Algorithm

Recently, Geem *et al.* [5] proposed a new metaheuristic algorithm, called HS, that was inspired by musical process of searching for "pleasant harmonies" through improvisation. For instance, when several notes from different musical instruments are played simultaneously on a random basis and this process is repeated, there is the possibility of finding better harmonies. In the HS methodology, these better harmonies are saved in a certain size of memory by replacing the worst harmony in the memory until the predefined maximum number of improvisations generating a new harmony is reached.

Its main difference from the others metaheuristic optimization algorithms is that HS does not get its main philosophy from a natural process, instead, gets from the musical improvisation which occurs when a group of musicians searches for a better state of harmony.

In the HS algorithm, musical performances seek a perfect state of harmony determined by aesthetic estimation, as the optimization algorithms seek a best state (i.e. global optimum) determined by objective function value. The optimization procedure of the HS algorithm can be synthesized in the following steps [8]:

Step 1. Initialize the optimization problem and HS algorithm parameters. First, the optimization problem is specified as follows:

$$\text{Minimize } f(x) \text{ subject to } x_i \in X_i, \quad i = 1,\ldots, N$$

where $f(x)$ is the objective function, x is the set of each decision variable (x_i); X_i is the set of the possible range of values for each design variable (continuous design variables), that is, $x_{i,lower} \le X_i \le x_{i,upper}$, where $x_{i,lower}$ and $x_{i,upper}$ are the lower and upper bounds for each decision variable; and N is the number of design variables. In this context, the HS algorithm parameters that are required to solve the optimization problem are also specified in this step. The number of solution vectors in harmony memory (HMS) that is the size of the harmony memory matrix, harmony memory considering rate (HMCR), pitch adjusting rate (PAR), and the maximum number of searches (stopping criterion) are selected in this step. Here, HMCR and PAR are parameters that are used to improve the solution vector. Both are defined in *Step 3*.

Step 2. Initialize the harmony memory. The harmony memory (HM) is a memory location where all the solution vectors (sets of decision variables) are stored. The HM matrix, shown in Eq. (11), is filled with randomly generated solution vectors using uniform distribution, where

$$\text{HM} = \begin{bmatrix} x_1^1 & x_2^1 & \cdots & x_{N\text{-}1}^1 & x_N^1 \\ x_1^2 & x_2^2 & \cdots & x_{N\text{-}1}^2 & x_N^2 \\ \vdots & \vdots & \vdots & \vdots & \vdots \\ x_1^{\text{HMS-1}} & x_2^{\text{HMS-1}} & \cdots & x_{N-1}^{\text{HMS-1}} & x_N^{\text{HMS-1}} \\ x_1^{\text{HMS}} & x_2^{\text{HMS}} & \cdots & x_{N-1}^{\text{HMS}} & x_N^{\text{HMS}} \end{bmatrix}. \tag{11}$$

Step 3. Improvise a new harmony from the HM. A new harmony vector, $x' = (x_1', x_2', \ldots, x_N')$, is generated based on three rules: i) memory consideration, ii) pitch adjustment, and iii) random selection. The generation of a new harmony is called 'improvisation'.

In the memory consideration, the value of the first decision variable (x_1') for the new vector is chosen from any value in the specified HM range $\left(x_1' - x_1^{HMS}\right)$. Values of the other decision variables $(x_2', \ldots, x_N')$ are chosen in the same manner. The HMCR, which varies between 0 and 1, is the rate of choosing one value from the historical values stored in the HM, while (1 - HMCR) is the rate of randomly selecting one value from the possible range of values.

$$x_i' \leftarrow \begin{cases} x_i' \in \left\{x_i^1, x_i^2, \ldots, x_i^{\text{HMS}}\right\} & \text{with probability HMCR} \\ x_i' \in X_i & \text{with probability (1 - HMCR).} \end{cases} \tag{12}$$

After, every component obtained by the memory consideration is examined to determine whether it should be pitch-adjusted. This operation uses the PAR parameter, which is the rate of pitch adjustment as follows:

$$\text{Pitch adjusting decision for } x_i^{'} \leftarrow \begin{cases} \text{Yes} & \text{with probability PAR} \\ \text{No} & \text{with probability (1 - PAR).} \end{cases} \quad (13)$$

The value of (1 - PAR) sets the rate of doing nothing. If the pitch adjustment decision for $x_i^{'}$ is Yes, $x_i^{'}$ is replaced as follows:

$$x_i^{'} \leftarrow x_i^{'} \pm r \cdot bw, \quad (14)$$

where *bw* is an arbitrary distance bandwidth, *r* is a random number generated using uniform distribution between 0 and 1. In *Step 3*, HM consideration, pitch adjustment or random selection is applied to each variable of the new harmony vector in turn.

Step 4. Update the HM. If the new harmony vector, $x' = (x_1^{'}, x_2^{'}, \ldots, x_N^{'})$ is better than the worst harmony in the HM, judged in terms of the objective function value, *F*, the new harmony is included in the HM and the existing worst harmony is excluded from the HM.

Step 5. Repeat *Steps 3* and *4* until the stopping criterion has been satisfied, usually a sufficiently good objective function or a maximum number of iterations (generations), t_{max}. Maximum number of iterations criterion is adopted in this work.

3.2 Proposed CHSDE Algorithm

Differential evolution is a population-based stochastic function minimizer (or maximizer) relating to evolutionary algorithms, whose simple yet powerful and straightforward features make it attractive for numerical optimization [11]. Inspired by mutation operation of classical differential evolution approach, the equation (14) is modified for

$$x_i^{'} \leftarrow x_i^{'} \pm \alpha \cdot r \cdot bw \cdot \left(x_{p_1}^{'} - x_{p_2}^{'} \right) \quad (15)$$

where p_1 and p_2 are mutually different integers and also different from the running index, *i*, randomly selected with uniform distribution from the set $\{1, 2, \cdots, i-1, i+1, \cdots, \text{HMS}\}$. In the CHSDE, the operation $r \cdot bw$ represents the mutation factor of classical differential evolution algorithm. In this work, it is adopted $\alpha = 50$.

Furthermore, other improvement in HS algorithm, using chaotic sequences combined with HS can be useful. Chaos is a kind of a feature of nonlinear dynamic system which exhibits bounded unstable dynamic behavior, ergodic, non-period behavior depended on initial condition and control parameters. In recent years, growing interests from physics, chemistry, biology and engineering have stimulated the studies of chaos for optimization problems [20-24].

In this paper, chaotic sequences using Logistic map [20] in HS approach are used to generate the values for HMCR, described in equation (12), is proposed. In this context, the equation (12) is modified to

$$x_i' \leftarrow \begin{cases} x_i' \in \{x_i^1, x_i^2, \ldots, x_i^{\text{HMS}}\} & \text{with probability HMCRchaotic}(t) \\ x_i' \in X_i & \text{with probability } (1 - \text{HMCRchaotic}(t)). \end{cases} \quad (16)$$

where HMCRchaotic is generated using the output (HMCRchaotic variable) of Logistic map, whose equation is given by:

$$\text{HMCRchaotic}(t) = \mu \cdot \text{HMCRchaotic}(t-1) \cdot [1 - \text{HMCRchaotic}(t-1)], \quad (17)$$

where t is the sample, and μ is a control parameter, $0 \le \mu \le 4$. The behavior of the system of equation (17) is greatly changed with the variation of μ. The value of μ determines whether HMCRchaotic stabilizes at a constant size, oscillates between a limited sequence of sizes, or behaves chaotically in an unpredictable pattern. A very small difference in the initial value of HMCRchaotic causes substantial differences in its long-time behavior. Equation (17) is deterministic, displaying chaotic dynamics when $\mu = 4$ and HMCRchaotic(1) $\notin$ {0, 0.25, 0.50, 0.75, 1}. In this paper, HMCRchaotic(t) is adopted with chaotic dynamics and distributed in the range between 0 and 1.

4 Simulation Results

Each individual of a population in tested HS and CHSDE approaches uses the variables vectors $\boldsymbol{n}$ and $\boldsymbol{r}$, where the boundaries are given in Section 2. During the evolution process, the integer variables n_i are treated as real variables; and in evaluating the objective function, the real values are transformed to the nearest integer values.

Each optimization method was implemented in MATLAB (MathWorks). All the programs were run under Windows XP on a 3.2 GHz Pentium IV processor with 2 GB of random access memory. To eliminate stochastic discrepancy, in each case study, it was adopted 50 independent runs for each optimization method. The total number of solution vectors in classical HS and CHSDE, i.e., the HMS was 25 and t_{max} = 200 generations. All tested HS approaches adopt 5,000 objective function evaluations in each run for the two benchmark problems.

In HS and CHSDE, it is necessary to tune a number of parameters to have good performance. The user-specified parameters of the HS and CHSDE algorithm were varied to establish the values most beneficial to the optimization process. Based on these simulations the values found to be most appropriate are: $bw = 0.01 \cdot (x_{i,upper} - x_{i,lower})$ and PAR = 0.3. Furthermore, in classical HS was adopted HMCR = 0.9.

In this work, the penalty-based method proposed in [21] was used in HS and CHSDE approaches for infeasible solutions (constraint violation). An approach is used to convert a constrained problem to an unconstrained one by modifying the search space. A penalty value is defined to take the constrained violation into account.

Table 3. Convergence results of $f(\boldsymbol{r},\boldsymbol{n})$ (50 runs) for the overspeed protection system using HS and CHSDE.

Optimization Method	$f(\boldsymbol{r},\boldsymbol{n})$			
	Minimum (Worst)	Mean	Maximum (Best)	Standard Deviation
HS	0.99960568	0.99993126	0.99995466	0.00005070
CHSDE	0.99994048	0.99994950	0.99995467	0.00000525

Table 4. Best result (50 runs) of HS and CHSDE approaches for the overspeed protection system.

Parameter	HS	CHSDE
$f(\boldsymbol{r},\boldsymbol{n})$	0.99995466	0.99995467
n_1	5	5
n_2	6	5
n_3	4	4
n_4	5	6
r_1	0.90124442	0.90165488
r_2	0.85037025	0.88821801
r_3	0.94848940	0.94807430
r_4	0.88792842	0.84996263
Slack (g_1)	55	55
Slack (g_2)	0.00000084	0.00934729
Slack (g_3)	24.80188272	15.36346308

Note: Slack is the unused resources.

Table 5. Comparison of result for the overspeed protection system using CHSDE with results in the literature.

Parameter	Dhingra [15]	Yokota *et al.* [16]	Chen [14]	CHSDE
$f(\boldsymbol{r},\boldsymbol{n})$	0.99961	0.999468	0.999942	0.99995467
n_1	6	3	5	5
n_2	6	6	5	5
n_3	3	3	5	4
n_4	5	5	5	6
r_1	0.81604	0.965593	0.903800	0.90165488
r_2	0.80309	0.760592	0.874992	0.88821801
r_3	0.98364	0.972646	0.919898	0.94807430
r_4	0.80373	0.804660	0.890609	0.84996263
MPI (%)	88.6333%	91.6673%	23.5689%	-
Slack (g_1)	65	92	50	55
Slack (g_2)	0.064	-70.733576	0.002152	0.00934729
Slack (g_3)	4.348	127.583189	28.803701	15.36346308

Note: Slack is the unused resources.

Maximum possible improvement (MPI):

$MPI(\%) = [R_s(\text{CHSDE}) - R_s(\text{other})]/[1 - R_s(\text{other})]$

The method proposed in [14] uses a procedure where the terms l are subtracted (maximization problem) from objective function $f(\boldsymbol{r},\boldsymbol{n})$ if $g(\boldsymbol{r},\boldsymbol{n}) > l$.

Table 3 shows the results over 50 independent runs for the overspeed protection system. CHSDE gives the best results, followed by the HS. By using the results in Table 3, in terms of best $f(\boldsymbol{r},\boldsymbol{n})$ result, the solutions of CHSDE are just slightly better than the solution found by HS for the overspeed protection system.

The best result obtained for the overspeed protection system using CHSDE was 0.99995467, as shown in Table 4. From Table 5, a best solution found by CHSDE for the overspeed protection system is significantly better than that obtained by Chen [14], Dhingra [15], and Yokota *et al.* [16].

The standard deviation is an important measure of algorithm robustness. From Table 6, for the HS and CHSDE approaches, the standard deviation in objective function values for the series-parallel system is very low. This implies that the proposed approach is robust and credible. The best result obtained for the series-parallel system using CHSDE was 0.99997664, as shown in Table 7.

Table 6. Convergence results of $f(\boldsymbol{r},\boldsymbol{n})$ (50 runs) for the series-parallel system using HS and CHSDE approaches.

Optimization	$f(\boldsymbol{r},\boldsymbol{n})$			
Method	Minimum (Worst)	Mean	Maximum (Best)	Standard Deviation
HS	0.99995030	0.99996483	0.99997654	0.00000583
CHSDE	0.99996423	0.99996782	0.99997664	0.00000481

Table 7. Best result (50 runs) of HS and CHSDE approaches for the series-parallel system.

Parameter	HS	CHSDE
$f(\boldsymbol{r},\boldsymbol{n})$	0.99997654	0.99997664
n_1	2	2
n_2	2	2
n_3	2	2
n_4	2	2
n_5	4	4
r_1	0.825317	0.819632
r_2	0.852423	0.844101
r_3	0.895007	0.895429
r_4	0.891666	0.895976
R_5	0.865721	0.868572
Slack (g_1)	40	40
Slack (g_2)	0.031580	0.000474
Slack (g_3)	1.609289	1.609289

Note: Slack is the unused resources.

Table 8 presents the corresponding results from [14],[17],[18]. As the results in Table 8 indicate, the solutions found by CHSDE outperformed the existing approaches.

Table 8. Comparison of result for the series-parallel system using CHSDE with results in the literature.

Parameter	Hikita *et al.* [18]	Hsieh *et al.* [17]	Chen [14]	CHSDE
$f(\boldsymbol{r},\boldsymbol{n})$	0.99996875	0.99997418	0.99997658	0.99997664
N_1	3	2	2	2
N_2	3	2	2	2
N_3	1	2	2	2
N_4	2	2	2	2
N_5	3	4	4	4
R_1	0.838193	0.785452	0.812485	0.819632
R_2	0.855065	0.842998	0.843155	0.844101
R_3	0.878859	0.885333	0.897385	0.895429
R_4	0.911402	0.917958	0.894516	0.895976
R_5	0.850355	0.870318	0.870590	0.868572
MPI (%)	25.248%	9.527%	0.256%	-
Slack (g_1)	53	40	40	40
Slack (g_2)	0.000000	1.194440	0.002627	0.000474
Slack (g_3)	7.110849	1.609289	1.609289	1.609289

Note: Slack is the unused resources.
Maximum possible improvement (MPI):
$MPI(\%) = [R_s(\text{CHSDE}) - R_s(\text{other})]/[1 - R_s(\text{other})]$

5 Conclusion and Future Research

The interest about metaheuristic optimization methods has grown for the last few years. Indeed, more and more papers about enhancements and modifications in meta-heuristics have been proposed. In this context, the HS is an emerging meta-heuristic optimization algorithm music-inspired, where each musical instrument corresponds to each decision variable, musical note corresponds to variable value, and, harmony corresponds to solution vector.

Modifications in classical HS can be useful for a balance between exploration and exploitation of the harmony memory. The aim of this paper is to investigate the performance of a new modification in classical HS, called CHSDE algorithm, proposed in this paper and applied to reliability-redundancy optimization.

The performance of the proposed CHSDE approach and the classical HS algorithm are evaluated in two benchmark problems in reliability-redundancy optimization. CHSDE has a slight advantage in terms of solution quality (maximum $f(\boldsymbol{r},\boldsymbol{n})$ value) over the other solvers for the two benchmark problems. As a consequence, CHSDE may be a promising and encouraging approach to deal with complex numerical optimization problems.

In future, the study of CHSDE can be extended to adopt diversity control mechanisms of harmony memory to deal with optimization reliability-redundancy problems in electrical power systems field.

Acknowledgments

This work was supported by the National Council of Scientific and Technologic Development of Brazil — CNPq — under Grants 568221/2008-7, 474408/2008-6, 302786/2008-2/PQ, 309646/2006-5/PQ and 478158/2009-2.

References

1. Kuo, W., Prasad, V.R.: An annotated overview of system-reliability optimization. IEEE Transactions on Reliability 49, 176–187 (2000)
2. Kuo, W., Prasad, V.R., Tillman, F., Hwang, C.L.: Optimization Reliability Design: Fundamentals and Applications. Cambridge University Press, Cambridge (2001)
3. Kuo, W., Wan, R.: Recent advances in optimal reliability allocation. IEEE Transactions on Systems, Man, and Cybernetics – Part A: Systems and Humans 37, 143–156 (2007)
4. Gen, M., Yun, Y.S.: Soft computing approach for reliability optimization: state-of-the-art survey. Reliability Engineering and System Safety 91, 1008–1026 (2006)
5. Geem, Z.J., Kim, H., Loganathan, G.V.: A new heuristic optimization algorithm: harmony search. Simulation 76, 60–68 (2001)
6. Saka, M.P.: Optimum design of steel sway frames to BS5950 using harmony search algorithm. Journal of Constructional Steel Research 65, 36–43 (2009)
7. Lee, K.S., Geem, Z.W.: A new meta-heuristic algorithm for continuous engineering optimization: harmony search theory and practice. Computer Methods in Applied Mechanics and Engineering 194, 3902–3933 (2005)
8. Coelho, L.S., Bernert, D.L.A.: An improved harmony search algorithm for synchronization of discrete-time chaotic systems. Chaos, Solitons & Fractals 41, 2526–2532 (2009)
9. Geem, Z.W.: Novel derivative of harmony search algorithm for discrete design variables. Applied Mathematics and Computation 199, 223–230 (2008)
10. Geem, Z.W.: Harmony search applications in industry. In: Prasad, B. (ed.) Soft Computing Applications in Industry, STUDFUZZ, vol. 226, pp. 117–134 (2008)
11. Storn, R., Price, K.V.: Differential evolution: a simple and efficient adaptive scheme for global optimization over continuous spaces, Technical Report TR-95-012, International Computer Science Institute, Berkeley, USA (1995)
12. Storn, R., Price, K.V.: Differential evolution a simple and efficient heuristic for global optimization over continuous spaces. Journal of Global Optimization 11, 341–359 (1997)
13. Price, K.V., Storn, R.M., Lampinen, J.A.: Differential Evolution: A Practical Approach to Global Optimization, 1st edn. Springer, Heidelberg (2005)
14. Chen, T.C.: IAs based approach for reliability redundancy allocation problems. Applied Mathematics and Computation 182, 1556–1567 (2006)
15. Dhingra, A.K.: Optimal apportionment of reliability & redundancy in series systems under multiple objectives. IEEE Transactions on Reliability 41, 576–582 (1992)
16. Yokota, T., Gen, M., Li, H.H.: Genetic algorithm for nonlinear mixed-integer programming problems and it's application. Computers and Industrial Engineering 30, 905–917 (1996)

17. Hsieh, Y.C., Chen, T.C., Bricker, D.L.: Genetic algorithm for reliability design problems. Microelectronic Reliability 38, 1599–1605 (1998)
18. Hikita, M., Nakagawa, Y., Harihisa, H.: Reliability optimization of systems by a surrogate constraints algorithm. IEEE Transactions on Reliability 41, 473–480 (1992)
19. Zhao, J.H., Liu, Z., Dao, M.T.: Reliability optimization using multiobjective ant colony system approaches. Reliability Engineering & System Safety 92, 109–120 (2007)
20. Caponetto, R., Fortuna, L., Fazzino, S., Xibilia, M.G.: Chaotic sequences to improve the performance of evolutionary algorithms. IEEE Transactions on Evolutionary Computation 7, 289–304 (2003)
21. Coelho, L.S., Mariani, V.C.: Combining of chaotic differential evolution and quadratic programming for economic dispatch optimization with valve-point effect. IEEE Transactions on Power Systems 21, 989–996 (2006)
22. Cong, L., Shaoqian, L.: Chaotic spreading sequences with multiple access performance better than random sequences. IEEE Transactions on Circuit and System-I, Fundamental Theory and Application 47, 394–397 (2000)
23. Zuo, X.Q., Fan, Y.S.: A chaos search immune algorithm with its application to neuro-fuzzy controller design. Chaos, Solitons & Fractals 30, 94–109 (2006)
24. Yang, D., Li, G., Cheng, G.: On the efficiency of chaos optimization algorithms for global optimization. Chaos, Solitons & Fractals 34, 1366–1375 (2007)

Gene Regulatory Network Identification from Gene Expression Time Series Data Using Swarm Intelligence

Debasish Datta, Amit Konar, Swagatam Das, and B.K. Panigrahi

A Gene Regulatory Network (GRN) usually is modelled as a directed graph, where the nodes represent genes and the directed arc from a given node i to node j represents the causal influence of gene i over gene j. The causal influence represented by an arc is enumerated by a signed weight associated with that arc. In this article, we model GRN by a recurrent fuzzy neural network, and attempt to identify the signed weights from the time response data of the gene micro-array. A cost function has been constructed to describe the weight identification as an optimization problem, and Particle Swarm Optimization algorithm has been used to optimize the cost function. The fuzzy membership distribution used to model network weights enhances search efficiency and hence computational overhead in the identification problem. Because of the nonlinearity in causal relationship between genes, there exist multiple solutions to the weight identification problem of GRN. In order to cater for the theoretical best solution, the identification problem has been decoupled into two sub-problems: i) determination of the existence/non-existence about the causal influence, and ii) determination of the sign and magnitude of the influence between any two genes of the network. The solutions obtained from these two sub-problems are then combined to accurately identify the both non-existing connections, and the sign and magnitude of weights to existing connections. Computer simulation reveals that the proposed realization outperforms the most recently reported work in this field in detecting the sign and magnitude and also the structure of the overall network.

1 Introduction

Genes are the key element in biological chromosomes that control the behaviour and characteristics of living creatures [46], [8]. Experiments undertaken by the biologists reveal that some genes of a species cannot function independently. There exist regulatory genes in De-oxy-ribo-nucleic-acid (DNA) that control the behaviour of the other genes. However, understanding the interaction of genes is not a simple problem as hardly any straightforward approach exists to determine their dependence until date.

Debasish Datta · Amit Konar · Swagatam Das
Department of Electronics and Telecommunication Engineering, Jadavpur University, WB, India

B.K. Panigrahi
Department of Electrical Engineering, IIT Delhi, India - 110016

B.K. Panigrahi, Y. Shi, and M.-H. Lim (Eds.): Handbook of Swarm Intelligence, ALO 8, pp. 517–542.
springerlink.com

1.1 Preliminaries

This regulatory mechanism is important from the point of view of genetic engineering [9], [49], as it provides insight into the interaction between different genes. Gene expression profiling or micro array analysis [45] has enabled us to measure thousands of genes in a single RNA (ribo-nucleic acid) sample. There are a variety of micro array platforms that have been developed to accomplish this gene expression profiling, and the basic idea for each is simple. A glass slide or membrane is spotted or "arrayed" with DNA fragments or oligonucleotides that represent a specific gene-coding region. Purified RNA is then fluorescently- or radioactively labelled, and hybridised to the slide/membrane. In some cases, hybridisation is done simultaneously with reference RNA to facilitate the comparison of data across multiple experiments. After thorough washing, the raw data is obtained by laser scanning or auto radiographic imaging. At this point, the data is entered into a database, called micro-array data, and analysed by a number of statistical/computational methods.

1.2 The Problem

This chapter proposes a novel approach to model GRN as a fuzzy recurrent neural network, and then attempts to determine the signed connection weights of the network by an evolutionary optimization algorithm. Since the dependence does not exist between any random pair of genes in a DNA, a neural net representation with exhaustive connectivity sometimes misleads the problem. In order to correctly determine the structure of a virtual GRN, we attempted to derive a binary solution to the identification problem where the two levels (different values e.g. 1 and 0) represent the presence or absence of connectivity between any two nodes of a neural net. Consequently causal dependence between any two genes in a GRN can be identified by this approach. However, the problem that still remains is to determine the magnitude and the sign of the weights of the neural net resembling the GRN. It won't be out of place to mention here that the sign indicates positive or negative influence, while the magnitude represents the strength of causal influence. Apparently, because of multiplicity of solution, the two sets of solution obtained for the weights sometimes do not have conformity. In other words a weight w_{ij} between neurons i and j, which was obtained zero, by one approach may occasionally yield a larger signed value by the other approach. To overcome this problem, we consider an aggregation of the results obtained from the above two sets of solutions, and finally provide a unique solution to this problem. A system validation is undertaken to validate the results against a well-known benchmark problem, and the results indicate a strong support to the algorithm with appreciable accuracy, outperforming the recently published results [7], [56] for this problem.

The distinctive features of this work lies in formulation of the problem by a recurrent neural net in order to represent circular dependence of gene interaction in the GRN. The fuzzy weights in the present context improve the search efficiency in the evolutionary algorithm, thereby reducing the computational overhead of the problem than the other realizations undertaken by the previous researchers. The Particle Swarm Optimization (PSO) algorithm [6], [39], has been employed here to minimize a cost (objective) function that represents the mean square error of the actual response with computational response of the genes. Overall computational overhead of the proposed technique discussed in this chapter thus is significantly low because of the incorporation of fuzzy parameters. The accuracies of the estimated network weights are ascertained through unrestricted connectionism among the nodes in a recurrent neural topology. Such structure with respective to specific cost function have no restriction on the solution in [- ∞, + ∞]. Lastly, the aggregation structural finding with computed values of signed weights eliminates the chances of missing nonexistent connections in the neural topology. Consequently independence between genes in the network in the GRN can be preserved by the proposed approach, thereby improving the quality of solution in the problem.

1.3 Review

Researchers are taking a keen interest in modelling GRN from this gene micro array data by soft computing techniques [49], [35], [34] and have attempted several approaches, including Boolean networks [2], [25], Linear differential model [5], [33] Bayesian networks [14], [20], linear additive regulation models [11], [48], and the like. A brief review of these models is given below.

Boolean networks consider two states of a gene: 1 for active and 0 for inactive. Boolean functions are used to describe the state change of a gene due to the interaction of other genes. The main drawback of Boolean network lies in ignorance of the effect of genes at intermediate levels, causing information loss because of binary quantization process. In addition, Boolean networks presume that the transition between genes' activation states are synchronous, which is biologically implausible. However, a study of dynamic behaviour modelling using the Boolean network is under investigation [16], [17].

Bayesian models have also been employed to represent GRN. A GRN under the Bayesian framework is considered a directed acyclic graph, where the vertices represent genes, and the edges represent the conditional dependence relations between a pair of genes. Bayesian networks are effective in handling noise, incompleteness, and stochastic nature of gene expression data. However, they are unable to capture the dynamical aspects of gene regulation. In recent times, dynamic Bayesian networks [10], [19] have been employed to model the temporal information of GRN. They are also capable of handling 'hidden variable', 'prior knowledge', and 'missing data'. Experiments with a large gene expression data set reveals that gene expressions in some time points are sometimes found missing [45]. Identification of the missing data in gene expressions thus is an open problem. The common approach to handle this problem is to employ Expectation-Maximizing (EM) algorithm [59]. Researchers used EM to determine the network parameters from incomplete data set when the structure of the network under consideration is known. But if the structure is unknown and the data is incomplete then EM is not a preferable method. To deal with this problem, researchers used biological knowledge such as transcription factor binding location data [59] along with gene expression to infer gene regulatory network. In [59], the authors proposed an expectation maximization based algorithm known as Structure Expectation Maximization (SEM), which make use of the transcriptional factor binding data in order to predict accurate structure from incomplete dataset.

Linear additive regulation models also are effective to describe the expression level of a gene at time t as a weighted sum of expression levels of other genes at time (t -1).

The main drawback of the linear additive network model lies in its incapability to capture nonlinear dynamics between gene regulations.

1.4 Sectional Classification

The chapter is organized as follows. In section 2, we propose a fuzzy extension to the existing framework of GRN model using recurrent neural network. Section 3 presents a scheme for generation of gene expression time series data for known GRN. We have generated the gene expressions of a 4-gene network, and graphically shown that they mimic the real gene expression time series data with some predefined values. In section 4, we propose a cost function, the minimization of which is needed to extract the network weights. In section 5, we describe the proposed inference method. In section 6, we presented the simulation results for a well-known 4-gene network, and for a larger network consisting of 10 genes, and in section 7 we used our model to simulate a GRN using real gene expression time series data of E.Coli. S.O.S DNA repair system.

2 The Proposed Model

GRN realized with recurrent neural network topology employs a gene dynamics given by (1).

$$T_i \frac{dg_i}{dt} = f(\sum_{j=1}^{Z} w_{ij} g_j - b_i) - k_i g_i \tag{1}$$

where,

g_i (t) is the expression of gene g_i at time t,

f(·) is a nonlinear sigmoid function , i.e. $f(x) = \frac{1}{1+e^{-x}}$,

k_i is a decay constant of ith gene,

T_i is the time constant for gene i,

b_i denotes the bias term for gene i, which is used to control the operating point in the linear region of the sigmoid function,

w_{ij} is the signed weight from gene i to gene j, representing the influence of gene j to gene i, and

Z is the number of genes present in the GRN under consideration.

Although apparent from section 1, we here emphasize the significance of fuzzy membership distribution to represent the weights and bias terms of the recurrent neural topology resembling the GRN. When the classical dynamics (1) is employed to determine the system parameters w_{ij} and b_i $(\forall i, j)$, the probability of selecting the right value of the parameter is insignificantly small for a wide selective range of the parameters. For example, if the range of w_{ij} is [-30, +30] then considering floating point weight with an interval of 0.1, the probability of selecting the right value of w_{ij} [= + 20 say] is 1/600. Naturally when evolutionary algorithm is used to adapt the weights, the probability that in the tth iteration the correct value of a weight w_{ij} will be determined is given by (2).

$$prob1={}^{N}C_t \times \left(\frac{1}{600}\right)^t \times \left(\frac{599}{600}\right)^{N-t} \tag{2}$$

where N = maximum number of iteration for which the program is run.

Alternatively consider a set $A=\{w_{ij}^1, w_{ij}^2, \ldots\ldots, w_{ij}^k\}$ where, w_{ij}^m (for m = 1 to k) denotes k distinct values of the weight w_{ij}. Then for w_{ij} = w_{ij}^m, we have a fuzzy membership $\mu_A(w_{ij}^m)$ in [0,1]. Now, if we presume that the $\mu_A(w_{ij}^m)$ are selected at a regular interval of 0.1, then the probability of correct selection of the membership $\mu_A(w_{ij}^m)$ (= 0.3 say) in the tth iteration of an evolutionary algorithm (to be introduced later) is given by $P(\mu_A(w_{ij}^m)) = {}^{N}C_t \times \left(\frac{1}{10}\right)^t \times \left(\frac{9}{10}\right)^{N-t}$, where N denotes maximum number of program iterations. Now the probability of selecting $\mu_A(w_{ij}^m)$ for all m (1 to k) together is given by (3).

$$\begin{aligned} prob2 &= \sum_{m=1}^{k} P(\mu_A(w_{ij}^m)) \\ &= k \times {}^{N}C_t \times \left(\frac{1}{10}\right)^t \times \left(\frac{9}{10}\right)^{N-t} \end{aligned} \tag{3}$$

It is apparent from (2) and (3) that prob2 is larger than prob1 by a factor of $k \times (60)^t$ approximately. This justifies the significance of fuzzy distribution of weights and bias term. Since a weight w_{ij} (or bias b_i) is distributed over a discrete set A, we need to de-fuzzyfy (decode) the weights (and bias term) by any typical de-fuzzyfication algorithm [24]. Here, we use the centroidal de-fuzzyfication algorithm where the de-fuzzyfication value of weight bias term denoted by $\overset{*}{w}_{ij}$ and $\overset{*}{b}_i$ are given by (4) and (5) respectively.

$$\overset{*}{w}_{ij} = \frac{\sum_{m=1}^{k} \left[w_{ij}^m \times \mu_A(w_{ij}^m, t)\right]}{\sum_{m=1}^{k} \mu_A(w_{ij}^m, t)} \tag{4}$$

$$\overset{*}{b}_i = \frac{\sum_{m=1}^{k} \left[b_i^m \times \mu_A(b_i^m, t)\right]}{\sum_{m=1}^{k} \mu_A(b_i^m, t)} \tag{5}$$

The additional term t in the argument of $\mu_A(w_{ij}^m, t)$ and $\mu_A(b_i^m, t)$ is used to represent the value of the parameter in iteration t of an evolutionary algorithm.

Example 1: De-fuzzyfication of weights

Let the weight w_{ij} is defined on a point set of three distinct values: 30, 0.001,-30. Consequently set A is given by $\{0.35 \mid 30, 0.89 \mid 0.001, 0.2 \mid -30\}$. Then on de-fuzzyfication by (4) we have

$$\overset{*}{w}_{ij} = \frac{30 \times 0.35 + 0.001 \times 0.89 - 30 \times 0.2}{0.35 + 0.89 + 0.2}$$
$$= 3.126$$

□

The ultimate model of the gene dynamics thus is restructured as

$$T_i \frac{dg_i}{dt} = f(\sum_{j=1}^{Z} \overset{*}{w}_{ij} g_j - \overset{*}{b}_i) - k_i g_i \tag{6}$$

where, $\overset{*}{w}_{ij}$ and $\overset{*}{b}_i$ represents the de-fuzzified value of weight and bias term respectively.

Since the available time series data usually are in discrete form we need to discretize the proposed gene dynamics (6). As in limiting case, the time derivative of g_i can be represented as

$$\frac{dg_i}{dt} = \lim_{\Delta t \to 0} \frac{g_i(t + \Delta t) - g_i(t)}{\Delta t} \tag{7}$$

the dynamics (6) on substitution of (7) reduces to

$$g_i(t+\Delta t) = \frac{\Delta t}{T_i} f(\sum_{j=1}^{Z} \overset{*}{w}_{ji} g_j(t) - \overset{*}{b}_i) + g_i(t)(1 - \frac{\Delta t}{T_i} k_i) \tag{8}$$

Equation (8) demonstrates the expression of a particular gene i at time $(t + \Delta t)$, as a function of its as well as other genes' response at time t.

Recurrent neural network (RNN) allows feedback from output of the neurons to their inputs. Such topological structure of RNN provides dynamic aspect which is most essential for Gene regulatory network. RNN model in the present context has been employed to represent the change of expression of genes; so it seems to be naturally equivalent with the real biologically observed Gene regulatory network. These are the main reasons of choosing RNN for the gene dynamics of this study.

In the vast literature of recurrent neural network, sigmoid function is a common choice as nonlinear function [4]. Some recent research works [7], [56] have shown that sigmoid function can successfully model the nonlinearity of gene regulation. We have chosen sigmoid function over other nonlinear function such as hill climbing nonlinearities, and tan hyperbolic functions due to the following reasons as indicated in [61]. Sigmoid function has a long history of using with neural network. It is a continuous function and is a solution to a particular differential equation. The function was also used as one of many possible smooth, monotonic "squashing" function defined on a bounded interval in many research works. Mainly due to its probabilistic properties related to classification it is chosen to use with neural networks. Hill function is described by the expression f(x) = 1- exp (-x) which provides same kind of output characteristics as of sigmoid function, except that the function produces zero while x becomes zero. This limitation make hill function infeasible for modelling gene dynamics. Tan hyperbolic function provides output in the range -1 to +1, but the real world gene expression data cannot be negative. Hence tan hyperbolic function is not feasible for gene dynamics modelling.

3 Gene Expression Generation for Known Network

In section 3, we generate artificial gene expression time series data to test the accuracy of our method. Using the proposed model in (8), we generate time series data with the parameter values of Table 1 for a 4-gene network;

Table 1. Parameter values of a 4-gene network

	Gene1	Gene2	Gene3	Gene4	b_i	T_i
Gene1	20.0	-20.0	0.0	0.0	0.0	10.0
Gene2	15.0	-10.0	0.0	0.0	-5.0	5.0
Gene3	0.0	-8.0	12.0	0.0	0.0	5.0
Gene4	0.0	0.0	8.0	-12.0	0.0	5.0

Parameters in Table 1 include gene-to-gene connection weights, bias terms and time constants. The first (4 x 4) sub-matrix refers to the connection weights. For example, box (1, 2) = -20.00 means that gene 2 has -20.00 unit of effect on gene1, i.e. we assume that the rows are regulated by the columns and hence they (columns) are regulator. We have chosen the same set of values as in [52] to compare the accuracy of our model with the results of the existing literature. The bias term b_i and time constants T_i refer to row-indexed genes' attributes. The time series data generated by expression (8) for the 4- genes are given in Fig.1.

In Fig.1 the x-axis represents the time observed points, and the y-axis represents gene expressions. It is seen that nearly after100 points, the expression of almost all the genes get saturated. Therefore this is the region from where we can extract maximum information. Because of this reason we have selected 100 data points for each gene profile. In our experiment, we have generated gene expression using five different time durations (i.e. five gene expressions time series data for each gene). The justification of using different time duration instead of taking more number of points from a single graph is that, there is some change in trajectory of the gene expression for different time duration, and taking values from graph having different trajectory convey more information than more point from a single graph.

Though in this section we presented the simulated gene expression of the 4- gene network, but we have tested our proposed model using a 10- gene network to show that it works well even for a large network. The procedure of the generation of the

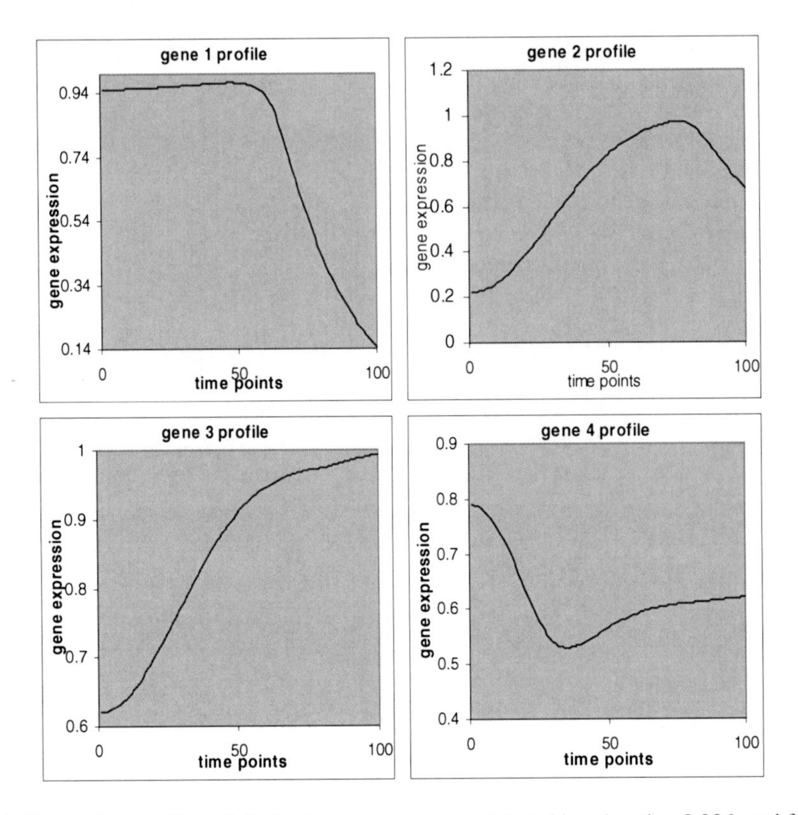

Fig. 1. Expression profiles of all the four genes generated by taking duration 0.006, and for 100 time points.

gene expression data for this 10-gene network is same as described for 4- gene network. For this reason we have not shown the simulated expression data. The used network of the 10-genes is shown in section 4.

4 Formulation as an Optimization Problem

In this section we present GRN identification as an optimization problem, and use an evolutionary algorithm to find the solution to this problem. This, however, requires construction of an objective function satisfying the following two fundamental objectives. The first objective ensures accuracy of the existing connection weights in the network, while the second is concerned with the accuracy in the skeletal structure of the simulated network. In other words, the first objective deals with the tuning of the network weights and the second is used to detect the sparse structure of the network topology.

Let $(g_{i,j}^{k})_{cal}$ and $(g_{i,j}^{k})_{ref}$ denotes the calculated and the reference gene expression. Then we define an error norm that minimizes the deviation of $(g_{i,j}^{k})_{cal}$ with $(g_{i,j}^{k})_{ref}$ for $\forall$ i , j, k. One possible way to represent this measure is to take the sum of the mean square deviation of error $\forall i \in [1,L]$, $\forall j \in [1,S]$ and $\forall k \in [1,M]$ where L is the number of gene, S is the number of time points in each time series gene expression data, and M is the number of time series used. Expression (9) is developed using the above idea.

$$error = \sum_{k=1}^{M} \sum_{j=1}^{S} \sum_{i=1}^{L} \{(g_{i,j}^{k})_{cal} - (g_{i,j}^{k})_{ref}\}^2 \qquad (9)$$

It may be noted that the evaluation of $(g_{i,j}^{k})_{cal}$ in (9) is evaluated by equation (8) by adding additional superscript k and subscripts j and cal to $g_i(t+\Delta t)$ in (8). The subscripts and superscripts of $(g_{i,j}^{k})_{ref}$ in (9) have the similar nomenclature to $(g_{i,j}^{k})_{cal}$ and thus needs no explanation.

The second objective that attempts to determine the sparse connection weights of the GRN is given by expression (10).

$$C = p \sum_{i=1}^{L} \sum_{j=1}^{L} \frac{|w_{ij}^{*}|}{1 + |w_{ij}^{*}|} \qquad (10)$$

The $\sum |w_{ij}|$ in the numerator of objective function (10) apparently ensures that some of the connection weights should be close enough to zero. However, when the second objective function (expression10) is added to first (expression 9), we should have a parity in their order of magnitude. This is achieved by attaching a denominator term $1+|w_{ij}^{*}|$ and a scale factor p to expression (10). Experimental studies with large scale data reveals that the best choice of p is 0.2 for the simulated gene expression data used in Table 1. The overall objective function is thus obtained by adding these two objective functions. This is shown in expression (11).

$$\begin{aligned} \text{cost} &= \text{error} + C \\ \text{cost} &= \sum_{k=1}^{M} \sum_{j=1}^{S} \sum_{i=1}^{L} \{(g_{i,j}^{k})_{cal} - (g_{i,j}^{k})_{ref}\}^2 \\ &+ p \sum_{i=1}^{L} \sum_{j=1}^{L} \frac{|w_{ij}^{*}|}{1 + |w_{ij}^{*}|} \end{aligned} \qquad (11)$$

The GRN identification thus remains as a minimization problem of the above cost function.

5 Solving the Optimisation Problem by Particle Swarm Optimization

In this section we briefly discuss the particle swarm optimization algorithm which is widely being used for optimization of nonlinear and rough objective function.

5.1 Brief Overview on Particle Swarm Optimization

Particle Swarm Optimization algorithm (PSO) [6] is inspired by the behaviour of the biological creatures, such as eagles to identify food in a given environment. It has been noted that an eagle joins a team with the largest eagle population in a forest with the help of having a good share of food. James Kennedy, Russell and Eberhart in 1995 employed this bio-inspired phenomenon to model the dynamics of stochastic particle with an ultimate aim to optimize non-linear functions of significant complexity.

PSO in principle is a parallel multi-agent search technique. Particles are conceptual entities which fly through the multi-dimensional search space. At any particular instant, each particle has a position and a velocity. The position vector of a particle with respect to the origin of the search space represents a trial solution of the search problem. At the beginning, a population of particles is initialized with random positions marked by vectors $\vec{X}_i$ and random velocities $\vec{V}_i$. The population of such particles is called a 'swarm' S. A neighbourhood relation N is defined in the swarm. N determines for any two particles P_i and P_j whether they are neighbours or not. Thus for any particle P, a neighbourhood can be assigned as N (P), containing all the neighbour of the particle. Different neighbourhood topologies and their effect on the swarm performance have been discussed in. The PSO used in this work, implicitly uses a so-called fully connected neighbourhood topology (or gbest). Every particle is a neighbour of every other particle.

Each particle P has two state variables:

- Its current position $\vec{x}(t)$
- Its current velocity $\vec{v}(t)$

And also a small memory comprising,

- Its previous best position $\vec{p}(t)$ i.e. personal best experience in terms of the objective function value $f(\vec{p}(t))$.

The best $\vec{p}(t)$ of all $P \in N$ (P): $\vec{g}(t)$ i.e. the best position found so far in the neighbourhood of the particle. The PSO scheme has the following algorithmic parameters:

- V_{max} or maximum velocity which restricts $\vec{V}_i(t)$ within the interval $[-v_{max}, v_{max}]$.
- An internal weight factor ω.
- Two uniformly distributed random numbers φ_1 and φ_2 which respectively determine the influence of $\vec{p}(t)$ and $\vec{g}(t)$ on the velocity update formula.
- Two constant multiplier terms C_1 and C_2 known as "self confidence" and "swarm confidence" respectively.

Initially the settings for $\vec{p}(t)$ and $\vec{g}(t)$ are $\vec{p}(0) = \vec{g}(0) = \vec{x}(0)$ for all particles.

Once the particles are initialized, the iterative optimization process begins where the position and velocities of all the particles are altered by the following recursive equation. The equations are presented for the d-th dimension of the position and velocity of the i-th particle.

$$V_{id}(t+1) = \omega V_{id}(t) + C_1\varphi_1.(P_d(t) - X_{id}(t)) + C_2\varphi_2.(P_d(t) - X_{id}(t))$$
$$X_{id}(t+1) = X_{id}(t) + V_{id}(t+1)$$

The first term in the velocity updating formula represents the initial velocity of the particle. The second term involving $\vec{p}(t)$ represents the personal experience of each particle and is referred to as "cognitive part". The last term of the relation is represented as the "social term" which represents how an individual particle is influenced by the other members of its velocity.

After having calculated the velocities and position for the next time steps t+1, the first iteration of the algorithm is completed. Typically, this process is iterated for a certain number of time steps, or until some acceptable solution has been found by the algorithm or until an upper limit of CPU usages has been reached. The algorithm can be summarized in the following pseudo code.

Procedure Particle_Swarm_optimization
Set t = 0;
Initialize φ_1 and φ_2, V_{max} and define N;
While (terminating_condition = FALSE)
{
$\forall p \in S$: Calculate $\vec{v}(t+1)$ and $\vec{x}(t+1)$ using the corresponding updating formula;
$\forall p \in S$: Update $\vec{p}(t+1)$ with $\vec{x}(t+1)$ if $f(\vec{x}(t+1))$ is better than $f(\vec{x}(t))$;
$\forall p \in S$: Update $\vec{g}(t+1)$ with the best $\vec{p}(t+1)$ in N (p);
}

Once the iterations are terminated, most of the particles are expected to converge to a small radius surrounding the global optima of the search space.

5.2 Solving the Gene Regulatory Network Identification by Particle Swarm Optimization

Here, we consider a population size of 70 where each particle vector includes fuzzified weights, fuzzified bias terms and time constants. The particle vectors are designated here as $\vec{X}_i(t)$ for i = [1, 2, 3, 4,……., NP]. Each particle represents a complete solution of the GRN under consideration. For example for a for a 4-gene network we have 16 weights, 4 bias terms, and 4 - time constants. Suppose each weight and bias term have a five point discrete fuzzy membership distribution, where the time constants are represented as nonfuzzy floating point numbers. Thus for the given 4-gene network of Fig.2, we have altogether 16×5+4×5+4=104 fields. The components of an individual solution (particle vector) of PSO are illustrated in Fig.3.

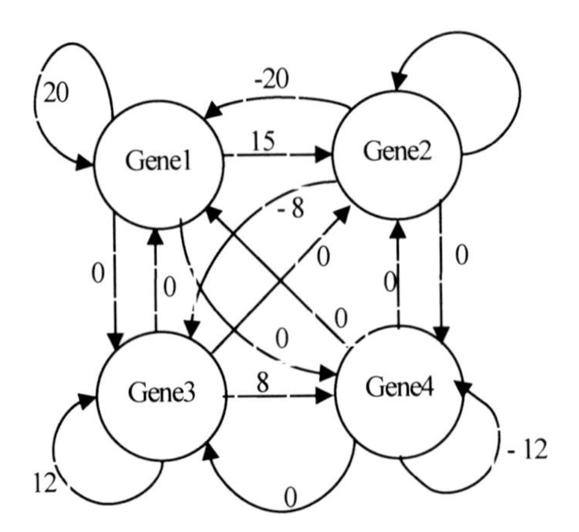

Fig. 2. 4-gene network of Table 1

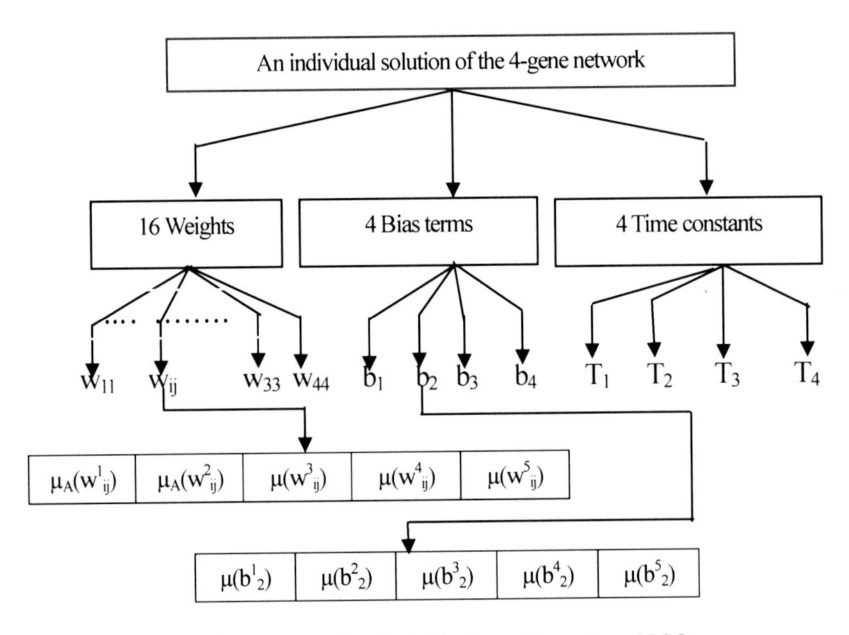

Fig. 3. Structure of an individual particle vector of PSO

It may be added here that the fuzzy set A in the present context is a five discrete point set with membership values in [0, 1]. The general form in fuzzy set A for a weight w_{ij} considered in this work is given below

$$A=\{\mu_A(w_{ij}^1)\,|\,w_{ij}^1, \mu_A(w_{ij}^2)\,|\,w_{ij}^2, \mu_A(w_{ij}^3)\,|\,w_{ij}^3, \mu_A(w_{ij}^4)\,|\,w_{ij}^4, \mu_A(w_{ij}^5)\,|\,w_{ij}^5\}$$
$$= \{-30\,|\,\mu_A(-30), -15\,|\,\mu_A(-15), 0.001\,|\,\mu_A(0.001), 15\,|\,\mu_A(15), 30\,|\,\mu_A(30)\}$$

In other words, we divided the interval [-30, 30] into five equal length, and thus obtained the point set {-30,-15, 0, 15, 30}. The discrete point zero is replaced by 0.001 as the numeric value zero has no effect on the result in the defuzzification process.

We initialise the membership value of each discrete point dividing a weight randomly in the interval [0,1]. After the initialization step is over, we calculate the velocity and the position of each individual particle in the swarm S. Update the individual and global best position as directed by the value of the fitness function.

The PSO program was executed for 5000 iterations, and the root mean square value of the error vector, obtained by taking the component wise difference between the present and previous position vectors for each particle, below a user-defined threshold is maintained. In case, all the position vectors satisfy the later condition, we terminate the algorithm once the condition is attained, even if 5000 iterations are not over yet. The resulting GRN obtained by the above algorithm for the best -fit particle is given in Fig.4.

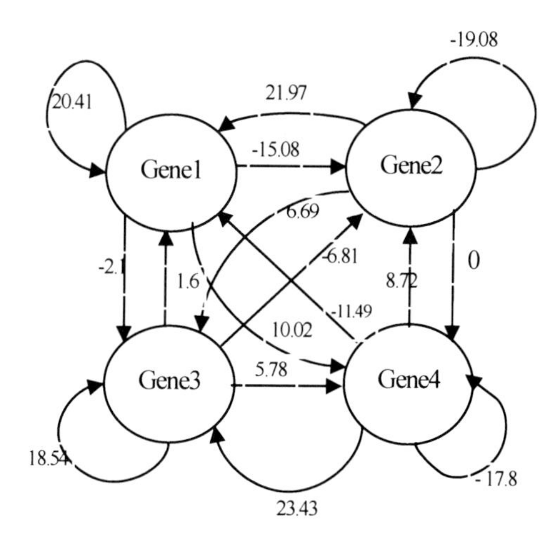

Fig. 4. Found best fit solution

5.3 Topology Finding Particle Swarm Optimization

It is apparent from Fig. 4 that some of the weights do not tally with the desired nonexistent connection of Fig. 2. Such situation occurs because of the possibility of multiplicity of GRN identification problem. One way to eliminate the nonexistent connection weights between any two genes is to binnarize the weights, where the binary value one indicates an existent connection and zero indicates non-existence of a weight. We used a binary version of PSO same as described in [22] (called topology finding PSO) in this section to accomplish this binarization of weights. The basic idea of introduction of the topology finding PSO lies in the detection of the nonexistent connections (zero connections) of the network. For this purpose, we first attempt a heuristic based local (hbl) search. In hbl, we set the values of each field of the best solution found using algorithm of the previous section (5.2) to zero one after another for the weights and bias terms. If that change improves the result (i.e. lower the value of the expression (11)), we allow the change, otherwise we disallow it. Then we repeat the experiment with the same hbl, but this time we set the value of two fields to zero simultaneously. In the second application of hbl, we select two fields consecutively first and then randomly. The process is repeated by selecting more number of fields simultaneously. But use of more number of fields simultaneously does not improve the solution. It should be mentioned here that steepest descent learning algorithm [53], [32] could also be used for this purpose. But we have not applied steepest descent algorithm here because of the fact that this algorithm fails when there is discontinuity in the used fitness function. Then we tried evolutionary algorithm like PSO for this purposed. We find that this technique produces the best result among the techniques we studied.

As discussed in the previous section (5.2) each particle vector includes three sets of information weights, bias terms, and time constants. Therefore the number of fields in a particle vector is $Z^2+2\times Z$ for a GRN consisting of Z number of genes. In the topology finding PSO, we used particle vectors having only (Z^2+Z) number of fields. We consider such an arrangement due to the reason that the last four fields of any particle vector corresponds to the time constants of the GRN under consideration, and time constants are used as denominator in the proposed gene dynamics (equation 8).

After the algorithm discussed in the previous section (5.2) converges, the best parameter vector is selected, which can be represented in a matrix form W as shown in Fig.5.

$W =$

From \ To	Connection weights (w_{ij})				bias and time constants	
	G1	G2	G3	G4	b	T
G1	21.41	−15.08	−2.1	10.2	2.99	14.83
G2	21.97	−19.08	6.69	0.0	−1.0	5.33
G3	1.6	−6.81	18.54	5.78	−222	4.08
G4	−11.49	8.72	23.43	−17.84	4.01	5.8

Fig. 5. An example of the best found solution of the algorithm of section 5.2

The best solution found from the algorithm discussed in section (5.2) is used here in topology detection PSO. In order to describe the topology detection PSO we will use a separate nomenclature. Here each particle $x_i = (x_{i1}, x_{i2}, \ldots\ldots, x_{id})$ $(i \in (1, Z))$ have velocity $v_i = (v_{i1}, v_{i2}, \ldots\ldots, v_{id})$ where $d=Z^2{}_{+}Z$. First Z^2 fields of position and velocity vector represents the connection weights among the genes (e.g. $w_{11}, w_{12}, w_{13}, \ldots, w_{1Z}, w_{21}, \ldots.. w_{2Z}, \ldots.. w_{ZZ}$) and last four field represents the bias terms (e.g. $b_1, b_2, \ldots\ldots, b_Z$). The main difference between classical PSO and topology finding PSO (used in this section) lies in the interpretation of the position vector of each particle as described in [23]. The position vector is updated using the formula described in (12).

$$x_{ij} = \begin{cases} 0 & \text{if } r_i(t) \geq fun(v_{ij}(t)) \\ 1 & \text{if } r_i(t) < fun(v_{ij}(t)) \end{cases} \quad (12)$$

where, $fun(v_{ij}(t)) = \dfrac{1}{1+e^{v_{ij}(t)}}$,

$x_{ij}(t)$ is the value j^{th} parameter of ith particle at iteration (time step) t.
$v_{ij}(t)$ is the corresponding velocity, and
$r_i(t) \sim U(0,1)$.

The traditional PSO velocity update equations remain same. A value one in x_{ij} represents the existence and zero value represents the nonexistence of a connection weight. Except the difference described above the algorithm of topology detection PSO is same as of the classical PSO described in section (5.1).

6 The Results

In this section, we provide the result of our inference method using the simulated gene expression data. First, we compare the results of our inference technique with and without the topology finding module of the 4-gene network introduced in section 3, to show that the addition of this topology finding module increase the accuracy of the whole inference technique. Secondly, we compare the result of our inference technique with the result of [56]. We have chosen [56] because in this paper, the authors performed a comprehensive study using the 4-gene network dataset, and provided a well-documented result. We found that our inference method improves the results both in detecting the existing connections, and the nonexistent connections. Sign of almost all the existent connection recovered correctly. Also the value of the existent connection is found with appreciable precision. Thirdly, experiments with 10 genes network reveals that our inference method works well even with large dataset. The bar chart in Fig. 6 shows that the average error (defined in equation (13)) reduces significantly due to incorporation of the topology detection module.

$$average\ error = \frac{\text{cost}}{Np} \tag{13}$$

In expression (13) cost is obtained from expression (11), and *Np* is the number of candidate solution present in the population pool of the proposed algorithm.

Table 2. Identified network without topology detection scheme, for population = 70, set cardinality = 5, average error for each individual solution = 2.37

	Gene1	Gene2	Gene3	Gene4	b_i	T_i
Gene1	21.41	-15.08	-2.1	10.02	2.99	14.83
Gene2	21.97	-19.08	6.69	0.0	-1.00	5.33
Gene3	1.6	-6.81	18.54	5.78	-22.2	4.08
Gene4	-11.49	8.72	23.43	-17.84	4.01	5.8

Table 3. Identified Network Using Frnn with Topology Detection PSO, for Population = 70, Set Cardinality = 5, Average Error for Each Individual Solution = 2.17, Average Error Reduction = 9.2%

	Gene1	Gene2	Gene3	Gene4	b_i	T_i
Gene1	21.41	-15.08	-2.1	10.02	2.99	14.83
Gene2	21.97	-19.08	6.69	0.00	-1.00	5.33
Gene3	0.00	-6.81	18.54	0.00	0.00	4.08
Gene4	-11.49	8.72	23.43	-17.84	4.01	5.8

As we can see that from Table 2, without the topology detection scheme, the inference algorithm identified one nonexistent connection and all the signs of the existent connections. Now, after applying the topology finding module, it identified two more nonexistent connections correctly along with one zero value bias term, as shown in Table 3. Also the average error for each candidate solution decreases considerably. If we consider only the nonexistent connection and the signs of the existent connections then the result can be understood more precisely. From Table 5 to Table 6 we have shown the results in that way. For better understanding we have shown the nonexistent connections and the signs of the existent connections of Table 1, in Table 4.

Table 4. Nonexistant connections and signs of existant connections of Table 1

	Gene1	Gene2	Gene3	Gene4	b_i	T_i
Gene1	+	-	0.0	0.0	0.0	10.0
Gene2	+	-	0.0	0.0	-5.0	5.0
Gene3	0.0	-	+	0.0	0.0	5.0
Gene4	0.0	0.0	+	-	0.0	5.0

Table 5. Nonexistant connections and signs of existant connections of Table 2

	Gene1	Gene2	Gene3	Gene4	b_i	T_i
Gene1	+	-	-2.1	10.02	2.99	14.83
Gene2	+	-	6.69	0.0	-1.00	5.33
Gene3	1.6	-	+	5.78	-22.2	4.08
Gene4	-11.49	8.72	+	-	4.01	5.8

Table 6. Nonexistant connections and signs of existant connections of Table 3

	Gene1	Gene2	Gene3	Gene4	b_i	T_i
Gene1	+	-	-2.1	10.02	2.99	14.83
Gene2	+	-	6.69	0.00	-1.00	5.33
Gene3	0.00	-	+	0.00	0.00	4.08
Gene4	-11.49	8.72	+	-	4.01	5.8

The result of Table 5 and Table 7 proves our claim. Also it can be seen from Table 2 to Table 3 that the identified existent connections are also detected with sufficient accuracy, which is a challenge of the current literature. For set cardinality 7 we are showing the result from Table 7 to Table 9.

Table 7. Identified network using FRNN without topology detection PSO, for population = 70, set cardinality = 7, average error for each individual solution = 1.39

	Gene1	Gene2	Gene3	Gene4	b_i	T_i
Gene1	15.42	-17.05	-2.74	-4.11	-6.16	10.84
Gene2	22.87	-16.18	2.62	8.7	-0.96	5.31
Gene3	-5.27	-0.50	1.2	17.7	-3.86	6.19
Gene4	8.2	16.37	20.92	-12.16	7.9	3.58

Table 8. Identified network using FRNN with topology detection scheme, for population = 70, set cardinality = 7, and average error for each individual solution = 1.13, average error reduction = 23%

	Gene1	Gene2	Gene3	Gene4	b_i	T_i
Gene1	15.42	-17.05	-2.74	-4.11	-6.16	10.84
Gene2	22.87	-16.18	2.62	8.7	-0.96	5.31
Gene3	0.00	-0.50	1.2	17.7	0.0	6.19
Gene4	0.00	16.37	20.92	-12.16	7.9	3.58

As can be seen from Table 8 the added topology finding module improves the solution quality largely for set cardinality 7. We show the nonexistent connection and the sign of the existent connection of Table 8 in Table 9.

Table 9. Nonexistant connection and sign of the existant connection of Table 8.

	Gene1	Gene2	Gene3	Gene4	b_i	T_i
Gene1	+	-	-2.74	-4.11	-6.16	10.84
Gene2	+	-	2.62	8.7	-0.96	5.31
Gene3	0.00	-	+	17.7	0.0	6.19
Gene4	0.00	16.37	+	-	7.9	3.58

For set cardinality 9, we have shown the result from Table 10 to Table 12.

Table 10. Identified network using FRNN with topology detection scheme, for population = 50, set cardinality = 9, and average error for each individual solution = 0.9466

	Gene1	Gene2	Gene3	Gene4	b_i	T_i
Gene1	19.83	-19.9	0.38	6.86	1.95	10.54
Gene2	20.37	-13.97	9.38	2.87	4.53	4.6
Gene3	-16.38	-16.16	10.55	18.87	-26.2	4.5
Gene4	0.78	-0.37	18.51	-28.9	-2.9	6.5

Table 11. Identified network using FRNN with topology detection scheme, for population = 50, set cardinality = 9, and average error for each individual solution = 0.8446, average error reduction = 11.84%

	Gene1	Gene2	Gene3	Gene4	b_i	T_i
Gene1	19.83	-19.9	0.00	6.86	1.95	10.54
Gene2	20.37	-13.97	9.38	2.87	4.53	4.6
Gene3	-16.38	-16.16	8.01	18.87	0.00	4.5
Gene4	0.00	0.00	18.51	-28.9	-2.9	6.5

The nonexistent connection and the sign of the existent connections of Table 11 are shown in Table 12.

Table 12. Nonexistant connection and the sign of the existant connection of Table 11.

	Gene1	Gene2	Gene3	Gene4	b_i	T_i
Gene1	+	-	0.00	6.86	1.95	10.54
Gene2	+	-	9.38	2.87	4.53	4.6
Gene3	-16.38	-	+	18.87	0.00	4.5
Gene4	0.00	0.00	+	-	-2.9	6.5

We have compared the result of our inference technique of Table 12 with the result of [56] as discussed above. For convenience we have shown the result obtained in [56] in Table 13

Table 13. Parameter obtain in [56]

	Gene1	Gene2	Gene3	Gene4
Gene1	+(29.0)	+(34.0)[wrong]	0.0(5.5)	0.0(7.0)
Gene2	+(83.5)	-(98.0)	0.0(3.5)	0.0(3.0)
Gene3	0.0(16.5)	-(60.5)	+(99.5)	0.0(15.0)
Gene4	0.0(10.0)	0.0(8.5)	+(100.0)	- (100.0)

The result shown in Table 13 was obtained using 5-time series data. In this table, the values in the parentheses indicate the percentage of occurrence of connections in the network within 200 runs. Clearly only two signs of the weights are determined with sufficient accuracy these are (Gene4, Gena3) and (Gene4, Gene4), and above all those values are among 200 runs. Therefore, no specific knowledge can be gained from that. On the contrary, the values of weights obtained in our method are due to execution of a single run, and from that one can easily have an idea about the nature of interaction among the genes. Comparing results of Table 12 with Table 13 it can be seen that our inference technique improves the result, and also the value of the detected existent connection are close to the original value.

In this chapter, we use a 10 - gene network to test our model. This database of Table 14 is chosen randomly. To the best of our knowledge testing of GRN

Table 14. Used 10 gene network

	G_1	G_2	G_3	G_4	G_5	G_6	G_7	G_8	G_9	G_{10}	b_i	T_i
G_1	0	0	20	0	0	-25	0	0	0	0	0	5
G_2	25	0	0	0	-30	0	0	0	0	0	0	3
G_3	0	0	-28	0	0	29	0	0	0	27	0	3
G_4	0	0	0	0	30	-20	0	23	22	25	0	13
G_5	28	0	0	0	-25	0	20	0	0	0	0	3
G_6	0	0	-10	13	12	0	0	0	0	13	7	5
G_7	15	0	0	0	-20	0	13	0	0	0	0	3
G_8	0	0	13	0	-7	0	0	0	19	0	4	3
G_9	-27	0	0	0	18	14	0	0	0	0	0	3
G_{10}	0	0	19	25	0	0	0	-21	0	0	11	3

Table 15. Identified network using the proposed model without topology detection pso, population size = 50, average errorper candidate= 17.77

	G_1	G_2	G_3	G_4	G_5	G_6	G_7	G_8	G_9	G_{10}	b_i	T_i
G_1	-3.09	26.27	6.99	-15.03	-3.77	-8.07	16.85	-6.17	-17.07	8.04	11.06	10.75
G_2	8.86	3.22	-5.7	-12.96	-8.18	-3.61	9.39	-11.07	-4.98	5.04	19.5	3.18
G_3	-1.82	4.12	8.0	-1.53	-0.88	21.72	-10.15	-4.48	-4.47	-2.93	-4.33	5.76
G_4	-21.29	2.95	4.02	-20.57	11.74	-22.12	-1.86	-4.92	9.7	19.03	-6.73	14.58
G_5	4.49	23.78	-0.64	-11.41	14.33	-13.53	7.22	16.15	11.85	-2.41	13.82	17.69
G_6	5.55	19.35	10.95	-12.13	5.86	-2.99	8.41	9.63	7.82	7.91	-10.1	4.32
G_7	8.17	2.82	6.69	-22.55	3.32	-8.76	12.95	-12.63	-3.59	-30.78	-12.37	20.58
G_8	8.74	10.25	14.19	-7.58	-22.29	5.35	6.35	2.74	14.69	16.5	-4.6	4.31
G_9	-5.25	18.45	4.12	3.78	2.72	10.99	-8.37	-8.74	3.76	13.02	-22.4	12.9
G_{10}	-10.62	-9.16	21.42	0.47	8.64	3.37	8.36	-1.23	4.85	3.68	-3.78	2.59

Table 16. Identified network using the proposed model with topology detection pso, population size = 50, average errorper candidate= 15.2, average error reduction = 17.9%

	G_1	G_2	G_3	G_4	G_5	G_6	G_7	G_8	G_9	G_{10}	b_i	T_i
G_1	0.00	26.27	6.99	0.00	-3.77	-8.07	0.00	-6.17	0.00	0.00	0.0	10.75
G_2	0.00	3.22	-5.7	0.00	-8.18	-3.61	9.39	-11.07	0.00	0.00	0.0	3.18
G_3	-1.82	4.12	8.0	0.00	-0.88	21.72	0.00	0.00	-4.47	0.00	-4.33	5.76
G_4	0.0	0.0	4.02	0.0	11.74	-22.12	-1.86	-4.92	9.7	0.0	-6.73	14.58
G_5	4.49	23.78	0.0	0.0	0.0	-13.53	7.22	0.0	11.85	0.0	0.0	17.69
G_6	0.0	0.0	0.0	-12.13	5.86	-2.99	0.0	0.0	0.0	7.91	-10.1	4.32
G_7	8.17	2.82	0.0	-22.55	0.0	-8.76	12.29	-12.63	-3.59	-30.78	-12.37	20.58
G_8	8.74	0.0	14.19	-7.58	0.0	0.0	0.0	2.74	14.69	16.5	-4.6	4.31
G_9	-5.25	0.0	4.12	0.0	2.72	0.0	-8.37	0.0	0.0	0.0	0.0	12.9
G_{10}	-10.62	0.0	21.42	0.47	8.64	0.0	8.36	-1.23	0.0	0.0	-3.78	2.59

inference method by recurrent neural using such a large database is not done in the existing literature. The network used is shown in Table 14. This experiment using such a large database truly reflects the efficiency and the accuracy of the proposed model. We have used our inference technique with population size of 50, and set cardinality = 7. We used 100 time points for each gene and 5 set of time series data for time durations of 0.006, 0.009, 0.011, 0.013, and 0.015. The testing is done on a laptop containing 1GB of RAM and Pentium dual - Core processor (1.76 GHz).

The simulation code is written in C language. We have not optimized our code but still the total inference technique takes nearly 4 hours, which is an improvement from the existent literature [34]. The result of the reverse engineering is shown from Table 15 to 18.

The nonexistent connections and the sign of the existent connections are shown in Table 17.

Table 17. Nonexistant connection and the sign of the existant connection of Table 16.

	G_1	G_2	G_3	G_4	G_5	G_6	G_7
G_1	0.00	26.27	+	0.00	-3.77	-	0.00
G_2	0.00(wrong)	3.22	-5.7	0.00	-	-3.61	9.39
G_3	-1.82	4.12	+(wrong)	0.00	-0.88	+	0.00
G_4	0.0	0.0	4.02	0.0	+	-	-1.86
G_5	+	23.78	0.0	0.0	0.0(wrong)	-13.53	+
G_6	0.0	0.0	0.0(wrong)	-(wrong)	+	-2.99	0.0
G_7	+	2.82	0.0	-22.55	0.0(wrong)	-8.76	+
G_8	8.74	0.0	+	-7.58	0.0(wrong)	0.0	0.0
G_9	-	0.0	4.12	0.0	+	0.0(wrong)	-8.37
G_{10}	-10.62	0.0	+	+	8.64	0.0	8.36

Remaining part of Table 17

G_8	G_9	G_{10}	b_i	T_i
-6.17	0.00	0.00	0.0	10.75
-11.07	0.00	0.00	0.0	3.18
0.00	-4.47	0.00(wrong)	-4.33	5.76
-(wrong)	+	0.0(wrong)	-6.73	14.58
0.0	11.85	0.0	0.0	17.69
0.0	0.0	+	-10.1	4.32
-12.63	-3.59	-30.78	-12.37	20.58
2.74	+	16.5	-4.6	4.31
0.0	0.0	0.0	0.0	12.9
-	0.0	0.0	-3.78	2.59

As can be seen from Table 17 that, out of 31 existent connections 20 existent connections are detected correctly; 36 out of 69 nonexistent connections are identified correctly. Therefore as a total 56 connections are identified correctly. This result is shown in Table 18.

Table 18. Summary of the result of the Table 17.

Existent connection			Nonexistent connection		
Total number connection	Detected	Accuracy	Total number connection	Detected	Accuracy
31	20	64.5%	69	36	52.2%

In Fig.6 we have shown the change of average error of each individual candidate solution of our inference technique with respect to set cardinality of the weights. We perform this test with set cardinality = 5, 7, 9 for number of individual in the population = 70, and same number of iterations.

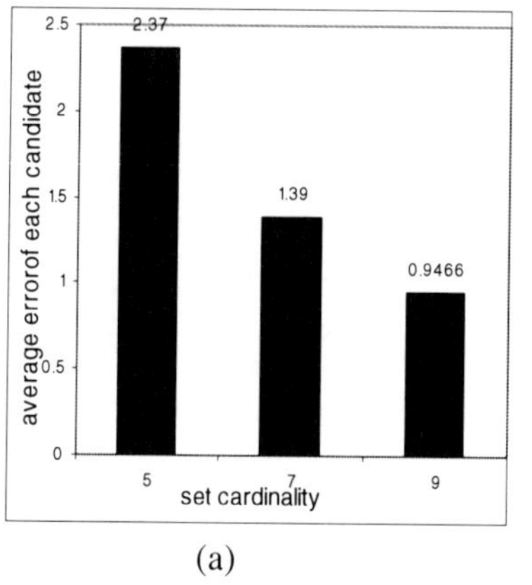

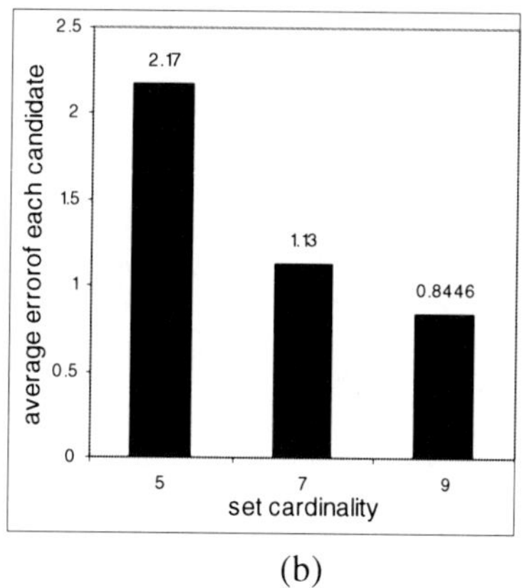

(a) (b)

Fig. 6. (a) Change of average error of each candidate with respect to weight distribution set cardinality without topology detection scheme, (b) Change of average error of each candidate with respect to weight distribution set cardinality with topology detection scheme.

It is clear from Fig. 6 that the average error decreases more rapidly in presence of proposed topology detection scheme. We calculated the average standard deviation (defined in expression (13)) of the weights obtained by our proposed technique over 30 runs for the 4-gene network of Section 3, and it comes out as 4.96. It proves that our proposed algorithm provides consistent result over several runs.

$$\text{average standard deviation} = \frac{\sum_{i=1}^{N}\sum_{j=1}^{N}\sigma_{ij}}{N^2} \qquad (14)$$

In expression (13) σ_{ij} represents the standard deviation of the weight w_{ij} over a specified number of runs of our proposed algorithm.

7 Identification of E.Coli. SOS DNA Repair Network

In this section, we infer the *E.Coli.* S.O.S DNA repair system using our proposed model. This is a well-studied network. This network consists of nearly 30 genes regulated at the transcription level. Four experiments have been conducted with different UV light intensities. Experiment 1, and 2 using UV= 5 jm^{-2}, and experiment 3, and 4 using UV=20jm^{-2}. Using these experiments, expressions of eight major genes have been documented. These genes are uvrD, lexA, umuD, recA, uvrA, uvrY, ruvA, and polB.

The function of the *E.Coli.* S.O.S. DNA repair network is as follows: lexA acts as a balancing factor for the whole network. In absence of any DNA damage it binds to the promoter region of the genes, suppressing the S.O.S genes in the network. When DNA damage occurs, recA (one of the S.O.S protein) becomes activated. It decreases

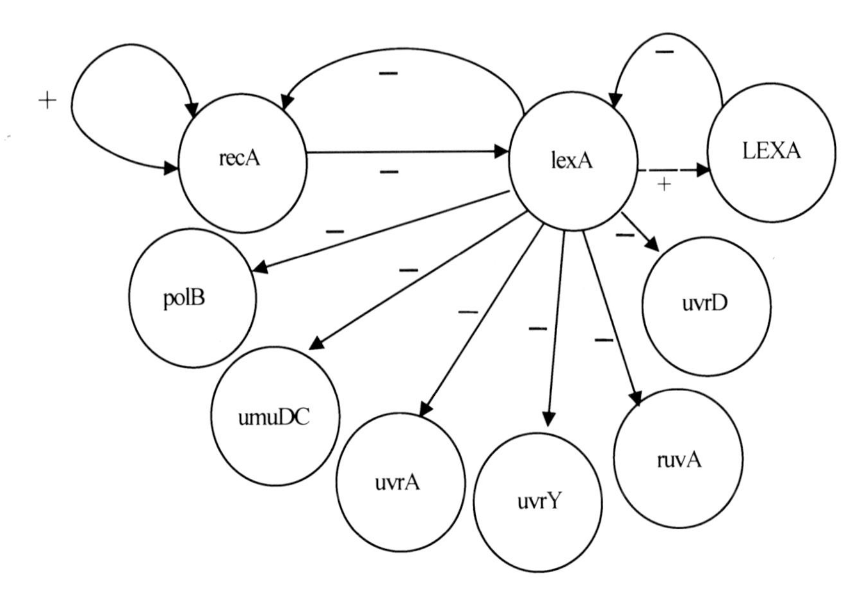

Fig. 7. E.Coli S.O.S. DNA repair network, activation is represented by '+' sign, and inhibitation by'-'; genes are written with small letter, and protein with capital letter

the level of lexA. As a result the S.O.S genes become activated. Once the damage has been repaired or bypassed, the levels of recA decrease. As a result, the level of lexA increases, and again deactivates the S.O.S genes. This regulatory mechanism is shown pictorially in Fig 7.

Gene profiles of *E.Coli* S.O.S DNA repair network consists of 50 data points [56], sampled every 6 minutes. The data set consists of four 8 × 50 matrix. Each column represents the observation of expression value at a particular time instant of eight genes, and each row represents the fifty expression value of a particular gene at

Table 19. Identified interaction matrix using proposed model, total time require for inference is 3 and half hour

	uvrD	lexA	umuDC	recA	uvrA	uvrY	ruvA	polB	b_i	T_i
uvrD	0.0	8.6	8.3	-5.1	0.0	0.0	0.0	0.0	0.0	15.4
lexA	-8.4	-14.2	-15.8	-16.23	0.0	- 4.4	9.4	8.4	-2.2	0.6
umuDC	0.0	19.7	0.74	0.0	0.0	0..1	0.0	13.3	9.9	4.6
recA	-10.1	2. 9	0,.54	-2. 4	0.0	0.0	-12	11.26	0.0	22.8
uvrA	0.0	0.0	0.0	-0.45	-7.6	-8	-3.4	0.0	-2.5	29.8
uvrY	-4.3	0.0	0.0	0.0	-14.1	0.0	0.0	0.0	0.0	0.63
ruvA	0.0	3.2	17	2	-2.5	0.0	-3.8	0.0	0.0	22.9
polB	-13.4	-11.5	0.0	0.0	0.0	0.0	-17.9	-6	-6.2	0.1

different time instants. It is one of the best data sets, which fits our model. We have conducted the same experiment, as with the above artificial data, and the identified interaction values between genes are represented in Table 19.

8 Conclusion

The work presented in this chapter is a major extension of the two research papers [56], [7]. The extension with respect to [56] is due to (i) incorporation of fuzzy weight and bias distribution, and (ii) topology detection scheme. The introduction of fuzzy distribution improves the accuracy in network weights without sacrificing computational cost. The topology detection scheme correctly identifies the nonexistent connections of network weights. A masking scheme has been adopted to keep the signed weights and eliminate the nonexistent weights by combining the results of fuzzy distribution based weight finding module and the topology detection module. The present work is also an extension of [7] due to the introduction of topology detection module and the masking scheme. Experiment with small gene network reveals that the proposed methodology of sign and nonexistent network weight extraction outperforms all reported works. Experiment with large 10-gene network further reveals that the signed weights and nonexistent connections can be retrieved to a level of 64.5 % and 52.2 % respectively. The principles of extraction of signed weights and nonexistent connections in an *E.Coli* S.O.S DNA repair network have been studied, and the results obtained support the biological evidences for the same system.

References

1. Akike, H.: Information Theory and an extension of the maximum likelihood Principle. In: Proc. Second int'l Symp. Information Theory, pp. 267–281 (1973)
2. Akutsu, T., Miyano, S., Kuhara, S.: Identification of genetic networks from a small number of gene expression patterns under the Boolean network model. In: Pac. Symp. Biocomput., pp. 17–28 (1999)
3. Bar-Joseph, Z.: Analyzing time series gene expression data. Bioinformatics 20(16), 2493–2503 (2004)
4. Bose, N.K., Liang, P.: Neural Network Fundamentals with Graphs. Algorithms, and Applications, p. 312. McGraw-Hill, New York (1996)
5. Chen, T., He, H.L., Church, G.M.: Modeling gene expression with differential equations. In: Pac. Symp. Biocomput., vol. 4, pp. 29–40 (1999)
6. Das, S., Abraham, A., Konar, A.: Metaheuristic Clustering, pp. 73–74. Springer, Heidelberg (2009)
7. Datta, D., Choudhuri, S.S., Konar, A., Nagar, A.K., Das, S.: A Recurrent Fuzzy Neural Model of a Gene Regulatory Network for Knowledge Extraction Using Differential Evolution. In: Proc. of IEEE Congress on Evolutionary Computation, Trondheim, Norway, May 18-21 (2009)
8. Dawkins, R.: The Selfish Gene. Oxford University Press, Oxford (1976)
9. De Jong, H.: Modeling and simulation of genetic regulatory systems: a literature review. Journal of Computational Biology 9, 67–103

10. D'haeseleer, P.: Reconstructing Gene Network from Large Scale Gene Expression Data. PhD dissertation, Univ. of New Mexico (2000)
11. D'haeseleer, P., Wen, X., Fuhrman, S., Somogyi, R.: Linear Modelling of mRNA Expression Levels during CNS Development and Injury. In: Proc. Pacific Symp. Bio. Computing, pp. 41–52 (1999)
12. Eberhart, R.C., Kennedy, J.: A new optimizer using particle swarm theory. In: Proc. of the Sixth Int. Symp. on Micro Machine and Human Science, Nayoga, Japan (1995)
13. Epinosa-soto, C., Padilla-Longoria, P., Alvarez-Buylla, E.R.: A Gene Regulatory Network Model for Cell-Fate Determination during Arabidopsis thaliana Flower Development That Is Robust and Recovers Experimental Gene Expression Profiles. In: The Plant Cell. American Society of Plant Biologists, vol. 16, pp. 2923–2939 (November 2004), http://www.plantcell.org
14. Friedman, N., Linial, M., Nachman, I., Pe'er, D.: Using Bayesian net work to analyze expression data. J. Comp. Biol. 7, 601–620 (2000)
15. Goldberg, D.E.: Genetic Algorithm in Search, Optimization and Machine Learning. Addison-Wesley, Reading (1989)
16. Hallinan, J., Wiles, J.: Evolving Genetic Regulatory Networks Using an Artificial Genome. In: Proc. Second Asia-Pacific Bioinformatics Conf., vol. 29, pp. 291–296 (2004)
17. Hallinan, J., Wiles, J.: Asynchronous Dynamics of an Artificial Genetic Regulatory Network. In: Proc. Ninth Int'l Conf. Simulation and Synthesis of Living Systems (2004)
18. Hassoun, M.H.: Fundamentals of Artificial neural network. MIT Press, Cambridge (1995)
19. Husmeier, D.: Sensitivity and Specificity of Inferring Genetic Regulatory Interactions from Micro array Experiments with Dynamic Bayesian Networks. Bioinformatics 19(17), 2271–2282 (2003)
20. Imoto, S., Gota, T., Miyano, S.: Estimation of genetic networks and functional structures between genes by using Bayesian networks and nonparametric regression. In: Pac. Symp. Biocomput., pp. 175–186 (2002)
21. Kennedy, J., Eberhart, R.C.: Swarm Intelligence. Morgan Kaufmann, San Francisco (2001)
22. Kennedy, J., Eberhart, R.C.: A discrete binary version of the particle swarm algorithm. In: Proc. Conf. on System, man, and Cybernetics, pp-, pp. 4104–4109 (1997)
23. Kim, J.-H., Lee, C.-H.: Multi-objective Evolutionary Process for Specific Personalities of artificial Creature. IEEE Computational Intelligence Magazine 3(1) (February 2008)
24. Kim, J.-H., Lee, K.-H., Kim, Y.-D., Park, I.-W.: Genetic Representation for Evolving Artificial Creature. In: Proc. of the IEEE Congress Evolutionary Computation, pp. 6838–6843 (2006)
25. Konar, A.: Computational Intelligence Principles, Techniques and Applications, pp. 119–120. Springer, Heidelberg (2009)
26. Liang, S., Fuhrman, S., Somogyi, R.: Reveal, a general reverse engineering algorithm for inference of genetic network architectures. In: Pac. Symp. Biocomput., pp. 18–29 (1998)
27. Li, S., Wunsch, D.C., O'Hair, E., Giesselman, M.G.: Extended Kalman filter training of neural network on SIMD parallel machine. Journal of Parallel and Distributed Computing 62, 544–562 (2002)
28. Li, X., Gi, Q.: Active Affective State Detection and User-Assistance with Dynamic Bayesian Networks. IEEE Trans. on Systems, Man and Cybernetics, Part-A: Systems and Humans 35(1) (January 2005)
29. Lng, C., Li, S.Q.: Chaotic spreading sequences with multiple access performance better than random sequences. IEEE transaction on Circuit and System -I, Fundamental Theory and Application 47(3), 394–397 (2000)

30. May, R.: Simple mathematical models with very complicated dynamics. Nature 261, 459–467 (1976)
31. Magnenat-Thalmann, N., Joslin, C., Berner, U.: Networked Virtual Park. In: Jain, L., Wilde, P.D. (eds.) Practical Applications of Computational Intelligence Techniques. Kluwer Academic, Dordrecht (2001)
32. Masri, S.F., Smyth, A.W., Chassiakos, A.G., Nakamura, M., Caughey, T.K.: Training Neural Networks By Adaptive Random Search Technique. Journal of Engineering Mechanics 125(2), 123–132 (1999)
33. Michael De Hoon, J.L., Imota, S., Kobayashi, K., Ogasawara, N., Miyano, S.: Inferring gene regulatory networks from time-ordered gene expression data of Bacillus subtills using differential equations. In: Pac. Symp. Biocomput., pp. 17–28 (2003)
34. Nasimul, N., Hitosi, I.: Inferring Gene Regulatory Networks Using Differential Evolution With Local Search heuristics. IEEE/ACM Transaction on computational biology and bioinformatics 4(4), 634–647 (2007)
35. Koduru, P., Dong, Z., Das, S., Welch, S.M., Roe, J.: Multi-Objective Evolutionary-Simplex Hybrid Approach for the Optimization of Differential Equation Models of Gene Networks. IEEE Transactions on Evolutionary Computation 12(5), 572–590 (2008)
36. Perrin, B., Ralaivola, L., Mazurie, A., Battani, S., Mallet, J., d'Alche-Buc, F.: Gene Networks Inference Using Dynamic Bayesian Networks. Bioinformatics 19, 138–148 (2003)
37. Roychowdhuri, P., Singh, Y.P., Chanskar, R.A.: Dynamic Tunneling Technique for efficient Training of Multilayer Perceptrons. IEEE transaction on Neural Networks 10(1) (January 1999)
38. Rumelhart, D.E., Hinton, G.E., Williams, R.J.: Learning representation by back propagation errors. Nature 323, 533–536 (1986)
39. Das, S., Morcos, K., Welch, S.M.: Combining Fuzzy Dominance Based PSO and Gradient Descent for Effective Parameter Estimation of Gene Regulatory Networks. In: Proceedings, IADIS Multi Conference on Computer Science and Information Systems, Algarve, Portugal (Ed. Antonio Palma dos Reis), pp. 3–10 (2009)
40. Schlitt, T., Brazma, A.: Current approaches to gene regulatory network modelling, BMC Bioinformatics, 8(Suppl 6):S9 (2007), doi:10.1186/1471-2105-8-S6-S9, This article is available from: `http://www.biomedcentral.com/1471-2105/8/S6/S9`
41. Somogyi, R., Sniegoski, C.A.: Modeling the complexity of genetic networks: understanding multigenic and pleiotropic regulation. Complexity 1, 45–63 (1996)
42. Storn, R., Price, K.: Differential Evolution –A Simple and Efficient Heuristic for Global Optimization over Continuous Spaces. Journal of Global Optimization 11, 341–359 (1997)
43. Storn, R., Price, K.: Differential Evolution – A simple evolution strategy for fast optimisation. Dr. Dobb's Journal 22(4), 18–24, 78(1997)
44. Storn, R., Price, K.: Differential Evolution- A Simple and Efficient Adaptive Scheme for Global Optimization over Continuous Spaces, Technical Report TR-95-012, Berkeley, CA (1995)
45. Spellman, P.T., Slerlock, G., Zhang, M.Q., Iyer, V.R., Anders, K., Eisen, M.B., Brown, P.O., Botstein, D., Futcher, B.: Comprehensive Identification of Cell Cycle-regulated Genes of the Yeast Saccharomyces cerevisiae by Microarray Hybridization. Molecular Biology of the Cell 9, 3273–3297 (1998)
46. The Modern Synthesis of Genetics and Evolution Copyright (1993-1997) by Laurence Moran `http://www.talkorigins.org/faqs/modern-synthesis.html`
47. VanBogelen, R.A., Greis, K.D., Blumenthal, R.M., Tani, T.H., Matthews, R.G.: Mapping regulatory networks in microbial cells. Trends Microbial. 7, 320–328 (1999)

48. Van Someren, E., Wessels, L., Reinders, M.: Linear Modeling of Genetic Networks from Experimental Data. In: Proc. Eighth Int'l Conf. Intelligent Systems for Molecular Biology, pp. 355–366 (2000)
49. Van Someren, E., Wessels, L., Reinders, M.: Genetic Network Models: A Comparative Study. In: Proc. SPIE, Micro-Arrays: Optical Technologies and Informatics, pp. 236–247 (2001)
50. Vohradsky, J.: Neural Network Model of Gene Expression. The FASEB journal 15, 354–846 (2001)
51. Wahde, M., Hartz, J.: Coarse-grained reverse engineering of genetic regulatory networks. Biosystems 55, 129–136 (2000)
52. Wahde, M., Hartz, J.: Modeling genetic regulatory dynamics in neural development. Journal of computational Biology 8, 429–442 (2001)
53. Werbos, P.: Back propagation through time: what it does and how to do it. Proceedings of IEEE 78(10), 1550–1560 (1990)
54. Wessels, L.F.A., Van Someren, E.P., Reinders, M.J.T.: A comparison of genetic network models. In: Pac. Symp. Biocomput., pp. 508–519 (2001)
55. Cai, X., Das, S., Welch, S.M., Koduru, P.: Simultaneous Structure Discovery and Parameter Estimation in Gene Networks Using a Multi-objective GP-PSO Hybrid Approach. International Journal of Bioinformatics Research and Applications 5(3), 254–268 (2009)
56. Xu, R., Wunsch II, D.C., Frank, R.L.: Inference of genetic regulatory networks with recurrent neural network models using particle swarm optimization. IEEE/ACM transaction on computational biology and bioinformatics 4(4), 681–692 (2007)
57. Xu, R., Hu, X., Wunsch, D.: Inference of genetic regulatory networks with recurrent neural network models. In: Proceedings of the 26th Annual International Conference on Engineering in Medicine and Biology Society, EMBC 2004, September 1-5, vol. 2, 4, pp. 2905–2908 (2004)
58. Xu, R., Venayagamoorthy, G.K., Wunsch II, D.C.: Modeling of gene regulatory networks with hybrid differential evolution and particle swarm optimization. Science Direct, Neural networks 20, 917–927 (2007), http://www.sciencedirect.com
59. Zhang, Y., Deng, Z., Jia, P.: A new dynamic Bayesian network for integrating multiple data in estimating gene networks. In: Third International Conference on Natural Computation (ICNC 2007) (2007)
60. Zitzler, E., Thiele, L.: Multiobjective evolutionary algorithms: A comparative case study and the strength Pareto approach. IEEE Trans. on Evolutionary Computation 3(4), 257–271 (1999)
61. Kros, J.F., Lin, M., Brown, M.L.: Effects of neural network s-Sigmoid function on KDD in the presence of imprecise data. Science Direct, Computer s & Operations Research 33(11), 3136–3194 (2006)

Author Index